Studies in Big Data

Volume 199

Series Editor

Janusz Kacprzyk, Polish Academy of Sciences, Warsaw, Poland

The series "Studies in Big Data" (SBD) publishes new developments and advances in the various areas of Big Data- quickly and with a high quality. The intent is to cover the theory, research, development, and applications of Big Data, as embedded in the fields of engineering, computer science, physics, economics and life sciences. The books of the series refer to the analysis and understanding of large, complex, and/or distributed data sets generated from recent digital sources coming from sensors or other physical instruments as well as simulations, crowd sourcing, social networks or other internet transactions, such as emails or video click streams and other. The series contains monographs, lecture notes and edited volumes in Big Data spanning the areas of computational intelligence including neural networks, evolutionary computation, soft computing, fuzzy systems, as well as artificial intelligence, data mining, modern statistics and Operations research, as well as self-organizing systems. Of particular value to both the contributors and the readership are the short publication timeframe and the world-wide distribution, which enable both wide and rapid dissemination of research output.

The books of this series are reviewed in a single blind peer review process.

Indexed by SCOPUS, EI Compendex, SCIMAGO and zbMATH.

All books published in the series are submitted for consideration in Web of Science.

Pradeep Singh • Balasubramanian Raman

Foundations of Machine Learning and AI

Geometry, Probability and Optimization

Springer

Pradeep Singh
Indian Institute of Information Technology
Surat
Surat, Gujarat, India

Balasubramanian Raman
Department of Computer Science and Engg
Indian Institute of Technology Roorkee
Roorkee, Uttarakhand, India

ISSN 2197-6503 ISSN 2197-6511 (electronic)
Studies in Big Data
ISBN 978-3-032-30335-6 ISBN 978-3-032-30336-3 (eBook)
https://doi.org/10.1007/978-3-032-30336-3

This Springer imprint is published by the registered company Springer Nature Switzerland AG
The registered company address is: Gewerbestrasse 11, 6330 Cham, Switzerland

To those who seek structure in uncertainty

Preface

The universe does not merely happen—it has shape.

We live inside a world that repeats itself. Not in the trivial sense that history loops, but in the deeper sense that *structure* reappears across scales: waves and spectra, symmetries and invariances, noise and signal, growth and decay, networks and flows. The same abstract forms show up in physics and finance, in biology and language, in images and genomes. Reality is generous with patterns.

Intelligence, in its widest sense, is the capacity to *meet* that structure: to compress experience into representations, to generalize beyond what was seen, to reason under uncertainty, and to act with purpose despite incomplete information. By this definition, intelligence is not confined to brains. It is a phenomenon that can emerge wherever a system can build models of the world and refine them through feedback. Life does it. Science does it. And now—machines do it too.

Machine Learning and AI are often introduced as technologies. They are that. But they are also something more enduring: a mathematical language for learning itself.

Why Foundations?

Modern AI systems can be astonishingly capable, yet they are also famously fragile: a slight shift in data distribution can collapse performance; an optimization choice can change a model from stable to chaotic; a "small" modeling assumption can become a large ethical or societal consequence. In practice, what separates robust understanding from accidental success is not access to more code—it is access to *principles*.

This book is written for readers who want that deeper grounding: not just how to run an algorithm, but why it works; not just how to train a model, but what training *means*; not just how to fit a curve, but how to think about geometry, uncertainty, and computation as a single story.

We begin with the language that makes the subject precise: the geometry of $\mathbb{R}^n$, linear maps, spectral ideas, and numerical stability. We build probability not as a list of distributions, but as a disciplined way of talking about uncertainty and evidence.

We connect estimation, testing, and resampling to the practical reality of learning from finite data. And then we turn to the engines that make learning possible at scale: linear models, kernels, high-dimensional methods, and optimization.

If you already use ML, these chapters will help you see familiar methods as consequences of a few recurring themes. If you are new to ML, the book aims to make the subject feel *coherent* rather than overwhelming.

Learning as Geometry, Uncertainty, and Search

At its core, learning is a three-part act.

- **Geometry:** We represent observations, features, and hypotheses inside spaces where we can measure similarity, distance, and direction. Vectors are not just arrays of numbers; they are points in a space where structure becomes visible. Inner products become notions of alignment. Norms become notions of size and robustness. Projections become the mathematics of approximation. Spectral decompositions become the mathematics of "important directions."
- **Uncertainty:** Data is partial, noisy, and finite. Probability gives a language for what we can and cannot infer. It tells us how evidence accumulates, how error behaves, and how confidence should be earned. Estimation and hypothesis testing are not bureaucratic rituals; they are the disciplined bridge between observations and claims.
- **Search:** Learning is optimization in disguise. Training is the process of navigating a landscape: sometimes convex and gentle, sometimes wild and jagged. Gradient methods are the workhorses not because they are magical, but because they exploit structure that appears across models. Regularization is not a hack; it is a way of encoding preference, simplicity, and stability.

These three perspectives are not separate topics. They are different faces of the same object. A least-squares problem is simultaneously a projection (geometry), an estimator (statistics), and a minimization (optimization). A kernel method is simultaneously an implicit embedding (geometry), a prior over functions (uncertainty, in a Bayesian reading), and a structured optimization problem (search). When you see these connections, the field stops being a collection of chapters and becomes a single map.

Wider Intelligence and the Universe of Structure

It is tempting to treat intelligence as a uniquely human spark. But intelligence, in a broad sense, can be viewed as a capacity that appears whenever a system can: *build internal representations of the world, compress experience into models, and use those models to predict and act.* By that definition, intelligence comes in degrees and in many forms: cells regulating themselves, animals navigating uncertain environments, societies coordinating knowledge, and machines learning from data.

This does not reduce intelligence to mere computation, nor does it claim that all forms of intelligence are the same. It simply recognizes a powerful idea: that *generalization*—the ability to go beyond what has been directly observed—is one of the most striking regularities in nature. The universe is not just a collection of events; it is a collection of *patterns*. And wherever patterns exist, there exists the possibility of learning them.

Mathematics is our way of naming those patterns with precision. Abstract structures are not detached from reality; they are reality's reusable skeletons. A vector space is not "just" an algebraic object; it is the simplest arena where combination, direction, and approximation make sense. A probability distribution is not "just" a formula; it is a compact description of uncertainty that can be updated by evidence. An optimization problem is not "just" a calculus exercise; it is a statement about preference and constraint, about what a system is trying to accomplish.

In this sense, studying ML is also a study of how structure appears: how high-dimensional geometry behaves, how noise and signal interact, how simple objectives produce complex behavior, and how abstract constraints shape what can be learned.

The Role of AI: A New Instrument for Understanding and Building

AI is often discussed as if it were only a set of products. But it is also an instrument—a new kind of scientific and creative apparatus. Just as the telescope extended vision into the cosmos and the microscope extended vision into the cell, learning systems extend our ability to detect structure in spaces too large, too noisy, or too complex for unaided intuition.

AI can help us find patterns in biology, climate, language, economics, and physics. It can help us compress and organize knowledge, simulate systems, and design artifacts. In that role, AI becomes a partner technology for discovery: not replacing human curiosity, but scaling its reach.

At the same time, the power to model and influence behavior comes with responsibilities. A model is never only a technical object; it is also a social object when deployed. Choices about data, objectives, evaluation, and constraints determine not just performance, but impact. Foundations matter here too. Understanding what an objective function actually optimizes, what "generalization" really guarantees, and how uncertainty should be communicated is part of building AI responsibly.

This book does not attempt to settle every philosophical question about intelligence or every societal question about AI. But it does aim to provide the conceptual tools required to think clearly: to separate what is provable from what is hopeful, to recognize when a method is stable or brittle, and to build models that are not only accurate, but understandable in their assumptions and limits.

How This Book Is Organized

The progression is intentional.

- **Chapters 1–5** build linear algebra and geometry with an ML mindset, including numerical stability. These chapters are not optional background; they are the language in which modern learning is written.
- **Chapters 6–9** develop probability, expectation, estimation, testing, and resampling. Here we learn to treat randomness not as a nuisance, but as a structure we can reason about.

- **Chapters 10–12** connect the mathematics to learning systems: linear and generalized linear models, kernels and high-dimensional methods, and the optimization machinery that trains modern models.

Each chapter is designed to be readable on its own, but the cumulative effect is the point: the reader should finish with a unified picture of ML as geometry + uncertainty + optimization.

How to Read (and How to Use) This Book

- If you prefer intuition first, read the opening sections of each chapter quickly, then return for the details.
- If you prefer rigor, follow the definitions carefully and treat the exercises as part of the text, not an appendix.
- If you are a practitioner, keep an eye on the recurring motifs: conditioning, invariances, bias–variance trade-offs, and the geometry of representations.

A practical note: the goal of foundations is not to memorize theorems. It is to internalize a way of thinking: *What is the space? What is the notion of closeness? What is the uncertainty model? What is being optimized? What assumptions make the claim true?* If you can ask those questions fluently, you will be able to learn new methods long after the current fashion changes.

An Invitation

The most exciting aspect of AI is not that it can imitate fragments of human output. It is that it forces us to confront a deeper question: *what does it mean for a system to form representations, to generalize, to reason under uncertainty, and to act effectively in the presence of constraints?*

Those are not only engineering questions. They are questions about knowledge, structure, and the space of possible intelligences.

The chapters ahead are a guided walk through the mathematics that makes these questions tractable. If you stay with the abstractions long enough for them to become familiar, you will find something surprising: the same few ideas return again and again, like constellations you start to recognize in different skies.

Welcome to *Foundations of Machine Learning & AI.*

Surat, India
Roorkee, India
March 25, 2026

Pradeep Singh
Balasubramanian Raman

Acknowledgments Acknowledgments The preparation of *Foundations of Machine Learning & AI* has been sustained by the support of institutions, colleagues, and communities to whom we owe a deep and lasting debt of gratitude. Foremost, we gratefully acknowledge the **Indian Institute of Information Technology Surat** and the **Indian Institute of Technology Roorkee** for providing the academic environments in which this work was conceived, developed, and completed. The intellectual culture, institutional encouragement, and scholarly resources available at these two institutes have been invaluable in shaping this book and in sustaining the broader research and teaching perspectives from which it emerged.

We gratefully acknowledge the support of the **Anusandhan National Research Foundation (ANRF)** and the **Department of Science and Technology (DST)**, Government of India, as well as the encouragement provided through fellowships and support associated with the **National Board of Higher Mathematics (NBHM)**, Department of Atomic Energy. Such support has been indispensable in enabling sustained engagement with fundamental research at the interface of mathematics, statistics, optimization, and artificial intelligence.

We are deeply thankful to our colleagues, collaborators, research scholars, and students whose questions, criticisms, and suggestions sharpened the exposition throughout this manuscript. Many ideas in these pages were refined through conversations in classrooms, seminars, offices, and informal discussions. Their intellectual companionship continually reminded us that a rigorous book must not only be correct, but also clear, motivated, and useful to learners entering the subject from diverse backgrounds.

A book on machine learning and AI is necessarily indebted to the broader scientific and open-source community. We acknowledge with gratitude the global communities that have built and maintained the mathematical, computational, and software foundations on which contemporary research and pedagogy depend. In particular, we thank the developers and maintainers of Python, NumPy, SciPy, SymPy, Matplotlib, scikit-learn, PyTorch, Jupyter, and many related libraries and platforms that have made experimentation, visualization, and reproducible scientific work significantly more accessible.

The first author (P.S.) would like to express heartfelt gratitude to *Prof. Anima Nagar* and *Prof. R. Balasubramanian* for their guidance, encouragement, and lasting influence on his intellectual journey. From them came not only mathematical insight and academic direction, but also a deeper appreciation for rigor, clarity, and the courage to pursue ambitious questions across disciplinary boundaries. Their mentorship has left a permanent imprint on the way this work was imagined and carried through.

Finally, we express our heartfelt gratitude to our families and friends for their patience, understanding, and unwavering support. Their quiet faith sustained the long hours of writing, revising, checking, and rethinking that a book of this scope demands. Whatever merit this volume may possess rests, in no small part, upon that constant support.

Competing Interests The authors declare that they have no competing interests relevant to the content of this manuscript.

Contents

About the Authors

Dr. Pradeep Singh is serving as an Assistant Professor (Grade–I) at IIIT Surat. He is a faculty member in the Department of Mathematics and Computational Sciences and joint faculty member in the Department of Computer Science & Engineering. His research spans Geometric Deep Learning, Neuro-symbolic AI, and Dynamical Systems. His research has appeared in premier venues, including *IEEE Transactions*; *Knowledge-Based Systems*; *Chaos, Solitons & Fractals*; *Neurocomputing*; and *ECAI*, contributing to the wider dialogue in AI and dynamical-systems research. He has authored three books: the Springer monographs *Deep Learning Through the Prism of Tensors* and *The Geometry of Intelligence: Foundations of Transformer Networks*, and the McGraw-Hill text *Machine Learning and Artificial Intelligence*. He earned his Ph.D. (2022) and master's degree from IIT Delhi, specializing in Symbolic Systems, and a bachelor's degree in Data Science from IIT Madras. He secured All-India Rank 1 in GATE 2020, JAM 2015, and CSIR-NET 2019. He has also received highly competitive awards, including the National Board for Higher Mathematics (NBHM) Master's, Doctoral, and Post-Doctoral Fellowships awarded by the Department of Atomic Energy, India. In 2019, he was one of only two researchers nationwide to receive the Dr Shyama Prasad Mukherjee (SPM) Fellowship in Mathematics from CSIR, India. He is also a recipient of the prestigious Human Frontier Science Program

Postdoctoral Fellowship, an international fellowship supporting frontier cross-disciplinary research.

Dr. Balasubramanian Raman (Senior Member, IEEE) is Professor (HAG) and Head of the Department of Computer Science & Engineering at the Indian Institute of Technology Roorkee, where he also holds a joint appointment with the Mehta Family School of Data Science & AI. He served as the iHUB Divyasampark Chair Professor from 2022 to 2025. He earned his Ph.D. from IIT Madras (2001). His research—spanning machine learning, computer vision, image/video processing, and pattern recognition—has produced more than 250 peer-reviewed publications and a Google-Scholar h-index of 53. He has undertaken post-doctoral and visiting appointments at Rutgers University, the University of Missouri–Columbia, Osaka Metropolitan University, Curtin University, and the University of Cyberjaya. To date, he has supervised 35 Ph.D. scholars (with a further 15 in progress) and regularly directs large, multi-institutional programs funded by leading national agencies and industry partners. An inventor of an Indian patent for a real-time fog-removal imaging system, he has also co-authored three books—*Deep Learning Through the Prism of Tensors* (Springer Nature, 2024), *The Geometry of Intelligence: Foundations of Transformer Networks in Deep Learning* (Springer Nature, 2025), and *Artificial Intelligence and Machine Learning* (McGraw-Hill, 2025). His honors include the DST BOYSCAST Fellowship, two IIT Roorkee Outstanding Teacher Awards, the Ramkumar Prize for Outstanding Teaching and Research, and the ICPC Coach Award, which recognized his teams' top-50 finishes at the ACM ICPC World Finals.

List of Common Abbreviations

Adam	Adaptive Moment Estimation
ADMM	Alternating Direction Method of Multipliers
AI	Artificial Intelligence
API	Application Programming Interface
ASR	Automatic Speech Recognition
AUC	Area Under the (ROC) Curve
CDF	Cumulative Distribution Function
CI	Confidence Interval
CNN	Convolutional Neural Network
CV	Computer Vision
DL	Deep Learning
ELBO	Evidence Lower BOund
ERM	Empirical Risk Minimization
EVD	Eigenvalue Decomposition
FDR	False Discovery Rate
FN	False Negative
FP	False Positive
FPR	False Positive Rate
FWER	Family-Wise Error Rate
GD	Gradient Descent
GMM	Gaussian Mixture Model
GPU	Graphics Processing Unit
GRU	Gated Recurrent Unit
HMM	Hidden Markov Model
IDE	Integrated Development Environment
i.i.d.	Independent and Identically Distributed
IR	Information Retrieval
IRLS	Iteratively Reweighted Least Squares
KKT	Karush–Kuhn–Tucker (conditions)
KL	Kullback–Leibler (divergence)
KNN	k-Nearest Neighbors

LBFGS	Limited-memory BFGS
LHS	Left-Hand Side
LLM	Large Language Model
LR	Learning Rate
LSTM	Long Short-Term Memory
LU	LU Factorization
MAE	Mean Absolute Error
MAML	Model-Agnostic Meta-Learning
MAP	Maximum A Posteriori
ML	Machine Learning
MLP	Multi-Layer Perceptron
MLE	Maximum Likelihood Estimation
MSE	Mean Squared Error
NLG	Natural Language Generation
NLL	Negative Log-Likelihood
NLP	Natural Language Processing
OCR	Optical Character Recognition
PAC	Probably Approximately Correct
PCA	Principal Component Analysis
PD	Positive Definite
PDF	Probability Density Function
PMF	Probability Mass Function
RNN	Recurrent Neural Network
PR	Precision–Recall
PSD	Positive Semidefinite
QR	QR Factorization
RAG	Retrieval-Augmented Generation
RHS	Right-Hand Side
RL	Reinforcement Learning
RMSE	Root Mean Squared Error
ROC	Receiver Operating Characteristic
RS	Recommender Systems
RV	Random Variable
SGD	Stochastic Gradient Descent
SL	Supervised Learning
SRM	Structural Risk Minimization
SSL	Self-Supervised Learning
SVD	Singular Value Decomposition
SVM	Support Vector Machine
TN	True Negative
TP	True Positive
TPR	True Positive Rate
TPU	Tensor Processing Unit
UL	Unsupervised Learning
VC	Vapnik–Chervonenkis (dimension)
ViT	Vision Transformer

Chapter 1
Mathematical Language and Notation

This chapter fixes the basic mathematical language that will be used throughout the book. Machine learning (ML) and artificial intelligence (AI) can be viewed as the study of *functions* learned from data, often under uncertainty. To talk precisely about data, models, and algorithms, we must be comfortable with: (i) *sets* as containers of objects (samples, features, hypotheses, parameters), (ii) *Cartesian products* and *tuples* to represent structured data, (iii) *functions* as maps that transform representations, and (iv) *relations* to formalize similarity, connectivity, and "belonging to the same class." We will repeatedly use two recurring ideas:

Remark 1.0.1 (Two Guiding Viewpoints)

1. **Data as elements of a product space.** A supervised dataset is naturally a finite subset (or multiset) of $\mathcal{X} \times \mathcal{Y}$, where $\mathcal{X}$ is the input (feature) space and $\mathcal{Y}$ is the output (label) space.
2. **Learning as choosing a function from a hypothesis class.** A model is often a function $h : \mathcal{X} \to \mathcal{Y}$ (or $h : \mathcal{X} \to \mathbb{R}$), chosen from a family $\mathcal{H}$ by minimizing a loss on data.

These two viewpoints already explain why sets, products, and maps are foundational objects (Halmos 1960) (cf. Fig. 1.1).

1.1 Sets, Functions, and Relations

1.1.1 Sets as Modeling Primitives

1.1.1.1 Basic Set Operations, Membership, and Subset Relations

Definition 1.1.1 (Set and Membership) A *set* is a collection of objects (called *elements*). If x is an element of a set A, we write $x \in A$; otherwise, we write $x \notin A$.

P. Singh, B. Raman, *Foundations of Machine Learning and AI*, Studies in Big Data 199, https://doi.org/10.1007/978-3-032-30336-3_1

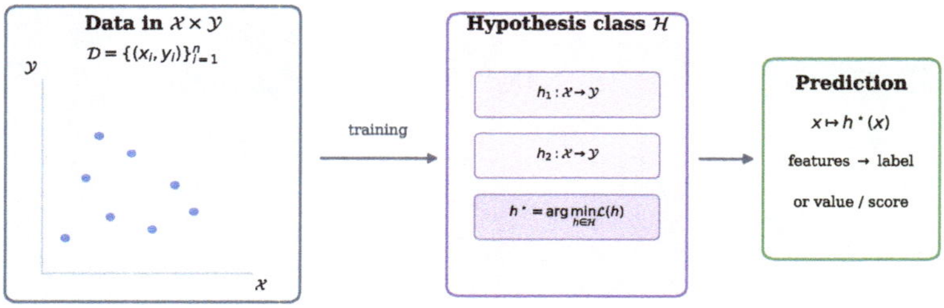

Fig. 1.1 Two guiding viewpoints for machine learning: data as a finite subset of a product space and learning as the selection of a function from a hypothesis class by minimizing loss on data

Two sets A and B are equal, written $A = B$, if they have exactly the same elements:

$$A = B \iff (\forall x)\,[x \in A \Leftrightarrow x \in B].$$

Definition 1.1.2 (Subset, Proper Subset, and Power Set) Let A and B be sets.

- A is a *subset* of B, written $A \subseteq B$, if every element of A is an element of B.
- A is a *proper subset* of B, written $A \subsetneq B$, if $A \subseteq B$ and $A \neq B$.
- The *power set* of A, denoted $\mathcal{P}(A)$, is the set of all subsets of A: $\mathcal{P}(A) = \{S : S \subseteq A\}$.

Remark 1.1.1 (Common Set-Builder Notation) A set may be specified by a predicate (property) $P(x)$ as

$$\{x \in A : P(x)\},$$

meaning the subset of A consisting of exactly those x for which $P(x)$ holds. In ML, one often uses set-builder notation to define hypothesis classes (e.g., linear predictors with bounded norm) or feasible sets of parameters.

Definition 1.1.3 (Union, Intersection, Difference, Complement) For sets $A, B \subseteq \Omega$ (where Ω is a universe under discussion),

$$\begin{aligned}
A \cup B &:= \{x \in \Omega : x \in A \text{ or } x \in B\},\\
A \cap B &:= \{x \in \Omega : x \in A \text{ and } x \in B\},\\
A \setminus B &:= \{x \in \Omega : x \in A \text{ and } x \notin B\},\\
A^c &:= \Omega \setminus A.
\end{aligned}$$

Proposition 1.1.1 (Basic Algebra of Sets) *For sets $A, B, C \subseteq \Omega$:*

1. *(Commutativity) $A \cup B = B \cup A$ and $A \cap B = B \cap A$.*
2. *(Associativity) $(A \cup B) \cup C = A \cup (B \cup C)$ and similarly for $\cap$.*

3. *(Distributivity)* $A \cap (B \cup C) = (A \cap B) \cup (A \cap C)$ *and* $A \cup (B \cap C) = (A \cup B) \cap (A \cup C)$.
4. *(De Morgan)* $(A \cup B)^c = A^c \cap B^c$ *and* $(A \cap B)^c = A^c \cup B^c$.

Remark 1.1.2 (Why Set Operations Matter in ML) Set operations appear whenever we talk about:

(i) *events* in probability (later chapters), e.g., $\{X \in A\} \cup \{X \in B\}$;
(ii) *decision regions* of classifiers, e.g., the positive class region R_+ and its complement $R_- = R_+^c$;
(iii) *dataset filtering*, e.g., selecting samples with certain attributes.

Example 1.1.1 (Dataset as a Finite Set) In supervised learning, let $\mathcal{X}$ be an input space and $\mathcal{Y}$ a label space. A dataset of n labeled examples can be written as

$$\mathcal{D} = \{(x_i, y_i)\}_{i=1}^n \subseteq \mathcal{X} \times \mathcal{Y}.$$

When duplicates are allowed, it is more precise to treat $\mathcal{D}$ as a *multiset* or as an indexed family $(x_i, y_i)_{i \in [n]}$ where $[n] := \{1, 2, \ldots, n\}$.

Remark 1.1.3 (Index Sets) Throughout the book, $[n] := \{1, 2, \ldots, n\}$ for integers $n \geq 1$. Index sets are convenient when data are best viewed as an ordered list rather than an unordered set.

1.1.1.2 Cartesian Products, Tuples, and Product Spaces for Datasets

Definition 1.1.4 (Ordered Pair and Cartesian Product) Given sets A and B, an *ordered pair* is written (a, b) with $a \in A$ and $b \in B$. The *Cartesian product* $A \times B$ is

$$A \times B := \{(a, b) : a \in A,\ b \in B\}.$$

More generally, for sets $A_1, \ldots, A_k$, the product $\prod_{j=1}^k A_j$ is the set of k-tuples $(a_1, \ldots, a_k)$ with $a_j \in A_j$ for each j.

Definition 1.1.5 (k-Tuples and Product Spaces) A *k-tuple* is an ordered list $(a_1, \ldots, a_k)$. If each A_j has additional structure (e.g., a vector space or a metric space), then $\prod_{j=1}^k A_j$ is called a *product space*, inheriting a natural componentwise structure.

Remark 1.1.4 (Data Matrices as Product Spaces) If each input $x_i \in \mathbb{R}^d$, then an n-sample dataset of inputs can be seen as an element of $(\mathbb{R}^d)^n$, which is naturally isomorphic to $\mathbb{R}^{n \times d}$ (a data matrix). This observation connects linear algebra directly to dataset representation (Axler 2015).

Example 1.1.2 (Multi-Modal Inputs as Products) Suppose an input consists of an image I and a text description T. If images live in $\mathcal{X}_{\text{img}}$ and text in $\mathcal{X}_{\text{text}}$, then the joint input space is

$$\mathcal{X} = \mathcal{X}_{\text{img}} \times \mathcal{X}_{\text{text}}.$$

A model may act on each component separately and then fuse representations; formally, this is a composition of maps on a product space.

Remark 1.1.5 (Labels, Targets, and Structured Outputs) The label space $\mathcal{Y}$ may be discrete (classification), continuous (regression), or structured (sequences, graphs, segmentations). In all cases, a labeled example is a pair in $\mathcal{X} \times \mathcal{Y}$.

1.1.2 Functions and Relations

1.1.2.1 Images, Preimages, and Inverse Images in Learning Pipelines

Definition 1.1.6 (Function (map)) A *function* (or *map*) from a set $\mathcal{X}$ to a set $\mathcal{Y}$ is a rule f that assigns to each $x \in \mathcal{X}$ a unique element $f(x) \in \mathcal{Y}$. We write $f : \mathcal{X} \to \mathcal{Y}$. The set $\mathcal{X}$ is the *domain*, and $\mathcal{Y}$ is the *codomain*.

Definition 1.1.7 (Image and Preimage) Let $f : \mathcal{X} \to \mathcal{Y}$ be a function.

- For a subset $S \subseteq \mathcal{X}$, the *image* of S under f is

$$f(S) := \{f(x) : x \in S\} \subseteq \mathcal{Y}.$$

- For a subset $T \subseteq \mathcal{Y}$, the *preimage* (also called *inverse image*) of T under f is

$$f^{-1}(T) := \{x \in \mathcal{X} : f(x) \in T\} \subseteq \mathcal{X}.$$

Remark 1.1.6 (Preimages Do Not Require Inverses) The notation $f^{-1}(T)$ denotes a *set* in $\mathcal{X}$, not necessarily the value of an inverse function. Even when f is not invertible, the preimage $f^{-1}(T)$ is always well-defined for any $T \subseteq \mathcal{Y}$. This is a crucial point in probability theory and ML, where events are often defined as preimages of measurable sets.

Proposition 1.1.2 (Preimage Preserves Set Operations) *Let $f : \mathcal{X} \to \mathcal{Y}$ and let $\{T_i\}_{i \in I}$ be subsets of $\mathcal{Y}$. Then:*

1. $f^{-1}\Big(\bigcup_{i \in I} T_i\Big) = \bigcup_{i \in I} f^{-1}(T_i)$,
2. $f^{-1}\Big(\bigcap_{i \in I} T_i\Big) = \bigcap_{i \in I} f^{-1}(T_i)$,
3. $f^{-1}(T^c) = (f^{-1}(T))^c$ *(complements taken in $\mathcal{X}$ and $\mathcal{Y}$).*

Remark 1.1.7 (Images Behave Differently) In general, $f(S \cap S') \subseteq f(S) \cap f(S')$ and equality need not hold. This asymmetry is one reason inverse images (preimages) are technically cleaner objects than images.

Definition 1.1.8 (Injective, Surjective, Bijective) A function $f : \mathcal{X} \to \mathcal{Y}$ is:

- *injective* (one-to-one) if $f(x) = f(x') \Rightarrow x = x'$,
- *surjective* (onto) if $f(\mathcal{X}) = \mathcal{Y}$,
- *bijective* if it is both injective and surjective.

If f is bijective, then there exists an inverse function $f^{-1} : \mathcal{Y} \to \mathcal{X}$.

Example 1.1.3 (Decision Regions are Preimages) Let $h : \mathbb{R}^d \to \{-1, +1\}$ be a binary classifier. The *positive decision region* is

$$R_+ := h^{-1}(\{+1\}) = \{x \in \mathbb{R}^d : h(x) = +1\},$$

and the negative region is $R_- := h^{-1}(\{-1\})$. Thus, a classifier partitions the input space into preimages of labels.

Example 1.1.4 (Feature Extraction as a Map) Let $\phi : \mathcal{X} \to \mathbb{R}^m$ be a feature map (e.g., embeddings from a neural network). If $B \subseteq \mathbb{R}^m$ is a set of "acceptable" representations (say, a ball around a prototype), then $\phi^{-1}(B)$ is the set of original inputs whose representations fall in that region. Understanding $\phi^{-1}(B)$ is central to interpretability and robustness.

Remark 1.1.8 (Composition and Pipelines) If $f : \mathcal{X} \to \mathcal{Y}$ and $g : \mathcal{Y} \to \mathcal{Z}$, then the composition $g \circ f : \mathcal{X} \to \mathcal{Z}$, $(g \circ f)(x) = g(f(x))$, formalizes ML pipelines: raw input $\to$ features $\to$ predictions. Preimages propagate *backward* through compositions: $(g \circ f)^{-1}(T) = f^{-1}(g^{-1}(T))$.

1.1.2.2 Equivalence Relations, Partitions, and Clustering Viewpoint

Definition 1.1.9 (Binary Relation) A *binary relation* R from a set A to a set B is any subset $R \subseteq A \times B$. If $(a, b) \in R$ we often write $a\,R\,b$. When $A = B$, we call $R \subseteq A \times A$ a relation *on* A.

Definition 1.1.10 (Equivalence Relation) A relation $\sim$ on a set A is an *equivalence relation* if for all $a, b, c \in A$:

1. (Reflexive) $a \sim a$,
2. (Symmetric) $a \sim b \Rightarrow b \sim a$,
3. (Transitive) $a \sim b$ and $b \sim c \Rightarrow a \sim c$.

Definition 1.1.11 (Equivalence Class and Quotient) Given an equivalence relation $\sim$ on A and an element $a \in A$, the *equivalence class* of a is

$$[a] := \{b \in A : b \sim a\}.$$

The set of all equivalence classes is denoted $A/\sim$ and is called the *quotient set*.

Definition 1.1.12 (Partition) A *partition* of a set A is a collection of nonempty, pairwise disjoint subsets $\{C_\alpha\}_{\alpha\in\mathcal{I}}$ such that $\bigcup_{\alpha\in\mathcal{I}} C_\alpha = A$. The sets C_α are called the *parts* (or *blocks*) of the partition.

Theorem 1.1.1 (Equivalence Relations $\Longleftrightarrow$ Partitions) *Let A be a set.*

1. *If $\sim$ is an equivalence relation on A, then the set of equivalence classes $\{[a] : a \in A\}$ forms a partition of A.*
2. *Conversely, if $\{C_\alpha\}_{\alpha\in\mathcal{I}}$ is a partition of A, then defining $a \sim b$ iff a and b belong to the same block (i.e. $\exists\alpha$ such that $a, b \in C_\alpha$) yields an equivalence relation on A.*

Proof

(1) Each $a \in A$ lies in $[a]$, so $\bigcup_{a\in A}[a] = A$. If $[a] \cap [b] \neq \emptyset$, pick c in the intersection; then $c \sim a$ and $c \sim b$. By symmetry and transitivity, $a \sim b$, hence $[a] = [b]$. Therefore, distinct classes are disjoint.

(2) Reflexivity holds since a belongs to some block and hence is in the same block as itself. Symmetry is immediate. Transitivity holds because if a and b are in the same block and b and c are in the same block, then all three are in that block, so $a \sim c$.

□

Remark 1.1.9 (Why Partitions Matter in ML) Many ML objects induce partitions:

- A classifier $h : \mathcal{X} \to \mathcal{Y}$ partitions $\mathcal{X}$ into $\{h^{-1}(\{y\})\}_{y\in\mathcal{Y}}$.
- A clustering algorithm outputs a partition of a dataset into clusters.
- Hashing or discretization maps inputs to bins, producing equivalence classes of "indistinguishable" items under that map.

Example 1.1.5 (Clustering as a Partition) Let $\mathcal{D} = \{x_1, \ldots, x_n\} \subseteq \mathcal{X}$ be an unlabeled dataset. A k-cluster output can be represented as a partition $\{C_1, \ldots, C_k\}$ of $\mathcal{D}$. Equivalently, it defines an equivalence relation on $\mathcal{D}$ by $x_i \sim x_j$ iff x_i and x_j are assigned to the same cluster.

Remark 1.1.10 (Similarity is Not Always an Equivalence Relation) In practice, "similarity" (e.g. cosine similarity exceeding a threshold) may fail transitivity. Thus similarity graphs are often modeled as general relations or as weighted graphs, not necessarily as equivalence relations. When we *force* transitivity (e.g. via connected components), we pass from a general relation to an equivalence relation.

Example 1.1.6 (Connected Components Induce Equivalence Classes) Let $G = (V, E)$ be an undirected graph on a dataset $V = \mathcal{D}$, where an edge $(x_i, x_j) \in E$ indicates "sufficient similarity". Define $x_i \sim x_j$ iff there exists a path in G between x_i and x_j. Then $\sim$ is an equivalence relation, and its equivalence classes are exactly the connected components of G. This construction underlies many graph-based clustering methods.

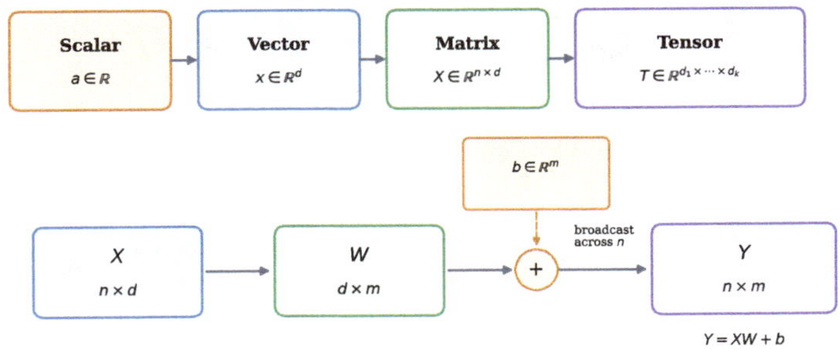

Fig. 1.2 A compact shape-calculus summary: scalar, vector, matrix, and tensor objects together with a batch affine map illustrating contraction over feature indices and broadcasting of the bias across the batch axis

1.2 Scalars, Vectors, Matrices, Tensors

Modern ML systems are ultimately built by composing linear-algebraic objects: scalars (parameters), vectors (feature representations), matrices (linear maps, Jacobians, attention weights), and tensors (multi-way arrays in deep learning). To avoid silent bugs and conceptual confusion, we fix *shape conventions* and *indexing rules* early. A consistent "shape calculus" also makes derivations (gradients, Jacobians, backpropagation) mechanically checkable (cf. Fig. 1.2).

Remark 1.2.1 (Convention Used Throughout This Book) Unless explicitly stated otherwise, **vectors are columns**. Thus $x \in \mathbb{R}^d$ is a $d \times 1$ column vector, and a linear model is written as $w^\top x$ with $w \in \mathbb{R}^d$. A batch of n samples is written as a data matrix $X \in \mathbb{R}^{n\times d}$ whose i-th row is $x_i^\top$.

1.2.1 Shapes, Indexing, and Data Layout

1.2.1.1 Vectors/Matrices as Structured Containers (Row vs. Column Conventions)

Definition 1.2.1 (Scalar, Vector, Matrix, Tensor)

- A *scalar* is a single number, typically in $\mathbb{R}$.
- A *vector* in $\mathbb{R}^d$ is an ordered d-tuple of scalars, written as a column $x = (x_1, \ldots, x_d)^\top \in \mathbb{R}^d$.
- A *matrix* $A \in \mathbb{R}^{m\times n}$ is a 2-way array with entries A_{ij}, $i \in [m]$, $j \in [n]$.
- A (finite-dimensional) *tensor* is a k-way array $T \in \mathbb{R}^{d_1\times\cdots\times d_k}$ with entries $T_{i_1,\ldots,i_k}$.

We call $(d_1, \ldots, d_k)$ the *shape* of T, and k its *order* (or rank in the array sense).

Remark 1.2.2 (Row vs. Column Conventions) Both conventions appear in the literature:

- **Column-vector convention:** $x \in \mathbb{R}^d$ (column), linear score $w^\top x$.
- **Row-vector convention:** $x^\top \in \mathbb{R}^{1\times d}$ (row), linear score $x^\top w$.

They are equivalent if used consistently. Most confusion arises when mixing them (e.g. writing Xw but defining X as $d \times n$ without transposing). We will use the column-vector convention for derivations, and store datasets as row-stacked matrices $X \in \mathbb{R}^{n\times d}$ for convenience.

Definition 1.2.2 (Dataset Layouts) Let $x_i \in \mathbb{R}^d$ for $i \in [n]$. Two common matrix layouts are:

- *Row-stacked (preferred here):* $X \in \mathbb{R}^{n\times d}$ with $X_{i:} = x_i^\top$.
- *Column-stacked:* $\widetilde{X} \in \mathbb{R}^{d\times n}$ with $\widetilde{X}_{:i} = x_i$.

They satisfy $\widetilde{X} = X^\top$.

Example 1.2.1 (Linear Prediction with Explicit Shapes) Let $w \in \mathbb{R}^d$ and $b \in \mathbb{R}$.

- Single example: $x \in \mathbb{R}^d$ and score $s = w^\top x + b \in \mathbb{R}$.
- Batch: $X \in \mathbb{R}^{n\times d}$ and score vector $s = Xw + b\mathbf{1} \in \mathbb{R}^n$, where $\mathbf{1} \in \mathbb{R}^n$ is the all-ones vector.

Remark 1.2.3 (Batches and Axes) In ML implementations, the first axis is often the *batch axis* (size n), followed by feature axes (size d) and sometimes time/spatial/channel axes. Making this explicit prevents common errors in matrix multiplications and tensor contractions.

1.2.1.2 Indexing Rules, Slicing, and Broadcasting Principles

Definition 1.2.3 (Indexing Notation) For a matrix $A \in \mathbb{R}^{m\times n}$:

- A_{ij} is the (i, j) entry.
- $A_{i:} \in \mathbb{R}^{1\times n}$ is the i-th row (row vector).
- $A_{:j} \in \mathbb{R}^m$ is the j-th column (column vector).

For a tensor $T \in \mathbb{R}^{d_1\times\cdots\times d_k}$, the entry is $T_{i_1,\ldots,i_k}$. Fixing some indices and leaving others free yields *slices*.

Remark 1.2.4 (Einstein Summation vs. Explicit Sums) In some texts, repeated indices imply summation (Einstein convention). We will *not* assume this by default, and will write sums explicitly unless the context is unmistakable. This improves readability for ML audiences and reduces ambiguity.

Definition 1.2.4 (Broadcasting (Shape Extension Principle)) Broadcasting is a rule for combining arrays of different shapes by implicitly replicating dimensions of size 1 to match the other operand. Formally, two shapes are *broadcast-compatible* if, when aligned from the trailing axes, each pair of axis lengths is either equal or one of them is 1. The result shape uses the maximum along each axis.

Example 1.2.2 (Broadcasting in Batch Computations) Let $X \in \mathbb{R}^{n\times d}$ and $\mu \in \mathbb{R}^d$. Centering data often uses

$$X_c = X - \mathbf{1}\mu^\top \in \mathbb{R}^{n\times d},$$

where $\mathbf{1} \in \mathbb{R}^n$. In array libraries, one can write $X - \mu^\top$ (or $X - \mu$ depending on convention), and broadcasting interprets μ as replicated across the batch axis.

Remark 1.2.5 (Broadcasting is a Convenience, Not a Theorem) Broadcasting is an implementation rule. It can hide unintended shape mismatches (e.g. subtracting a $(d,)$ vector from a (n, d) matrix is correct, but subtracting a $(n,)$ vector might silently broadcast along the wrong axis). When in doubt, reshape explicitly (e.g. $\mu^\top$ vs. μ vs. $\mu\mathbf{1}^\top$).

Proposition 1.2.1 (Trace and Frobenius Inner Product Identity) *For matrices $A, B \in \mathbb{R}^{m\times n}$, the Frobenius inner product satisfies*

$$\langle A, B\rangle_F := \sum_{i=1}^{m}\sum_{j=1}^{n} A_{ij}B_{ij} = tr(A^\top B).$$

Remark 1.2.6 (Why This Identity Appears Often) The identity $\langle A, B\rangle_F = \mathrm{tr}(A^\top B)$ is a "bridge" between coordinate sums and compact matrix calculus. It will appear later in least squares, PCA, and gradient derivations.

1.2.2 Block Structures and Modular Notation

1.2.2.1 Block Vectors/Matrices for Feature Groups and Multi-view Data

Definition 1.2.5 (Block Vector and Block Matrix) A *block vector* is a vector formed by stacking sub-vectors:

$$x = \begin{bmatrix} x^{(1)} \\ \vdots \\ x^{(p)} \end{bmatrix}, \qquad x^{(r)} \in \mathbb{R}^{d_r}, \ \sum_{r=1}^{p} d_r = d.$$

A *block matrix* is a matrix partitioned into submatrices:

$$A = \begin{bmatrix} A_{11} & \cdots & A_{1q} \\ \vdots & \ddots & \vdots \\ A_{p1} & \cdots & A_{pq} \end{bmatrix}, \qquad A_{rs} \in \mathbb{R}^{d_r \times e_s}.$$

Remark 1.2.7 (Feature Groups and Structured Regularization) Block decompositions formalize feature groups (demographics, text features, sensor channels, etc.). Many regularizers and priors act at the group level (e.g. encouraging sparsity by blocks), and many architectures process blocks separately before fusing representations.

Example 1.2.3 (Two-View Learning as a Block Vector) Suppose each example has two feature views: $x^{(1)} \in \mathbb{R}^{d_1}$ (text embedding) and $x^{(2)} \in \mathbb{R}^{d_2}$ (image embedding). Then

$$x = \begin{bmatrix} x^{(1)} \\ x^{(2)} \end{bmatrix} \in \mathbb{R}^{d_1+d_2}, \qquad w = \begin{bmatrix} w^{(1)} \\ w^{(2)} \end{bmatrix}.$$

A linear score decomposes as

$$w^\top x = (w^{(1)})^\top x^{(1)} + (w^{(2)})^\top x^{(2)}.$$

This makes clear how each modality contributes additively.

Proposition 1.2.2 (Block Matrix Multiplication (Compatibility Check)) *Let A be a block matrix with blocks A_{rs} and let x be a block vector with blocks $x^{(s)}$. If dimensions are compatible, then the r-th block of Ax is*

$$(Ax)^{(r)} = \sum_s A_{rs}\, x^{(s)}.$$

Remark 1.2.8 (Where This Shows up Later) Block identities are used in: (i) least squares with partitioned features, (ii) multitask learning, (iii) Gaussian models with block covariance matrices, and (iv) neural networks with concatenation and residual paths.

1.2.2.2 Tensor Notation for Multi-way Arrays and Neural Layers

Definition 1.2.6 (Tensor Axes Used in Deep Learning) A common tensor layout for a mini-batch of images is

$$X \in \mathbb{R}^{n \times c \times h \times w},$$

where n is batch size, c is number of channels, and h, w are spatial dimensions. For sequences, a common layout is

$$X \in \mathbb{R}^{n \times T \times d},$$

where T is sequence length and d is embedding dimension.

Remark 1.2.9 (Order Matters: Same Numbers, Different Meaning) The shapes (n, T, d) and (T, n, d) contain the same number of scalars but represent different semantics. Always track *which axis means what* (batch, time, features, channels, height, width). Axis conventions differ across libraries; our analysis will specify them explicitly.

Definition 1.2.7 (Mode-k Fibers and Unfolding (Matricization)) Let $T \in \mathbb{R}^{d_1 \times \cdots \times d_k}$. Fixing all indices except the r-th produces a *mode-r fiber*. *Unfolding* (or matricization) reshapes T into a matrix $T_{(r)} \in \mathbb{R}^{d_r \times (d_1 \cdots d_{r-1} d_{r+1} \cdots d_k)}$ by stacking mode-r fibers as columns (exact ordering is a convention).

Remark 1.2.10 (Why Unfolding is Useful) Many tensor operations can be expressed as matrix multiplications after unfolding. This is central in analyzing neural layers (convolutions, attention) and in deriving gradients, because matrix calculus is simpler than general tensor calculus (Kolda and Bader 2009).

Definition 1.2.8 (Tensor Contraction: A Controlled Summation) A *tensor contraction* is a generalization of matrix multiplication obtained by summing over one or more shared indices. For example, if $A \in \mathbb{R}^{m \times n}$ and $B \in \mathbb{R}^{n \times p}$ then

$$(AB)_{ip} = \sum_{j=1}^{n} A_{ij} B_{jp},$$

which is a contraction over the index j.

Example 1.2.4 (Affine Layer as a Tensor Contraction) Let a batch of embeddings be $X \in \mathbb{R}^{n \times d}$ and weights $W \in \mathbb{R}^{d \times m}$. The output $Y \in \mathbb{R}^{n \times m}$ of an affine layer is

$$Y = XW + b\mathbf{1}^\top,$$

where $b \in \mathbb{R}^m$. Entrywise: $Y_{i\ell} = \sum_{j=1}^{d} X_{ij} W_{j\ell} + b_\ell$, which is a contraction over the feature index j.

Example 1.2.5 (Self-Attention Shapes (Minimal Bookkeeping)) For a batch of token embeddings $X \in \mathbb{R}^{n \times T \times d}$, define linear maps $W_Q, W_K, W_V \in \mathbb{R}^{d \times d_h}$ producing $Q, K, V \in \mathbb{R}^{n \times T \times d_h}$. Attention scores are (conceptually) a batched matrix product:

$$S = QK^\top \in \mathbb{R}^{n \times T \times T},$$

where the transpose is applied on the last two axes of K (turning $(T \times d_h)$ into $(d_h \times T)$ per batch). Softmax is applied along the last axis, and the output is

$$Y = \mathrm{softmax}(S)\, V \in \mathbb{R}^{n \times T \times d_h}.$$

The point is not the full attention theory here, but that *correctness is largely a shape argument.*

Remark 1.2.11 (Shape Errors Are Conceptual Errors) Many implementation mistakes (wrong transpose, wrong axis for normalization) are actually mathematical mistakes about which indices are being contracted or broadcast. A reliable habit is to annotate every intermediate quantity with its shape before computing.

1.3 Asymptotics (O, Θ, o)

Asymptotic notation provides a compact language to describe how quantities scale when a problem becomes *large*: number of samples $n \to \infty$, dimension $d \to \infty$, model size $p \to \infty$, time horizon $T \to \infty$, or desired accuracy $\varepsilon \downarrow 0$. In ML/AI, asymptotics serves two complementary purposes:

1. **Algorithmic scaling:** how runtime, memory, or iteration complexity grows with n, d, p, T.
2. **Statistical scaling:** how error, risk, or generalization bounds shrink with more data and how they depend on dimension/complexity.

We first formalize O, Θ, o precisely, then discuss scaling laws that repeatedly reappear in learning theory (Cormen et al. 2022).

1.3.1 Growth Rates and Scaling Laws

1.3.1.1 Big-O, little-o, and Tight Bounds in Algorithmic Analysis

Definition 1.3.1 (Big-O, Big-Ω, Big-Θ) Let $f, g : \mathbb{N} \to \mathbb{R}$ (or $f, g : (0, \infty) \to \mathbb{R}$) be functions with $g(n) > 0$ for all sufficiently large n.

- $f(n) = O(g(n))$ if there exist constants $C > 0$ and n_0 such that

$$|f(n)| \le C\, g(n) \qquad \forall n \ge n_0.$$

- $f(n) = \Omega(g(n))$ if there exist constants $c > 0$ and n_0 such that

$$|f(n)| \ge c\, g(n) \qquad \forall n \ge n_0.$$

- $f(n) = \Theta(g(n))$ if $f(n) = O(g(n))$ and $f(n) = \Omega(g(n))$. Equivalently, there exist constants $c, C > 0$ and n_0 such that

$$c\, g(n) \le |f(n)| \le C\, g(n) \qquad \forall n \ge n_0.$$

Remark 1.3.1 (Asymptotics Ignores Constant Factors (By Design)) When we write runtime $T(n) = O(n^2)$, we assert that for large n the runtime is bounded by a constant multiple of n^2. Different implementations may differ by large constants, but the *growth class* remains quadratic.

Definition 1.3.2 (Little-o and Little-ω) Let f, g be as above with $g(n) > 0$ eventually.

- $f(n) = o(g(n))$ if

$$\lim_{n\to\infty} \frac{f(n)}{g(n)} = 0.$$

- $f(n) = \omega(g(n))$ if

$$\lim_{n\to\infty} \frac{f(n)}{g(n)} = \infty.$$

Remark 1.3.2 (Little-o is Strictly Smaller Order) If $f = o(g)$, then $f = O(g)$, but not conversely. For example, $n = O(n)$ but $n \ne o(n)$, while $n = o(n \log n)$.

Proposition 1.3.1 (Limit Characterization of Θ) *Assume $g(n) > 0$ for all large n and $f(n) \ge 0$ for all large n. If the limit $L := \lim_{n\to\infty} f(n)/g(n)$ exists and satisfies $0 < L < \infty$, then $f(n) = \Theta(g(n))$. If $L = 0$ then $f(n) = o(g(n))$; if $L = \infty$ then $f(n) = \omega(g(n))$.*

Example 1.3.1 (Common Runtime Orders) For $n \to \infty$:

$$\log n = o(n^\alpha) \quad \text{for any } \alpha > 0, \qquad n^\alpha = o(a^n) \quad \text{for any } a > 1,$$
$$a^n = o(n!) \quad \text{for any } a > 1.$$

Thus, logarithmic < polynomial < exponential < factorial growth.

Remark 1.3.3 (Landau Notation is About a Specific Asymptotic Regime) Always state (at least implicitly) *what is going to infinity*. In ML, we may consider $n \to \infty$ with fixed d, or high-dimensional regimes where $d = d(n) \to \infty$ and $d/n \to \gamma \in (0, \infty)$, or accuracy-driven regimes $\varepsilon \downarrow 0$. The same algorithm may behave very differently under different regimes.

Definition 1.3.3 (Asymptotic Upper Bounds for Algorithms) If an algorithm on input size n runs in time $T(n)$ (worst-case), we say it runs in *time* $O(g(n))$ if

$T(n) = O(g(n))$. Similarly for memory $M(n)$. When randomness is involved, one may specify expected time $\mathbb{E}[T(n)]$ or high-probability bounds.

Remark 1.3.4 (Worst-Case vs. Average-Case vs. Amortized) Algorithmic bounds can be worst-case (guaranteed), average-case (under an input distribution), or amortized (average over a sequence of operations). ML systems often have distributional structure, so both worst-case and typical-case perspectives can matter.

1.3.1.2 Log Factors, Poly(d) vs. Exponential Dependence

Definition 1.3.4 (Polylog and Polynomial) We say $f(n)$ is *polylogarithmic* if $f(n) = (\log n)^{O(1)}$. We say $f(n)$ is *polynomial* if $f(n) = n^{O(1)}$. We say $f(n)$ is *exponential* if $f(n) = a^{\Theta(n)}$ for some $a > 1$.

Remark 1.3.5 (Why Log Factors are Often "Harmless") A runtime $n^2 \log n$ is asymptotically larger than n^2, but in many practical ranges it behaves similarly. In learning theory, log factors often arise from union bounds, covering numbers, or confidence parameters δ (e.g. $\log(1/\delta)$). One often sees bounds stated "up to log factors".

Proposition 1.3.2 (Logarithms Are Dominated By Any Polynomial) *For any* $\alpha > 0$,

$$(\log n)^k = o(n^\alpha) \qquad \textit{for every fixed } k \in \mathbb{N}.$$

Example 1.3.2 (Poly(d) vs. exp(d) in High Dimension) Suppose a method scales as $O(d^3)$ in dimension d (e.g. naive matrix inversion), while an alternative scales as $O(2^d)$ (e.g. exhaustive subset search). For $d = 50$, $d^3 = 125000$ is feasible, but $2^{50} \approx 1.13 \times 10^{15}$ is not. Thus, when d is moderately large, exponential dependence on d is typically prohibitive.

Remark 1.3.6 (Common ML Meanings of "Polynomial Time") In theoretical CS, "efficient" often means polynomial in the input size. In ML, the input size may involve both n (samples) and d (features), and sometimes model size p. A method polynomial in n but exponential in d may be unusable in high-dimensional settings. Always record which parameters appear in the exponent.

Definition 1.3.5 (Soft-O Notation) We write $\widetilde{O}(g(n))$ to mean $O(g(n) \cdot (\log n)^{O(1)})$; i.e. we hide polylogarithmic factors. Similarly, $\widetilde{\Omega}$ and $\widetilde{\Theta}$ hide polylog factors.

Remark 1.3.7 (When $\widetilde{O}$ is Appropriate) Soft-O is convenient when log factors are not central to the discussion (e.g. when comparing n^2 vs. n^3). However, in tight comparisons (or when logs carry meaning such as confidence δ), log factors should be stated explicitly.

1.3.2 *Asymptotics in Statistical Learning*

1.3.2.1 Sample Complexity vs. Dimension: Informal Scaling Heuristics

A central question in ML is: *How many samples n are needed to learn a model to a desired accuracy?* This is the domain of *sample complexity*. Precise answers depend on assumptions and the complexity of the hypothesis class, but several robust scaling heuristics recur.

Remark 1.3.8 (A Prototype Decomposition: Approximation + Estimation) In many settings, the expected prediction error can be informally decomposed as

$$\text{(total error)} \approx \text{(approximation error)} + \text{(estimation error)} + \text{(optimization error)}.$$

- Approximation error decreases with richer model classes.
- Estimation error typically decreases with more data, often like $O(1/\sqrt{n})$ up to complexity factors.
- Optimization error depends on computation/iterations and may dominate if training is truncated.

This decomposition is conceptual, but it guides how we read asymptotic statements.

Definition 1.3.6 (A Minimal Notion of Risk (Informal, for Scaling)) Let (X, Y) be a random pair on $\mathcal{X} \times \mathcal{Y}$ and let $\ell(\hat{y}, y) \geq 0$ be a loss. For a predictor h, its (population) risk is $R(h) := \mathbb{E}[\ell(h(X), Y)]$. Given data $\mathcal{D}$, an estimator produces $\widehat{h} = \widehat{h}(\mathcal{D})$ and we are interested in how $R(\widehat{h}) - \inf_{h \in \mathcal{H}} R(h)$ scales with n, d, and model complexity.

Remark 1.3.9 (Typical $1/\sqrt{n}$ Scaling) In many classical settings with bounded variance (and without strong structural assumptions), estimation error exhibits $O(1/\sqrt{n})$ behavior (e.g. due to concentration of empirical averages). This scaling is a statistical analogue of CLT-type behavior and often appears (up to constants/logs) in generalization bounds.

Example 1.3.3 (Parametric Heuristic: Effective Degrees of Freedom) For a well-specified p-parameter model, a common heuristic is

$$\text{estimation error} \sim \sqrt{\frac{p}{n}} \quad \text{(up to constants and logs).}$$

This suggests that if p grows with n, then learning requires n to dominate p in the sense $p/n \to 0$ for vanishing estimation error. While the precise statement depends on the model and loss, the scaling captures a broad truth: *more degrees of freedom require more data.*

Remark 1.3.10 (High-Dimensional Regimes) When d (or p) is comparable to n, classical asymptotics (fixed dimension, $n \to \infty$) can fail. Modern analyses often consider $d = d(n) \to \infty$ with $d/n \to \gamma$ or sparsity regimes where only $s \ll d$ coordinates matter. The key message is: the *relevant* complexity is not always raw dimension d, but an effective complexity such as sparsity, rank, margin, or norm constraints.

Remark 1.3.11 (Curse and Blessing of Dimensionality) **Curse:** Without structure, high-dimensional spaces require exponentially many samples to cover. **Blessing:** With structure (sparsity, low-rank, compositionality), the effective complexity can be much smaller than d, allowing sample complexity to scale with the structured parameter (e.g. s or rank r) rather than d. This theme will recur in chapters on regularization, kernels, and representation learning.

1.3.2.2 Computational vs. Statistical Tradeoffs (Time–Accuracy Regimes)

In ML, we rarely get unlimited computation. We therefore care about two questions simultaneously:

1. How does error scale with data (statistics)?
2. How does error scale with computation (optimization/algorithms)?

Their interaction yields *time–accuracy* tradeoffs (cf. Fig. 1.3).

Definition 1.3.7 (Three Sources of Error: A Computational Lens) Let $\widehat{R}_n(h)$ denote an empirical objective minimized in training and let $\widehat{h}_T$ be the predictor

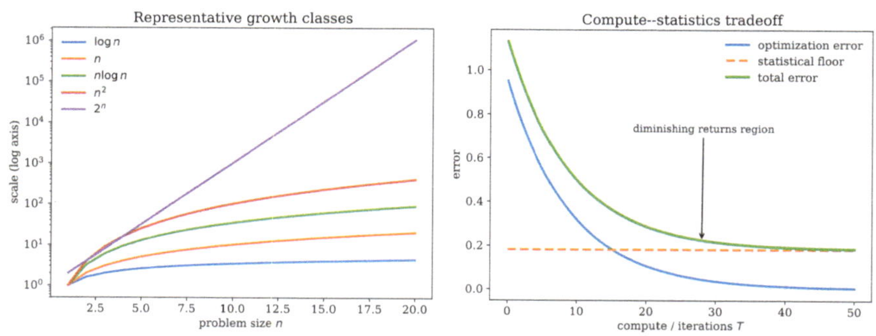

Fig. 1.3 Left: representative asymptotic growth classes. Right: a conceptual compute–statistics picture in which optimization error decreases with more computation until a statistical floor dominates further gains

obtained after T optimization steps. Then we may informally write

$$R(\widehat{h}_T) - \inf_{h\in\mathcal{H}} R(h) \approx \underbrace{\left(\inf_{h\in\mathcal{H}} R(h) - R^\star\right)}_{\text{approximation (model bias)}} + \underbrace{\left(R(\widehat{h}_n) - \inf_{h\in\mathcal{H}} R(h)\right)}_{\text{estimation (generalization)}} + \underbrace{\left(R(\widehat{h}_T) - R(\widehat{h}_n)\right)}_{\text{optimization (compute)}},$$

where $R^\star$ is the Bayes risk and $\widehat{h}_n$ denotes an (ideal) exact empirical minimizer. The terms are not always separable, but the decomposition is a useful organizing principle.

Remark 1.3.12 (Optimization Error Can Dominate Early, Estimation Error Dominates Late) In early training, optimization error is typically large and dominates. As T increases, optimization error decreases until it becomes comparable to the estimation error, beyond which additional computation yields diminishing returns unless we change the model class, data, or regularization.

Example 1.3.4 (Accuracy-Driven Complexity in Convex Optimization) For many smooth convex objectives, first-order methods reach ε-suboptimality in $T = O(1/\varepsilon)$ (or $O(\log(1/\varepsilon))$ under strong convexity) iterations. If each iteration costs $O(nd)$, the total computational cost may scale as $O(nd/\varepsilon)$ or $O(nd\log(1/\varepsilon))$. Thus, even when statistics says "more data helps", computation may become the bottleneck.

Remark 1.3.13 (Statistical Accuracy Sets a Sensible Optimization Target) It is often wasteful to optimize far beyond the statistical precision allowed by the data. If generalization error is on the order of $\sqrt{p/n}$, driving optimization error far below that scale may not improve test performance. This insight motivates early stopping and approximate solvers.

Remark 1.3.14 (Sublinear-Time and Streaming Perspectives) For massive datasets, one aims for algorithms whose cost is sublinear in n per pass (e.g. mini-batch SGD), or that use one-pass/streaming computation. In such regimes, the key asymptotic question becomes: *how does test error scale with the number of gradient evaluations or processed samples?*

Remark 1.3.15 (Practical Reading Guide) When you see a bound or runtime claim, ask:

- Which variable is asymptotic ($n, d, p, T, \varepsilon^{-1}, \delta^{-1}$)?
- Are constants suppressed benign (implementation constants) or meaningful (noise level, margin, condition number)?
- Are there hidden log factors (e.g. $\log(1/\delta)$ for confidence)?
- Does the dependence on d appear in an exponent?

These questions prevent misinterpretation and guide correct algorithm/model choices.

1.4 Common Symbols and Conventions

This section fixes notation that will be used consistently across the book. The goal is not to be exhaustive, but to provide a reliable reference so that later derivations can be read unambiguously. Whenever a convention differs across communities (statistics vs. ML systems vs. numerical linear algebra), we state what we adopt here.

1.4.1 Optimization and Estimation Notation

1.4.1.1 arg min / arg max, Constraints, and Feasible Sets

Definition 1.4.1 (Minimizers, Maximizers, and the Arg Operators) Let $f : \mathcal{S} \to \mathbb{R}$ be an objective defined on a set $\mathcal{S}$.

- The *minimum value* (infimum) is $\inf_{x\in\mathcal{S}} f(x)$ and, if attained, the *minimum* is $\min_{x\in\mathcal{S}} f(x)$.
- The *set of minimizers* is

$$\arg\min_{x\in\mathcal{S}} f(x) := \{x^\star \in \mathcal{S} : f(x^\star) = \inf_{x\in\mathcal{S}} f(x)\}.$$

- Analogously,

$$\arg\max_{x\in\mathcal{S}} f(x) := \{x^\star \in \mathcal{S} : f(x^\star) = \sup_{x\in\mathcal{S}} f(x)\}.$$

Remark 1.4.1 (Argmin Returns a Set) The expression arg min returns a *set* (possibly empty, possibly containing many points). In informal writing one often selects an arbitrary element and writes $x^\star \in \arg\min_{x\in\mathcal{S}} f(x)$. We will follow this convention when uniqueness is not guaranteed.

Definition 1.4.2 (Constrained Optimization and Feasible Set) A *constrained optimization problem* has the form

$$\min_{x\in\mathcal{S}} f(x) \quad \text{subject to } x \in \mathcal{C},$$

where $\mathcal{S}$ is the ambient domain (often $\mathbb{R}^d$) and $\mathcal{C} \subseteq \mathcal{S}$ encodes constraints. The *feasible set* is $\mathcal{F} := \mathcal{S} \cap \mathcal{C}$, and the problem is equivalently $\min_{x\in\mathcal{F}} f(x)$.

Example 1.4.1 (Regularization as Constrained vs. Penalized Form) For a loss $L(w)$ and norm $\|\cdot\|$, the two forms

$$\min_{w\in\mathbb{R}^d} L(w) + \lambda\|w\| \qquad \text{and} \qquad \min_{w\in\mathbb{R}^d} L(w) \text{ s.t. } \|w\| \le r$$

are closely related (often equivalent for suitable correspondence between λ and r under convexity). This dual viewpoint will appear repeatedly (ridge, lasso, margin constraints) (Boyd and Vandenberghe 2004).

Remark 1.4.2 (Standard Symbols) Throughout:

$$n = \text{number of samples}, \quad d = \text{dimension of features}, \quad p = \text{number of parameters},$$

$$T = \text{iterations/steps}, \quad \varepsilon = \text{accuracy tolerance}, \quad \delta = \text{failure probability}.$$

1.4.1.2 Indicator Functions, Support, and Event Notation

Definition 1.4.3 (Indicator Function) For an event or condition E, the *indicator* $\mathbf{1}\{E\}$ (also written $\mathbb{I}\{E\}$) is

$$\mathbf{1}\{E\} := \begin{cases} 1, & \text{if } E \text{ is true}, \\ 0, & \text{otherwise}. \end{cases}$$

For a set A, the indicator of membership is $\mathbf{1}\{x \in A\}$.

Remark 1.4.3 (Indicators Turn Logic into Algebra) Indicators obey useful algebraic identities, e.g.

$$\mathbf{1}\{E \cap F\} = \mathbf{1}\{E\}\mathbf{1}\{F\}, \qquad \mathbf{1}\{E \cup F\} \le \mathbf{1}\{E\} + \mathbf{1}\{F\}.$$

These are frequently used in probability bounds and learning-theoretic arguments.

Definition 1.4.4 (Support: Discrete and Continuous)

- For a vector $x \in \mathbb{R}^d$, its *support* is

$$\operatorname{supp}(x) := \{i \in [d] : x_i \neq 0\}.$$

- For a discrete random variable X with pmf $p(x) = \mathbb{P}(X = x)$, its support is $\operatorname{supp}(X) := \{x : p(x) > 0\}$.
- For a density p on $\mathbb{R}^d$, one common notion is $\operatorname{supp}(p) := \overline{\{x : p(x) > 0\}}$ (closure of the set where the density is positive).

Remark 1.4.4 (Support in ML Practice) Sparse modeling is about controlling $|\operatorname{supp}(w)|$ (number of nonzeros) or approximating it. In probabilistic modeling, support matters for validity of likelihoods, change-of-variables, and constraints (e.g. simplex-valued variables).

Definition 1.4.5 (Event Notation) Given a random variable X and a set A, the event $\{X \in A\}$ means $\{\omega : X(\omega) \in A\}$. For real-valued X, common events are

$$\{X \le t\}, \{X > t\}, \{|X| \ge t\}, \{\|X\| \le r\}.$$

We use $\mathbb{P}(E)$ for the probability of an event E and $\mathbb{E}[\cdot]$ for expectation.

Remark 1.4.5 (Naming Conventions) We typically use uppercase letters for random variables (X, Y), lowercase for realizations (x, y), calligraphic letters for sets/spaces $(\mathcal{X}, \mathcal{H})$, and boldface for vectors when needed $(\mathbf{x})$. However, in many ML contexts, X is also used for the data matrix; context will disambiguate, and we will explicitly state shapes when ambiguity is possible.

1.4.2 *Linear-Algebraic Conventions*

1.4.2.1 Inner Products, Norms, Orthogonality, and Projections

Definition 1.4.6 (Euclidean Inner Product) For $x, y \in \mathbb{R}^d$, the (standard) inner product is

$$\langle x, y \rangle := x^\top y = \sum_{i=1}^{d} x_i y_i.$$

Definition 1.4.7 (Norms) For $x \in \mathbb{R}^d$, the standard norms are

$$\|x\|_2 := \sqrt{\langle x, x \rangle}, \qquad \|x\|_1 := \sum_{i=1}^{d} |x_i|, \qquad \|x\|_\infty := \max_{i \in [d]} |x_i|.$$

For a matrix $A \in \mathbb{R}^{m \times n}$, the Frobenius norm is

$$\|A\|_F := \sqrt{\sum_{i=1}^{m} \sum_{j=1}^{n} A_{ij}^2} = \sqrt{\mathrm{tr}(A^\top A)}.$$

Definition 1.4.8 (Orthogonality) Vectors $x, y \in \mathbb{R}^d$ are *orthogonal* if $\langle x, y \rangle = 0$. A matrix $Q \in \mathbb{R}^{d \times d}$ is *orthogonal* if $Q^\top Q = I$ (equivalently $Q^{-1} = Q^\top$).

Remark 1.4.6 (Geometric Meaning) Orthogonality formalizes "no linear correlation" in Euclidean geometry. In least squares, the residual is orthogonal to the fitted subspace; in PCA, principal components are orthogonal directions.

Definition 1.4.9 (Projection Matrix) A matrix $P \in \mathbb{R}^{d \times d}$ is a *projection* if $P^2 = P$ (idempotent). It is an *orthogonal projection* if additionally $P^\top = P$.

Proposition 1.4.1 (Key Properties of Orthogonal Projections) *If P is an orthogonal projection, then:*

1. $\mathbb{R}^d = range(P) \oplus null(P)$ *(orthogonal direct sum),*
2. $\|Px\|_2 \le \|x\|_2$ *for all* x,
3. $x - Px$ *is orthogonal to* $range(P)$.

Remark 1.4.7 (Notation for Subspaces) We use $\text{span}\{v_1, \ldots, v_k\}$ for the span, $\text{range}(A)$ for the column space, and $\text{null}(A)$ for the null space. For a subspace $\mathcal{U}$, the orthogonal complement is $\mathcal{U}^\perp$.

1.4.2.2 Transpose, Adjoint, Trace, and Standard Matrix Identities

Definition 1.4.10 (Transpose and Adjoint) For a real matrix $A \in \mathbb{R}^{m\times n}$, $A^\top \in \mathbb{R}^{n\times m}$ denotes the transpose. For complex matrices, the *adjoint* (conjugate transpose) is $A^* := \overline{A}^\top$. Unless stated otherwise, this book works over $\mathbb{R}$, so $A^* = A^\top$.

Definition 1.4.11 (Trace) For a square matrix $A \in \mathbb{R}^{d\times d}$, the trace is

$$\text{tr}(A) := \sum_{i=1}^{d} A_{ii}.$$

Proposition 1.4.2 (Trace Identities) *Let A, B be matrices of compatible sizes.*

1. *Linearity:* $tr(A + B) = tr(A) + tr(B)$ *and* $tr(cA) = c\,tr(A)$.
2. *Cyclicity: if products are defined,* $tr(AB) = tr(BA)$. *More generally,* $tr(ABC) = tr(BCA) = tr(CAB)$.
3. *Frobenius inner product: for* $A, B \in \mathbb{R}^{m\times n}$,

$$\langle A, B\rangle_F = tr(A^\top B).$$

Remark 1.4.8 (Cyclicity is Not Commutativity) The identity $\text{tr}(AB) = \text{tr}(BA)$ does *not* mean $AB = BA$. It only states equality of traces when the products are defined.

Proposition 1.4.3 (Standard Transpose Rules) *For compatible matrices/vectors:*

$$(A^\top)^\top = A, \quad (A+B)^\top = A^\top + B^\top, \quad (AB)^\top = B^\top A^\top, \quad (x^\top Ay) = y^\top A^\top x.$$

If A is invertible, then $(A^{-1})^\top = (A^\top)^{-1}$.

Definition 1.4.12 (Identity and Zero Matrices) I_d denotes the $d \times d$ identity matrix, and 0 denotes a zero vector/matrix whose shape is clear from context.

Remark 1.4.9 (Symmetry and Positive Semidefiniteness) A matrix $A \in \mathbb{R}^{d\times d}$ is *symmetric* if $A^\top = A$. It is *positive semidefinite* (PSD), written $A \succeq 0$, if $x^\top Ax \ge 0$

for all $x \in \mathbb{R}^d$. PSD matrices arise as covariance matrices, Gram matrices, and Hessians of convex quadratics.

Example 1.4.2 (A Compact Quadratic Form Identity) For $x \in \mathbb{R}^d$ and $A \in \mathbb{R}^{d\times d}$,

$$x^\top Ax = \mathrm{tr}(x^\top Ax) = \mathrm{tr}(Axx^\top).$$

This simple identity is frequently used in deriving gradients and connecting quadratic forms to covariance-like objects.

Remark 1.4.10 (Notational Warnings)

- $\|A\|$ without a subscript may denote different norms in different contexts (spectral, Frobenius, operator). We will specify $\|A\|_F$ (Frobenius) and $\|A\|_2$ (spectral/operator) when needed.
- X may denote a random variable or a data matrix; context and shape annotations resolve ambiguity.
- $\mathbb{1}$ denotes an all-ones vector; its dimension will be clear from context.

1.5 Exercises

Unless explicitly stated otherwise, assume the **column-vector convention** and use the symbols and rules fixed in this chapter.

Exercise 1.5.1 (Set Algebra and Preimages under a Map) Let $\mathcal{X} = \{1, 2, 3, 4, 5, 6, 7, 8, 9, 10\}$ and define $f : \mathcal{X} \to \mathbb{Z}$ by $f(x) = x \bmod 4$ (so $f(x) \in \{0, 1, 2, 3\}$). Let $A = \{1, 3, 4, 7, 9, 10\}$ and $B = \{2, 3, 6, 7, 8, 10\}$. Compute explicitly:

(i) $A \cup B$, $A \cap B$, $A \setminus B$, and $(A \cup B)^c$ (complement in $\mathcal{X}$);
(ii) $f(A)$ and $f(B)$;
(iii) $f^{-1}(\{1, 3\})$, $f^{-1}(f(A))$, and $f^{-1}(\{0\}) \cap (A \cup B)$; and
(iv) verify directly on these sets the identities in Proposition 1.1.2 for $T_1 = \{1\}$ and $T_2 = \{3\}$ by computing both sides.

Exercise 1.5.2 (Decision Regions as Preimages; Partition Induced by a Classifier) Let $\mathcal{X} = \mathbb{R}^2$ and define the classifier $h : \mathcal{X} \to \{-1, +1\}$ by

$$h(x) = \begin{cases} +1, & \text{if } w^\top x + b \ge 0, \\ -1, & \text{if } w^\top x + b < 0, \end{cases} \qquad w = \begin{bmatrix} 2 \\ -1 \end{bmatrix}, \qquad b = -3.$$

Compute the decision regions $R_+ = h^{-1}(\{+1\})$ and $R_- = h^{-1}(\{-1\})$ as explicit subsets of $\mathbb{R}^2$ described by inequalities. Then, for the finite dataset $\mathcal{D} = \{x_1, \ldots, x_6\}$ with

$$x_1 = \begin{bmatrix} 0 \\ 0 \end{bmatrix}, \ x_2 = \begin{bmatrix} 2 \\ 1 \end{bmatrix}, \ x_3 = \begin{bmatrix} 1 \\ -1 \end{bmatrix}, \ x_4 = \begin{bmatrix} 3 \\ 0 \end{bmatrix}, \ x_5 = \begin{bmatrix} 2 \\ 3 \end{bmatrix}, \ x_6 = \begin{bmatrix} 4 \\ 2 \end{bmatrix},$$

compute $h(x_i)$ for each i and show that the family $\{R_+ \cap \mathcal{D},\ R_- \cap \mathcal{D}\}$ is a partition of $\mathcal{D}$ in the sense of Theorem 1.1.1.

Exercise 1.5.3 (Equivalence Relation from a Graph; Connected Components) Let $V = \{1, 2, 3, 4, 5, 6, 7, 8\}$ and let $G = (V, E)$ be an undirected graph with edge set

$$E = \{(1, 2), (2, 3), (3, 1), (4, 5), (5, 6), (6, 4), (6, 7)\}.$$

Define a relation $\sim$ on V by $i \sim j$ iff there exists a path between i and j in G. Compute all equivalence classes $[i]$ and the quotient set $V/\sim$, and verify explicitly that $\sim$ is an equivalence relation by checking reflexivity, symmetry, and transitivity from the path definition (do not appeal to known theorems without argument).

Exercise 1.5.4 (Shapes and Consistent Matrix Calculus) Let $n = 4$ and $d = 3$. Consider the data matrix $X \in \mathbb{R}^{n\times d}$ and vectors $w \in \mathbb{R}^d$, $b \in \mathbb{R}$ given by

$$X = \begin{bmatrix} 1 & 0 & 2 \\ -1 & 3 & 1 \\ 0 & 2 & -2 \\ 4 & 1 & 0 \end{bmatrix}, \qquad w = \begin{bmatrix} 2 \\ -1 \\ 1 \end{bmatrix}, \qquad b = -2.$$

(i) Compute the score vector $s = Xw + b\mathbf{1} \in \mathbb{R}^n$ explicitly.
(ii) Define $\mu = \frac{1}{n}\sum_{i=1}^{n} X_{i:}^\top \in \mathbb{R}^d$ and compute μ and the centered matrix $X_c = X - \mathbf{1}\mu^\top$.
(iii) Compute $X_c^\top \mathbf{1}$ and interpret the result.
(iv) Let $A = X_c^\top X_c \in \mathbb{R}^{d\times d}$; compute A and verify that A is symmetric.

Exercise 1.5.5 (Broadcasting and Explicit Replication) Let $X \in \mathbb{R}^{3\times 4}$ and $v \in \mathbb{R}^4$ be

$$X = \begin{bmatrix} 2 & -1 & 0 & 3 \\ 1 & 4 & -2 & 0 \\ -3 & 2 & 1 & 5 \end{bmatrix}, \qquad v = \begin{bmatrix} 1 \\ 0 \\ -1 \\ 2 \end{bmatrix}.$$

Define $Y = X - \mathbf{1}v^\top \in \mathbb{R}^{3\times 4}$ and compute Y explicitly. Then compute the row-wise maxima $m \in \mathbb{R}^3$ defined by $m_i = \max_{j\in[4]} Y_{ij}$. Finally, define $Z \in \mathbb{R}^{3\times 4}$ by $Z_{ij} = Y_{ij} - m_i$ (row-wise centering) and compute Z.

Exercise 1.5.6 (Block Vectors and Block Matrices: Multi-View Fusion) Let $x^{(1)} \in \mathbb{R}^2$, $x^{(2)} \in \mathbb{R}^3$ and define the block vector

$$x = \begin{bmatrix} x^{(1)} \\ x^{(2)} \end{bmatrix} \in \mathbb{R}^5, \qquad x^{(1)} = \begin{bmatrix} 1 \\ -2 \end{bmatrix}, \quad x^{(2)} = \begin{bmatrix} 0 \\ 3 \\ 1 \end{bmatrix}.$$

Let $A \in \mathbb{R}^{4\times 5}$ be block-partitioned as $A = [A_1 \;\; A_2]$ with

$$A_1 = \begin{bmatrix} 1 & 0 \\ -1 & 2 \\ 0 & 3 \\ 2 & -2 \end{bmatrix} \in \mathbb{R}^{4\times 2}, \qquad A_2 = \begin{bmatrix} 0 & 1 & -1 \\ 2 & 0 & 1 \\ -1 & 1 & 0 \\ 3 & -2 & 2 \end{bmatrix} \in \mathbb{R}^{4\times 3}.$$

(i) Compute Ax using the block multiplication identity in Proposition 1.2.2 by first computing $A_1x^{(1)}$ and $A_2x^{(2)}$.

(ii) Let $r \in \mathbb{R}^4$ be the residual $r = Ax - y$ with $y = \begin{bmatrix} 1 \\ 0 \\ -1 \\ 2 \end{bmatrix}$; compute r and its ℓ_2 norm.

(iii) Define a block-regularized objective $J(x) = \|Ax - y\|_2^2 + \lambda_1\|x^{(1)}\|_2^2 + \lambda_2\|x^{(2)}\|_2^2$ with $\lambda_1 = 1$, $\lambda_2 = 2$; compute $J(x)$.

Exercise 1.5.7 (Tensor Indexing and Unfolding) Let $T \in \mathbb{R}^{2\times 3\times 2}$ be defined by its entries

$$T_{i,j,k} = \begin{cases} i + j + k, & \text{if } (i, j, k) \in \{1, 2\} \times \{1, 2, 3\} \times \{1, 2\}, \end{cases}$$

where indices start at 1.

(i) Write out all entries of T explicitly (as two 2×3 "slices" for $k = 1$ and $k = 2$).

(ii) Form the mode-1 unfolding $T_{(1)} \in \mathbb{R}^{2\times 6}$ by stacking mode-1 fibers as columns in lexicographic order of (j, k):

$$(j, k) = (1, 1), (2, 1), (3, 1), (1, 2), (2, 2), (3, 2).$$

Compute $T_{(1)}$ explicitly.

(iii) Let $u = \begin{bmatrix} 1 \\ -1 \end{bmatrix}$ and define $S \in \mathbb{R}^{3\times 2}$ by the contraction $S_{j,k} = \sum_{i=1}^{2} u_i\, T_{i,j,k}$. Compute S explicitly.

Exercise 1.5.8 (Trace, Frobenius Inner Product, and a Quadratic Identity) Let

$$A = \begin{bmatrix} 2 & -1 & 0 \\ -1 & 3 & 1 \\ 0 & 1 & 4 \end{bmatrix}, \qquad B = \begin{bmatrix} 1 & 2 & -1 \\ 0 & -1 & 3 \\ 2 & 1 & 0 \end{bmatrix}.$$

(i) Compute $\langle A, B\rangle_F = \sum_{i,j} A_{ij} B_{ij}$ directly and compute $\mathrm{tr}(A^\top B)$; verify Proposition 1.4.2(3).

(ii) For $x = \begin{bmatrix} 1 \\ -2 \\ 1 \end{bmatrix}$ compute $x^\top Ax$ and verify the identity $x^\top Ax = \mathrm{tr}(Axx^\top)$ by computing $xx^\top$ and then $Axx^\top$.

Exercise 1.5.9 (Orthogonal Projection via a Given Basis) Let $U \in \mathbb{R}^{3\times 2}$ have columns

$$u_1 = \begin{bmatrix} 1 \\ 1 \\ 0 \end{bmatrix}, \qquad u_2 = \begin{bmatrix} 1 \\ -1 \\ 2 \end{bmatrix}, \qquad U = \begin{bmatrix} u_1 \; u_2 \end{bmatrix}.$$

Assume u_1 and u_2 are linearly independent and define the orthogonal projector onto range(U) by

$$P := U(U^\top U)^{-1}U^\top.$$

(i) Compute $U^\top U$, $(U^\top U)^{-1}$, and hence P explicitly.
(ii) Verify directly that $P^\top = P$ and $P^2 = P$.
(iii) For $x = \begin{bmatrix} 2 \\ 0 \\ 1 \end{bmatrix}$ compute Px and the residual $r = x - Px$, and check that $U^\top r = 0$.

Exercise 1.5.10 (Asymptotics: Compare Scaling Laws Quantitatively) Consider the following candidate runtimes as functions of n:

$$T_1(n) = n^2 \log n, \qquad T_2(n) = n^{2.2}, \qquad T_3(n) = n \cdot 2^{\sqrt{\log_2 n}}.$$

(i) Determine (with justification using Definitions 1.3.1 and 1.3.2) which of the relations $T_i(n) = o(T_j(n))$ hold as $n \to \infty$.
(ii) Evaluate $T_1(n)$, $T_2(n)$, $T_3(n)$ numerically at $n \in \{10^2, 10^4, 10^8\}$ (use natural logs for log and base-2 logs for $\log_2$) and comment briefly on whether the finite-n ordering matches the asymptotic ordering.

Exercise 1.5.11 (Indicator Algebra and Event Manipulation) Let $\Omega = \{1, 2, \ldots, 12\}$ be equally likely outcomes, and define events

$$E = \{2, 4, 6, 8, 10, 12\}, \qquad F = \{1, 2, 3, 4, 5, 6\}, \qquad G = \{3, 6, 9, 12\}.$$

(i) Compute $\mathbf{1}\{E \cap F\}$, $\mathbf{1}\{E \cup G\}$, and $\mathbf{1}\{F \setminus G\}$ as functions on Ω by listing their values on each $\omega \in \Omega$.

(ii) Using indicator identities, verify the equality

$$\mathbf{1}\{E \cup G\} = \mathbf{1}\{E\} + \mathbf{1}\{G\} - \mathbf{1}\{E \cap G\}$$

pointwise on Ω. (iii) Compute $\mathbb{P}(E)$, $\mathbb{P}(F)$, $\mathbb{P}(G)$, $\mathbb{P}(E \cap G)$, and $\mathbb{P}(E \cup G)$.

Chapter 2
Geometry of $\mathbb{R}^n$: Norms, Inner Products, Projections

Many learning problems are geometric at their core. Data points form *clouds* in an ambient Euclidean space, models carve the space into *decision regions*, and training algorithms move through parameter space by following *gradients* whose meaning depends on an underlying inner product. Regularization, robustness, generalization, and optimization can often be understood by asking a single question: *what is the geometry of the objects we are manipulating?* This chapter develops a precise language for that geometry—norms, inner products, orthogonality, and projections—and shows how these concepts already explain common design choices in ML.these concepts already explain common design choices in ML (Strang 2019) (cf. Fig. 2.1).

2.1 Why Geometry Matters in Machine Learning

2.1.1 Geometric Objects Behind Learning Problems

2.1.1.1 Data Clouds, Hypothesis Classes, and Decision Regions

Definition 2.1.1 (Data Cloud and Dataset Geometry) A (finite) dataset of inputs is a set (or indexed family) $\mathcal{D} = \{x_i\}_{i=1}^n \subseteq \mathbb{R}^d$. We call $\mathcal{D}$ a *data cloud* in the ambient space $\mathbb{R}^d$. Associated geometric summaries include the empirical mean

$$\bar{x} := \frac{1}{n}\sum_{i=1}^{n} x_i \in \mathbb{R}^d,$$

and the centered cloud $\{x_i - \bar{x}\}_{i=1}^n$.

Remark 2.1.1 (Geometry is a Model of Similarity) When we embed objects (images, texts, graphs) into $\mathbb{R}^d$, we are implicitly declaring that *closeness in* $\mathbb{R}^d$

P. Singh, B. Raman, *Foundations of Machine Learning and AI*, Studies in Big Data 199, https://doi.org/10.1007/978-3-032-30336-3_2

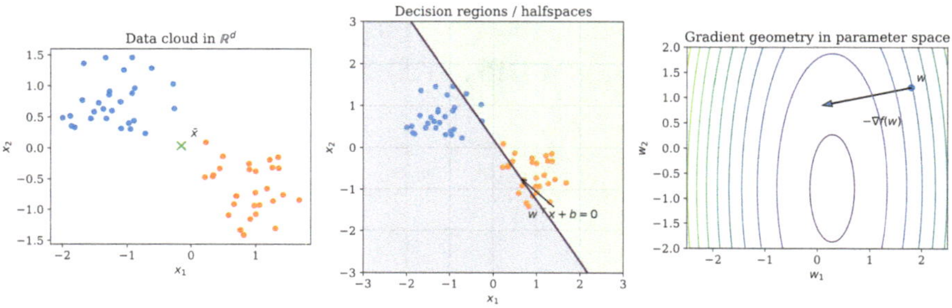

Fig. 2.1 Three geometric primitives behind learning: a data cloud in ambient space, a linear decision boundary partitioning the input space, and a parameter-space objective whose gradient direction depends on the chosen geometry

reflects *semantic similarity* in the original domain. Thus, the geometry of the embedding space is part of the modeling choice, not merely a computational convenience.

Definition 2.1.2 (Hypothesis Class and Parameterization) Let $\mathcal{X} \subseteq \mathbb{R}^d$ and $\mathcal{Y}$ be an output space. A *hypothesis class* is a set of functions $\mathcal{H} \subseteq \{h : \mathcal{X} \to \mathcal{Y}\}$. Often $\mathcal{H}$ is parameterized by a set Θ and a map $\Phi : \Theta \to \mathcal{H}$ so that $h_\theta := \Phi(\theta)$. The geometry of Θ (e.g., Euclidean, constrained, manifold) influences both optimization and generalization.

Definition 2.1.3 (Decision Regions as Preimages) Let $h : \mathcal{X} \to \mathcal{Y}$ be a predictor. For $y \in \mathcal{Y}$, its *decision region* is

$$R_y := h^{-1}(\{y\}) = \{x \in \mathcal{X} : h(x) = y\}.$$

The family $\{R_y\}_{y\in\mathcal{Y}}$ partitions $\mathcal{X}$ into regions of constant prediction (ignoring boundary conventions when $\mathcal{Y}$ is finite).

Example 2.1.1 (Linear Classification Induces Halfspaces) For binary classification on $\mathcal{X} = \mathbb{R}^d$, let

$$h(x) = \mathrm{sign}(w^\top x + b), \qquad w \in \mathbb{R}^d, \ b \in \mathbb{R}.$$

Then the decision regions are the halfspaces

$$R_+ = \{x : w^\top x + b \geq 0\}, \qquad R_- = \{x : w^\top x + b < 0\},$$

separated by the affine hyperplane $\{x : w^\top x + b = 0\}$. Training a linear classifier is therefore the problem of choosing a hyperplane that separates (or best fits) the data.

Example 2.1.2 (Nearest-Neighbor Classification and Voronoi Cells) Given labeled points $\{(x_i, y_i)\}_{i=1}^n \subseteq \mathbb{R}^d \times \mathcal{Y}$ and a distance $d(\cdot, \cdot)$ on $\mathbb{R}^d$, the 1-NN rule predicts $h(x) = y_{i^\star}$ where $i^\star \in \arg\min_{i\in[n]} d(x, x_i)$. The sets

$$V_i := \{x \in \mathbb{R}^d : d(x, x_i) \leq d(x, x_j)\ \forall j \in [n]\}$$

form a Voronoi-type partition of $\mathbb{R}^d$ (ties occur on boundaries). Thus, NN methods are explicitly geometric: they carve space into cells based on the chosen distance.

Remark 2.1.2 (Geometry Links Representation and Prediction) Examples 2.1.1–2.1.2 illustrate a key message: once an embedding $\mathcal{X} \subseteq \mathbb{R}^d$ and a notion of distance/angle is fixed, many models are simply different ways of partitioning $\mathcal{X}$.

2.1.1.2 Distances, Similarities, and Invariances

Definition 2.1.4 (Metric and Similarity) A *metric* on $\mathcal{X} \subseteq \mathbb{R}^d$ is a function $d : \mathcal{X} \times \mathcal{X} \to [0, \infty)$ satisfying: (i) $d(x, y) = 0 \Leftrightarrow x = y$, (ii) symmetry $d(x, y) = d(y, x)$, (iii) triangle inequality $d(x, z) \leq d(x, y) + d(y, z)$. A *similarity* is any function $s : \mathcal{X} \times \mathcal{X} \to \mathbb{R}$ intended to be larger for "more similar" pairs. Often $s(x, y) = \langle x, y\rangle$ or $s(x, y) = \frac{\langle x,y\rangle}{\|x\|_2\|y\|_2}$ (cosine similarity) once an inner product is fixed.

Remark 2.1.3 (Choice of Metric is a Modeling Choice) Different metrics can yield radically different partitions (cf. Example 2.1.2). In high-dimensional ML, the distinction between ℓ_1-, ℓ_2-, and ℓ_∞-geometries is not cosmetic: it encodes assumptions about sparsity, robustness to outliers, and the structure of perturbations.

Definition 2.1.5 (Group Action and Invariance) Let G be a group and $\mathcal{X} \subseteq \mathbb{R}^d$. A (left) *group action* is a map $\cdot : G \times \mathcal{X} \to \mathcal{X}$ such that $e \cdot x = x$ and $(g_1 g_2) \cdot x = g_1 \cdot (g_2 \cdot x)$ for all $g_1, g_2 \in G$. A predictor $h : \mathcal{X} \to \mathcal{Y}$ is *G-invariant* if

$$h(g \cdot x) = h(x) \qquad \forall g \in G,\ \forall x \in \mathcal{X}.$$

Remark 2.1.4 (Orbits and Quotients) The *orbit* of x is $\mathrm{Orb}(x) := \{g \cdot x : g \in G\}$. If h is G-invariant, then h is constant on each orbit, so the prediction depends only on the equivalence class of x under the relation $x \sim x'$ iff $x' \in \mathrm{Orb}(x)$ (cf. Chapter 1 on equivalence relations and partitions). Geometrically, invariance means "collapse along symmetry directions."

Example 2.1.3 (Translation Invariance as a Geometric Constraint) For images represented as vectors in $\mathbb{R}^d$, let G be a set of discrete translations acting on pixels. A classifier that is (approximately) invariant under G should assign the same label to an image and its translated version. Convolutional architectures hard-code such approximate invariances; in representation space, this corresponds to identifying points along translation orbits.

2.1.2 A "Shape Calculus" for Reasoning About Models

2.1.2.1 Ambient Space vs. Intrinsic Structure (Manifolds, Subspaces)

Definition 2.1.6 (Ambient vs. Intrinsic Dimension (Informal)) The *ambient space* is the vector space in which data are represented, typically $\mathbb{R}^d$. The *intrinsic structure* refers to the possibility that data effectively lie near a lower-dimensional set, such as a k-dimensional subspace, affine subspace, or (more generally) a k-dimensional manifold with $k \ll d$.

Definition 2.1.7 (Linear and Affine Models of Intrinsic Structure) A *linear subspace* $\mathcal{U} \subseteq \mathbb{R}^d$ is a subset closed under addition and scalar multiplication. An *affine subspace* is a translate $a + \mathcal{U} := \{a + u : u \in \mathcal{U}\}$ for some $a \in \mathbb{R}^d$.

Example 2.1.4 (Low-Dimensional Structure via Centering) If points $\{x_i\}_{i=1}^n$ lie near an affine subspace $a + \mathcal{U}$, then the centered points $x_i - \bar{x}$ lie near the linear subspace $\mathcal{U}$ (when $\bar{x}$ is close to a). This is why centering is a canonical preprocessing step before PCA.

Definition 2.1.8 (Embedded Manifold (Minimal Working Definition)) A subset $\mathcal{M} \subseteq \mathbb{R}^d$ is a *(smooth) k-dimensional embedded manifold* if for every $x \in \mathcal{M}$ there exists a neighborhood $U \subseteq \mathbb{R}^d$ of x and a smooth bijection between $\mathcal{M} \cap U$ and an open set in $\mathbb{R}^k$ with smooth inverse (a local coordinate chart).

Remark 2.1.5 (Why Intrinsic Structure Matters) When data concentrate near a low-dimensional structure, distances, angles, and projections can behave differently than in generic $\mathbb{R}^d$. Many techniques (PCA, manifold learning, representation learning) can be interpreted as attempts to recover or exploit intrinsic structure to improve generalization and computational efficiency.

2.1.2.2 Geometry as a Guide to Regularization and Optimization

Definition 2.1.9 (Regularization as Geometric Control) Given an empirical objective $\widehat{L}(w)$, a regularized problem has the form

$$\min_{w \in \mathbb{R}^d} \widehat{L}(w) + \lambda\, \Omega(w),$$

where $\Omega : \mathbb{R}^d \to [0, \infty)$ is a regularizer (often a norm or semi-norm). Geometrically, level sets $\{w : \Omega(w) \leq r\}$ constrain the "shape" of admissible solutions.

Remark 2.1.6 (Penalized vs. Constrained View) Under suitable conditions (notably convexity), the penalized form above is closely related to the constrained form

$$\min_{w \in \mathbb{R}^d} \widehat{L}(w) \quad \text{s.t.} \quad \Omega(w) \leq r.$$

Thus, choosing Ω is choosing a *geometry* in parameter space: Euclidean balls (ℓ_2), cross-polytopes (ℓ_1), or mixed norms for group structure.

Example 2.1.5 (Level Sets of ℓ_2 vs. ℓ_1) In $\mathbb{R}^2$, the ℓ_2 ball $\{w : \|w\|_2 \le r\}$ is a disk, while the ℓ_1 ball $\{w : \|w\|_1 \le r\}$ is a diamond. When minimizing a convex loss over these sets, the different boundary geometry explains why ℓ_1 constraints tend to produce sparse solutions (solutions often land on corners).

Definition 2.1.10 (Directional Derivative and Gradient (Geometric Characterization)) Let $f : \mathbb{R}^d \to \mathbb{R}$ be differentiable at w. The directional derivative in direction u is

$$Df(w)[u] := \lim_{t \to 0} \frac{f(w + tu) - f(w)}{t}.$$

Given an inner product $\langle \cdot, \cdot \rangle$, the *gradient* $\nabla f(w)$ is the unique vector satisfying

$$Df(w)[u] = \langle \nabla f(w), u \rangle \qquad \forall u \in \mathbb{R}^d.$$

Proposition 2.1.1 (Steepest Ascent Depends on the Norm) *Let f be differentiable at w and fix a norm $\|\cdot\|$ with dual norm $\|\cdot\|_*$. Then*

$$\max_{\|u\| \le 1} Df(w)[u] = \|\nabla f(w)\|_*,$$

and any maximizer $u^\star$ is a steepest-ascent direction *under $\|\cdot\|$. In particular, under the Euclidean norm $\|\cdot\|_2$, the steepest ascent direction is $u^\star = \nabla f(w)/\|\nabla f(w)\|_2$.*

Remark 2.1.7 (Optimization is Geometry in Motion) Gradient descent chooses steps by a geometric rule: move against the steepest ascent direction measured in a chosen norm/inner product. Changing the geometry (via preconditioning, normalization, or natural gradients) changes the meaning of "steepest" and can dramatically improve conditioning and convergence.

Example 2.1.6 (Quadratic Level Sets and Conditioning) Consider $f(w) = \frac{1}{2} w^\top A w - b^\top w$ with $A \succ 0$. Its level sets are ellipsoids aligned with eigenvectors of A. If A is ill-conditioned (eigenvalues vary widely), then the ellipsoids are elongated, and vanilla gradient descent may zig-zag. Preconditioning corresponds to changing coordinates so that the ellipsoids become closer to spheres.

2.2 Vector Spaces and Subspaces

The geometry of Euclidean learning problems rests on linear structure. Subspaces describe families of features, directions of variation, constraints, and latent representations. A large portion of ML can be viewed as selecting (or learning) an appropriate subspace: PCA chooses a low-dimensional subspace capturing variance;

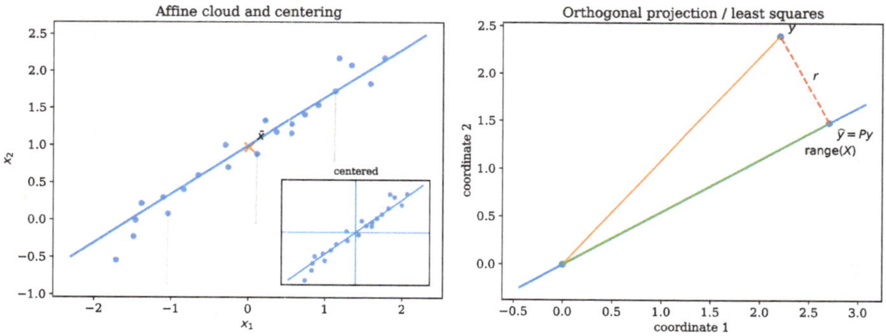

Fig. 2.2 Left: an affine data cloud, its centered version, and the associated low-dimensional subspace model. Right: orthogonal projection geometry underlying least squares, with fitted vector in the column space and residual orthogonal to it

linear regression projects onto a column space; regularization controls which directions in parameter space are used. We therefore develop the basic calculus of spans, bases, dimension, rank, and orthonormal coordinates (Axler 2015) (cf. Fig. 2.2).

2.2.1 *Span, Basis, and Dimension*

2.2.1.1 Linear Independence and Coordinate Representations

Definition 2.2.1 (Vector Space and Subspace) A *vector space* over $\mathbb{R}$ is a set V equipped with addition and scalar multiplication satisfying the usual axioms. A subset $\mathcal{U} \subseteq V$ is a *subspace* if: (i) $0 \in \mathcal{U}$, (ii) $u, v \in \mathcal{U} \Rightarrow u + v \in \mathcal{U}$, (iii) $u \in \mathcal{U}$, $\alpha \in \mathbb{R} \Rightarrow \alpha u \in \mathcal{U}$. In this chapter, $V = \mathbb{R}^d$ unless stated otherwise.

Definition 2.2.2 (Span) Given vectors $v_1, \ldots, v_k \in \mathbb{R}^d$, their *span* is

$$\operatorname{span}\{v_1, \ldots, v_k\} := \left\{ \sum_{i=1}^{k} \alpha_i v_i \,:\, \alpha_1, \ldots, \alpha_k \in \mathbb{R} \right\}.$$

It is the smallest subspace containing all v_i.

Definition 2.2.3 (Linear Independence and Dependence) Vectors $v_1, \ldots, v_k \in \mathbb{R}^d$ are *linearly independent* if the only solution to

$$\sum_{i=1}^{k} \alpha_i v_i = 0$$

is $\alpha_1 = \cdots = \alpha_k = 0$. Otherwise, they are *linearly dependent*.

Remark 2.2.1 (Independence as Uniqueness of Coordinates) If $v_1, \dots, v_k$ are linearly independent, then every $x \in \text{span}\{v_1, \dots, v_k\}$ has a *unique* coordinate representation $x = \sum_{i=1}^{k} \alpha_i v_i$. If they are dependent, coordinates are not unique.

Definition 2.2.4 (Basis and Dimension) A *basis* of a subspace $\mathcal{U} \subseteq \mathbb{R}^d$ is a set of vectors $\{u_1, \dots, u_k\} \subseteq \mathcal{U}$ that is linearly independent and spans $\mathcal{U}$. The *dimension* of $\mathcal{U}$, written $\dim(\mathcal{U})$, is the number of vectors in any basis of $\mathcal{U}$.

Theorem 2.2.1 (Steinitz Exchange/Basis Size is Well-Defined) *Any two bases of a finite-dimensional subspace $\mathcal{U} \subseteq \mathbb{R}^d$ have the same number of elements. Hence $\dim(\mathcal{U})$ is well-defined.*

Definition 2.2.5 (Coordinate Map (with Respect to a Basis)) Let $\mathcal{B} = (u_1, \dots, u_k)$ be an ordered basis of a k-dimensional subspace $\mathcal{U}$. Every $x \in \mathcal{U}$ can be written uniquely as $x = \sum_{i=1}^{k} \alpha_i u_i$. The coordinate vector of x in basis $\mathcal{B}$ is

$$[x]_{\mathcal{B}} := (\alpha_1, \dots, \alpha_k)^\top \in \mathbb{R}^k.$$

Example 2.2.1 (Coordinates via a Basis Matrix) Let $U = [u_1 \ \cdots \ u_k] \in \mathbb{R}^{d \times k}$ have basis vectors as columns. Then $x \in \mathcal{U}$ iff there exists $\alpha \in \mathbb{R}^k$ with $x = U\alpha$. If columns of U are linearly independent, then $U^\top U$ is invertible and the coordinates are

$$\alpha = (U^\top U)^{-1} U^\top x \quad \text{whenever } x \in \mathcal{U}.$$

This formula is the normal-equations solution and foreshadows projections and least squares.

2.2.1.2 Rank as Data Dimension: Feature-Span Interpretation

Definition 2.2.6 (Rank) For a matrix $X \in \mathbb{R}^{n \times d}$, the *rank* is

$$\text{rank}(X) := \dim(\text{range}(X)) = \dim(\text{range}(X^\top)),$$

i.e., the dimension of the column space (equivalently, the row space).

Remark 2.2.2 (Rank as Intrinsic Linear Dimension of a Dataset) If X is a data matrix with rows $x_i^\top$, then:

- $\text{rank}(X)$ is the maximum number of linearly independent columns, i.e., independent *features*.
- $\text{rank}(X)$ is also the maximum number of linearly independent rows, i.e., independent *samples*.

Thus, $\text{rank}(X)$ measures the intrinsic linear dimension of the data in the ambient feature space. When $\text{rank}(X) \ll d$, the data lie in a low-dimensional subspace, and many problems simplify.

Proposition 2.2.1 (Rank Bounds) *For any $X \in \mathbb{R}^{n\times d}$,*

$$\operatorname{rank}(X) \leq \min\{n, d\}.$$

Moreover, $\operatorname{rank}(X) = d$ *iff the columns are linearly independent (full column rank), and* $\operatorname{rank}(X) = n$ *iff the rows are linearly independent (full row rank).*

Example 2.2.2 (Feature Span and Identifiability) Consider linear regression with design matrix $X \in \mathbb{R}^{n\times d}$. If $\operatorname{rank}(X) < d$, then there exist nonzero directions $v \in \mathbb{R}^d$ with $Xv = 0$ (nontrivial null space), meaning w and $w + v$ yield identical predictions $Xw = X(w+v)$. Thus parameters are not identifiable without additional structure (e.g., regularization).

2.2.2 Subspaces Used in ML

2.2.2.1 Column/Row/Null Spaces as Modeling Objects

Definition 2.2.7 (Fundamental Subspaces) For $X \in \mathbb{R}^{n\times d}$:

$$\operatorname{range}(X) = \{Xw : w \in \mathbb{R}^d\} \subseteq \mathbb{R}^n \quad \text{(column space)},$$

$$\operatorname{range}(X^\top) = \{X^\top v : v \in \mathbb{R}^n\} \subseteq \mathbb{R}^d \quad \text{(row space)},$$

$$\operatorname{null}(X) = \{w \in \mathbb{R}^d : Xw = 0\} \subseteq \mathbb{R}^d \quad \text{(null space)}.$$

Theorem 2.2.2 (Rank–Nullity) *For $X \in \mathbb{R}^{n\times d}$,*

$$\operatorname{rank}(X) + \dim(\operatorname{null}(X)) = d.$$

Remark 2.2.3 (Interpretations in Learning)

- $\operatorname{range}(X)$ is the set of all *possible prediction vectors* Xw for linear models; least squares searches for the closest point in $\operatorname{range}(X)$ to a target y.
- $\operatorname{null}(X)$ contains parameter perturbations that do not change predictions on the training inputs. This is central to underdetermined systems and to implicit bias of optimization algorithms.
- $\operatorname{range}(X^\top)$ is the span of features that are "activated" by linear combinations of samples and appears in gradients of empirical losses.

Example 2.2.3 (Least Squares as Projection Onto $\operatorname{range}(X)$**)** Given labels $y \in \mathbb{R}^n$, the least squares predictor solves

$$\min_{w\in\mathbb{R}^d} \|Xw - y\|_2^2.$$

Geometrically, Xw ranges over range$(X) \subseteq \mathbb{R}^n$, and the solution picks the point in range(X) closest to y in Euclidean distance. Thus, the fitted vector $\widehat{y} = X\widehat{w}$ is the orthogonal projection of y onto range(X) (proved in the projections section).

2.2.2.2 Affine Subspaces: Centering and Translation of Data Clouds

Definition 2.2.8 (Affine Subspace and Affine Span) An *affine subspace* of $\mathbb{R}^d$ is a set of the form $a + \mathcal{U} = \{a + u : u \in \mathcal{U}\}$ where $a \in \mathbb{R}^d$ and $\mathcal{U}$ is a linear subspace. Given points $x_1, \ldots, x_n \in \mathbb{R}^d$, their *affine span* is

$$\text{aff}\{x_1, \ldots, x_n\} := \left\{ \sum_{i=1}^{n} \alpha_i x_i : \sum_{i=1}^{n} \alpha_i = 1 \right\}.$$

Remark 2.2.4 (Centering Converts Affine to Linear) Let $\bar{x} = \frac{1}{n}\sum_{i=1}^{n} x_i$ and define centered points $z_i = x_i - \bar{x}$. Then $\text{aff}\{x_i\} = \bar{x} + \text{span}\{z_1, \ldots, z_n\}$. Thus, studying affine structure of a cloud reduces to studying linear structure after centering. This is why PCA is usually performed on centered data.

Example 2.2.4 (Affine Constraints in Regression with Intercept) A linear model with intercept predicts $w^\top x + b$. In matrix form with $X \in \mathbb{R}^{n \times d}$, this is $Xw + b\mathbf{1}$, i.e., predictions lie in the affine subspace

$$\mathcal{A} := \text{range}(X) + \text{span}\{\mathbf{1}\} \subseteq \mathbb{R}^n.$$

Thus, fitting an intercept enlarges the prediction set from a subspace to an affine subspace.

2.2.3 *Orthonormal Bases and Geometric Coordinates*

2.2.3.1 Gram–Schmidt and Orthonormalization (Conceptual)

Definition 2.2.9 (Orthonormal Set and Orthonormal Basis) Vectors $q_1, \ldots, q_k \in \mathbb{R}^d$ are *orthonormal* if

$$\langle q_i, q_j \rangle = \begin{cases} 1, & i = j, \\ 0, & i \neq j. \end{cases}$$

An *orthonormal basis* of a subspace $\mathcal{U}$ is an orthonormal set that spans $\mathcal{U}$.

Proposition 2.2.2 (Orthonormal Coordinates are Inner Products) *If $q_1, \ldots, q_k$ is an orthonormal basis of $\mathcal{U}$ and $x \in \mathcal{U}$, then*

$$x = \sum_{i=1}^{k} \langle q_i, x \rangle \, q_i, \qquad [x]_{\mathcal{Q}} = \big(\langle q_1, x \rangle, \ldots, \langle q_k, x \rangle\big)^\top,$$

where $\mathcal{Q} = (q_1, \ldots, q_k)$ is the ordered orthonormal basis.

Remark 2.2.5 (Why Orthonormal Bases are Special) In a general basis, coordinates require solving a linear system (often via $(U^\top U)^{-1} U^\top$), which can be numerically unstable when the basis vectors are nearly dependent. In an orthonormal basis, coordinates are obtained by simple inner products, which is stable and geometrically transparent.

Definition 2.2.10 (Gram–Schmidt (Algorithmic Statement)) Let $v_1, \ldots, v_k \in \mathbb{R}^d$ be linearly independent. The Gram–Schmidt procedure constructs an orthonormal set $q_1, \ldots, q_k$ spanning the same subspace by

$$\tilde{q}_1 = v_1, \qquad q_1 = \frac{\tilde{q}_1}{\|\tilde{q}_1\|_2},$$

and for $j = 2, \ldots, k$,

$$\tilde{q}_j = v_j - \sum_{i=1}^{j-1} \langle q_i, v_j \rangle \, q_i, \qquad q_j = \frac{\tilde{q}_j}{\|\tilde{q}_j\|_2}.$$

Remark 2.2.6 (Geometric Meaning of Gram–Schmidt) At each step, Gram–Schmidt subtracts from v_j its projection onto the span of earlier vectors, leaving a component orthogonal to that span. Normalization then produces unit vectors. Thus, it is a constructive proof that every finite-dimensional subspace admits an orthonormal basis.

2.2.3.2 Coordinates Under Orthonormal Bases and Stability Intuition

Definition 2.2.11 (Basis Matrix and Orthonormality) Let $Q = [q_1 \;\cdots\; q_k] \in \mathbb{R}^{d \times k}$. Then $\{q_i\}_{i=1}^k$ is orthonormal iff

$$Q^\top Q = I_k.$$

If additionally $k = d$, then Q is an orthogonal matrix and also satisfies $Q Q^\top = I_d$.

Proposition 2.2.3 (Stable Coordinate Extraction) *If $Q \in \mathbb{R}^{d \times k}$ has orthonormal columns, then for any $x \in \mathbb{R}^d$, the coefficient vector $\alpha := Q^\top x \in \mathbb{R}^k$ satisfies*

$$\|Q^\top x\|_2 \le \|x\|_2, \qquad \|Q\alpha\|_2 = \|\alpha\|_2,$$

and $Q^\top$ acts as a norm-nonexpansive map while Q acts as an isometric embedding of $\mathbb{R}^k$ into $\mathbb{R}^d$.

Remark 2.2.7 (Stability Intuition) In numerical computations, instability often arises from amplifying small perturbations. Orthonormal transformations preserve Euclidean norm and angles, hence they do not amplify noise. This is why QR-based methods and orthonormal bases are preferred in least squares and PCA computations.

Example 2.2.5 (Coordinates vs. Solving Normal Equations) If $U \in \mathbb{R}^{d \times k}$ has non-orthonormal columns, coordinates of $x \in \text{range}(U)$ are found by solving $U\alpha = x$, or using $\alpha = (U^\top U)^{-1} U^\top x$. When $U^\top U$ is ill-conditioned, small perturbations in x can cause large perturbations in α. If instead $U = Q$ has orthonormal columns, then $\alpha = Q^\top x$ requires no inversion and is stable.

2.3 Affine Geometry, Hyperplanes, and Halfspaces

Many foundational ML models are *affine* rather than purely linear: they include offsets (intercepts), thresholds, and biases. The geometry of affine sets explains why centering matters, why hyperplanes are canonical decision boundaries, and why constraints and regularizers carve out polytopes and other feasible regions. This section develops the algebra and geometry of affine combinations, hyperplanes, and halfspaces, with an emphasis on computationally checkable formulas (distances, margins, and constraint sets) (Boyd and Vandenberghe 2004).

2.3.1 Affine Combinations and Affine Hull

2.3.1.1 Affine Independence and Barycentric Coordinates

An affine set is a translated subspace, so it remembers an "origin" only up to translation. This is the right language for datasets in $\mathbb{R}^d$, because clouds rarely pass through the origin. Affine combinations impose the constraint that coefficients sum to one, ensuring invariance under translation. The most important computational consequence is *barycentric coordinates*: representing a point relative to a set of affinely independent anchors. In ML, such coordinates appear in simplex-based methods, mixture models, and in understanding convex hulls of prototypes (e.g., nearest convex hull classification). Affine independence is the precise condition that guarantees uniqueness of barycentric coordinates.

Definition 2.3.1 (Affine Combination and Affine Hull) Let $x_0, \ldots, x_k \in \mathbb{R}^d$. An *affine combination* is any linear combination $\sum_{i=0}^{k} \alpha_i x_i$ with

$$\sum_{i=0}^{k} \alpha_i = 1.$$

The *affine hull* of $\{x_0, \ldots, x_k\}$ is

$$\operatorname{aff}\{x_0, \ldots, x_k\} := \left\{ \sum_{i=0}^{k} \alpha_i x_i : \sum_{i=0}^{k} \alpha_i = 1 \right\}.$$

Definition 2.3.2 (Affine Independence) Points $x_0, \ldots, x_k \in \mathbb{R}^d$ are *affinely independent* if the vectors

$$x_1 - x_0,\ x_2 - x_0,\ \ldots,\ x_k - x_0$$

are linearly independent in $\mathbb{R}^d$. Equivalently, if $\sum_{i=0}^{k} \alpha_i x_i = 0$ and $\sum_{i=0}^{k} \alpha_i = 0$ imply $\alpha_0 = \cdots = \alpha_k = 0$.

Proposition 2.3.1 (Affine Hull as Translation of a Span) *For any* $x_0, \ldots, x_k \in \mathbb{R}^d$,

$$\operatorname{aff}\{x_0, \ldots, x_k\} = x_0 + \mathit{span}\{x_1 - x_0, \ldots, x_k - x_0\}.$$

Definition 2.3.3 (Barycentric Coordinates) Let $x_0, \ldots, x_k \in \mathbb{R}^d$ be affinely independent. Every $x \in \operatorname{aff}\{x_0, \ldots, x_k\}$ can be written uniquely as

$$x = \sum_{i=0}^{k} \alpha_i x_i, \qquad \sum_{i=0}^{k} \alpha_i = 1,$$

and the coefficients $(\alpha_0, \ldots, \alpha_k)$ are called the *barycentric coordinates* of x relative to $(x_0, \ldots, x_k)$.

Proposition 2.3.2 (Computing Barycentric Coordinates) *Let* $x_0, \ldots, x_k \in \mathbb{R}^d$ *be affinely independent, and define the matrix*

$$U := \begin{bmatrix} x_1 - x_0 & x_2 - x_0 & \cdots & x_k - x_0 \end{bmatrix} \in \mathbb{R}^{d \times k}.$$

Then for any $x \in \operatorname{aff}\{x_0, \ldots, x_k\}$, *the vector* $\beta \in \mathbb{R}^k$ *defined by*

$$\beta := (\alpha_1, \ldots, \alpha_k)^\top$$

is the unique solution of the linear system

$$U\beta = x - x_0,$$

and $\alpha_0 = 1 - \sum_{j=1}^{k} \alpha_j$.

Remark 2.3.1 (Translation Invariance of Barycentric Coordinates) If we translate all anchor points by the same vector t, i.e., replace x_i by x_i+t and x by $x+t$, then the barycentric coordinates remain unchanged. This is exactly why the constraint $\sum_i \alpha_i = 1$ is the correct affine analogue of linear coordinates.

2.3.1.2 Simplex Geometry and Convex Combinations

Convex combinations are affine combinations with nonnegative coefficients. They encode "mixtures" and "soft assignments" and are ubiquitous in ML: probability vectors lie in simplices, mixture models combine components by convex weights, attention weights are convex coefficients over tokens, and convex hulls describe the region obtainable by interpolating prototypes. The simplex provides a clean geometric object whose facets, vertices, and barycentric coordinates admit explicit computations.

Definition 2.3.4 (Convex Combination, Convex Hull) A *convex combination* of $x_0, \dots, x_k \in \mathbb{R}^d$ is an affine combination

$$x = \sum_{i=0}^{k} \alpha_i x_i, \qquad \alpha_i \geq 0, \; \sum_{i=0}^{k} \alpha_i = 1.$$

The *convex hull* is the set of all convex combinations:

$$\operatorname{conv}\{x_0, \dots, x_k\} := \left\{ \sum_{i=0}^{k} \alpha_i x_i \; : \; \alpha_i \geq 0, \; \sum_{i=0}^{k} \alpha_i = 1 \right\}.$$

Definition 2.3.5 (Simplex) If $x_0, \dots, x_k$ are affinely independent, then the set

$$\Delta(x_0, \dots, x_k) := \operatorname{conv}\{x_0, \dots, x_k\}$$

is a *k-simplex* embedded in $\mathbb{R}^d$. Its vertices are $x_0, \dots, x_k$, and its affine hull has dimension k.

Remark 2.3.2 (Probability Simplex) The standard probability simplex in $\mathbb{R}^{k+1}$ is

$$\Delta^k := \left\{ \alpha \in \mathbb{R}^{k+1} \; : \; \alpha_i \geq 0, \; \sum_{i=0}^{k} \alpha_i = 1 \right\}.$$

Any convex combination $\sum_{i=0}^{k} \alpha_i x_i$ uses coefficients $\alpha \in \Delta^k$. In ML, softmax outputs lie in Δ^k, and categorical distributions are parameterized by points in Δ^k.

Proposition 2.3.3 (Convex Hull as Intersection of Halfspaces (Informal Preview)) *A convex polytope can be represented either as a convex hull of finitely many points (a "V-representation") or as an intersection of finitely many halfspaces (an "H-representation"). We will make the halfspace viewpoint precise later in this section.*

2.3.2 Hyperplanes and Linear Decision Boundaries

2.3.2.1 Normal Vectors, Offsets, and Signed Distances

A hyperplane is the set of points with a fixed dot product against a normal vector. In ML, hyperplanes appear as decision boundaries of linear classifiers and as level sets of affine functions. The normal vector w is not merely a parameter: it specifies both the orientation of the boundary and the direction along which the classifier's score changes most rapidly. The signed distance to a hyperplane is a fundamental computational quantity: it determines which side of the boundary a point lies on and measures confidence. Importantly, scaling (w, b) scales the raw score but not the boundary, and the distance formula precisely separates "geometry" from "parameterization."

Definition 2.3.6 (Affine Hyperplane) Let $w \in \mathbb{R}^d$ be nonzero and $b \in \mathbb{R}$. The *(affine) hyperplane* with normal w and offset b is

$$H(w, b) := \{x \in \mathbb{R}^d : w^\top x + b = 0\}.$$

Remark 2.3.3 (Scaling Invariance of Hyperplanes) For any $\lambda \neq 0$,

$$H(w, b) = H(\lambda w, \lambda b).$$

Thus, (w, b) is not unique: only the *direction* of w and the ratio between w and b matters for the set H.

Definition 2.3.7 (Halfspaces Associated with a Hyperplane) Given (w, b) with $w \neq 0$, define the closed halfspaces

$$\mathcal{H}_+(w, b) := \{x : w^\top x + b \geq 0\}, \qquad \mathcal{H}_-(w, b) := \{x : w^\top x + b \leq 0\}.$$

Then $H(w, b) = \mathcal{H}_+(w, b) \cap \mathcal{H}_-(w, b)$.

Definition 2.3.8 (Signed Distance to a Hyperplane) Let $H(w, b)$ be a hyperplane with $w \neq 0$. The *signed distance* from $x \in \mathbb{R}^d$ to $H(w, b)$ is

$$\operatorname{dist}_{\pm}(x, H(w, b)) := \frac{w^\top x + b}{\|w\|_2}.$$

The (unsigned) Euclidean distance is $\operatorname{dist}(x, H) = \frac{|w^\top x+b|}{\|w\|_2}$.

Proposition 2.3.4 (Closest Point on a Hyperplane) *Let $H(w, b)$ be a hyperplane with $w \neq 0$. For any $x \in \mathbb{R}^d$, the unique closest point $x^\star \in H(w, b)$ is*

$$x^\star = x - \frac{w^\top x + b}{\|w\|_2^2}\, w,$$

and $\|x - x^\star\|_2 = \frac{|w^\top x+b|}{\|w\|_2}$.

Proof Let $x^\star = x - \alpha w$. Enforce the constraint $w^\top x^\star + b = 0$: $w^\top(x - \alpha w) + b = 0$, hence $\alpha = (w^\top x + b)/\|w\|_2^2$. Then $x^\star \in H(w, b)$ and $x - x^\star$ is parallel to w, which is orthogonal to the hyperplane, so this is the orthogonal projection and hence the closest point. □

Remark 2.3.4 (Interpretation of the Score) For a linear score function $s(x) = w^\top x + b$, the normalized score $s(x)/\|w\|_2$ is exactly the signed distance to the decision boundary. Thus, normalization by $\|w\|_2$ produces a geometrically meaningful confidence measure.

2.3.2.2 Margins, Separability, and Geometric Classification

Geometric classification is about separating point clouds by a hyperplane. When separation is possible, there are infinitely many separating hyperplanes; the key is to prefer those with large *margin*, i.e., those that keep all training points far from the boundary in signed-distance units. Margins connect geometry to robustness: a larger margin implies that small perturbations of the input are less likely to cross the boundary. The margin concept also clarifies why scaling (w, b) does not matter for classification but does matter for optimization objectives, and why norm constraints/regularization naturally appear in max-margin formulations.

Definition 2.3.9 (Linear Separability) A labeled dataset $\{(x_i, y_i)\}_{i=1}^n \subseteq \mathbb{R}^d \times \{-1, +1\}$ is *linearly separable* if there exist $w \in \mathbb{R}^d$, $b \in \mathbb{R}$ such that

$$y_i(w^\top x_i + b) > 0 \qquad \forall i \in [n].$$

Definition 2.3.10 (Functional Margin and Geometric Margin) Fix (w, b) with $w \neq 0$ and labeled examples (x_i, y_i) with $y_i \in \{-1, +1\}$.

- The *functional margin* of (w, b) on example i is

$$\gamma_i^{\text{fun}} := y_i(w^\top x_i + b).$$

- The *geometric margin* of (w, b) on example i is

$$\gamma_i^{\text{geo}} := \frac{y_i(w^\top x_i + b)}{\|w\|_2}.$$

- The (dataset) margins are $\gamma^{\text{fun}} := \min_i \gamma_i^{\text{fun}}$ and $\gamma^{\text{geo}} := \min_i \gamma_i^{\text{geo}}$.

Remark 2.3.5 (Functional Margin Scales; Geometric Margin is Invariant) If we replace (w, b) by $(\lambda w, \lambda b)$ with $\lambda > 0$, then γ^{fun} scales by λ, but γ^{geo} is unchanged. Thus, geometric margin captures the intrinsic separation of the classifier, independent of parameter scaling.

Proposition 2.3.5 (Geometric Margin Equals Minimum Signed Distance) *For $y_i \in \{-1, +1\}$, the geometric margin satisfies*

$$\gamma^{\text{geo}} = \min_{i \in [n]} \operatorname{dist}\big(x_i, H(w, b)\big),$$

where distance is Euclidean and $H(w, b)$ is the decision hyperplane.

Proof By Definition 2.3.8, the signed distance of x_i to $H(w, b)$ is $(w^\top x_i + b)/\|w\|_2$. Multiplying by y_i flips the sign so that correctly classified points yield positive distance. Taking the minimum over i yields the claimed equality. □

Remark 2.3.6 (Margin as Robustness Radius) If $\gamma^{\text{geo}} > 0$, then any perturbation δ with $\|\delta\|_2 < \gamma^{\text{geo}}$ cannot change the label: $y_i(w^\top(x_i + \delta) + b) > 0$ remains true. Hence, margin lower-bounds the ℓ_2-robustness radius on the training set.

2.3.3 Halfspaces and Polytopes

2.3.3.1 Intersections of Halfspaces and Feasible Regions

Many ML objectives are solved under linear constraints: nonnegativity, simplex constraints for probabilities, box constraints, or margin constraints for classifiers. Each linear inequality defines a halfspace; the feasible set is an intersection of halfspaces, hence a convex polyhedron (and, if bounded, a polytope). This viewpoint unifies geometry and optimization: feasibility is a geometric property, and algorithms often move by projecting onto or navigating within these polyhedral regions.

Definition 2.3.11 (Halfspace and Polyhedron) A (closed) halfspace in $\mathbb{R}^d$ is a set of the form

$$\{x \in \mathbb{R}^d : a^\top x \leq c\}$$

for some $a \in \mathbb{R}^d$, $c \in \mathbb{R}$ with $a \neq 0$. A *polyhedron* is an intersection of finitely many halfspaces:

$$\mathcal{P} := \{x \in \mathbb{R}^d : Ax \leq b\},$$

where $A \in \mathbb{R}^{m \times d}$, $b \in \mathbb{R}^m$, and the inequality is componentwise. If $\mathcal{P}$ is bounded, it is called a *polytope*.

Proposition 2.3.6 (Convexity of Halfspaces and Polyhedra) *Every halfspace is convex, and any intersection of convex sets is convex. Hence every polyhedron* $\{x : Ax \leq b\}$ *is convex.*

Remark 2.3.7 (Convexity is Algorithmic Leverage) Convex feasible regions enable global optimality for convex objectives and make projections and separation arguments tractable. Even for nonconvex objectives (e.g., deep networks), convex constraints (simplex/box) remain computationally manageable.

Example 2.3.1 (Probability Simplex as a Polytope) The probability simplex in $\mathbb{R}^k$ is

$$\Delta^{k-1} = \left\{ p \in \mathbb{R}^k : p_i \geq 0 \ \forall i, \ \sum_{i=1}^{k} p_i = 1 \right\}.$$

This is a polytope: it is the intersection of k halfspaces $p_i \geq 0$ (equivalently $-p_i \leq 0$) with the affine hyperplane $\sum_i p_i = 1$.

2.3.3.2 Geometry of Linear Constraints in Learning Problems

Linear constraints encode structure and prior knowledge. They express feasibility of probabilities, budgets, fairness constraints, monotonicity constraints, and margin constraints. In regularization, norm balls can be viewed as (possibly infinite) intersections of halfspaces via duality, explaining why different norms impose different "shapes." From a computational perspective, the critical operation is checking feasibility and computing projections onto constraint sets. Understanding constraints geometrically also clarifies Lagrange multipliers as normal vectors to active constraints at an optimum.

Definition 2.3.12 (Feasible Region and Active Constraints) Let $\mathcal{F} = \{x \in \mathbb{R}^d : Ax \leq b\}$ be a polyhedron with rows $a_i^\top$ of A. A constraint $a_i^\top x \leq b_i$ is *active* at

$x \in \mathcal{F}$ if $a_i^\top x = b_i$. The set of active constraints determines the local "face" of the polyhedron containing x.

Example 2.3.2 (Box Constraints and Clipping) The box $\mathcal{B} = [\ell_1, u_1] \times \cdots \times [\ell_d, u_d] \subseteq \mathbb{R}^d$ is the polytope

$$\mathcal{B} = \{x : x_i \geq \ell_i, \ x_i \leq u_i \ \forall i\}.$$

Projection onto $\mathcal{B}$ is coordinatewise clipping:

$$(\Pi_{\mathcal{B}}(z))_i = \min\{u_i, \max\{\ell_i, z_i\}\}.$$

Such constraints appear in bounded-parameter models and in adversarial perturbation sets.

Example 2.3.3 (Margin Constraints as Halfspaces) Given labeled data $(x_i, y_i) \in \mathbb{R}^d \times \{-1, +1\}$, the constraints

$$y_i(w^\top x_i + b) \geq 1 \qquad (i \in [n])$$

are linear inequalities in (w, b). Thus, the set of separators with functional margin at least 1 is a polyhedron in parameter space. Max-margin classifiers can be viewed as finding a point in this feasible region with minimal norm.

Remark 2.3.8 (Dual Norms and Halfspace Views of Norm Balls) For a norm $\|\cdot\|$ with dual norm $\|\cdot\|_*$, the unit ball $\{w : \|w\| \leq 1\}$ can be characterized as

$$\{w : \langle w, u \rangle \leq 1 \text{ for all } u \text{ with } \|u\|_* \leq 1\}.$$

Thus, norm constraints are (possibly infinite) intersections of halfspaces. This geometric viewpoint will be useful when we discuss regularization and robustness.

2.4 Norms and Metrics

Norms and metrics formalize the notion of *size* and *closeness* in $\mathbb{R}^d$. In ML, these choices are never neutral: they encode inductive biases, determine which perturbations are considered small, and control the geometry of feasible regions in regularization. This section develops standard norms, their unit balls, norm equivalence (why all norms are comparable in finite dimension yet can behave very differently), and weighted geometries such as Mahalanobis distance that adapt to covariance and conditioning (cf. Fig. 2.3).

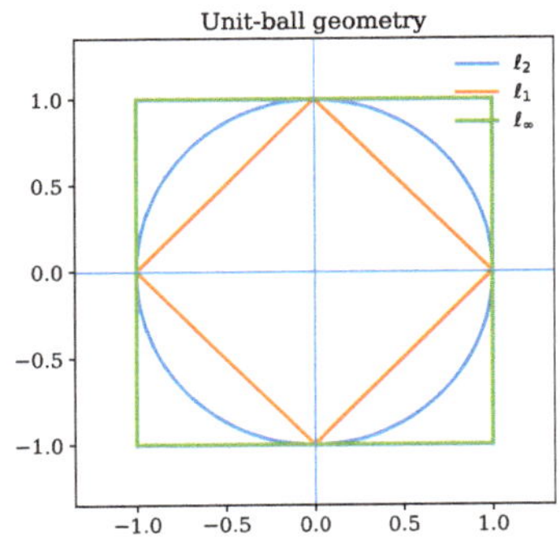

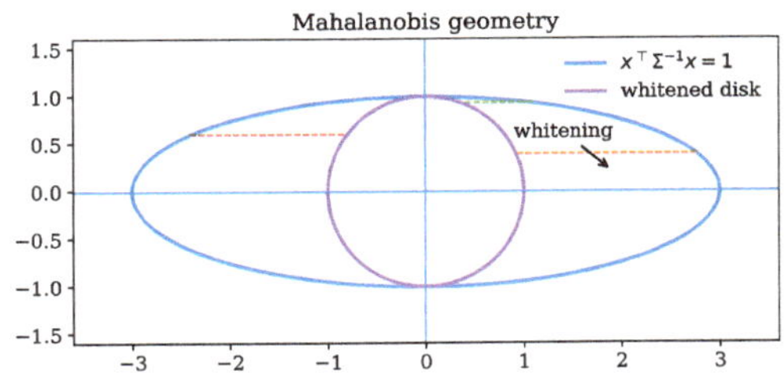

Fig. 2.3 Left: the ℓ_1, ℓ_2, and ℓ_∞ unit balls in two dimensions, illustrating different inductive biases. Right: a Mahalanobis ellipsoid and its whitening transformation into a Euclidean disk

2.4.1 Standard Norms and Geometry

2.4.1.1 $\ell_1, \ell_2, \ell_\infty$: Unit Balls and Inductive Bias

A norm is a geometric lens: it turns vectors into lengths and defines a unit ball. Different unit-ball shapes produce different "preferred" solutions when used as constraints or penalties. The ℓ_2 ball is rotationally symmetric and thus treats all directions equally; ℓ_1 has sharp corners aligned with coordinate axes, favoring sparse coordinate solutions; ℓ_∞ produces box-like constraints that limit the maximum coordinate magnitude. These geometries underpin why ridge regression shrinks uniformly, lasso induces sparsity, and ℓ_∞-constraints appear naturally in worst-case coordinate-bounded perturbation models.

Definition 2.4.1 (Norm) A function $\|\cdot\| : \mathbb{R}^d \to [0, \infty)$ is a *norm* if for all $x, y \in \mathbb{R}^d$ and $\alpha \in \mathbb{R}$:

1. (Positive definiteness) $\|x\| = 0 \Leftrightarrow x = 0$,
2. (Homogeneity) $\|\alpha x\| = |\alpha|\,\|x\|$,
3. (Triangle inequality) $\|x + y\| \leq \|x\| + \|y\|$.

Definition 2.4.2 (Standard ℓ_p Norms) For $x \in \mathbb{R}^d$:

$$\|x\|_1 := \sum_{i=1}^{d} |x_i|, \qquad \|x\|_2 := \left(\sum_{i=1}^{d} x_i^2\right)^{1/2}, \qquad \|x\|_\infty := \max_{i\in[d]} |x_i|.$$

Definition 2.4.3 (Unit Ball) Given a norm $\|\cdot\|$, its *unit ball* is

$$B_{\|\cdot\|}(1) := \{x \in \mathbb{R}^d : \|x\| \leq 1\}.$$

More generally, the radius-r ball is $B_{\|\cdot\|}(r) := \{x : \|x\| \leq r\}$.

Remark 2.4.1 (Geometry of the Three Unit Balls) In $\mathbb{R}^2$:

- $B_{\|\cdot\|_2}(1)$ is a disk (rotationally invariant).
- $B_{\|\cdot\|_1}(1)$ is a diamond with corners on the coordinate axes.
- $B_{\|\cdot\|_\infty}(1)$ is an axis-aligned square.

In higher dimensions, these become sphere, cross-polytope, and hypercube, respectively.

Remark 2.4.2 (Inductive Bias via Constraints) When solving $\min_w \widehat{L}(w)$ subject to $\|w\| \le r$, the feasible set is the norm ball. If the loss level sets are smooth and the constraint boundary has corners (as for ℓ_1), solutions often occur at corners, which correspond to sparse coordinates. Smooth boundaries (as for ℓ_2) produce smoother shrinkage across coordinates.

Proposition 2.4.1 (ℓ_p Inequalities for $p \in \{1, 2, \infty\}$) *For all $x \in \mathbb{R}^d$,*

$$\|x\|_\infty \le \|x\|_2 \le \|x\|_1, \qquad \|x\|_1 \le \sqrt{d}\,\|x\|_2, \qquad \|x\|_2 \le \sqrt{d}\,\|x\|_\infty.$$

Example 2.4.1 (Same Perturbation, Different Notions of "Small") Let $x = (1, 1, \ldots, 1)^\top \in \mathbb{R}^d$. Then $\|x\|_1 = d$, $\|x\|_2 = \sqrt{d}$, and $\|x\|_\infty = 1$. Thus, the same vector can be "tiny" in ℓ_∞ but "large" in ℓ_1. In robustness, choosing ℓ_∞ vs. ℓ_2 changes which perturbation patterns are allowed: ℓ_∞ bounds each coordinate, while ℓ_2 bounds total energy.

2.4.1.2 Norm Equivalence in Finite Dimensions and Robustness Intuition

In finite-dimensional spaces, all norms are equivalent: none is infinitely stronger than another. This is a foundational fact ensuring that convergence and continuity notions do not depend on the chosen norm. However, the equivalence constants often scale with dimension (typically like $\sqrt{d}$ or d), so in high-dimensional ML the choice of norm can have major quantitative consequences. This tension explains why we can safely say "all norms define the same topology," yet we must still be careful because the constants matter for robustness bounds, generalization, and conditioning.

Theorem 2.4.1 (Norm Equivalence in Finite Dimensions) *Let $\|\cdot\|_a$ and $\|\cdot\|_b$ be two norms on $\mathbb{R}^d$. There exist constants $0 < c \le C < \infty$ (depending on d and the norms) such that for all $x \in \mathbb{R}^d$,*

$$c\,\|x\|_a \le \|x\|_b \le C\,\|x\|_a.$$

Remark 2.4.3 (High-Dimensional Caution) The constants c, C may grow with d. For example, by Proposition 2.4.1, $\|x\|_1 \le \sqrt{d}\,\|x\|_2$ and $\|x\|_2 \le \sqrt{d}\,\|x\|_\infty$. Thus an ℓ_2-ball of radius 1 contains an ℓ_∞-ball of radius $1/\sqrt{d}$. In large d, these are very different robustness radii.

Proposition 2.4.2 (Dual Norms (Useful for Robustness and Margins)) *For a norm* $\|\cdot\|$ *on* $\mathbb{R}^d$*, its* dual norm *is*

$$\|u\|_* := \sup_{\|x\|\leq 1} \langle u, x\rangle.$$

In particular,

$$(\|\cdot\|_1)^* = \|\cdot\|_\infty, \qquad (\|\cdot\|_2)^* = \|\cdot\|_2, \qquad (\|\cdot\|_\infty)^* = \|\cdot\|_1.$$

Remark 2.4.4 (Robustness and Dual Norms) For a linear score $s(x) = w^\top x + b$, the worst-case change under perturbations $\|\delta\| \leq \varepsilon$ is

$$\sup_{\|\delta\|\leq\varepsilon} |w^\top \delta| = \varepsilon\, \|w\|_*.$$

Thus, the dual norm governs how sensitive the model is to perturbations measured in the primal norm. This relation will reappear in margin-based generalization and adversarial robustness.

2.4.2 Distances Induced by Norms

2.4.2.1 Metric Properties and Geometric Meaning of "Closeness"

A norm induces a metric by measuring the length of the difference vector. This immediately gives translation invariance: distances depend only on relative displacement not on absolute location. In ML, this matters because learning tasks are often invariant to reparameterizations such as shifting the origin (e.g., centered data), and because distances should reflect semantic changes. The triangle inequality encodes that detours cannot be shorter than direct paths, a property used in nearest-neighbor search, clustering, and covering arguments in generalization theory.

Definition 2.4.4 (Metric Induced by a Norm) Let $\|\cdot\|$ be a norm on $\mathbb{R}^d$. Define

$$d_{\|\cdot\|}(x, y) := \|x - y\|.$$

Then $d_{\|\cdot\|}$ is a metric on $\mathbb{R}^d$.

Proposition 2.4.3 (Metric Properties) *For all* $x, y, z \in \mathbb{R}^d$*,* $d_{\|\cdot\|}$ *satisfies:*

1. $d(x, y) \geq 0$ *and* $d(x, y) = 0 \Leftrightarrow x = y$*,*
2. $d(x, y) = d(y, x)$*,*
3. $d(x, z) \leq d(x, y) + d(y, z)$*.*

Remark 2.4.5 (Geometric Meaning) The set $\{y : d_{\|\cdot\|}(x, y) \leq r\}$ is exactly the translated norm ball $x + B_{\|\cdot\|}(r)$. Thus, the geometry of neighborhoods (local closeness) is controlled by the unit ball shape.

Example 2.4.2 (Nearest Neighbors Depend on the Norm) Given points $x_1, \ldots, x_n$, the nearest neighbor of a query q is $i^\star \in \arg\min_i \|q - x_i\|$. Changing $\|\cdot\|$ changes the Voronoi partition (Example 2.1.2) and can change $i^\star$, especially in high dimensions where different norms emphasize different coordinate patterns.

2.4.2.2 Feature Scaling and Normalization as Metric Design

Feature scaling is not a cosmetic preprocessing step; it is a geometric choice that changes distances and angles. If one feature has units in thousands and another in decimals, Euclidean distance will be dominated by the larger-scale feature. Standardization and normalization effectively *reweight* axes, changing the norm and hence the induced metric. From a modeling standpoint, scaling encodes our belief about which directions matter and how to trade off deviations across features.

Definition 2.4.5 (Weighted ℓ_2 Norm (Diagonal Scaling)) Let $s_1, \ldots, s_d > 0$ and let $D = \operatorname{diag}(s_1, \ldots, s_d)$. Define

$$\|x\|_D := \|Dx\|_2 = \left(\sum_{i=1}^{d} s_i^2 x_i^2\right)^{1/2}.$$

Then $\|\cdot\|_D$ is a norm, and the induced metric is $d_D(x, y) = \|D(x - y)\|_2$.

Remark 2.4.6 (Standardization as Diagonal Weighting) If features have empirical standard deviations σ_i, then scaling by $s_i = 1/\sigma_i$ makes each feature have unit variance (approximately). This corresponds to measuring distances in units of "standard deviations" along each axis.

Example 2.4.3 (Why Scaling Changes Nearest Neighbors) Let $x = (0, 0)$, $y = (1, 0)$, $z = (0, 100)$ in $\mathbb{R}^2$ and query $q = (0, 1)$. Under ℓ_2, $\|q - y\|_2 = \sqrt{(1-0)^2 + (0-1)^2} = \sqrt{2}$ while $\|q - z\|_2 = \sqrt{0^2 + (100-1)^2} = 99$, so y is nearest. If we scale the second coordinate by $1/100$ (i.e., $D = \operatorname{diag}(1, 1/100)$), then $d_D(q, z) = \|(0, -99/100)\|_2 = 0.99$ and $d_D(q, y) = \|(1, -1/100)\|_2 \approx 1.005$, so the nearest neighbor changes from y to z. This illustrates that scaling encodes feature importance.

2.4.3 *Weighted Geometries*

2.4.3.1 Mahalanobis Distance and Whitening Geometry

A diagonal weighting adjusts axes independently, but real data exhibit correlated features. Mahalanobis distance generalizes Euclidean distance by using a positive definite matrix to weight directions, shrinking directions with high variance and expanding directions with low variance, *after accounting for correlations.* Whitening is the coordinate transformation that makes the covariance identity, so Mahalanobis distance becomes Euclidean in whitened space. This is central in Gaussian models, LDA/QDA, and in understanding conditioning of optimization problems (Bhatia 2007).

Definition 2.4.6 (Positive Definite Matrices) A symmetric matrix $M \in \mathbb{R}^{d\times d}$ is *positive semidefinite* (PSD), written $M \succeq 0$, if $x^\top Mx \geq 0$ for all $x \in \mathbb{R}^d$, and *positive definite* (PD), written $M \succ 0$, if $x^\top Mx > 0$ for all nonzero x.

Definition 2.4.7 (Mahalanobis Norm and Distance) Let $M \in \mathbb{R}^{d\times d}$ with $M \succ 0$. Define the *Mahalanobis norm* and induced distance by

$$\|x\|_M := \sqrt{x^\top Mx}, \qquad d_M(x, y) := \|x - y\|_M = \sqrt{(x - y)^\top M(x - y)}.$$

Proposition 2.4.4 (Mahalanobis Distance is Euclidean After a Linear Map) *Let $M \succ 0$. Then there exists an invertible matrix L such that $M = L^\top L$ (e.g., Cholesky factor). For such L,*

$$\|x\|_M = \|Lx\|_2, \qquad d_M(x, y) = \|L(x - y)\|_2.$$

Hence, Mahalanobis distance is Euclidean distance in the transformed coordinates $z = Lx$.

Definition 2.4.8 (Whitening) Let random $X \in \mathbb{R}^d$ have mean μ and covariance $\Sigma = \mathbb{E}[(X - \mu)(X - \mu)^\top]$ with $\Sigma \succ 0$. A *whitening transform* is any invertible linear map W such that

$$\mathrm{Cov}\big(W(X - \mu)\big) = I_d.$$

One canonical choice is $W = \Sigma^{-1/2}$ (defined via eigendecomposition).

Remark 2.4.7 (Mahalanobis Distance from Covariance) If $\Sigma \succ 0$ is a covariance matrix, then the distance

$$d_{\Sigma^{-1}}(x, y) = \sqrt{(x - y)^\top \Sigma^{-1}(x - y)}$$

measures closeness in *units of standard deviation accounting for correlation*. In the whitened space $z = \Sigma^{-1/2}(x - \mu)$, this becomes ordinary Euclidean distance.

2.4.3.2 Anisotropy, Covariance Shape, and Conditioning

Anisotropy means "direction-dependent scale." In data space, anisotropy is captured by covariance eigenvalues: large-variance directions form elongated ellipsoids, and low-variance directions are tight. In optimization, anisotropy appears as ill-conditioned Hessians: gradients point in directions scaled by curvature. Weighted geometries (preconditioning, whitening) aim to reduce anisotropy, turning elongated ellipsoids into near-spherical sets, thereby improving both statistical comparisons (fair distances) and computational behavior (faster convergence).

Definition 2.4.9 (Covariance Ellipsoid) Let $\Sigma \succ 0$. The set

$$\mathcal{E}_\Sigma(r) := \{x \in \mathbb{R}^d : x^\top \Sigma^{-1} x \le r^2\}$$

is an ellipsoid. If $\Sigma = Q \Lambda Q^\top$ with eigenvalues $\lambda_1, \dots, \lambda_d > 0$, then $\mathcal{E}_\Sigma(r)$ has principal axes q_i with semi-axis lengths $r\sqrt{\lambda_i}$.

Definition 2.4.10 (Condition Number (Spectral)) For $A \succ 0$, the (spectral) condition number is

$$\kappa(A) := \frac{\lambda_{\max}(A)}{\lambda_{\min}(A)}.$$

Large $\kappa(A)$ indicates strong anisotropy (elongated level sets) and often slow convergence for basic first-order methods.

Remark 2.4.8 (Geometry of Conditioning) For the quadratic $f(w) = \frac{1}{2} w^\top A w - b^\top w$ with $A \succ 0$, level sets are ellipsoids shaped by A. If $\kappa(A) \gg 1$, the ellipsoids are highly elongated and gradient descent may zig-zag. Preconditioning corresponds to changing variables $w = Pz$ so that the transformed curvature matrix becomes better conditioned.

Example 2.4.4 (A Concrete Anisotropy Calculation) Let

$$\Sigma = \begin{bmatrix} 9 & 0 \\ 0 & 1 \end{bmatrix}.$$

Then $\mathcal{E}_\Sigma(1) = \{x : x_1^2/9 + x_2^2/1 \le 1\}$ is an ellipse with semi-axes 3 and 1. Whitening with $W = \Sigma^{-1/2} = \mathrm{diag}(1/3, 1)$ maps this ellipse to the unit disk. Hence Euclidean distance in whitened coordinates corresponds to Mahalanobis distance in the original coordinates.

2.5 Dual Norms, Inequalities, and Geometric Bounds

Norm duality and a small set of core inequalities are the "inequality engine" behind much of learning theory: generalization bounds, stability estimates, margin arguments, and robustness guarantees. Geometrically, the dual norm describes the support function of the primal unit ball and governs worst-case linear fluctuations under bounded perturbations. We present dual norms and their computation for standard pairs, then record Cauchy–Schwarz, Hölder, and Minkowski, followed by practical norm-conversion bounds. Finally, we connect these tools to the geometry of regularization penalties and structured sparsity.

2.5.1 Dual Norms and Support Functions

2.5.1.1 Duality: $\|x\| = \sup_{\|y\|_*\leq 1}\langle x, y\rangle$

Linear functionals $\langle x, y\rangle$ measure how much a vector x aligns with a "probe" direction y. Bounding such alignment is central in ML: a classifier's score is a linear functional, gradients are linear functionals, and adversarial perturbations maximize linear functionals. The dual norm is defined precisely to control this interaction. The identity $\|x\| = \sup_{\|y\|_*\leq 1}\langle x, y\rangle$ says that every norm can be recovered from its dual as a support function of the dual unit ball. This makes duality a bridge between algebra (suprema) and geometry (supporting hyperplanes).

Definition 2.5.1 (Dual Norm) Let $\|\cdot\|$ be a norm on $\mathbb{R}^d$. Its *dual norm* $\|\cdot\|_*$ is

$$\|y\|_* := \sup_{\|x\|\leq 1} \langle x, y\rangle.$$

Proposition 2.5.1 (Dual Norm is a Norm) *For any norm $\|\cdot\|$ on $\mathbb{R}^d$, the mapping $\|\cdot\|_*$ defined in Definition 2.5.1 is itself a norm.*

Proposition 2.5.2 (Generalized Cauchy–Schwarz Inequality) *For all $x, y \in \mathbb{R}^d$,*

$$\langle x, y\rangle \leq \|x\|\,\|y\|_*.$$

Moreover, $\|y\|_* = \sup_{\|x\|\leq 1}\langle x, y\rangle$ is the smallest constant C such that $\langle x, y\rangle \leq C\|x\|$ holds for all x.

Definition 2.5.2 (Support Function) For a nonempty set $K \subseteq \mathbb{R}^d$, its *support function* is

$$h_K(y) := \sup_{x\in K}\langle x, y\rangle.$$

Remark 2.5.1 (Dual Norm as a Support Function) Let $B = \{x : \|x\| \leq 1\}$ be the primal unit ball. Then $\|y\|_* = h_B(y)$. Thus, dual norms encode supporting hyperplanes of the primal unit ball and quantify the maximum response of B in direction y.

Theorem 2.5.1 (Duality Representation of a Norm) *Let $\|\cdot\|$ be a norm with dual $\|\cdot\|_*$. Then for all $x \in \mathbb{R}^d$,*

$$\|x\| = \sup_{\|y\|_* \leq 1} \langle x, y \rangle.$$

Proof First, by Proposition 2.5.2, for any y with $\|y\|_* \leq 1$,

$$\langle x, y \rangle \leq \|x\| \, \|y\|_* \leq \|x\|,$$

so $\sup_{\|y\|_* \leq 1} \langle x, y \rangle \leq \|x\|$. For the reverse inequality, if $x = 0$ the claim is trivial. For $x \neq 0$, consider the primal unit ball $B = \{u : \|u\| \leq 1\}$. By convexity of B and the fact that $x/\|x\| \in \partial B$, there exists a supporting hyperplane at $x/\|x\|$ with normal y satisfying $\langle x/\|x\|, y \rangle = \sup_{u \in B} \langle u, y \rangle = \|y\|_*$. Scaling y so that $\|y\|_* = 1$ yields $\langle x, y \rangle = \|x\|$. Hence $\sup_{\|y\|_* \leq 1} \langle x, y \rangle \geq \|x\|$. □

2.5.1.2 Dual Pairs: ℓ_1–ℓ_∞ and ℓ_2–ℓ_2

Computing dual norms for standard ℓ_p norms yields formulas that directly control worst-case linear deviations. The dual pair (ℓ_1, ℓ_∞) explains why sparsity regularization (ℓ_1) corresponds to coordinate-bounded perturbations (ℓ_∞), and why ℓ_∞ adversaries are naturally bounded by ℓ_1 weights. The self-duality of ℓ_2 is the familiar Cauchy–Schwarz world: Euclidean geometry where angles and projections behave cleanly.

Proposition 2.5.3 (ℓ_2 is Self-Dual) *For $y \in \mathbb{R}^d$,*

$$\|y\|_{2,*} = \sup_{\|x\|_2 \leq 1} \langle x, y \rangle = \|y\|_2.$$

Moreover, the supremum is attained at $x = y/\|y\|_2$ when $y \neq 0$.

Proposition 2.5.4 (ℓ_1 and ℓ_∞ are Dual) *For $y \in \mathbb{R}^d$,*

$$\sup_{\|x\|_1 \leq 1} \langle x, y \rangle = \|y\|_\infty, \qquad \sup_{\|x\|_\infty \leq 1} \langle x, y \rangle = \|y\|_1.$$

Hence $(\|\cdot\|_1)^* = \|\cdot\|_\infty$ and $(\|\cdot\|_\infty)^* = \|\cdot\|_1$.

Proof For $\|x\|_1 \leq 1$, we have $\langle x, y \rangle = \sum_i x_i y_i \leq \sum_i |x_i| \, |y_i| \leq \|y\|_\infty \sum_i |x_i| \leq \|y\|_\infty$, so the supremum is at most $\|y\|_\infty$. Let $j \in \arg\max_i |y_i|$ and choose $x = \operatorname{sign}(y_j) e_j$ (the j-th basis vector with matching sign); then $\|x\|_1 = 1$ and $\langle x, y \rangle =$

$|y_j| = \|y\|_\infty$, achieving the bound. The second identity is analogous: $\langle x, y\rangle \le \sum_i |x_i|\,|y_i| \le \|x\|_\infty \|y\|_1 \le \|y\|_1$, with equality at $x_i = \text{sign}(y_i)$. □

Remark 2.5.2 (Worst-Case Linear Deviation) For any perturbation set $\{\delta : \|\delta\| \le \varepsilon\}$ and any weight vector w,

$$\sup_{\|\delta\|\le\varepsilon} |w^\top \delta| = \varepsilon\, \|w\|_*.$$

Thus, choosing an ℓ_∞ perturbation model makes $\|w\|_1$ the controlling sensitivity, while choosing an ℓ_2 perturbation model makes $\|w\|_2$ the controlling sensitivity.

2.5.2 *Core Inequalities in* $\mathbb{R}^n$

2.5.2.1 Cauchy–Schwarz, Hölder, and Minkowski

These three inequalities form the backbone of Euclidean and ℓ_p geometry. Cauchy–Schwarz controls inner products by lengths (angles), Hölder generalizes this to dual ℓ_p–ℓ_q pairs, and Minkowski is the triangle inequality for ℓ_p norms. In ML proofs, these inequalities bound losses, gradients, and error terms, often reducing complicated expressions to norm products that can be controlled.

Theorem 2.5.2 (Cauchy–Schwarz) *For all* $x, y \in \mathbb{R}^d$,

$$|\langle x, y\rangle| \le \|x\|_2\, \|y\|_2.$$

Equality holds iff x and y are linearly dependent.

Theorem 2.5.3 (Hölder Inequality) *Let $p, q \in [1, \infty]$ satisfy $\frac{1}{p} + \frac{1}{q} = 1$ (with $1/\infty := 0$). Then for all $x, y \in \mathbb{R}^d$,*

$$|\langle x, y\rangle| \le \|x\|_p\, \|y\|_q.$$

In particular, $\|\cdot\|_q$ is the dual norm of $\|\cdot\|_p$.

Theorem 2.5.4 (Minkowski Inequality) *For $p \in [1, \infty]$ and all $x, y \in \mathbb{R}^d$,*

$$\|x + y\|_p \le \|x\|_p + \|y\|_p.$$

Remark 2.5.3 (A Common Proof Pattern) In optimization and learning-theoretic derivations, one repeatedly encounters terms like $\langle g, w - w^\star\rangle$ where g is a gradient-like vector. Hölder (or generalized Cauchy–Schwarz) turns this into $\|g\|_* \|w - w^\star\|$, separating a "data/gradient" term from a "parameter distance" term.

2.5.2.2 Norm Inequalities and Conversion Bounds (ℓ_p vs. ℓ_q)

Norm conversion bounds quantify how the same vector can look different under different geometries. These inequalities are particularly important in high dimensions: constants can scale as powers of d, and thus switching norms can dramatically change robustness radii and regularization strength. The ℓ_p–ℓ_q conversion bounds also justify using whichever norm is algebraically convenient, at the cost of dimension-dependent factors.

Proposition 2.5.5 (ℓ_p Monotonicity and Conversion) *Let $x \in \mathbb{R}^d$ and $1 \le p \le q \le \infty$. Then*

$$\|x\|_q \le \|x\|_p \le d^{\left(\frac{1}{p}-\frac{1}{q}\right)}\|x\|_q.$$

In particular:

$$\|x\|_\infty \le \|x\|_2 \le \|x\|_1, \qquad \|x\|_1 \le \sqrt{d}\,\|x\|_2, \qquad \|x\|_2 \le \sqrt{d}\,\|x\|_\infty.$$

Remark 2.5.4 (Tightness and Dimension Dependence) The factor $d^{(1/p-1/q)}$ is sharp in general (up to constants), attained for vectors with equal-magnitude coordinates. Hence, while all norms are equivalent, the equivalence can be quantitatively severe when d is large.

Example 2.5.1 (A Conversion Bound in Robustness) If a model is certified robust to all ℓ_2 perturbations of size ε, then it is also robust to all ℓ_∞ perturbations of size at most $\varepsilon/\sqrt{d}$ (since $\|\delta\|_2 \le \sqrt{d}\|\delta\|_\infty$). The $\sqrt{d}$ factor can be substantial in high-dimensional inputs.

2.5.3 Geometry of Regularization Penalties

2.5.3.1 ℓ_1 Corners and Sparsity; ℓ_2 Smoothness and Shrinkage

Regularization chooses a geometry in parameter space. The minimizer of a loss under an ℓ_1 constraint often lands on a corner of the ℓ_1 ball, yielding zeros in many coordinates (sparsity). By contrast, the ℓ_2 ball has no corners and treats directions smoothly, leading to uniform shrinkage rather than exact zeros. This is not merely intuition: it is a geometric consequence of how level sets of the loss meet the boundary of the constraint set.

Remark 2.5.5 (ℓ_1 vs. ℓ_2 via Unit-Ball Geometry) In $\mathbb{R}^d$, the ℓ_1 unit ball is a cross-polytope with many sharp vertices aligned with coordinate axes. A typical convex loss has smooth level sets. When minimizing over the ℓ_1 ball, tangency often occurs at a vertex, forcing many coordinates to be exactly zero. The ℓ_2 ball is strictly convex (round), so tangency occurs on a smooth boundary, typically producing dense solutions with small but nonzero coordinates.

Proposition 2.5.6 (A One-Line Sparsity Certificate) *Let $S \subseteq [d]$ and define the coordinate projection $(x_S)_i = x_i$ if $i \in S$, else 0. If $\|x\|_1 = \|x_S\|_1$ for some S with $|S| = s$, then x has at most s nonzero entries. Conversely, if x is s-sparse, then $\|x\|_1 = \|x_{\mathrm{supp}(x)}\|_1$ with $|\operatorname{supp}(x)| \leq s$.*

Remark 2.5.6 (Shrinkage vs. Thresholding) Ridge (ℓ_2) penalization shrinks all coordinates continuously. Lasso (ℓ_1) penalization introduces thresholding effects: small coordinates are set exactly to zero in many regimes. This behavior will be formalized later when we derive proximal operators.

2.5.3.2 Group Norms and Block-Structured Sparsity Intuition

In many applications, features come in groups (multi-modal blocks, polynomial expansions, one-hot groups, sensor channels). We may want to select or drop entire groups rather than individual coordinates. Group norms achieve this by combining an ℓ_2 norm within each group (to treat group directions smoothly) with an ℓ_1 aggregation across groups (to create corners at group level). The resulting unit ball has block-aligned corners, encouraging group-sparse solutions.

Definition 2.5.3 (Group Partition and Group Norms) Let $\{G_1, \ldots, G_m\}$ be a partition of $[d]$ into disjoint nonempty groups. For $x \in \mathbb{R}^d$, write $x_{G_j} \in \mathbb{R}^{|G_j|}$ for the subvector on indices in G_j. Define the *group ℓ_1/ℓ_2 norm* by

$$\|x\|_{1,2} := \sum_{j=1}^{m} \|x_{G_j}\|_2.$$

More generally, for $p, q \geq 1$, one may define $\|x\|_{p,q} := \left(\sum_{j=1}^{m} \|x_{G_j}\|_q^p\right)^{1/p}$.

Remark 2.5.7 (Geometry: Corners at the Group Level) The unit ball $\{x : \|x\|_{1,2} \leq 1\}$ is convex and has non-smooth points when some group subvectors are exactly zero. Thus, optimizing with $\|x\|_{1,2}$ encourages many entire groups to vanish, producing *block-sparse* solutions. Within an active group, the ℓ_2 norm treats directions smoothly and avoids coordinate-level sparsity bias.

Proposition 2.5.7 (Dual of the Group ℓ_1/ℓ_2 Norm) *Let $\|\cdot\|_{1,2}$ be as in Definition 2.5.3. Its dual norm is*

$$\|y\|_* = \max_{j \in [m]} \|y_{G_j}\|_2.$$

Remark 2.5.8 (Interpretation of the Group Dual) The dual says: the worst-case alignment with a group-sparse unit ball is achieved by concentrating on the single group with largest ℓ_2 magnitude. This mirrors the ℓ_1–ℓ_∞ duality at the group level and is central in group-sparsity generalization and robustness bounds.

2.6 Inner Products and Angles

Inner products endow $\mathbb{R}^d$ with notions of *angle*, *orthogonality*, and *projection*. These are not merely geometric decorations: cosine similarity is a standard tool in embeddings, orthogonal decompositions underpin least squares and PCA, and orthogonal transformations describe invariances that many models should respect (or at least behave stably under). This section formalizes these ideas and provides computationally useful identities.

2.6.1 Orthogonality and Similarity

2.6.1.1 Cosine Similarity and Correlation Geometry

In many ML pipelines, the absolute scale of an embedding is less meaningful than its direction. Cosine similarity compares directions by normalizing vectors, making it invariant to positive rescaling. This is why cosine similarity is common in information retrieval and representation learning. When embeddings are centered, cosine similarity becomes closely related to correlation: it measures linear association after removing mean effects. In high dimensions, cosine similarity also connects to concentration phenomena: random vectors tend to be nearly orthogonal, which has practical implications for nearest-neighbor search and embedding design.

Definition 2.6.1 (Inner Product and Induced Norm) On $\mathbb{R}^d$, the standard inner product is

$$\langle x, y\rangle := x^\top y = \sum_{i=1}^{d} x_i y_i,$$

and the induced Euclidean norm is $\|x\|_2 := \sqrt{\langle x, x\rangle}$.

Definition 2.6.2 (Cosine Similarity and Angle) For nonzero $x, y \in \mathbb{R}^d$, the *cosine similarity* is

$$\cos(x, y) := \frac{\langle x, y\rangle}{\|x\|_2\,\|y\|_2} \in [-1, 1].$$

The (acute/obtuse) *angle* $\theta \in [0, \pi]$ between x and y is defined by

$$\cos\theta = \cos(x, y).$$

Proposition 2.6.1 (Scale Invariance of Cosine Similarity) *For any $\alpha, \beta > 0$ and nonzero x, y,*

$$\cos(\alpha x, \beta y) = \cos(x, y).$$

If α or β is negative, the cosine may flip sign accordingly.

Remark 2.6.1 (Cosine vs. Euclidean Distance After Normalization) If $\|x\|_2 = \|y\|_2 = 1$, then

$$\|x - y\|_2^2 = 2 - 2\langle x, y\rangle = 2\big(1 - \cos(x, y)\big).$$

Thus, on the unit sphere, maximizing cosine similarity is equivalent to minimizing Euclidean distance.

Definition 2.6.3 (Centered Correlation Geometry for Data Vectors) For $u, v \in \mathbb{R}^n$ (e.g., two feature columns across n samples), define the centering operator

$$C := I_n - \frac{1}{n}\mathbf{1}\mathbf{1}^\top, \qquad \tilde{u} := Cu,\ \tilde{v} := Cv.$$

The (sample) correlation coefficient can be written as

$$\operatorname{corr}(u, v) := \frac{\langle \tilde{u}, \tilde{v}\rangle}{\|\tilde{u}\|_2\,\|\tilde{v}\|_2},$$

i.e., cosine similarity of centered vectors.

Remark 2.6.2 (Interpretation) Correlation is direction similarity after removing the mean (translation component along $\mathbf{1}$). This makes clear why correlation is invariant to adding constants to u or v.

2.6.1.2 Angles Between Subspaces: Principal-Angle Intuition

In learning, we often compare not just individual vectors but *subspaces*: feature subspaces, representation subspaces, tangent spaces of manifolds, or the span of learned filters. The correct notion of "angle between subspaces" is given by principal angles, which quantify the best (and worst) possible alignment between directions in the two subspaces. Principal angles provide a stable geometric diagnostic: if two subspaces have small principal angles, then information expressed in one can be transferred to the other with little distortion. This is relevant in domain adaptation, subspace tracking, and comparing learned representations.

Definition 2.6.4 (Principal Angles (Definition-Level Statement)) Let $\mathcal{U}, \mathcal{V} \subseteq \mathbb{R}^d$ be subspaces with $\dim(\mathcal{U}) = k$ and $\dim(\mathcal{V}) = \ell$, and let $m := \min\{k, \ell\}$.

The *principal angles* $0 \leq \theta_1 \leq \cdots \leq \theta_m \leq \pi/2$ are defined recursively by

$$\cos\theta_j := \max_{\substack{u\in\mathcal{U},\, v\in\mathcal{V}\\ \|u\|_2=\|v\|_2=1\\ u\perp u_1,\ldots,u_{j-1}\\ v\perp v_1,\ldots,v_{j-1}}} \langle u, v\rangle,$$

where (u_j, v_j) are maximizers at step j (called principal vectors).

Remark 2.6.3 (Key Intuition) θ_1 is the smallest angle between any unit vector in $\mathcal{U}$ and any unit vector in $\mathcal{V}$ (best alignment), while θ_m captures the worst alignment among the best-matched orthogonal directions. If $\theta_1 = 0$, the subspaces intersect nontrivially. If all $\theta_j = 0$ (and $k = \ell$), then $\mathcal{U} = \mathcal{V}$. If all $\theta_j = \pi/2$, the subspaces are orthogonal.

Remark 2.6.4 (Computational Preview) If $Q_U \in \mathbb{R}^{d\times k}$ and $Q_V \in \mathbb{R}^{d\times \ell}$ have orthonormal columns spanning $\mathcal{U}$ and $\mathcal{V}$, then the singular values of $Q_U^\top Q_V$ are $\cos\theta_j$. This connects principal angles to SVD (developed later).

2.6.2 Geometric Decompositions

2.6.2.1 Pythagorean Theorem and Orthogonal Decompositions

Orthogonal decompositions are the algebraic form of "splitting signal into independent parts." When a vector is decomposed into orthogonal components, squared norm decomposes additively. This is the basis of least squares (signal + residual) and variance decomposition (explained + unexplained). In ML, such decompositions justify objective functions that sum squared errors and motivate projections onto subspaces.

Theorem 2.6.1 (Pythagorean Theorem in Inner-Product Spaces) *If $a, b \in \mathbb{R}^d$ satisfy $\langle a, b\rangle = 0$ (orthogonality), then*

$$\|a + b\|_2^2 = \|a\|_2^2 + \|b\|_2^2.$$

Definition 2.6.5 (Orthogonal Decomposition) Let $\mathcal{U} \subseteq \mathbb{R}^d$ be a subspace. A decomposition $x = u + r$ with $u \in \mathcal{U}$ and $r \in \mathcal{U}^\perp$ is called an *orthogonal decomposition* of x relative to $\mathcal{U}$.

Proposition 2.6.2 (Existence and Uniqueness via Orthogonal Projection) *For any subspace $\mathcal{U} \subseteq \mathbb{R}^d$ and any $x \in \mathbb{R}^d$, there exists a unique pair (u, r) with $u \in \mathcal{U}$ and $r \in \mathcal{U}^\perp$ such that $x = u + r$. The vector u is the orthogonal projection of x onto $\mathcal{U}$ and $r = x - u$ is the residual.*

Remark 2.6.5 (Geometric Meaning) The projection u is the closest point in $\mathcal{U}$ to x under $\|\cdot\|_2$, and the residual r points orthogonally away from $\mathcal{U}$. Thus, orthogonal decompositions are "best approximations" in Euclidean geometry.

2.6.2.2 Energy/Variance Interpretations in Learning Pipelines

In signal processing and statistics, $\|x\|_2^2$ is often called the *energy* of a signal. In ML, squared norms and squared errors are ubiquitous because they admit Pythagorean decompositions and because they connect to variance. For centered data, total variance is the sum of squared distances to the mean; projections onto subspaces split this total variance into explained variance (captured by the subspace) and residual variance (orthogonal complement). This geometric view anticipates PCA, bias–variance tradeoffs, and error decompositions.

Definition 2.6.6 (Total Sum of Squares (TSS) for Centered Data) Let $x_1, \ldots, x_n \in \mathbb{R}^d$ and $\bar{x} = \frac{1}{n}\sum_{i=1}^{n} x_i$. The total sum of squares is

$$\mathrm{TSS} := \sum_{i=1}^{n} \|x_i - \bar{x}\|_2^2.$$

If we form the centered data matrix $X_c \in \mathbb{R}^{n\times d}$ with rows $(x_i - \bar{x})^\top$, then

$$\mathrm{TSS} = \|X_c\|_F^2.$$

Proposition 2.6.3 (Energy Split Under Projection) *Let $\mathcal{U} \subseteq \mathbb{R}^d$ be a subspace and write $x = u + r$ with $u \in \mathcal{U}$ and $r \in \mathcal{U}^\perp$. Then*

$$\|x\|_2^2 = \|u\|_2^2 + \|r\|_2^2.$$

For a dataset $\{x_i\}$, summing over i yields an additive decomposition of total energy into energy captured by $\mathcal{U}$ and residual energy orthogonal to $\mathcal{U}$.

Remark 2.6.6 (Explained Variance Intuition) If $\mathcal{U}$ is a low-dimensional representation subspace (e.g., PCA subspace), $\sum_i \|u_i\|_2^2$ measures variance explained by the representation, while $\sum_i \|r_i\|_2^2$ measures information discarded. This provides a geometric criterion for choosing representation dimension.

2.6.3 Isometries and Invariances

2.6.3.1 Orthogonal Matrices as Distance-Preserving Maps

Orthogonal transformations (rotations and reflections) preserve inner products, norms, and hence distances and angles. They capture a core invariance of Euclidean geometry: changing coordinates by an orthogonal matrix does not change the problem. In ML, orthogonal transformations appear as benign reparameterizations (e.g., whitening + rotation), as data augmentations, and as symmetry operations. They also underpin numerically stable algorithms: QR factorization relies on orthogonal matrices precisely because they do not amplify errors.

Definition 2.6.7 (Orthogonal Matrix) A matrix $Q \in \mathbb{R}^{d \times d}$ is *orthogonal* if

$$Q^\top Q = I_d.$$

Equivalently, $Q^{-1} = Q^\top$ and the columns (and rows) of Q form an orthonormal basis.

Proposition 2.6.4 (Inner Product and Norm Invariance) *If Q is orthogonal, then for all $x, y \in \mathbb{R}^d$,*

$$\langle Qx, Qy \rangle = \langle x, y \rangle, \qquad \|Qx\|_2 = \|x\|_2, \qquad \|Qx - Qy\|_2 = \|x - y\|_2.$$

Proof $\langle Qx, Qy \rangle = x^\top Q^\top Q y = x^\top y$. Setting $y = x$ yields $\|Qx\|_2^2 = \|x\|_2^2$. Finally, apply the norm identity to $x - y$. □

Definition 2.6.8 (Isometry) A map $\phi : \mathbb{R}^d \to \mathbb{R}^d$ is an *isometry* (with respect to $\|\cdot\|_2$) if $\|\phi(x) - \phi(y)\|_2 = \|x - y\|_2$ for all x, y. Every orthogonal linear map $x \mapsto Qx$ is an isometry.

2.6.3.2 Rotations/Reflections and Invariance of Norms and Angles

Rotations and reflections are the fundamental rigid motions of Euclidean space. They do not change lengths or angles, so any geometric statement expressed purely in terms of $\|\cdot\|_2$ and $\langle \cdot, \cdot \rangle$ is invariant under such transformations. In ML, this matters when we interpret embeddings: if only relative geometry matters, then embeddings are identifiable only up to an orthogonal transformation. It also matters for algorithm design: enforcing orthogonality constraints can preserve information and stabilize training (e.g., orthogonal initialization, orthogonal recurrent weights).

Definition 2.6.9 (Rotation and Reflection) An orthogonal matrix Q is a *rotation* if $\det(Q) = +1$ and a *reflection* (or improper rotation) if $\det(Q) = -1$.

Proposition 2.6.5 (Angle Invariance) *Let $x, y \in \mathbb{R}^d$ be nonzero and Q orthogonal. Then the angle between Qx and Qy equals the angle between x and y:*

$$\frac{\langle Qx, Qy \rangle}{\|Qx\|_2 \|Qy\|_2} = \frac{\langle x, y \rangle}{\|x\|_2 \|y\|_2}.$$

Hence cosine similarity is invariant under orthogonal transformations.

Remark 2.6.7 (Identifiability of Embeddings) If an embedding method is trained using only inner products or distances (e.g., certain contrastive objectives), then the solution is typically identifiable only up to an orthogonal transformation of the embedding space, since such transformations preserve all pairwise distances and angles.

Remark 2.6.8 (Orthogonality as a Numerical Safeguard) Multiplication by an orthogonal matrix does not amplify Euclidean norm, so it does not magnify perturbations. This is why orthogonal factors (QR, Householder reflections) are central to stable numerical linear algebra and to robust coordinate transforms in ML pipelines.

2.7 Gram Matrices and Positive Semidefinite Geometry

Gram matrices encode geometry: they are the algebraic record of all pairwise inner products among a set of vectors. This makes them fundamental in ML. In kernel methods, one replaces explicit feature vectors by an implicit Gram matrix computed via a kernel function. In statistics, covariance matrices are PSD and define uncertainty ellipsoids. In optimization, PSD constraints define the PSD cone, a central object in convex geometry and semidefinite programming. This section develops Gram matrices, positive semidefiniteness, the PSD cone viewpoint, and the conversion between inner products and distances (via centering).

2.7.1 Gram Matrices

2.7.1.1 Gram Matrix as Inner-Product Table and Kernel Precursor

Given vectors $x_1, \dots, x_n$, the Gram matrix stores $\langle x_i, x_j \rangle$ for all pairs. Geometrically, this table determines angles and lengths of projections, and (up to rigid motions) it determines the point configuration. Computationally, it turns many problems into linear algebra on an $n \times n$ matrix. In kernels, we do not even need explicit coordinates in $\mathbb{R}^d$; it suffices to know inner products in a (possibly infinite-dimensional) feature space. The Gram matrix is thus the bridge between Euclidean geometry and similarity-based learning.

Definition 2.7.1 (Gram Matrix) Let $x_1, \dots, x_n \in \mathbb{R}^d$ and form the data matrix $X \in \mathbb{R}^{n \times d}$ with rows $x_i^\top$. The *Gram matrix* of $\{x_i\}_{i=1}^n$ is

$$G := XX^\top \in \mathbb{R}^{n \times n}, \qquad G_{ij} = \langle x_i, x_j \rangle.$$

Remark 2.7.1 (Two Gram Viewpoints: Sample–Sample vs. Feature–Feature) For $X \in \mathbb{R}^{n\times d}$, both

$$G_s := XX^\top \in \mathbb{R}^{n\times n} \quad \text{and} \quad G_f := X^\top X \in \mathbb{R}^{d\times d}$$

are Gram-type matrices: G_s compares samples via inner products, G_f compares features via inner products. Both are PSD and share the same nonzero eigenvalues (up to multiplicity), a fact central in PCA/SVD.

Definition 2.7.2 (Kernel and Kernel Matrix) A function $k : \mathcal{X} \times \mathcal{X} \to \mathbb{R}$ is a *kernel* if there exists a feature map $\phi : \mathcal{X} \to \mathcal{H}$ into an inner-product space $(\mathcal{H}, \langle\cdot,\cdot\rangle_{\mathcal{H}})$ such that

$$k(x, z) = \langle\phi(x), \phi(z)\rangle_{\mathcal{H}}.$$

Given data points $x_1, \dots, x_n \in \mathcal{X}$, the *kernel matrix* is $K \in \mathbb{R}^{n\times n}$ with $K_{ij} = k(x_i, x_j)$.

Remark 2.7.2 (Why Gram Matrices Enable Kernelization) Many algorithms can be written using only inner products $\langle x_i, x_j\rangle$. Replacing these by $k(x_i, x_j)$ yields a nonlinear method in the original space but linear in the feature space $\phi(\mathcal{X})$.

2.7.1.2 Positive Semidefiniteness and Geometric Feasibility

Not every symmetric matrix is a Gram matrix of real vectors. The exact feasibility condition is positive semidefiniteness: a symmetric matrix is Gram iff it is PSD. This provides a clean geometric characterization: PSD means that all quadratic forms are nonnegative. In ML, this becomes a sanity check: if you propose a similarity matrix intended to be inner products, it must be PSD (or else it corresponds to an indefinite geometry). This is also the basis of "PSD correction" methods (eigenvalue clipping) for noisy similarity matrices.

Definition 2.7.3 (Positive Semidefinite (PSD)) A symmetric matrix $A \in \mathbb{R}^{n\times n}$ is *positive semidefinite*, written $A \succeq 0$, if

$$z^\top Az \geq 0 \qquad \forall z \in \mathbb{R}^n.$$

It is *positive definite*, written $A \succ 0$, if $z^\top Az > 0$ for all $z \neq 0$.

Proposition 2.7.1 (Gram Matrices are PSD) *If $G = XX^\top$ is a Gram matrix, then $G \succeq 0$.*

Proof For any $z \in \mathbb{R}^n$,

$$z^\top Gz = z^\top XX^\top z = \|X^\top z\|_2^2 \geq 0.$$

□

Theorem 2.7.1 (Geometric Feasibility: PSD $\iff$ Gram) *A symmetric matrix $G \in \mathbb{R}^{n\times n}$ is a Gram matrix of some vectors in $\mathbb{R}^d$ (for some $d \leq n$) iff $G \succeq 0$. Equivalently, $G \succeq 0$ iff there exist vectors $x_1, \ldots, x_n \in \mathbb{R}^n$ such that $G_{ij} = \langle x_i, x_j\rangle$.*

Proof ($\Rightarrow$) Follows from Proposition 2.7.1. ($\Leftarrow$) If $G \succeq 0$, then by spectral decomposition $G = Q\Lambda Q^\top$ with $\Lambda = \operatorname{diag}(\lambda_1, \ldots, \lambda_n)$, $\lambda_i \geq 0$. Let $\Lambda^{1/2} = \operatorname{diag}(\sqrt{\lambda_1}, \ldots, \sqrt{\lambda_n})$ and set $X := \Lambda^{1/2}Q^\top$. Then $G = Q\Lambda Q^\top = (Q\Lambda^{1/2})(Q\Lambda^{1/2})^\top = X^\top X$, i.e., G is a Gram matrix (up to transpose convention). □

Remark 2.7.3 (Rank and Embedding Dimension) If $G \succeq 0$ and $\operatorname{rank}(G) = r$, then the vectors realizing G can be chosen in $\mathbb{R}^r$. Thus, PSD plus low rank implies a low-dimensional geometric realization.

2.7.2 PSD Cones and Geometry

2.7.2.1 Eigenvalue Nonnegativity and Quadratic Forms

The PSD condition is equivalent to "all eigenvalues are nonnegative." This equivalence is more than a fact: it explains how PSD constraints behave geometrically. Eigenvalues represent curvature along orthogonal directions. PSD means no direction has negative curvature, so quadratic forms act like squared lengths in some transformed coordinates. In ML optimization, PSD matrices appear as covariance matrices and as Hessians of convex functions; eigenvalues control conditioning and stability.

Theorem 2.7.2 (PSD $\iff$ Nonnegative Eigenvalues) *For a symmetric matrix $A \in \mathbb{R}^{n\times n}$, the following are equivalent:*

1. *$A \succeq 0$.*
2. *All eigenvalues of A are nonnegative.*
3. *There exists a matrix B such that $A = B^\top B$ (or $A = BB^\top$).*

Remark 2.7.4 (PSD Cone) The set $\mathbb{S}^n_+ := \{A \in \mathbb{R}^{n\times n} : A = A^\top,\ A \succeq 0\}$ is a convex cone: if $A, B \succeq 0$ and $\alpha, \beta \geq 0$, then $\alpha A + \beta B \succeq 0$. This cone is the feasible region of semidefinite programs and encodes "matrix-valued" nonnegativity.

Remark 2.7.5 (Quadratic Forms as Generalized Squared Lengths) If $A \succeq 0$, then $x^\top Ax = \|Bx\|_2^2$ for some B by Theorem 2.7.2(3). Thus, PSD matrices define weighted geometries and appear naturally in Mahalanobis distances and energy functions.

2.7.2.2 Covariance Matrices and Geometric Uncertainty Ellipsoids

Covariance matrices are PSD by construction and encode the shape of uncertainty. In a Gaussian model, contours of equal density are ellipsoids given by Mahalanobis distance; more generally, covariance describes anisotropy of data clouds. This geometric view explains why whitening and PCA align with covariance eigenvectors: they rotate into principal axes and rescale by standard deviations. In learning pipelines, such transformations can dramatically improve conditioning and reveal low-dimensional structure.

Definition 2.7.4 (Covariance Matrix) Let $X \in \mathbb{R}^d$ be a random vector with mean $\mu = \mathbb{E}[X]$. Its covariance matrix is

$$\Sigma := \mathbb{E}\big[(X-\mu)(X-\mu)^\top\big] \in \mathbb{R}^{d\times d}.$$

Proposition 2.7.2 (Covariance is PSD) *Every covariance matrix Σ is PSD.*

Proof For any $v \in \mathbb{R}^d$,

$$v^\top \Sigma v = \mathbb{E}\big[v^\top (X-\mu)(X-\mu)^\top v\big] = \mathbb{E}\big[(v^\top (X-\mu))^2\big] \geq 0.$$

□

Definition 2.7.5 (Uncertainty Ellipsoid) Let $\Sigma \succ 0$. The ellipsoid

$$\mathcal{E}_\Sigma(r) := \left\{x \in \mathbb{R}^d : x^\top \Sigma^{-1} x \leq r^2\right\}$$

is the set of points within Mahalanobis radius r from the origin (or from μ after shifting).

Remark 2.7.6 (Eigen-Geometry of Covariance) If $\Sigma = Q\Lambda Q^\top$ with eigenvalues $\lambda_i > 0$, then $\mathcal{E}_\Sigma(r)$ has principal axes q_i with semi-axis lengths $r\sqrt{\lambda_i}$. Large variance directions correspond to long axes, small variance directions to short axes.

2.7.3 From Gram Matrices to Distances

2.7.3.1 Recovering Squared Distances from Inner Products (Centering Trick)

Inner products and distances are two sides of the same geometry. If we know all inner products (Gram matrix), we can recover all squared distances. Conversely, if we know all squared distances between points, we can recover a centered Gram matrix. The subtlety is translation: distances are translation-invariant, while Gram matrices depend on the choice of origin. Centering removes this ambiguity. This "centering trick" is the algebraic heart of classical multidimensional scaling (MDS) and appears implicitly in kernel PCA and distance-based embedding methods.

Proposition 2.7.3 (Distances from a Gram Matrix) *Let $x_1, \dots, x_n \in \mathbb{R}^d$ with Gram matrix G where $G_{ij} = \langle x_i, x_j \rangle$. Define squared distances $D^2 \in \mathbb{R}^{n \times n}$ by $D^2_{ij} = \|x_i - x_j\|_2^2$. Then*

$$D^2_{ij} = G_{ii} + G_{jj} - 2G_{ij}.$$

Proof $\|x_i - x_j\|_2^2 = \langle x_i - x_j, x_i - x_j \rangle = \langle x_i, x_i \rangle + \langle x_j, x_j \rangle - 2\langle x_i, x_j \rangle$. □

Definition 2.7.6 (Centering Matrix) Let $\mathbf{1} \in \mathbb{R}^n$ be the all-ones vector and define

$$H := I_n - \frac{1}{n}\mathbf{1}\mathbf{1}^\top.$$

Then H is symmetric and idempotent ($H^\top = H$, $H^2 = H$), and Hx subtracts the mean of the components of x.

Proposition 2.7.4 (Double-Centering: Gram from Squared Distances) *Let $x_1, \dots, x_n \in \mathbb{R}^d$ and let $X \in \mathbb{R}^{n \times d}$ be the row-stacked data matrix. Let $X_c := HX$ be the centered data matrix (rows $x_i - \bar{x}$), and let $G_c := X_c X_c^\top$ be the centered Gram matrix. Let D^2 be the squared-distance matrix with $D^2_{ij} = \|x_i - x_j\|_2^2$. Then*

$$G_c = -\frac{1}{2} H D^2 H.$$

Remark 2.7.7 (What Centering Accomplishes) Distances do not determine the absolute location of points, only their relative configuration. Centering fixes a canonical translation by forcing the mean to be zero. The matrix G_c encodes this centered configuration.

2.7.3.2 Kernel Distances and Similarity-Based Learning Viewpoint

In kernel methods, similarity is represented by a PSD kernel matrix K that acts like a Gram matrix in a feature space. This makes it natural to define distances *in that feature space* even when coordinates are implicit. Such kernel distances unify many algorithms: nearest neighbors in feature space, kernel k-means, and kernel PCA all rely on distance computations derived from K. Importantly, PSD ensures that these distances behave like squared Euclidean distances in some (possibly high-dimensional) embedding, preserving geometric consistency.

Definition 2.7.7 (Kernel-Induced Squared Distance) Let k be a kernel with feature map ϕ into an inner-product space $\mathcal{H}$. Define the squared distance in feature space by

$$d_k(x, z)^2 := \|\phi(x) - \phi(z)\|_{\mathcal{H}}^2 = \langle \phi(x), \phi(x) \rangle_{\mathcal{H}} + \langle \phi(z), \phi(z) \rangle_{\mathcal{H}} - 2\langle \phi(x), \phi(z) \rangle_{\mathcal{H}}.$$

Using $k(x, z) = \langle \phi(x), \phi(z) \rangle_{\mathcal{H}}$, this becomes

$$d_k(x, z)^2 = k(x, x) + k(z, z) - 2k(x, z).$$

Remark 2.7.8 (Kernel Distance from the Kernel Matrix) Given data $x_1, \ldots, x_n$ and kernel matrix K, the pairwise squared distances in feature space satisfy

$$d_k(x_i, x_j)^2 = K_{ii} + K_{jj} - 2K_{ij}.$$

Thus, one can run distance-based algorithms in implicit feature space without explicitly computing ϕ.

Remark 2.7.9 (Similarity Matrices Must be PSD to be Gram-Like) If a proposed similarity matrix is not PSD, it cannot be a Gram matrix of real vectors, so distances derived from $K_{ii} + K_{jj} - 2K_{ij}$ may fail to satisfy Euclidean consistency. In practice, one may "PSD-project" a noisy similarity matrix by setting negative eigenvalues to zero (eigenvalue clipping), yielding the closest PSD matrix in Frobenius norm.

2.8 Orthogonal Projections

Orthogonal projection is the canonical "best approximation" operator in Euclidean geometry. It underlies least squares, PCA, and the geometry of residuals, and it is the algebraic mechanism behind many error decompositions. This section develops projection operators, distance-to-subspace formulas, and practical projection matrices, with careful attention to numerical stability.

2.8.1 Projection Operators

2.8.1.1 Projectors, Idempotence, and Geometric Decompositions

A projection is an operator that leaves points in a target set unchanged and maps all other points to their "shadow" on that set. Algebraically, a linear projector is characterized by idempotence $P^2 = P$. Geometrically, orthogonal projectors are characterized by symmetry $P^\top = P$ and induce orthogonal decompositions $x = Px + (I - P)x$ into a component in the subspace and a residual orthogonal to it. This decomposition is the template for "signal + noise" or "explained + unexplained" splits throughout ML.

Definition 2.8.1 (Linear Projector and Orthogonal Projector) A matrix $P \in \mathbb{R}^{d\times d}$ is a *(linear) projector* if it is idempotent:

$$P^2 = P.$$

It is an *orthogonal projector* if additionally P is symmetric:

$$P^\top = P.$$

Proposition 2.8.1 (Range/Null Characterization) *If P is a projector, then*

$$\text{range}(P) = \{Px : x \in \mathbb{R}^d\}, \qquad \text{null}(P) = \{x : Px = 0\},$$

and every $x \in \mathbb{R}^d$ admits a decomposition

$$x = Px + (I - P)x,$$

where $Px \in$ range(P) and $(I - P)x \in$ null(P). If P is orthogonal, then null$(P) =$ range$(P)^\perp$.

Proof Idempotence implies $P(Px) = Px$, so Px is fixed by P and lies in the range. Also, $P(I - P)x = Px - P^2x = 0$, hence $(I - P)x$ lies in the null space. If $P^\top = P$, then for $u = Pv \in$ range(P) and $r \in$ null(P), $\langle u, r\rangle = \langle Pv, r\rangle = \langle v, P^\top r\rangle = \langle v, Pr\rangle = 0$, so null$(P) \subseteq$ range$(P)^\perp$. Dimension counting yields equality. □

Remark 2.8.1 (Geometric Meaning of Symmetry) Idempotence alone yields a projection onto a subspace along some complementary subspace (possibly oblique). Symmetry $P^\top = P$ is exactly the condition that the projection is orthogonal (closest in ℓ_2).

2.8.1.2 Pythagoras Theorem and Error Decomposition

Orthogonal projection yields a Pythagorean decomposition of squared error. This is the geometric origin of least squares optimality and of variance decompositions. The key point is orthogonality: if $x = Px + r$ with $r \perp$ range(P), then $\|x\|_2^2 = \|Px\|_2^2 + \|r\|_2^2$. Summed over samples, this becomes a clean split of total energy into explained and residual energy.

Theorem 2.8.1 (Pythagorean Theorem for Orthogonal Projections) *Let P be an orthogonal projector. For any $x \in \mathbb{R}^d$, define $u := Px$ and $r := x - u =$*

$(I - P)x$. Then $u \perp r$ and

$$\|x\|_2^2 = \|u\|_2^2 + \|r\|_2^2.$$

Proof Orthogonality follows from Proposition 2.8.1: $u \in \text{range}(P)$ and $r \in \text{null}(P) = \text{range}(P)^{\perp}$. Then apply the Euclidean Pythagorean theorem (Theorem 2.6.1). □

Remark 2.8.2 (Error Decomposition Template) If u is the "model component" and r is the residual, then squared error decomposes additively. This is why squared loss admits especially clean geometry: it is compatible with orthogonality.

2.8.2 Distance to Subspaces

2.8.2.1 Least-Squares Motivation via Closest-Point Projection

Least squares is projection in disguise. The vector of model predictions lies in a subspace (the column space), and fitting chooses the closest point to the observed labels. This viewpoint avoids calculus: optimality is simply the nearest-point property of orthogonal projection. It also clarifies uniqueness: if the subspace is fixed, the projected point is unique; non-uniqueness of parameters occurs only when the mapping from parameters to the subspace point is not injective (i.e., when the design matrix has a nontrivial null space).

Definition 2.8.2 (Distance from a Point to a Subspace) Let $\mathcal{U} \subseteq \mathbb{R}^d$ be a subspace. The distance from $x \in \mathbb{R}^d$ to $\mathcal{U}$ is

$$\text{dist}(x, \mathcal{U}) := \inf_{u \in \mathcal{U}} \|x - u\|_2.$$

Theorem 2.8.2 (Closest-Point Projection Onto a Subspace) *For any subspace $\mathcal{U} \subseteq \mathbb{R}^d$ and any $x \in \mathbb{R}^d$, there exists a unique $u^\star \in \mathcal{U}$ such that*

$$\|x - u^\star\|_2 = \text{dist}(x, \mathcal{U}).$$

Moreover, writing $r^\star := x - u^\star$, we have $r^\star \in \mathcal{U}^{\perp}$.

Remark 2.8.3 (Least Squares as Projection in $\mathbb{R}^n$) Given $X \in \mathbb{R}^{n \times d}$ and $y \in \mathbb{R}^n$, the least squares problem

$$\min_{w \in \mathbb{R}^d} \|Xw - y\|_2^2$$

chooses $\widehat{y} := X\widehat{w}$ as the orthogonal projection of y onto $\text{range}(X) \subseteq \mathbb{R}^n$. Thus, $\widehat{y}$ is unique even when $\widehat{w}$ is not (when $\text{null}(X) \neq \{0\}$).

2.8.2.2 Residual Geometry and Orthogonality Conditions

The defining optimality condition of least squares is orthogonality of the residual to the model subspace. This is both a geometric statement (residual is perpendicular to the column space) and an algebraic statement (normal equations). In ML terms, it means that the residual has zero correlation with every feature direction representable by the model. This property is central to diagnostics, interpretability, and to understanding what information the model fails to capture.

Proposition 2.8.2 (Orthogonality Condition for Best Approximation) *Let $\mathcal{U} \subseteq \mathbb{R}^d$ be a subspace and let $u^\star$ be the closest point to x in $\mathcal{U}$. Then for all $u \in \mathcal{U}$,*

$$\langle x - u^\star,\ u \rangle = 0.$$

Equivalently, $x - u^\star \in \mathcal{U}^\perp$.

Proposition 2.8.3 (Normal Equations for Least Squares) *Let $X \in \mathbb{R}^{n\times d}$ and $y \in \mathbb{R}^n$. A vector $\widehat{w} \in \mathbb{R}^d$ minimizes $\|Xw - y\|_2^2$ iff it satisfies the normal equations*

$$X^\top (X\widehat{w} - y) = 0.$$

Proof Let $\widehat{y} = X\widehat{w}$ be the projection of y onto range(X) and $r = y - \widehat{y}$ the residual. By Proposition 2.8.2, $r \perp$ range(X), i.e., $\langle r, Xv\rangle = 0$ for all v. Equivalently, $v^\top X^\top r = 0$ for all v, hence $X^\top r = 0$, i.e., $X^\top (X\widehat{w} - y) = 0$. □

Remark 2.8.4 (Statistical Interpretation) The condition $X^\top r = 0$ says each feature column is uncorrelated with the residual (under the sample inner product). In linear regression, this is why residual plots against fitted features are diagnostic for misspecification.

2.8.3 Projection Matrices in Practice

2.8.3.1 Projecting onto Spans Using $P = A(A^\top A)^{-1}A^\top$ (Full-Rank Case)

To compute projections concretely, we represent a subspace as the span of columns of a matrix A. When A has full column rank, the orthogonal projector onto range(A) has a closed form: $P = A(A^\top A)^{-1}A^\top$. This is the matrix that maps any vector to its least-squares fit in the subspace. In practice, one rarely forms $(A^\top A)^{-1}$ explicitly (for numerical reasons), but the formula is conceptually fundamental and will reappear in regression, PCA, and kernel methods.

Theorem 2.8.3 (Projection Onto a Column Space (Full Column Rank)) *Let $A \in \mathbb{R}^{d\times k}$ have full column rank* (rank$(A) = k$)*, and define*

$$P := A(A^\top A)^{-1}A^\top \in \mathbb{R}^{d\times d}.$$

Then:

1. *P is an orthogonal projector onto* range(A)*;*
2. *for any $x \in \mathbb{R}^d$, Px is the unique minimizer of* $\min_{u \in \text{range}(A)} \|x - u\|_2$*;*
3. *writing $x = Px + r$ yields $r \in$* range$(A)^\perp$ *and* $\|x\|_2^2 = \|Px\|_2^2 + \|r\|_2^2$.

Proof

(1) Symmetry: $P^\top = A(A^\top A)^{-1}A^\top = P$ since $(A^\top A)^{-1}$ is symmetric. Idempotence:

$$P^2 = A(A^\top A)^{-1}A^\top A(A^\top A)^{-1}A^\top = A(A^\top A)^{-1}(A^\top A)(A^\top A)^{-1}A^\top = P.$$

Moreover, for any x, $Px \in$ range(A), and if $x = Az$ then $P(Az) = Az$ (fixed), so P projects onto range(A).

(2) Any $u \in$ range(A) has the form $u = Az$. Minimizing $\|x - Az\|_2^2$ yields normal equations $A^\top(Az - x) = 0$, so $z = (A^\top A)^{-1}A^\top x$ and $u^\star = Az = Px$.

(3) Follows from orthogonal projection properties (Theorem 2.8.1).

□

Remark 2.8.5 (If A is Not Full Rank) If $A^\top A$ is singular, then $(A^\top A)^{-1}$ does not exist and the formula must be replaced by a pseudoinverse: $P = AA^\dagger$ (or $P = A(A^\top A)^\dagger A^\top$). This will be discussed in later chapters on numerical linear algebra.

2.8.3.2 Numerical/Conditioning Intuition (Why Geometry Can Become Unstable)

Projection is a geometric operation, but its computation can be numerically unstable when the basis matrix A is ill-conditioned. The culprit is the inversion of $A^\top A$: if columns of A are nearly linearly dependent, then $A^\top A$ is nearly singular, and small perturbations in data (or floating-point errors) can produce large changes in the computed projection coefficients. This is why stable implementations use orthonormal bases (QR decomposition) rather than normal equations. The geometric message is that "almost the same subspace" can have a poorly conditioned coordinate system.

Remark 2.8.6 (Conditioning of Normal Equations) If $\kappa(\cdot)$ denotes the spectral condition number, then typically

$$\kappa(A^\top A) = \kappa(A)^2.$$

Thus, solving least squares via $A^\top A$ squares the conditioning and can dramatically amplify numerical error. Orthonormalization (QR) avoids this squaring and is preferred in practice.

Remark 2.8.7 (Stable Alternative via Orthonormal Bases) If $A = QR$ is a QR factorization with Q having orthonormal columns spanning range(A), then the projector is

$$P = QQ^\top,$$

and $Px = QQ^\top x$ is computed without forming $(A^\top A)^{-1}$. This connects numerical stability directly to geometry: orthonormal coordinates preserve lengths and do not amplify noise.

2.9 Projections onto Convex Sets and Proximal Geometry

Orthogonal projection onto subspaces is linear and admits explicit formulas. Projection onto a general *convex set* is nonlinear, yet it remains geometrically well-behaved: the Euclidean projection exists and is unique, it is non-expansive, and it satisfies Pythagorean-type inequalities. These facts are the geometric foundation of many constrained optimization algorithms. Moreover, the *proximal operator* generalizes projection by allowing a penalty term, capturing shrinkage and thresholding effects (central in sparse learning). We give a careful preview here; the algorithmic consequences will be developed in the optimization chapter.

2.9.1 Convex Sets in $\mathbb{R}^n$

2.9.1.1 Convex Combinations, Convex Hull, and Extreme Points

Convexity formalizes "no holes" under interpolation: the line segment between two feasible points stays feasible. This property enables global reasoning with local tools (supporting hyperplanes, projections, subgradients). Convex hulls describe all mixtures of a finite set of prototypes, and extreme points represent the "atoms" of convex sets. In ML, probabilities form simplices, feasible sets of linear constraints form polytopes, and many regularizers correspond to convex balls. Understanding convex combinations and extreme points provides a geometric intuition for why some constraints induce sparsity (extreme points aligned with axes or groups).

Definition 2.9.1 (Convex Set) A set $C \subseteq \mathbb{R}^d$ is *convex* if for all $x, y \in C$ and all $\theta \in [0, 1]$,

$$\theta x + (1 - \theta) y \in C.$$

Definition 2.9.2 (Convex Hull) For a set $S \subseteq \mathbb{R}^d$, the convex hull is

$$\operatorname{conv}(S) := \left\{ \sum_{i=1}^{m} \alpha_i s_i \;:\; m \in \mathbb{N},\; s_i \in S,\; \alpha_i \geq 0,\; \sum_{i=1}^{m} \alpha_i = 1 \right\}.$$

Definition 2.9.3 (Extreme Point) Let $C \subseteq \mathbb{R}^d$ be convex. A point $x \in C$ is an *extreme point* if whenever $x = \theta y + (1 - \theta)z$ for some $y, z \in C$ and $\theta \in (0, 1)$, it follows that $y = z = x$.

Remark 2.9.1 (Why Extreme Points Matter) For many convex optimization problems, optimal solutions occur at extreme points when the objective is linear. In regularization, geometry of extreme points explains sparsity and group selection.

2.9.1.2 Balls, Simplices, and Polytopes as Constraint Sets

The most common constraint sets in ML are convex: norm balls (regularization), simplices (probabilities/attention weights), and polytopes (linear inequalities). These sets are chosen because they admit efficient projections or proximal operators, and because their geometry matches prior beliefs (bounded energy, nonnegativity, budgets). Even when the learning objective is nonconvex, using convex constraints keeps subproblems tractable.

Example 2.9.1 (Canonical Convex Constraint Sets)

- ℓ_2-ball: $B_2(r) = \{x : \|x\|_2 \leq r\}$.
- ℓ_1-ball: $B_1(r) = \{x : \|x\|_1 \leq r\}$.
- Simplex: $\Delta^{k-1} = \{p \in \mathbb{R}^k : p_i \geq 0, \sum_i p_i = 1\}$.
- Polytope: $\{x \in \mathbb{R}^d : Ax \leq b\}$.

All are convex, and each corresponds to a distinct geometric bias.

2.9.2 Projection Theorem for Convex Sets

2.9.2.1 Existence/Uniqueness of Euclidean Projections

The projection onto a convex set is the closest feasible point. Convexity is exactly the property that guarantees uniqueness: in nonconvex sets, multiple closest points can exist. Existence follows when the set is closed and nonempty (in Euclidean space). This theorem is the geometric cornerstone for projected gradient methods and for interpreting constraints as "hard regularization."

Definition 2.9.4 (Euclidean Projection) Let $C \subseteq \mathbb{R}^d$ be nonempty. The *Euclidean projection* of $x \in \mathbb{R}^d$ onto C is

$$\Pi_C(x) \in \arg\min_{z \in C} \|x - z\|_2.$$

Theorem 2.9.1 (Projection Onto Closed Convex Sets) *If $C \subseteq \mathbb{R}^d$ is nonempty, closed, and convex, then for every $x \in \mathbb{R}^d$, the minimization problem* $\min_{z \in C} \|x - z\|_2$ *has a* unique *solution, denoted $\Pi_C(x)$.*

Remark 2.9.2 (Closedness is Essential for Existence) If C is not closed, the infimum distance may not be attained (e.g., $C = (0, 1) \subset \mathbb{R}$ and $x = 0$). Convexity is essential for uniqueness.

2.9.2.2 Non-Expansiveness and Pythagorean-Type Properties

Euclidean projection onto a closed convex set is a *non-expansive* map: it cannot increase distances. This is a powerful stability property: projecting noisy iterates does not amplify the noise. Additionally, projection satisfies a Pythagorean-type inequality that generalizes orthogonal decomposition: the error vector from x to its projection is orthogonal (in a generalized sense) to feasible directions. These properties are the basis for convergence proofs of projected gradient and proximal methods.

Theorem 2.9.2 (Characterization and Pythagorean Inequality) *Let $C \subseteq \mathbb{R}^d$ be nonempty, closed, and convex, and let $p = \Pi_C(x)$. Then:*

*1. (*Variational inequality*) For all $z \in C$,*

$$\langle x - p, \ z - p \rangle \leq 0.$$

*2. (*Pythagorean-type inequality*) For all $z \in C$,*

$$\|x - z\|_2^2 \geq \|x - p\|_2^2 + \|p - z\|_2^2.$$

Proof

(1) Consider the function $\varphi(t) = \|x - (p + t(z - p))\|_2^2$ for $t \in [0, 1]$. Since C is convex, $p + t(z - p) \in C$ and minimality of p implies $\varphi(t) \geq \varphi(0)$. Thus $\varphi'(0) \geq 0$. Computing

$$\varphi(t) = \|x - p - t(z - p)\|_2^2 = \|x - p\|_2^2 - 2t\langle x - p, \ z - p \rangle + t^2\|z - p\|_2^2,$$

gives $\varphi'(0) = -2\langle x - p, \ z - p \rangle \geq 0$, hence the variational inequality.

(2) Expand:

$$\|x - z\|_2^2 = \|(x - p) + (p - z)\|_2^2 = \|x - p\|_2^2 + \|p - z\|_2^2 + 2\langle x - p, \ p - z \rangle,$$

and apply (1) with z replaced by z to get $\langle x - p, \ p - z \rangle \geq 0$, yielding the inequality.

□

Theorem 2.9.3 (Non-Expansiveness) *Let C be nonempty, closed, and convex. Then for all $x, y \in \mathbb{R}^d$,*

$$\|\Pi_C(x) - \Pi_C(y)\|_2 \le \|x - y\|_2.$$

Remark 2.9.3 (Stability Message) Projection is "safe": it cannot blow up perturbations. This is one reason projected/proximal methods are robust numerically.

2.9.3 Proximal Viewpoint (Preview for Optimization)

2.9.3.1 Prox Operators as Generalized Projections

Projection solves $\min_{z \in C} \frac{1}{2}\|z - x\|_2^2$, i.e., it balances closeness to x with the hard constraint $z \in C$. The proximal operator replaces the hard constraint by a soft penalty $\lambda\,\Omega(z)$. This unifies regularization and geometry: prox is "projection in a penalized geometry." Prox operators are the building blocks of modern optimization algorithms for nonsmooth regularizers (lasso, group lasso, TV norms).

Definition 2.9.5 (Proximal Operator) Let $\Omega : \mathbb{R}^d \to (-\infty, \infty]$ be a proper function (often convex, possibly taking $+\infty$ to encode constraints), and let $\lambda > 0$. The *proximal operator* is

$$\operatorname{prox}_{\lambda\Omega}(x) := \arg\min_{z \in \mathbb{R}^d} \left\{ \frac{1}{2}\|z - x\|_2^2 + \lambda\,\Omega(z) \right\},$$

when the minimizer exists and is unique.

Remark 2.9.4 (Projection is a Special Prox) If Ω is the indicator of a set C,

$$\iota_C(z) := \begin{cases} 0, & z \in C, \\ +\infty, & z \notin C, \end{cases}$$

then $\operatorname{prox}_{\lambda\iota_C}(x) = \Pi_C(x)$ for any $\lambda > 0$. Thus, constraints are regularizers in disguise.

2.9.3.2 Geometry Behind Soft-Thresholding and Shrinkage

The most iconic proximal effect is *soft-thresholding*: the prox of the ℓ_1 norm. Geometrically, this is the consequence of ℓ_1 having corners: the minimizer often hits a coordinate axis, creating exact zeros. By contrast, the prox of ℓ_2^2 produces smooth shrinkage without sparsity. These two behaviors correspond precisely to the different boundary geometries of the ℓ_1 and ℓ_2 balls discussed earlier, now expressed as explicit coordinatewise formulas.

Proposition 2.9.1 (Soft-Thresholding (Prox of ℓ_1)) *Let $\Omega(z) = \|z\|_1$. Then for $x \in \mathbb{R}^d$,*

$$\operatorname{prox}_{\lambda\|\cdot\|_1}(x) = \big(\operatorname{sign}(x_i)\max\{|x_i| - \lambda, 0\}\big)_{i=1}^{d}.$$

Remark 2.9.5 (Geometric Reading) Each coordinate is pulled toward 0 by λ and clipped at 0 if it crosses. This is the precise algebraic form of "corners induce sparsity."

Proposition 2.9.2 (Ridge Shrinkage (Prox of Squared ℓ_2)) *Let $\Omega(z) = \frac{1}{2}\|z\|_2^2$. Then*

$$\operatorname{prox}_{\lambda\Omega}(x) = \frac{1}{1+\lambda}\,x.$$

Remark 2.9.6 (Smoothness vs. Sparsity) The ℓ_2^2 prox scales all coordinates uniformly and never produces exact zeros (unless $x_i = 0$). Thus, ℓ_2-type penalties shrink but do not threshold.

2.10 High-Dimensional Geometry: Phenomena Behind Modern ML

High-dimensional Euclidean geometry behaves in ways that strongly influence modern ML. As dimension grows, random vectors become nearly orthogonal, norms and distances concentrate sharply, and volume migrates toward thin shells. These effects both empower and complicate learning: they justify random projections and certain initialization heuristics but also warn against naive distance-based intuition (e.g., nearest neighbors) in large d (Vershynin 2018) (cf. Fig. 2.4).

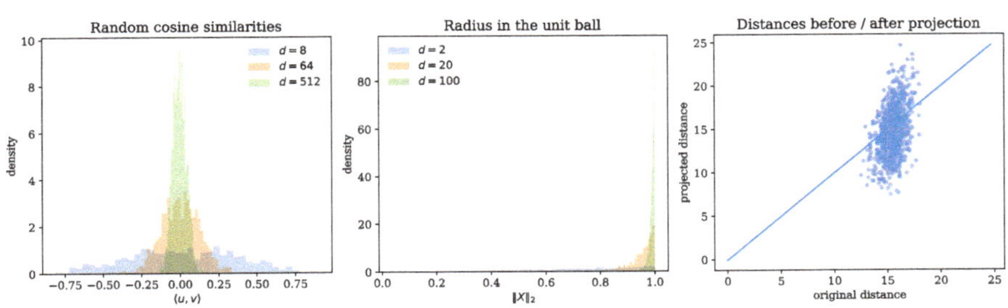

Fig. 2.4 Three high-dimensional effects relevant to ML: random directions become nearly orthogonal, mass in the Euclidean ball concentrates near a thin shell, and random projections approximately preserve pairwise distances

2.10.1 Concentration and Near-Orthogonality

2.10.1.1 Angles Between Random Vectors and Cosine Similarity Behavior

In embedding-based ML (retrieval, contrastive learning), cosine similarity is often the primary notion of similarity. A key high-dimensional fact is that if vectors are "generic" (e.g., isotropic noise or random directions), then their cosine similarity concentrates near 0 as d increases. This means (i) random embeddings look mutually orthogonal, (ii) meaningful similarity must rise above a small baseline noise floor, and (iii) dot products can be treated as approximately Gaussian fluctuations around 0 after normalization. These principles justify why normalized embeddings are stable and why small cosine differences can still be statistically significant when d is large.

Definition 2.10.1 (Uniform Random Direction on the Sphere) A random vector $u \in \mathbb{R}^d$ is *uniform on the unit sphere* $\mathbb{S}^{d-1} = \{x : \|x\|_2 = 1\}$ if it is rotationally invariant, i.e., $Qu \stackrel{d}{=} u$ for every orthogonal matrix Q. Equivalently, if $g \sim \mathcal{N}(0, I_d)$, then $u = g/\|g\|_2$ is uniform on $\mathbb{S}^{d-1}$.

Proposition 2.10.1 (Near-Orthogonality of Random Directions) *Let u, v be independent and uniform on $\mathbb{S}^{d-1}$. Then*

$$\mathbb{E}\big[\langle u, v\rangle\big] = 0, \qquad \mathbb{E}\big[\langle u, v\rangle^2\big] = \frac{1}{d}.$$

Moreover, for any $\varepsilon \in (0, 1)$,

$$\mathbb{P}\Big(|\langle u, v\rangle| \geq \varepsilon\Big) \leq 2\exp\Big(-\frac{(d-1)\varepsilon^2}{2}\Big).$$

Remark 2.10.1 (Cosine Similarity Baseline) For unit vectors, $\cos(u, v) = \langle u, v\rangle$. Proposition 2.10.1 implies typical cosine similarities are of order $1/\sqrt{d}$. Thus, in $d = 768$ (common embedding sizes), a random cosine magnitude is typically only a few percent.

2.10.1.2 Distance Concentration and Nearest-Neighbor Caveats

Distance-based methods (nearest neighbors, k-means) rely on meaningful contrasts between "close" and "far." In high dimension, many natural distributions exhibit *distance concentration*: pairwise distances are all nearly the same. This causes nearest-neighbor rankings to become unstable and sensitive to noise, and it explains why practitioners often (i) reduce dimension, (ii) learn a metric, or (iii) use cosine similarity on normalized embeddings. A useful mental model is: if distances concentrate, then the geometry loses contrast unless structure (clusters/manifolds/sparsity) is present.

Proposition 2.10.2 (Distance Concentration on the Sphere) *Let u, v be independent uniform on $\mathbb{S}^{d-1}$. Then*

$$\|u - v\|_2^2 = 2 - 2\langle u, v\rangle,$$

and hence $\|u - v\|_2$ concentrates near $\sqrt{2}$ as d grows. In particular, for any $\varepsilon \in (0, 1)$,

$$\mathbb{P}\Big(\big|\|u - v\|_2 - \sqrt{2}\big| \geq \varepsilon\Big) \;\leq\; 2\exp\Big(-c\,d\,\varepsilon^2\Big)$$

for a universal constant $c > 0$.

Remark 2.10.2 (Nearest-Neighbor Contrast Heuristic) Suppose all pairwise distances in a dataset lie in $[\mu(1 - \eta), \mu(1 + \eta)]$ for some small η. Then the ratio between the farthest and nearest neighbor distances from any query is at most

$$\frac{1 + \eta}{1 - \eta} \approx 1 + 2\eta,$$

so the distance contrast is weak. In such regimes, NN decisions can flip under small perturbations or minor scaling changes.

Example 2.10.1 (Why Normalization Helps) If embeddings are normalized to $\|x\|_2 = 1$, Euclidean distance and cosine similarity are linked by $\|x - y\|_2^2 = 2(1 - \langle x, y\rangle)$. Thus, cosine similarity directly controls distance, and the geometry becomes "angular" rather than "radial," reducing one major source of variability (norm differences).

2.10.2 Volumes and Intuition in High Dimensions

2.10.2.1 Sphere vs. Ball: Where the Mass Lives

High-dimensional volume behaves counterintuitively: most of the volume of a Euclidean ball lies near its boundary, in a thin shell. This has practical implications: random initialization in a ball tends to have nearly fixed norm; random perturbations tend to be almost orthogonal; and constraints like $\|x\|_2 \leq r$ effectively behave like $\|x\|_2 \approx r$ for typical samples. This "thin shell" phenomenon is one of the geometric reasons that norm-based regularization and normalization produce predictable behavior in large d.

Proposition 2.10.3 (Radius Distribution in the Unit Ball) *Let X be uniformly distributed in the unit Euclidean ball $B_2(1) = \{x \in \mathbb{R}^d : \|x\|_2 \leq 1\}$ and set*

$R = \|X\|_2$. *Then for* $r \in [0, 1]$,

$$\mathbb{P}(R \le r) = r^d, \qquad \textit{and hence} \qquad \mathbb{P}(R \ge 1 - \varepsilon) = 1 - (1 - \varepsilon)^d.$$

In particular, for fixed $\varepsilon \in (0, 1)$ *and large* d, $\mathbb{P}(R \ge 1 - \varepsilon) \approx 1 - e^{-d\varepsilon}$.

Remark 2.10.3 (Thin-Shell Intuition) Proposition 2.10.3 implies that for large d, a uniform point in the unit ball has radius extremely close to 1 with high probability. Thus, "uniform in the ball" behaves like "almost uniform on the sphere" in high dimensions.

2.10.2.2 Implications for Sampling, Initialization, and Regularization

The thin-shell effect and concentration guide several ML design heuristics. Random weights drawn from isotropic distributions have norms that concentrate, so scale can be controlled by variance choices. Regularizers based on norms constrain solutions to shells whose geometry determines inductive bias: ℓ_2 constraints are rotationally symmetric (direction-agnostic), ℓ_1 constraints concentrate on axis-aligned corners (sparsity), and group norms impose block-level sparsity. In optimization, these geometries affect conditioning and step-size stability.

Remark 2.10.4 (Initialization and Typical Norms) If $w \sim \mathcal{N}(0, \sigma^2 I_d)$, then $\|w\|_2^2/\sigma^2$ concentrates near d. Thus, choosing $\sigma^2 \propto 1/d$ keeps $\|w\|_2$ of order 1, stabilizing early activations in linear layers. This is a geometric (norm-scaling) explanation behind variance-preserving initializations.

Remark 2.10.5 (Regularization as Controlling Typical Fluctuations) If perturbations are modeled as $\|\delta\|_2 \le \varepsilon$, then the worst-case linear deviation is governed by $\|w\|_2$. If perturbations are modeled as $\|\delta\|_\infty \le \varepsilon$, the sensitivity is governed by $\|w\|_1$ (dual norm effect). Hence, the choice of regularizer and the choice of robustness model are geometrically linked.

2.10.3 Random Projections (Geometric Preview)

2.10.3.1 Distance Preservation Intuition (Toward JL-type Results)

Random projections exploit concentration: a random linear map into $\mathbb{R}^m$ (with $m \ll d$) can approximately preserve pairwise distances among a finite set of points. The key is that for a fixed vector x, $\|Ax\|_2^2$ concentrates tightly around $\|x\|_2^2$ when A is suitably randomized. A union bound over all pairs then yields the Johnson–Lindenstrauss (JL) phenomenon. This is foundational for dimensionality reduction, approximate nearest neighbors, and randomized numerical linear algebra.

Proposition 2.10.4 (Single-Vector Concentration under Gaussian Projection) *Let $A \in \mathbb{R}^{m\times d}$ have i.i.d. entries $A_{ij} \sim \mathcal{N}(0, 1/m)$, and fix $x \in \mathbb{R}^d$. Then $\mathbb{E}\|Ax\|_2^2 = \|x\|_2^2$, and for any $\varepsilon \in (0, 1)$,*

$$\mathbb{P}\Big(\big|\|Ax\|_2^2 - \|x\|_2^2\big| \geq \varepsilon\|x\|_2^2\Big) \;\leq\; 2\exp\big(-c\,m\,\varepsilon^2\big)$$

for a universal constant $c > 0$.

Theorem 2.10.1 (JL-Type Statement (Finite Set, Informal Parameterization)) *Let $x_1, \dots, x_N \in \mathbb{R}^d$ and let $\varepsilon \in (0, 1)$. If A is as in Proposition 2.10.4 and*

$$m \;\gtrsim\; \varepsilon^{-2}\log N,$$

then with high probability,

$$(1-\varepsilon)\|x_i - x_j\|_2^2 \leq \|A(x_i - x_j)\|_2^2 \leq (1+\varepsilon)\|x_i - x_j\|_2^2 \qquad \forall i, j \in [N].$$

Remark 2.10.6 (Geometric Punchline) A random map to dimension $m = O(\log N)$ (up to ε^{-2}) preserves geometry of N points. Thus, "intrinsic dataset size" (via $\log N$) can matter more than ambient dimension d for many geometric tasks.

2.10.3.2 Sketching Intuition for Large-Scale Learning

Sketching uses random projections to speed up large linear-algebra subroutines (least squares, covariance estimation, PCA). Instead of forming $X^\top X$ on the full dataset, we compress rows with a sketch matrix $S \in \mathbb{R}^{m\times n}$ and solve a smaller problem involving SX. When S approximately preserves norms on the relevant subspace (e.g., the column space of X), the sketched solution approximates the original solution with provable guarantees. This provides a geometric route to scalable learning: preserve the geometry that matters, ignore the rest.

Remark 2.10.7 (Sketched Least Squares (Conceptual Template)) Given $X \in \mathbb{R}^{n\times d}$ and $y \in \mathbb{R}^n$, the original problem is $\min_w \|Xw - y\|_2^2$. A sketched problem is

$$\min_w \|S(Xw - y)\|_2^2 = \min_w \|(SX)w - Sy\|_2^2,$$

where $S \in \mathbb{R}^{m\times n}$ is random with $m \ll n$. If S is a subspace embedding for range(X), then $\|SXv\|_2 \approx \|Xv\|_2$ for all v, so the sketched objective preserves the geometry of residuals induced by X.

2.11 Geometry in Core ML Tasks

The geometric concepts developed in this chapter (norms, inner products, hyperplanes, Gram/PSD geometry, projections) recur in essentially every ML task. Here we summarize the geometric backbone of classification, regression, and representation learning, emphasizing computations and the role of geometry in stability and robustness.

2.11.1 *Classification Geometry*

2.11.1.1 Margins, Separating Hyperplanes, and Robustness to Perturbations

For linear classifiers, the margin is a geometric distance to the decision boundary, and it directly controls robustness: a perturbation must be large enough (in the chosen norm) to cross the boundary. This yields simple, exam-level quantitative robustness guarantees using dual norms. In practice, the same geometry explains why scaling (w, b) does not change the decision boundary, why normalized margins are meaningful, and why regularization of w interacts with adversarial sensitivity.

Proposition 2.11.1 (Margin-Based Robustness via Dual Norms) *Let* $h(x) = \operatorname{sign}(w^\top x + b)$ *with* $w \neq 0$ *and labels* $y \in \{-1, +1\}$. *Assume a point* (x, y) *satisfies the functional margin condition*

$$y(w^\top x + b) \geq \gamma \quad \text{for some } \gamma > 0.$$

Let $\|\cdot\|$ *be any norm with dual* $\|\cdot\|_*$. *Then for any perturbation* δ *with*

$$\|\delta\| \leq \frac{\gamma}{\|w\|_*},$$

the label is unchanged: $y(w^\top(x + \delta) + b) \geq 0$.

Proof By Hölder (Theorem 2.5.3),

$$y(w^\top(x + \delta) + b) = y(w^\top x + b) + y\, w^\top \delta \geq \gamma - |w^\top \delta| \geq \gamma - \|w\|_* \|\delta\| \geq 0.$$

□

Remark 2.11.1 (Euclidean Specialization) For $\|\cdot\| = \|\cdot\|_2$, we have $\|w\|_* = \|w\|_2$ and the robustness radius becomes $\gamma/\|w\|_2$, matching the geometric margin interpretation.

2.11.1.2 Decision Regions, Voronoi Cells, and Nearest-Centroid Classifiers

Nearest-neighbor and nearest-centroid classifiers are purely geometric: they partition space by distance. Voronoi cells describe nearest-neighbor regions; nearest-centroid partitions become hyperplane arrangements under Euclidean distance. This is an important computational fact: even though nearest-centroid uses distances, its boundaries are linear in $\mathbb{R}^d$. Thus, geometry reveals when distance-based classifiers reduce to linear decision rules and how normalization changes these partitions.

Definition 2.11.1 (Voronoi Cells) Given sites $x_1, \ldots, x_n \in \mathbb{R}^d$ and a metric $d(\cdot, \cdot)$, the Voronoi cell of x_i is

$$V_i := \{x \in \mathbb{R}^d : d(x, x_i) \le d(x, x_j)\ \forall j \in [n]\}.$$

The collection $\{V_i\}$ (up to boundary ties) partitions $\mathbb{R}^d$.

Proposition 2.11.2 (Nearest-Centroid Boundaries are Hyperplanes under ℓ_2) *Let class centroids be $\mu_1, \ldots, \mu_K \in \mathbb{R}^d$ and define the nearest-centroid rule $h(x) \in \arg\min_k \|x - \mu_k\|_2$. For two classes $a \neq b$, the decision boundary between a and b is the hyperplane*

$$\{x : \|x - \mu_a\|_2^2 = \|x - \mu_b\|_2^2\} = \{x : (\mu_b - \mu_a)^\top x = \tfrac{1}{2}(\|\mu_b\|_2^2 - \|\mu_a\|_2^2)\}.$$

Proof Expand squared norms and cancel $\|x\|_2^2$: $\|x - \mu_a\|_2^2 - \|x - \mu_b\|_2^2 = -2\mu_a^\top x + \|\mu_a\|_2^2 + 2\mu_b^\top x - \|\mu_b\|_2^2$, and set equal to 0. □

2.11.2 *Regression Geometry*

2.11.2.1 Least Squares as Orthogonal Projection: Bias–Variance Intuition

Least squares is the archetype of projection geometry: the fitted vector is the orthogonal projection of y onto the column space. This yields the hat matrix P and the residual-maker $I - P$, providing an exact algebraic error decomposition. In statistical regression with noise, the same matrices determine bias and variance: P describes how noise is transmitted into predictions, and conditioning of $X^\top X$ controls how noise is amplified in parameter estimates.

Definition 2.11.2 (Hat Matrix and Residual-Maker) Let $X \in \mathbb{R}^{n \times d}$ have full column rank. Define

$$P := X(X^\top X)^{-1} X^\top \in \mathbb{R}^{n \times n}, \qquad R := I_n - P.$$

Then P is the orthogonal projector onto range(X) and R projects onto range$(X)^\perp$.

Proposition 2.11.3 (Projection Decomposition of Least Squares) *For any $y \in \mathbb{R}^n$, the least-squares fitted vector is $\widehat{y} = Py$ and the residual is $r = Ry$. Moreover, $\widehat{y} \perp r$ and*

$$\|y\|_2^2 = \|\widehat{y}\|_2^2 + \|r\|_2^2.$$

Remark 2.11.2 (Bias–Variance (Preview in the Linear Model)) If $y = X\beta + \varepsilon$ with $\mathbb{E}[\varepsilon] = 0$ and $\mathrm{Cov}(\varepsilon) = \sigma^2 I_n$, then $\widehat{\beta} = (X^\top X)^{-1}X^\top y$ satisfies $\mathbb{E}[\widehat{\beta}] = \beta$ (unbiased), and $\mathrm{Cov}(\widehat{\beta}) = \sigma^2(X^\top X)^{-1}$, showing directly how geometry of $X^\top X$ controls estimator variance.

2.11.2.2 Collinearity, Subspace Angles, and Instability

Collinearity means some features lie nearly in the span of others, so the design matrix is close to rank-deficient. Geometrically, this implies the column space has directions with extremely small singular values, making coordinates ill-conditioned. Consequently, small perturbations in y (noise) can cause large changes in $\widehat{\beta}$, even if fitted values $\widehat{y}$ remain stable. Subspace angles provide an additional diagnostic: if a feature subspace is nearly aligned with another, separation of their effects becomes unstable.

Definition 2.11.3 (Condition Number of a Full-Rank Design) For $X \in \mathbb{R}^{n\times d}$ with full column rank, define $\kappa(X) := \sigma_{\max}(X)/\sigma_{\min}(X)$, where $\sigma_{\max}, \sigma_{\min}$ are the largest and smallest singular values.

Proposition 2.11.4 (Sensitivity of Least Squares to Perturbations) *Let X have full column rank and let $\widehat{\beta}(y) = (X^\top X)^{-1}X^\top y$. For any perturbation $\Delta y \in \mathbb{R}^n$,*

$$\|\widehat{\beta}(y+\Delta y) - \widehat{\beta}(y)\|_2 \;\le\; \|(X^\top X)^{-1}X^\top\|_2\,\|\Delta y\|_2 \;=\; \frac{1}{\sigma_{\min}(X)}\,\|\Delta y\|_2.$$

Thus, small $\sigma_{\min}(X)$ (collinearity) implies large parameter sensitivity.

Remark 2.11.3 (Fitted Values Can Be Stable Even When Parameters Are Not) Although $\widehat{\beta}$ may be unstable under collinearity, the fitted vector $\widehat{y} = Py$ depends on the projection P, which can remain relatively stable as a subspace operator. This is one reason why prediction can be reliable while interpretability of coefficients is not.

2.11.3 Representation Geometry

2.11.3.1 Embeddings, Similarity Search, and Metric Learning Objectives

Representation learning maps objects into a geometric space where similarity and task structure become linear or distance-based. Once embeddings are fixed, core tasks reduce to geometric primitives: nearest-neighbor search, clustering, margin separation, and retrieval by cosine similarity. Metric learning makes this explicit by optimizing objectives that push similar points close and dissimilar points far, often under normalization that constrains embeddings to spheres or cones.

Definition 2.11.4 (Embedding Map and Induced Similarity) An *embedding* is a map $\phi : \mathcal{X} \to \mathbb{R}^d$. Given a similarity s on $\mathbb{R}^d$ (e.g., $s(u, v) = \langle u, v \rangle$ or cosine similarity), one induces a similarity on $\mathcal{X}$ via $s_\phi(x, z) = s(\phi(x), \phi(z))$.

Definition 2.11.5 (Triplet Loss (Metric Learning Template)) Given an anchor a, a positive p (similar), and a negative n (dissimilar), and a margin $\alpha > 0$, the triplet loss based on squared Euclidean distance is

$$\mathcal{L}_{\text{tri}}(a, p, n) := \max\left\{0,\ \|\phi(a) - \phi(p)\|_2^2 - \|\phi(a) - \phi(n)\|_2^2 + \alpha\right\}.$$

Remark 2.11.4 (Geometric Effect of the Triplet Constraint) The constraint $\|\phi(a)-\phi(p)\|_2^2+\alpha \leq \|\phi(a)-\phi(n)\|_2^2$ creates a margin-separated annulus structure: negatives must lie outside a ball around the anchor whose radius is determined by positives plus margin.

2.11.3.2 Normalization Layers and Geometry of Feature Spaces

Normalization is geometry engineering: it removes nuisance degrees of freedom (mean, scale) and stabilizes training by controlling norms. ℓ_2-normalization projects features onto a sphere, turning dot products into cosine similarity. BatchNorm/LayerNorm apply affine transformations to enforce unit variance (and often zero mean) along chosen axes, which can be interpreted as moving from an anisotropic geometry to a more isotropic one. These operations change distances and angles, so they should be understood as deliberate geometric design choices, not mere tricks.

Definition 2.11.6 (ℓ_2 Normalization) For $x \in \mathbb{R}^d$ with $x \neq 0$, define the normalized vector

$$\text{norm}(x) := \frac{x}{\|x\|_2} \in \mathbb{S}^{d-1}.$$

Remark 2.11.5 (Dot Product Becomes Cosine Similarity under Normalization) If $\tilde{x} = x/\|x\|_2$ and $\tilde{y} = y/\|y\|_2$, then $\langle \tilde{x}, \tilde{y} \rangle = \cos(x, y)$. Thus, many retrieval systems normalize embeddings so that ranking by dot product equals ranking by cosine similarity.

Definition 2.11.7 (Per-Sample Centering and Scaling (LayerNorm-Style Abstraction)) For a feature vector $x \in \mathbb{R}^d$, define its mean and variance

$$\mu(x) := \frac{1}{d}\sum_{i=1}^{d} x_i, \qquad \sigma^2(x) := \frac{1}{d}\sum_{i=1}^{d}(x_i - \mu(x))^2.$$

A normalized feature is

$$\mathrm{LN}(x) := \frac{x - \mu(x)\mathbf{1}}{\sqrt{\sigma^2(x) + \epsilon}},$$

with a small $\epsilon > 0$ for numerical stability (and optional learned affine re-scaling later).

Remark 2.11.6 (Geometric Reading) Layerwise normalization removes translation along $\mathbf{1}$ and rescales by the within-feature standard deviation, effectively mapping features to a standardized shell. This reduces anisotropy and stabilizes gradient magnitudes in deep networks.

2.12 Exercises

Unless stated otherwise, work in Euclidean space with the standard inner product $\langle x, y\rangle = x^\top y$ and norm $\|\cdot\|_2$.

Exercise 2.12.1 (Norm Balls, Inclusions, and Geometric Consequences) In $\mathbb{R}^3$, consider the vectors

$$x = \begin{bmatrix} 2 \\ -1 \\ 0 \end{bmatrix}, \qquad y = \begin{bmatrix} 1 \\ 1 \\ -1 \end{bmatrix}, \qquad z = \begin{bmatrix} 0 \\ 3 \\ 4 \end{bmatrix}.$$

Compute $\|x\|_1, \|x\|_2, \|x\|_\infty$ (and similarly for y, z), and verify explicitly the inequalities $\|v\|_\infty \le \|v\|_2 \le \|v\|_1$ and $\|v\|_1 \le \sqrt{3}\|v\|_2$ for each $v \in \{x, y, z\}$; then determine the largest $\varepsilon > 0$ such that the ℓ_∞-ball $B_\infty(\varepsilon)$ is contained in the ℓ_2-ball $B_2(1)$ in $\mathbb{R}^3$ (i.e., find ε so that $\|\delta\|_\infty \le \varepsilon \Rightarrow \|\delta\|_2 \le 1$).

Exercise 2.12.2 (Mahalanobis Geometry and Whitening) Let $\Sigma = \begin{bmatrix} 4 & 2 \\ 2 & 3 \end{bmatrix}$ and assume $\Sigma \succ 0$. (i) Compute Σ^{-1} explicitly and hence compute the Mahalanobis distance $d_{\Sigma^{-1}}(a,b) = \sqrt{(a-b)^\top \Sigma^{-1}(a-b)}$ for

$$a = \begin{bmatrix}1\\0\end{bmatrix}, \qquad b = \begin{bmatrix}3\\2\end{bmatrix}, \qquad c = \begin{bmatrix}0\\1\end{bmatrix}.$$

(ii) Compute the eigen-decomposition $\Sigma = Q\Lambda Q^\top$ and from it compute $\Sigma^{-1/2} = Q\Lambda^{-1/2}Q^\top$. (iii) Verify numerically that for the whitening map $W = \Sigma^{-1/2}$,

$$d_{\Sigma^{-1}}(u,v) = \|W(u-v)\|_2$$

for $(u,v) \in \{(a,b),(a,c),(b,c)\}$.

Exercise 2.12.3 (Affine Independence and Barycentric Coordinates) Let $x_0 = (0,0,0)^\top$, $x_1 = (2,0,0)^\top$, $x_2 = (0,1,0)^\top$, $x_3 = (0,0,3)^\top$ in $\mathbb{R}^3$. (i) Prove that x_0, x_1, x_2, x_3 are affinely independent by checking linear independence of $\{x_1 - x_0, x_2 - x_0, x_3 - x_0\}$. (ii) For

$$x = \begin{bmatrix}1\\ \frac{1}{2}\\ 1\end{bmatrix},$$

compute the barycentric coordinates $(\alpha_0,\alpha_1,\alpha_2,\alpha_3)$ such that $x = \sum_{i=0}^{3}\alpha_i x_i$ and $\sum_i \alpha_i = 1$, and determine whether x lies in the simplex $\Delta(x_0,x_1,x_2,x_3)$ by checking $\alpha_i \geq 0$.

Exercise 2.12.4 (Hyperplanes, Signed Distances, and Closest Points) Let $w = \begin{bmatrix}2\\-1\\2\end{bmatrix}$, $b = -3$, and define the hyperplane $H = \{x \in \mathbb{R}^3 : w^\top x + b = 0\}$. (i) Compute the signed distance $\operatorname{dist}_\pm(p,H)$ and Euclidean distance $\operatorname{dist}(p,H)$ for

$$p = \begin{bmatrix}1\\0\\1\end{bmatrix}, \qquad q = \begin{bmatrix}0\\3\\0\end{bmatrix}.$$

(ii) Using Proposition 2.3.4, compute the closest points $p^\star, q^\star \in H$ to p and q. (iii) Verify that $p - p^\star$ and $q - q^\star$ are parallel to w and that $w^\top p^\star + b = 0$ and $w^\top q^\star + b = 0$.

Exercise 2.12.5 (Margins and Robustness Radii via Dual Norms) Consider a binary classifier $h(x) = \operatorname{sign}(w^\top x + b)$ in $\mathbb{R}^2$ with

$$w = \begin{bmatrix}3\\4\end{bmatrix}, \qquad b = -5.$$

Let labeled points be (x_i, y_i) for $i \in [4]$ where

$$x_1 = \begin{bmatrix} 2 \\ 2 \end{bmatrix}, \; y_1 = +1; \quad x_2 = \begin{bmatrix} 1 \\ 0 \end{bmatrix}, \; y_2 = -1; \quad x_3 = \begin{bmatrix} 3 \\ 1 \end{bmatrix}, \; y_3 = +1;$$

$$x_4 = \begin{bmatrix} 0 \\ 2 \end{bmatrix}, \; y_4 = -1.$$

(i) Compute the functional margins $\gamma_i^{\text{fun}} = y_i(w^\top x_i + b)$ and the geometric margins $\gamma_i^{\text{geo}} = \gamma_i^{\text{fun}}/\|w\|_2$. (ii) Compute the largest ε_2 such that all four points are robustly classified under ℓ_2 perturbations $\|\delta\|_2 \leq \varepsilon_2$. (iii) Compute the largest ε_∞ such that all four points are robustly classified under ℓ_∞ perturbations $\|\delta\|_\infty \leq \varepsilon_\infty$, using the dual norm rule $\sup_{\|\delta\|_\infty \leq \varepsilon} |w^\top \delta| = \varepsilon \|w\|_1$.

Exercise 2.12.6 (Polyhedral Feasible Region and Vertices) Let $\mathcal{P} \subseteq \mathbb{R}^2$ be the polytope defined by the intersection of halfspaces

$$\mathcal{P} = \{x \in \mathbb{R}^2 : \; Ax \leq b\}, \quad A = \begin{bmatrix} 1 & 0 \\ 0 & 1 \\ -1 & 0 \\ 0 & -1 \\ 1 & 1 \end{bmatrix}, \quad b = \begin{bmatrix} 2 \\ 2 \\ 0 \\ 0 \\ 3 \end{bmatrix}.$$

(i) Sketch $\mathcal{P}$ and list all its vertices explicitly by intersecting the relevant boundary lines. (ii) For the linear objective $f(x) = c^\top x$ with $c = \begin{bmatrix} 2 \\ 1 \end{bmatrix}$, compute $\arg\max_{x \in \mathcal{P}} f(x)$ by evaluating f on the vertices. (iii) Identify the set of active constraints at each vertex (Definition 2.3.12).

Exercise 2.12.7 (Orthonormal Basis via Gram–Schmidt and Coordinate Extraction) Let $v_1 = \begin{bmatrix} 1 \\ 1 \\ 0 \end{bmatrix}$, $v_2 = \begin{bmatrix} 1 \\ 0 \\ 1 \end{bmatrix}$, $v_3 = \begin{bmatrix} 0 \\ 1 \\ 1 \end{bmatrix}$ in $\mathbb{R}^3$. (i) Apply Gram–Schmidt (Definition 2.2.10) to produce an orthonormal basis q_1, q_2, q_3. (ii) Form $Q = [q_1 \; q_2 \; q_3]$ and verify $Q^\top Q = I_3$. (iii) For $x = \begin{bmatrix} 2 \\ -1 \\ 3 \end{bmatrix}$, compute the orthonormal coordinates $\alpha = Q^\top x$ and reconstruct x as $Q\alpha$.

Exercise 2.12.8 (Principal-Angle Computation in a Concrete Case) Let $\mathcal{U} = \operatorname{span}\{u_1, u_2\}$ and $\mathcal{V} = \operatorname{span}\{v_1, v_2\}$ in $\mathbb{R}^3$, where

$$u_1 = \begin{bmatrix} 1 \\ 0 \\ 0 \end{bmatrix}, \quad u_2 = \begin{bmatrix} 0 \\ 1 \\ 0 \end{bmatrix}, \qquad v_1 = \frac{1}{\sqrt{2}} \begin{bmatrix} 1 \\ 1 \\ 0 \end{bmatrix}, \quad v_2 = \begin{bmatrix} 0 \\ 0 \\ 1 \end{bmatrix}.$$

(i) Construct orthonormal basis matrices Q_U, Q_V for $\mathcal{U}, \mathcal{V}$. (ii) Compute $M = Q_U^\top Q_V$ and its singular values $\sigma_1 \geq \sigma_2$. (iii) Compute the principal angles $\theta_i = \arccos(\sigma_i)$ and interpret geometrically whether $\mathcal{U}$ and $\mathcal{V}$ intersect and whether they are orthogonal.

Exercise 2.12.9 (Gram Matrix Feasibility and Reconstruction) Consider the symmetric matrix

$$G = \begin{bmatrix} 4 & 2 & 0 \\ 2 & 2 & 1 \\ 0 & 1 & 1 \end{bmatrix}.$$

(i) Verify that G is PSD by checking that all eigenvalues are nonnegative (compute them explicitly) *or* by verifying that all leading principal minors are nonnegative. (ii) Construct vectors $x_1, x_2, x_3 \in \mathbb{R}^3$ such that $G_{ij} = \langle x_i, x_j \rangle$ by producing a factorization $G = XX^\top$. (iii) Using Proposition 2.7.3, compute the squared distances $\|x_i - x_j\|_2^2$ for all pairs (i, j).

Exercise 2.12.10 (Recovering a Centered Gram Matrix from Distances (Double-Centering)) Let the squared-distance matrix $D^2 \in \mathbb{R}^{4\times 4}$ be

$$D^2 = \begin{bmatrix} 0 & 1 & 4 & 9 \\ 1 & 0 & 1 & 4 \\ 4 & 1 & 0 & 1 \\ 9 & 4 & 1 & 0 \end{bmatrix}.$$

(i) Form the centering matrix $H = I_4 - \frac{1}{4}\mathbf{1}\mathbf{1}^\top$ and compute the centered Gram matrix

$$G_c = -\frac{1}{2} H D^2 H.$$

(ii) Verify $G_c \succeq 0$ by computing its eigenvalues. (iii) Construct a realization $x_1, \ldots, x_4 \in \mathbb{R}^r$ (for minimal r) such that $\langle x_i, x_j \rangle = (G_c)_{ij}$ and $\sum_{i=1}^4 x_i = 0$.

Exercise 2.12.11 (Orthogonal Projection Onto a Span and Residual Orthogonality) Let $A \in \mathbb{R}^{4\times 2}$ have columns

$$a_1 = \begin{bmatrix} 1 \\ 0 \\ 1 \\ 0 \end{bmatrix}, \quad a_2 = \begin{bmatrix} 0 \\ 1 \\ 1 \\ 1 \end{bmatrix}, \quad A = \begin{bmatrix} a_1 & a_2 \end{bmatrix},$$

and let $x = \begin{bmatrix} 2 \\ -1 \\ 1 \\ 3 \end{bmatrix}$. (i) Compute the orthogonal projector $P = A(A^\top A)^{-1}A^\top$ and then compute $u = Px$ and residual $r = x - u$. (ii) Verify $A^\top r = 0$ and compute $\|x\|_2^2 - \|u\|_2^2 - \|r\|_2^2$ to confirm the Pythagorean identity. (iii) Compute $\text{dist}(x, \text{range}(A))$ and verify it equals $\|r\|_2$.

Exercise 2.12.12 (Projection Onto a Simplex) Let $z = \begin{bmatrix} 0.6 \\ -0.2 \\ 1.4 \end{bmatrix} \in \mathbb{R}^3$ and consider the probability simplex

$$\Delta^2 = \{p \in \mathbb{R}^3 : p_i \geq 0, \sum_{i=1}^{3} p_i = 1\}.$$

Compute the Euclidean projection $p^\star = \Pi_{\Delta^2}(z)$ by solving the KKT conditions for $\min_{p \in \Delta^2} \|p - z\|_2^2$ and report $p^\star$ exactly. Then verify the variational inequality characterization (Theorem 2.9.2) by checking that $\langle z - p^\star, \ p - p^\star \rangle \leq 0$ holds for the three vertices e_1, e_2, e_3 of Δ^2.

Exercise 2.12.13 (Non-Expansiveness of Convex Projection in a Concrete Set) Let $C = [-1, 1]^2 \subseteq \mathbb{R}^2$ and define $x = \begin{bmatrix} 2 \\ -3 \end{bmatrix}$, $y = \begin{bmatrix} -4 \\ 0.5 \end{bmatrix}$. (i) Compute $\Pi_C(x)$ and $\Pi_C(y)$ explicitly (coordinatewise clipping). (ii) Compute $\|\Pi_C(x) - \Pi_C(y)\|_2$ and $\|x - y\|_2$ and verify numerically the non-expansiveness inequality $\|\Pi_C(x) - \Pi_C(y)\|_2 \leq \|x - y\|_2$ (Theorem 2.9.3). (iii) Compute $\|x - \Pi_C(x)\|_2$ and interpret it as the distance from x to C.

Exercise 2.12.14 (Proximal Operators: Soft-Thresholding and Ridge Shrinkage) Let $x = \begin{bmatrix} 3 \\ -0.7 \\ 1.2 \\ -4 \end{bmatrix}$ and $\lambda = 1.5$. (i) Compute $\text{prox}_{\lambda\|\cdot\|_1}(x)$ using the soft-thresholding formula in Proposition 2.9.1. (ii) Compute $\text{prox}_{\lambda(\frac{1}{2}\|\cdot\|_2^2)}(x)$ using Proposition 2.9.2. (iii) For each prox output $z^\star$, compute the objective value $\frac{1}{2}\|z^\star - x\|_2^2 + \lambda\Omega(z^\star)$ and compare it against the value at $z = 0$ to verify improvement.

Exercise 2.12.15 (High-Dimensional Near-Orthogonality Numerics) Let $d = 20$ and define two vectors $u, v \in \mathbb{R}^{20}$ by

$$u_i = \frac{i}{\sqrt{\sum_{j=1}^{20} j^2}}, \qquad v_i = \frac{(-1)^i}{\sqrt{20}}, \qquad i \in [20].$$

(i) Verify $\|u\|_2 = \|v\|_2 = 1$ and compute $\langle u, v\rangle$ and the angle $\theta = \arccos(\langle u, v\rangle)$. (ii) Compute $\|u-v\|_2$ and verify $\|u-v\|_2^2 = 2-2\langle u, v\rangle$. (iii) Interpret the magnitude of $\langle u, v\rangle$ relative to the heuristic scale $1/\sqrt{d}$.

Exercise 2.12.16 (Random Projection Computation (Fixed Matrix)) Let $A \in \mathbb{R}^{3\times 5}$ be

$$A = \frac{1}{\sqrt{3}} \begin{bmatrix} 1 & 0 & -1 & 2 & 0 \\ 0 & 2 & 1 & 0 & -1 \\ 1 & 1 & 0 & -1 & 2 \end{bmatrix}.$$

Let points $x_1, x_2, x_3 \in \mathbb{R}^5$ be

$$x_1 = \begin{bmatrix} 1 \\ 0 \\ 2 \\ -1 \\ 1 \end{bmatrix}, \quad x_2 = \begin{bmatrix} 0 \\ 1 \\ -1 \\ 2 \\ 0 \end{bmatrix}, \quad x_3 = \begin{bmatrix} 2 \\ -1 \\ 0 \\ 1 \\ -2 \end{bmatrix}.$$

(i) Compute all pairwise distances $\|x_i - x_j\|_2$. (ii) Compute the projected points $y_i = Ax_i \in \mathbb{R}^3$ and all pairwise distances $\|y_i - y_j\|_2$. (iii) For each pair (i, j) compute the distortion ratio $\|y_i - y_j\|_2/\|x_i - x_j\|_2$ and comment whether distances are approximately preserved.

Exercise 2.12.17 (Nearest-Centroid Boundaries as Hyperplanes) In $\mathbb{R}^2$, let class centroids be

$$\mu_1 = \begin{bmatrix} 0 \\ 0 \end{bmatrix}, \quad \mu_2 = \begin{bmatrix} 2 \\ 1 \end{bmatrix}, \quad \mu_3 = \begin{bmatrix} 1 \\ 3 \end{bmatrix}.$$

(i) Derive the three pairwise decision boundaries between classes using Proposition 2.11.2. (ii) Determine the predicted class for each query point

$$q_1 = \begin{bmatrix} 1 \\ 1 \end{bmatrix}, \quad q_2 = \begin{bmatrix} 2 \\ 3 \end{bmatrix}, \quad q_3 = \begin{bmatrix} -1 \\ 1 \end{bmatrix}$$

by comparing squared distances to the centroids. (iii) Identify the region (as inequalities) in which class 2 is selected.

Chapter 3
Linear Maps and Matrix Algebra

Linear maps are the fundamental compositional units of ML: they encode feature extraction, readouts, attention mixing, convolutions (as structured linear maps), whitening, projections, and local behavior of nonlinear networks via Jacobians. This chapter develops a rigorous "operator-first" view (linear maps between vector spaces) and then connects it to matrices as coordinate representations. The main theme is that *geometry and information flow are controlled by kernels, images, and rank*, while *computation and stability are controlled by coordinate choices and conditioning*.

3.1 Why Linear Maps Are the Core Abstraction in ML

3.1.1 Linear Maps as Modules in Learning Systems

3.1.1.1 Features, Embeddings, and Linear Readouts

Most modern pipelines can be written as: a nonlinear representation map ϕ (an "encoder") followed by a linear map (a "readout"). The readout is the simplest trainable module: it preserves convexity of many objectives (e.g., least squares, logistic regression), admits closed-form or well-conditioned solvers, and provides a clean interface between geometry (what the representation makes linearly separable) and learning (what the readout can generalize). This is why linear probes are ubiquitous: they test the separability/information content of representations (Strang 2019) (cf. Fig. 3.1).

Definition 3.1.1 (Linear Readout/Linear Probe) Let $\phi : \mathcal{X} \to \mathbb{R}^m$ be a feature/embedding map. A *linear readout* is a function $h : \mathcal{X} \to \mathbb{R}^k$ of the form

$$h(x) = W\,\phi(x) + b, \qquad W \in \mathbb{R}^{k\times m},\ b \in \mathbb{R}^k.$$

P. Singh, B. Raman, *Foundations of Machine Learning and AI*, Studies in Big Data 199, https://doi.org/10.1007/978-3-032-30336-3_3

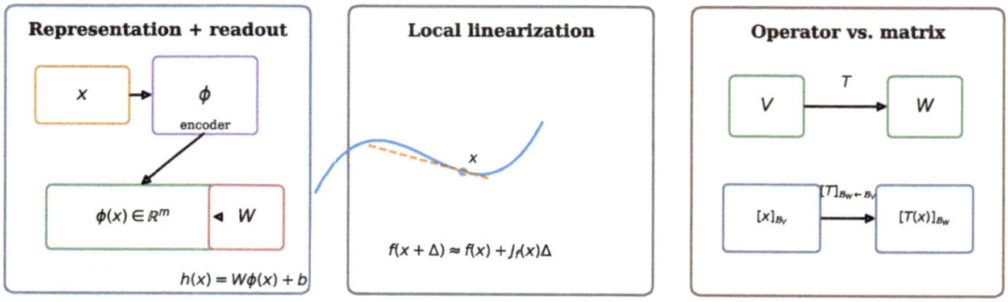

Fig. 3.1 An operator-first view of learning: a nonlinear encoder produces a representation, a linear readout acts on that representation, and the Jacobian gives the local linear map of a nonlinear model around a point

For binary classification one often uses $k = 1$ and takes $\text{sign}(h(x))$; for multiclass one uses $k = K$ and applies arg max.

Remark 3.1.1 (When Linear Readouts Are "Enough") If $\phi(\mathcal{X})$ organizes data so that classes are separable by hyperplanes in $\mathbb{R}^m$, then a linear readout suffices for classification. Thus, representation learning can be viewed as learning a map ϕ that makes the task linearly solvable.

Example 3.1.1 (Reservoir/Feature Map + Linear Regression) Let $\phi(x) \in \mathbb{R}^m$ be a fixed nonlinear feature map (e.g., random features or a reservoir state). Given training pairs (x_i, y_i) with $y_i \in \mathbb{R}$, the linear readout solves

$$\min_{w \in \mathbb{R}^m,\, b \in \mathbb{R}} \sum_{i=1}^{n} (w^\top \phi(x_i) + b - y_i)^2,$$

a least-squares problem. The learning happens in the linear map even though the representation is nonlinear.

3.1.1.2 Jacobian Viewpoint: Local Linearization of Nonlinear Models

Even when a model is highly nonlinear, it behaves *approximately linearly* at small scales. The Jacobian captures the best linear approximation and governs sensitivity, stability, and backpropagation. In optimization, gradients are obtained by composing Jacobians (chain rule), which is why linear maps (and their matrix representations) sit at the computational core of deep learning.

Definition 3.1.2 (Jacobian) Let $f : \mathbb{R}^d \to \mathbb{R}^k$ be differentiable at x. The *Jacobian* $J_f(x) \in \mathbb{R}^{k \times d}$ is the matrix whose entries are

$$(J_f(x))_{ij} := \frac{\partial f_i}{\partial x_j}(x).$$

Theorem 3.1.1 (First-Order (Linear) Approximation) *If $f : \mathbb{R}^d \to \mathbb{R}^k$ is differentiable at x, then for small $\Delta \in \mathbb{R}^d$,*

$$f(x + \Delta) = f(x) + J_f(x)\,\Delta + o(\|\Delta\|_2).$$

Remark 3.1.2 (Sensitivity as an Operator Norm) The Jacobian is a linear map $\Delta \mapsto J_f(x)\Delta$. Thus, local sensitivity is controlled by matrix norms such as $\|J_f(x)\|_2$ (Lipschitz-type behavior).

Example 3.1.2 (Jacobian of an Affine Map) For $f(x) = Ax + b$ with $A \in \mathbb{R}^{k\times d}$, the Jacobian is constant: $J_f(x) = A$ for all x. Thus, every affine layer in a network contributes a linear map to the overall Jacobian product.

3.1.2 Coordinate-Free vs. Coordinate-Based Thinking

3.1.2.1 Linear Operators vs. Matrices: What Changes, What Does Not

A linear map is a coordinate-free object: it acts on vectors in a vector space regardless of how we write those vectors. A matrix is a coordinate-dependent representation of that map in chosen bases (Axler 2015). Confusing the two leads to conceptual errors: changing the basis changes the matrix, but the underlying transformation is the same. The correct invariant quantities (trace, determinant, rank, eigenvalues up to similarity) capture properties of the operator itself.

Definition 3.1.3 (Linear Map/Linear Operator) Let V, W be real vector spaces. A map $T : V \to W$ is *linear* if for all $u, v \in V$ and $\alpha, \beta \in \mathbb{R}$,

$$T(\alpha u + \beta v) = \alpha T(u) + \beta T(v).$$

Definition 3.1.4 (Matrix Representation of a Linear Map) Let $\mathcal{B}_V = (v_1, \ldots, v_d)$ be a basis of V and $\mathcal{B}_W = (w_1, \ldots, w_k)$ a basis of W. There exists a unique matrix $[T]_{\mathcal{B}_W \leftarrow \mathcal{B}_V} \in \mathbb{R}^{k\times d}$ such that for every $x \in V$,

$$[T(x)]_{\mathcal{B}_W} = [T]_{\mathcal{B}_W \leftarrow \mathcal{B}_V}\,[x]_{\mathcal{B}_V}.$$

Remark 3.1.3 (Same Operator, Different Matrices) If we change bases, the coordinate vectors and the representing matrix both change, but T does not. Thus, statements that depend only on T should be invariant under basis change (e.g., rank).

Example 3.1.3 (Coordinate Change Without Changing the Map) Let $T : \mathbb{R}^2 \to \mathbb{R}^2$ be "rotate by 90°." In the standard basis, its matrix is $\begin{bmatrix} 0 & -1 \\ 1 & 0 \end{bmatrix}$. In a rotated basis, the representing matrix changes, but T still rotates vectors by 90° in the plane.

3.1.2.2 Choosing Coordinates: Interpretability and Numerical Stability

Coordinates are a design choice. In interpretability, we prefer coordinates aligned with meaningful features (original units, principal components, disentangled factors). In computation, we prefer coordinates that stabilize algorithms: orthonormal bases preserve lengths, reduce conditioning issues, and avoid squaring condition numbers in normal equations. Thus, "choosing a basis" is simultaneously about semantics and numerical robustness.

Remark 3.1.4 (Orthonormal Bases and Stability) If Q is orthogonal, then $\|Qx\|_2 = \|x\|_2$ and $\|Q^\top x\|_2 = \|x\|_2$ (isometries). Algorithms expressed in orthonormal coordinates do not amplify Euclidean perturbations, which is why QR-based methods are numerically stable.

Remark 3.1.5 (Conditioning Depends on Coordinates) The matrix representing T can have very different condition numbers under different bases. Poorly chosen bases can make an intrinsically benign operator appear ill-conditioned and destabilize computations.

3.2 Linear Maps and Matrices

3.2.1 Linear Maps as Transformations

3.2.1.1 Linearity, Composition, and Matrix Representation

Linearity is the algebraic expression of superposition: outputs for mixtures are mixtures of outputs. This property enables modularity: composition of linear maps is linear, and in coordinates, composition is matrix multiplication. For ML, this is the backbone of forward propagation through linear layers and of structured linear operators (e.g., convolutions as sparse Toeplitz-like matrices). Shape checks are not optional: they are the formal proof that a composition is well-defined.

Proposition 3.2.1 (Composition of Linear Maps is Linear) *If $T : V \to W$ and $S : W \to U$ are linear maps, then $S \circ T : V \to U$ is linear.*

Proposition 3.2.2 (Matrix of a Composition) *Let $T : V \to W$ and $S : W \to U$ be linear maps with chosen bases. Then the matrix of the composition satisfies*

$$[S \circ T] = [S]\,[T],$$

with dimensions compatible (matrix multiplication encodes composition).

Example 3.2.1 (Shape Check) If $A \in \mathbb{R}^{k\times d}$ and $B \in \mathbb{R}^{m\times k}$, then $BA \in \mathbb{R}^{m\times d}$. Thus, $x \mapsto Ax$ maps $\mathbb{R}^d \to \mathbb{R}^k$ and then $y \mapsto By$ maps $\mathbb{R}^k \to \mathbb{R}^m$, so the composed map is $x \mapsto (BA)x : \mathbb{R}^d \to \mathbb{R}^m$.

3.2.1.2 Kernels, Images, and Rank as Information Flow

Every linear map discards information along its kernel and preserves information along directions not in the kernel. The image (range) describes what outputs are achievable, and the rank measures the effective output dimension. In ML terms, a linear layer compresses representations when rank is small and expands when rank is large (up to min dimensions). Rank-nullity is the accounting identity of information flow: input dimension splits into transmitted degrees of freedom (rank) and lost degrees of freedom (nullity).

Definition 3.2.1 (Kernel and Image) For a linear map $T : V \to W$,

$$\ker(T) := \{v \in V : T(v) = 0\}, \qquad \operatorname{im}(T) := \{T(v) : v \in V\} \subseteq W.$$

Definition 3.2.2 (Rank and Nullity) If V is finite-dimensional, define

$$\operatorname{rank}(T) := \dim(\operatorname{im}(T)), \qquad \operatorname{nullity}(T) := \dim(\ker(T)).$$

Theorem 3.2.1 (Rank–Nullity) *If $T : V \to W$ is linear and* $\dim(V) = d < \infty$, *then*

$$\operatorname{rank}(T) + \operatorname{nullity}(T) = d.$$

Remark 3.2.1 (Information-Flow Interpretation) $\operatorname{nullity}(T)$ counts how many independent directions are annihilated (information lost), while $\operatorname{rank}(T)$ counts how many independent output directions survive (information transmitted).

Example 3.2.2 (Non-Identifiability from a Nontrivial Kernel) Let $X \in \mathbb{R}^{n \times d}$ represent a design matrix and consider the map $T(w) = Xw$. If $\ker(T) = \operatorname{null}(X) \neq \{0\}$, then w and $w + v$ yield identical predictions for any $v \in \operatorname{null}(X)$. Thus, parameters are not identifiable without additional constraints or regularization.

3.2.2 Change of Basis and Coordinate Dependence

3.2.2.1 Similarity Transforms and Invariants (Trace, Determinant, Eigenvalues)

Changing coordinates changes the matrix representation by a similarity transform. This is the precise algebra behind "same linear operator, different matrices." Quantities invariant under similarity are intrinsic to the operator: trace and determinant (in any basis) are the same, and eigenvalues are preserved. These invariants act as coordinate-free summaries: trace measures total expansion along axes (in an averaged sense), determinant measures volume scaling, and eigenvalues describe invariant directions and dynamical behavior.

Proposition 3.2.3 (Change of Basis for Endomorphisms) *Let $T : V \to V$ be linear on a d-dimensional space. Let $\mathcal{B}$ and $\mathcal{B}'$ be two bases, and let S be the change-of-coordinates matrix satisfying $[x]_{\mathcal{B}} = S\,[x]_{\mathcal{B}'}$ for all $x \in V$. Then the matrix representations satisfy the similarity relation*

$$[T]_{\mathcal{B}'} = S^{-1}\,[T]_{\mathcal{B}}\,S.$$

Theorem 3.2.2 (Similarity Invariants) *If A and B are similar, i.e., $B = S^{-1}AS$ for some invertible S, then:*

$$\operatorname{tr}(B) = \operatorname{tr}(A), \qquad \det(B) = \det(A),$$

and A and B have the same multiset of eigenvalues (with algebraic multiplicity).

Remark 3.2.2 (Why Invariants Matter in ML) In dynamical systems (e.g., linearized dynamics), eigenvalues determine stability. In optimization, eigenvalues of Hessians determine curvature and conditioning. Because eigenvalues are similarity invariants, they are properties of the underlying operator, not artifacts of coordinates.

3.2.2.2 Orthonormal Bases and Isometries

Among all coordinate systems, orthonormal bases are special: they preserve Euclidean geometry. Changing from one orthonormal basis to another corresponds to multiplication by an orthogonal matrix, an isometry. This means dot products, norms, and angles are unchanged, so computations are stable and interpretations are geometric. In practice, many algorithms first convert a basis to an orthonormal one (QR/Gram–Schmidt) precisely to operate in an isometric coordinate system.

Definition 3.2.3 (Orthogonal Similarity) Matrices A and B are *orthogonally similar* if $B = Q^{\top}AQ$ for some orthogonal Q. This corresponds to changing between orthonormal bases.

Proposition 3.2.4 (Orthogonal Changes Preserve Euclidean Structure) *If Q is orthogonal, then for all x, y,*

$$\langle Qx, Qy\rangle = \langle x, y\rangle, \qquad \|Qx\|_2 = \|x\|_2.$$

Hence orthogonal basis changes do not distort lengths or angles.

Remark 3.2.3 (Numerical Advantage) Orthogonal transformations have condition number 1 in ℓ_2. Thus, rewriting computations in orthonormal coordinates avoids amplification of floating-point errors.

3.2.3 *Matrix Algebra Essentials*

3.2.3.1 Transpose Rules, Associativity, Distributivity, and Shape Checks

Matrix algebra is the syntax of linear maps in coordinates. To prevent silent errors, one must treat dimensions as part of the statement. Many ML bugs are dimension bugs. Associativity enables reparenthesization (important for computational cost), distributivity enables linear expansion, and transpose rules encode how dual spaces and inner products interact. These are the basic rewriting rules used throughout optimization and backpropagation.

Proposition 3.2.5 (Algebraic Rules) *Let A, B, C be matrices of compatible sizes and let $\alpha \in \mathbb{R}$.*

1. *(Associativity)* $(AB)C = A(BC)$.
2. *(Distributivity)* $A(B + C) = AB + AC$ *and* $(A + B)C = AC + BC$.
3. *(Transpose rules)* $(A + B)^\top = A^\top + B^\top$, $(\alpha A)^\top = \alpha A^\top$, *and* $(AB)^\top = B^\top A^\top$.

Remark 3.2.4 (Shape Checks as Proof Obligations) An expression like AB is defined only if the number of columns of A equals the number of rows of B. Writing shapes above intermediate terms is a rigorous way to verify correctness of derivations and implementations.

3.2.3.2 Trace Tricks and Frobenius Inner Products

Trace identities convert sums into compact algebra and expose cyclic invariances. They are the backbone of matrix calculus: many gradients are derived by rewriting objectives in trace form and applying cyclicity. The Frobenius inner product identifies matrices with vectors in a Hilbert space, enabling "Euclidean" geometry for matrices. In ML, these tools appear in least squares, PCA, covariance estimation, and regularization.

Definition 3.2.4 (Trace and Frobenius Inner Product) For $A \in \mathbb{R}^{d\times d}$, $\operatorname{tr}(A) = \sum_{i=1}^{d} A_{ii}$. For $A, B \in \mathbb{R}^{m\times n}$, the Frobenius inner product is

$$\langle A, B\rangle_F := \sum_{i=1}^{m}\sum_{j=1}^{n} A_{ij}B_{ij} = \operatorname{tr}(A^\top B).$$

Proposition 3.2.6 (Cyclicity of Trace) *If products are defined, then*

$$\operatorname{tr}(AB) = \operatorname{tr}(BA), \qquad \operatorname{tr}(ABC) = \operatorname{tr}(BCA) = \operatorname{tr}(CAB).$$

Remark 3.2.5 (A Common ML Rewrite) Squared Frobenius norms can be written as

$$\|A\|_F^2 = \langle A, A \rangle_F = \mathrm{tr}(A^\top A),$$

which is often the starting point for deriving gradients and for connecting objectives to eigen-structure.

Example 3.2.3 (Quadratic Forms as Traces) For $x \in \mathbb{R}^d$ and $M \in \mathbb{R}^{d\times d}$,

$$x^\top Mx = \mathrm{tr}(x^\top Mx) = \mathrm{tr}(Mxx^\top).$$

This identity is ubiquitous in covariance/PSD geometry and matrix calculus.

3.3 Fundamental Subspaces and Rank Structure

A matrix A is not just an array of numbers: it is a linear map between Euclidean spaces. The most important geometric objects attached to A are its *four fundamental subspaces*: column space, row space, null space, and left null space. Together with rank–nullity, they explain existence/uniqueness of solutions to linear systems, identifiability of parameters in linear models, and why redundancy in data features leads to instability. We also preview low-rank structure as an inductive bias in modern ML.

3.3.1 The Four Fundamental Subspaces

3.3.1.1 Column Space, Row Space, Null Space, Left Null Space

For $A \in \mathbb{R}^{m\times n}$, the map $x \mapsto Ax$ sends $\mathbb{R}^n$ to $\mathbb{R}^m$. Its image is the column space, the set of attainable outputs. The null space is the set of input directions that map to zero, i.e., degrees of freedom invisible to the output. The row space is the column space of $A^\top$, living in input space and capturing which input directions are "seen" by A. The left null space is the null space of $A^\top$ and captures which output directions are orthogonal to everything Ax can produce. In learning, these subspaces directly describe predictability, constraints, and residual structure.

Definition 3.3.1 (Four Fundamental Subspaces) Let $A \in \mathbb{R}^{m\times n}$. Define

$$\mathcal{C}(A) := \mathrm{range}(A) \subseteq \mathbb{R}^m \quad \text{(column space)},$$

$$\mathcal{C}(A^\top) := \mathrm{range}(A^\top) \subseteq \mathbb{R}^n \quad \text{(row space)},$$

$$\mathcal{N}(A) := \{x \in \mathbb{R}^n : Ax = 0\} \subseteq \mathbb{R}^n \quad \text{(null space)},$$

$$\mathcal{N}(A^\top) := \{y \in \mathbb{R}^m : A^\top y = 0\} \subseteq \mathbb{R}^m \quad \text{(left null space)}.$$

Proposition 3.3.1 (Orthogonal Decompositions) *For any $A \in \mathbb{R}^{m\times n}$,*

$$\mathbb{R}^n = \mathcal{C}(A^\top) \oplus \mathcal{N}(A), \qquad \mathbb{R}^m = \mathcal{C}(A) \oplus \mathcal{N}(A^\top),$$

where $\oplus$ denotes an orthogonal direct sum with respect to the standard inner product.

Remark 3.3.1 (Why These Decompositions Matter) In input space $\mathbb{R}^n$, every vector splits into a part that affects Ax (row space) and a part that is invisible (null space). In output space $\mathbb{R}^m$, every vector splits into a producible component (column space) and an orthogonal residual component (left null space). This is the exact geometric content of least squares.

3.3.1.2 Geometric Meaning in Regression and Classification

In linear regression, the set of all possible prediction vectors is exactly the column space $\mathcal{C}(X)$ of the design matrix X. Fitting is projecting y onto $\mathcal{C}(X)$; the residual lies in the left null space $\mathcal{N}(X^\top)$. In classification, a linear classifier depends on the row space directions of the feature map: if a direction lies in the null space of the feature transformation, it cannot influence the score. Thus, the four subspaces explain what information can be represented, what is ignored, and why redundancies (nontrivial null space) lead to non-identifiability of parameters.

Example 3.3.1 (Regression Geometry) Let $X \in \mathbb{R}^{n\times d}$ and $w \in \mathbb{R}^d$. Predictions $\widehat{y} = Xw$ range over $\mathcal{C}(X) \subseteq \mathbb{R}^n$. The least-squares residual $r = y - \widehat{y}$ satisfies $X^\top r = 0$, hence $r \in \mathcal{N}(X^\top)$.

Example 3.3.2 (Classification and Feature Null Space) Consider a feature map $A : \mathbb{R}^n \to \mathbb{R}^m$ (matrix $A \in \mathbb{R}^{m\times n}$) and a linear score $s(x) = w^\top(Ax) + b$. If $v \in \mathcal{N}(A)$, then $A(x+v) = Ax$, so $s(x+v) = s(x)$ for all x. Thus, directions in $\mathcal{N}(A)$ are invariant directions of the classifier: they are invisible to the representation.

3.3.2 Rank–Nullity Theorem

3.3.2.1 Statement and Consequences for Identifiability

Rank–nullity is the dimension bookkeeping identity behind linear models. If A maps an n-dimensional input space to an m-dimensional output space, then n degrees of freedom split into $\operatorname{rank}(A)$ degrees that survive and $\operatorname{nullity}(A)$ degrees that are annihilated. In regression, this determines identifiability: if $\mathcal{N}(X) \neq \{0\}$, multiple parameter vectors yield the same predictions, and additional structure (regularization, priors) is needed to select a unique solution.

Theorem 3.3.1 (Rank–Nullity) *For $A \in \mathbb{R}^{m \times n}$,*

$$\operatorname{rank}(A) + \dim(\mathcal{N}(A)) = n, \qquad \operatorname{rank}(A) + \dim(\mathcal{N}(A^\top)) = m.$$

Remark 3.3.2 (Identifiability Criterion for Linear Prediction) In linear regression $y \approx Xw$, the mapping $w \mapsto Xw$ is injective iff $\mathcal{N}(X) = \{0\}$, equivalently iff $\operatorname{rank}(X) = d$ (full column rank). If $\operatorname{rank}(X) < d$, then parameters are not identifiable from predictions.

3.3.2.2 Rank as Intrinsic Data Dimension and Redundancy

For a data matrix $X \in \mathbb{R}^{n \times d}$, rank measures the effective linear dimension of the dataset: how many independent directions exist among features (columns) or samples (rows). Redundancy appears as low rank: some features are linear combinations of others or lie nearly so (approximate low rank), leading to collinearity and instability. In practice, singular values quantify this continuously: small singular values correspond to nearly redundant directions.

Definition 3.3.2 (Intrinsic Linear Dimension) Given a data matrix $X \in \mathbb{R}^{n \times d}$, the intrinsic linear dimension of the column space is $\operatorname{rank}(X)$. If $\operatorname{rank}(X) = r \ll d$, then all samples lie in an r-dimensional subspace of feature space (after centering if needed).

Remark 3.3.3 (Redundancy in Features) If two columns of X are identical (or proportional), rank decreases. More subtly, if a column is nearly a linear combination of others, rank may remain full but the smallest singular value is small, leading to poor conditioning. This is the geometric origin of multicollinearity.

3.3.3 Low-Rank Structure

3.3.3.1 Rank Constraints as Inductive Bias

Low rank is a structural prior: it asserts that a matrix is generated by a small number of latent factors. In ML, low rank appears in collaborative filtering, embedding tables, linear dynamical systems, and representation learning (bottlenecks). A rank constraint reduces degrees of freedom from $O(nd)$ to $O(r(n + d))$, which can dramatically improve sample efficiency and reduce overfitting when the assumption is valid.

Definition 3.3.3 (Rank-r Matrices) A matrix $M \in \mathbb{R}^{m \times n}$ has rank at most r if it can be factorized as

$$M = UV^\top$$

for some $U \in \mathbb{R}^{m \times r}$ and $V \in \mathbb{R}^{n \times r}$.

Remark 3.3.4 (Degrees of Freedom) A rank-r matrix is specified (up to non-uniqueness) by U and V, requiring $r(m + n)$ parameters rather than mn. This is one reason low-rank models scale well and generalize.

3.3.3.2 Connections to Matrix Completion and Representation Learning

Matrix completion assumes a partially observed matrix is low rank and seeks to infer missing entries. Representation learning can be seen as learning low-rank structure: embeddings are low-rank factorizations of co-occurrence or interaction matrices. Even when not stated explicitly, many architectures implicitly prefer low-rank solutions due to bottlenecks and regularization. This preview sets the stage for later chapters on SVD, PCA, and optimization.

Remark 3.3.5 (Preview: Matrix Completion Template) Given observed entries $\{(i, j, M_{ij}) : (i, j) \in \Omega\}$, one seeks a low-rank X matching those entries:

$$\min_{X:\, \mathrm{rank}(X) \le r} \sum_{(i,j)\in\Omega} (X_{ij} - M_{ij})^2.$$

This is nonconvex in rank, but admits tractable relaxations and factorization approaches.

3.4 Linear Systems and Least Squares Geometry

Solving $Ax = b$ is the basic computational problem behind fitting linear models. When a solution exists, the set of solutions is an affine space; when it does not, the least-squares solution is defined by projecting b onto the column space of A. Conditioning determines reliability: near-singularity amplifies noise (Trefethen and Bau 1997). We develop existence/uniqueness, the geometric meaning of consistency, least squares via projections, and ridge/Tikhonov regularization as stabilized inversion.

3.4.1 Solving $Ax = b$: Existence and Uniqueness

3.4.1.1 Consistency, Degrees of Freedom, and Null-Space Solutions

A system is consistent iff b lies in the column space. When consistent, solutions form an affine translate of the null space: $x = x_0 + v$ with $v \in \mathcal{N}(A)$. Thus, degrees of freedom in solutions are exactly the nullity. When A is full column rank (injective), the solution is unique (if it exists). When A is full row rank (surjective onto $\mathbb{R}^m$ when $m \leq n$), solutions exist for every b but may not be unique.

Theorem 3.4.1 (Consistency Criterion) *The linear system $Ax = b$ has a solution iff $b \in \mathcal{C}(A)$.*

Proposition 3.4.1 (Solution Set as Affine Space) *If x_0 is one solution to $Ax = b$, then the set of all solutions is*

$$\{x : Ax = b\} = x_0 + \mathcal{N}(A).$$

Remark 3.4.1 (Degrees of Freedom) If $\dim(\mathcal{N}(A)) > 0$, there are infinitely many solutions: each null-space direction can be added without changing Ax. This is the linear-algebraic source of non-identifiability in underdetermined regression.

3.4.1.2 Conditioning Intuition: When Solutions Are Unreliable

Even when a unique solution exists, it may be unreliable numerically and statistically. If A is ill-conditioned (small $\sigma_{\min}(A)$), then small perturbations in b or rounding errors can cause large changes in x. In ML, this means fitted parameters can have high variance and be unstable under resampling, even if predictions remain stable. Conditioning is therefore a geometric property of how A stretches space: it measures anisotropy of the mapping.

Definition 3.4.1 (Condition Number (Spectral)) If $A \in \mathbb{R}^{m \times n}$ has full column rank, define

$$\kappa_2(A) := \frac{\sigma_{\max}(A)}{\sigma_{\min}(A)}.$$

Remark 3.4.2 (Sensitivity Bound (Informal)) For full-rank square A, $\|A^{-1}\|_2 = 1/\sigma_{\min}(A)$. Thus, $x = A^{-1}b$ satisfies $\|\Delta x\|_2 \lesssim \|A^{-1}\|_2 \|\Delta b\|_2$: small $\sigma_{\min}(A)$ amplifies perturbations in b.

3.4.2 Least Squares as Projection

3.4.2.1 Normal Equations and Geometric Derivation

When $Ax = b$ is inconsistent, the best approximation chooses x so that Ax is the closest point in $\mathcal{C}(A)$ to b. This is an orthogonal projection problem in $\mathbb{R}^m$. Algebraically, the orthogonality of the residual to the column space yields the normal equations $A^\top(Ax - b) = 0$. Geometrically, this means the residual lies in the left null space.

Definition 3.4.2 (Least Squares) Given $A \in \mathbb{R}^{m\times n}$ and $b \in \mathbb{R}^m$, a *least-squares solution* is any minimizer of

$$\min_{x\in\mathbb{R}^n} \|Ax - b\|_2^2.$$

Proposition 3.4.2 (Normal Equations) *A vector* $\widehat{x}$ *minimizes* $\|Ax - b\|_2^2$ *iff it satisfies*

$$A^\top(A\widehat{x} - b) = 0.$$

Remark 3.4.3 (Projection Viewpoint) The fitted vector $\widehat{b} := A\widehat{x}$ is the orthogonal projection of b onto $\mathcal{C}(A)$, hence is unique. The coefficient vector $\widehat{x}$ is unique iff $\mathcal{N}(A) = \{0\}$ (full column rank).

3.4.2.2 Residual Orthogonality and Fitted Subspaces

The residual $r = b - A\widehat{x}$ is orthogonal to every column of A. Equivalently, it is orthogonal to $\mathcal{C}(A)$, hence lies in $\mathcal{N}(A^\top)$. This leads to the Pythagorean decomposition $\|b\|_2^2 = \|\widehat{b}\|_2^2 + \|r\|_2^2$ once b is decomposed into its components in $\mathcal{C}(A)$ and $\mathcal{N}(A^\top)$ (after centering if needed). In regression diagnostics, $A^\top r = 0$ expresses that residuals have zero sample correlation with each feature column.

Proposition 3.4.3 (Residual Orthogonality) *If* $\widehat{x}$ *is a least-squares solution and* $r = b - A\widehat{x}$, *then*

$$r \in \mathcal{N}(A^\top) \qquad \text{and} \qquad \langle r, Au\rangle = 0 \;\; \forall u \in \mathbb{R}^n.$$

Remark 3.4.4 (Geometric Decomposition of b**)** There exists a unique decomposition $b = \widehat{b}+r$ with $\widehat{b} \in \mathcal{C}(A)$ and $r \in \mathcal{N}(A^\top)$. This is the orthogonal decomposition $\mathbb{R}^m = \mathcal{C}(A) \oplus \mathcal{N}(A^\top)$.

3.4.3 Regularized Linear Systems

3.4.3.1 Ridge/Tikhonov as Stabilized Inversion

Ridge (Tikhonov) regularization replaces a potentially ill-conditioned or singular system by a well-conditioned one. Instead of solving $A^\top Ax = A^\top b$, we solve $(A^\top A + \lambda I)x = A^\top b$. Geometrically, ridge trades exact fit for stability by penalizing large coefficients, shrinking solutions toward the origin. Spectrally, it shifts eigenvalues away from zero, improving conditioning and reducing variance amplification.

Definition 3.4.3 (Ridge/Tikhonov Problem) Given $A \in \mathbb{R}^{m\times n}$, $b \in \mathbb{R}^m$, and $\lambda > 0$, the ridge solution is

$$\widehat{x}_\lambda \in \arg\min_{x\in\mathbb{R}^n} \|Ax - b\|_2^2 + \lambda\|x\|_2^2.$$

Proposition 3.4.4 (Closed Form via Regularized Normal Equations) *The ridge objective is strictly convex, and its unique minimizer satisfies*

$$(A^\top A + \lambda I)\widehat{x}_\lambda = A^\top b, \qquad \textit{hence} \qquad \widehat{x}_\lambda = (A^\top A + \lambda I)^{-1}A^\top b.$$

Remark 3.4.5 (Why $\lambda > 0$ Stabilizes) If $A^\top A$ is singular or ill-conditioned, adding λI makes it positive definite and improves the smallest eigenvalue: $\lambda_{\min}(A^\top A + \lambda I) \geq \lambda$, so inversion becomes stable.

3.4.3.2 Bias–Variance and Conditioning Tradeoffs

Ridge introduces bias (solutions are shrunk) but reduces variance (less sensitivity to noise), the classic bias–variance tradeoff. Geometrically, ridge discourages moving far in parameter space, which damps the effect of nearly-null directions of A. Spectrally, ridge acts as a filter on singular directions: directions with small singular values are suppressed most strongly. This is why ridge can improve generalization and numerical robustness, especially under collinearity.

Remark 3.4.6 (Spectral Filtering Intuition) If $A = U\Sigma V^\top$ is an SVD, then ridge effectively scales components along V by factors $\sigma_i/(\sigma_i^2 + \lambda)$ in the solution (derived later). Small σ_i directions are strongly damped, improving stability.

Remark 3.4.7 (Choosing λ) Large λ improves conditioning but increases bias (underfitting). Small λ reduces bias but may inherit ill-conditioning (overfitting/instability). In practice, λ is tuned by cross-validation; theoretically, it interacts with noise level and effective dimension.

3.5 Affine Maps and Homogeneous Coordinates

Most ML pipelines are not purely linear: they include translations (bias terms), centering, and normalization. These are *affine* transformations. Treating them as operators clarifies their algebra (composition), their geometric effects (changing origin, rescaling directions, whitening correlations), and their impact on conditioning. Homogeneous coordinates provide a standard trick: represent affine maps as linear maps in one higher dimension, so that composition becomes matrix multiplication.

3.5.1 *Affine Geometry in Data Preprocessing*

3.5.1.1 Centering, Scaling, and Whitening as Operators

Centering and scaling are linear/affine operators on data matrices: they can be written as left/right multiplication by structured matrices. This viewpoint makes it clear which axes are being normalized (samples vs. features) and how these operations commute (or do not). Whitening is a particularly important linear operator: it rotates into covariance eigenvectors and rescales by inverse standard deviations, transforming anisotropic clouds into isotropic ones. In learning, these operations reduce numerical issues and align the geometry so that distances and gradients behave predictably.

Definition 3.5.1 (Centering Operator on a Dataset) Let $X \in \mathbb{R}^{n\times d}$ be a data matrix with rows $x_i^\top$ and let $\mathbf{1} \in \mathbb{R}^n$ be all ones. Define the empirical feature-mean (as a row vector) by

$$\mu^\top := \frac{1}{n}\mathbf{1}^\top X \in \mathbb{R}^{1\times d}, \qquad \mu \in \mathbb{R}^d.$$

The centered data matrix is

$$X_c := X - \mathbf{1}\mu^\top.$$

Equivalently, with $H := I_n - \frac{1}{n}\mathbf{1}\mathbf{1}^\top$, we have $X_c = HX$.

Definition 3.5.2 (Diagonal Feature Scaling) Let $D = \mathrm{diag}(s_1, \ldots, s_d)$ with $s_i > 0$. Feature scaling is the linear map

$$X \mapsto XD,$$

which multiplies feature i by s_i.

Definition 3.5.3 (Whitening Transform (Data-Matrix Form)) Let $X_c \in \mathbb{R}^{n\times d}$ be centered data and define the empirical covariance

$$\widehat{\Sigma} := \frac{1}{n} X_c^\top X_c \in \mathbb{R}^{d\times d}.$$

Assume $\widehat{\Sigma} \succ 0$ and define $\widehat{\Sigma}^{-1/2}$ via eigendecomposition. The whitened data are

$$X_w := X_c\, \widehat{\Sigma}^{-1/2},$$

so that $\frac{1}{n} X_w^\top X_w = I_d$.

Remark 3.5.1 (Operator View of Preprocessing Pipelines) Centering acts on the *sample axis* via left multiplication (HX), while scaling/whitening act on the *feature axis* via right multiplication (XD or $X\widehat{\Sigma}^{-1/2}$). Thus, the algebra of preprocessing is matrix multiplication with a clear geometric meaning.

3.5.1.2 Feature Standardization and Conditioning Implications

Standardization (subtract mean, divide by standard deviation) changes the metric of feature space: it measures distance in "standard deviation units" per coordinate. This reduces dominance by large-scale features and often improves conditioning of $X^\top X$, stabilizing least squares and gradient-based optimization. However, standardization is not always benign: if a feature has near-zero variance, scaling can blow up noise. Hence, conditioning analysis motivates adding ϵ stabilizers and careful handling of constant/rare features.

Definition 3.5.4 (Standardization) Let $X \in \mathbb{R}^{n\times d}$ and let $\mu \in \mathbb{R}^d$ be the empirical mean. Let $\sigma \in \mathbb{R}^d$ have entries

$$\sigma_j := \sqrt{\frac{1}{n}\sum_{i=1}^{n}(X_{ij} - \mu_j)^2}.$$

Define $D_\sigma := \operatorname{diag}(\sigma_1^{-1}, \ldots, \sigma_d^{-1})$ (with an ϵ-stabilizer if needed). The standardized data are

$$X_s := (X - \mathbf{1}\mu^\top) D_\sigma.$$

Remark 3.5.2 (Conditioning Link (Qualitative)) If columns have vastly different scales, then $X^\top X$ can be ill-conditioned because some directions dominate. Standardization attempts to equalize scales, often reducing $\kappa(X^\top X)$ and improving numerical stability of solvers.

3.5.2 Homogeneous Coordinates

3.5.2.1 Embedding Affine Maps into Linear Maps

Affine maps $x \mapsto Ax + b$ are linear maps plus translation. Homogeneous coordinates convert them into pure linear maps in dimension $d + 1$ by appending a constant 1 to the vector. This is a powerful representational trick: composition of affine maps becomes ordinary matrix multiplication, allowing preprocessing pipelines and neural affine layers to be treated uniformly as linear algebra.

Definition 3.5.5 (Affine Map) An *affine map* $f : \mathbb{R}^d \to \mathbb{R}^k$ is

$$f(x) = Ax + b, \qquad A \in \mathbb{R}^{k\times d}, \ b \in \mathbb{R}^k.$$

Definition 3.5.6 (Homogeneous Lifting) Define the homogeneous embedding $\iota : \mathbb{R}^d \to \mathbb{R}^{d+1}$ by

$$\iota(x) := \begin{bmatrix} x \\ 1 \end{bmatrix}.$$

Given an affine map $f(x) = Ax + b$, define the homogeneous matrix

$$\widetilde{A} := \begin{bmatrix} A & b \\ 0^\top & 1 \end{bmatrix} \in \mathbb{R}^{(k+1)\times(d+1)}.$$

Then

$$\widetilde{A}\,\iota(x) = \begin{bmatrix} Ax + b \\ 1 \end{bmatrix}.$$

Remark 3.5.3 (Bias Terms Become Weights) In homogeneous coordinates, the bias vector b becomes an extra column of the linear map, and the constant 1 becomes an extra feature. This is why linear models with intercept can be written as linear models on augmented features.

3.5.2.2 Composing Preprocessing Steps as Matrix Products

A preprocessing pipeline often applies multiple affine transformations: center, scale, rotate, shift, etc. Homogeneous coordinates provide a clean algebra: if each step is represented by a homogeneous matrix, the entire pipeline is the product of these matrices. This is valuable both conceptually (understanding which operations commute) and computationally (precomputing a single transform).

Proposition 3.5.1 (Composition of Affine Maps via Homogeneous Matrices) *Let $f(x) = A_1x + b_1$ and $g(x) = A_2x + b_2$ with compatible shapes. Let $\widetilde{A}_1, \widetilde{A}_2$ be their homogeneous matrices. Then the composition satisfies*

$$\iota\big((g \circ f)(x)\big) = \widetilde{A}_2\,\widetilde{A}_1\,\iota(x).$$

In particular, the composed affine map is $x \mapsto (A_2A_1)x + (A_2b_1 + b_2)$.

Example 3.5.1 (Center then Scale) Let $f(x) = x - \mu$ (centering) and $g(x) = Dx$ (scaling). Then $(g \circ f)(x) = D(x - \mu) = Dx - D\mu$. Homogeneous representation gives

$$\widetilde{A}_f = \begin{bmatrix} I & -\mu \\ 0^\top & 1 \end{bmatrix}, \qquad \widetilde{A}_g = \begin{bmatrix} D & 0 \\ 0^\top & 1 \end{bmatrix}, \qquad \widetilde{A}_g\widetilde{A}_f = \begin{bmatrix} D & -D\mu \\ 0^\top & 1 \end{bmatrix}.$$

3.5.3 *Augmentations as Transformations*

3.5.3.1 Data Augmentation as Group Action (Informal)

Data augmentation applies transformations (rotations, translations, color jitter, permutations) to generate new samples. Many such transformations form (approximately) a group acting on the input space. This viewpoint connects to invariance: if the label is preserved under the action, then a good predictor should be invariant. It also connects to equivariance: intermediate representations may transform predictably under the action. We keep the treatment informal here; the main geometric lesson is to recognize augmentations as structured transformations.

Definition 3.5.7 (Group Action (Recall)) A group G acts on $\mathcal{X}$ if there is a map $(g, x) \mapsto g \cdot x$ such that $e \cdot x = x$ and $(g_1g_2) \cdot x = g_1 \cdot (g_2 \cdot x)$.

Remark 3.5.4 (Augmentation Distributions) In practice, augmentations sample $g \sim \pi$ from a distribution on G (or a semigroup of transformations) and replace (x, y) by $(g \cdot x, y)$ when label is invariant.

3.5.3.2 Invariances and Equivariances

Invariance means output does not change under transformations; equivariance means representation changes in a controlled way. Linear maps are the simplest equivariant objects: if a representation is linear and the group action is linear, equivariance constraints become commutation relations $T(gx) = \rho(g)T(x)$. This idea is foundational in convolutional networks (translation equivariance) and modern equivariant architectures (rotations, permutations).

Definition 3.5.8 (Invariance and Equivariance) Let G act on $\mathcal{X}$ and let $\rho : G \to \mathrm{GL}(k)$ be a representation acting on $\mathbb{R}^k$. A map $f : \mathcal{X} \to \mathbb{R}^k$ is

invariant if $f(g \cdot x) = f(x)\ \forall g, x,$ *equivariant* if $f(g \cdot x) = \rho(g) f(x)\ \forall g, x.$

3.6 Block Matrices and Modular Model Structure

Modern ML systems are modular: they process multiple modalities, share parameters across tasks, and couple submodels through structured interactions. Block matrices provide the algebra for this modularity. They enable decomposed computations, efficient updates, and clear interpretations of sharing vs. separation. Schur complements provide a powerful tool for eliminating blocks and understanding conditional structure, including a bridge to Gaussian conditioning.

3.6.1 Block Matrices for Multi-view and Multi-task Learning

3.6.1.1 Concatenation, Fusion, and Block-Structured Predictors

Multi-view data combines heterogeneous feature groups (text/image/tabular) and multi-task learning shares representations across outputs. Concatenation yields block vectors; linear predictors on concatenated features become block-structured. This makes it transparent how each view contributes and how cross-view interactions appear as off-diagonal blocks. The same notation clarifies architectures with separate encoders and a fusion head.

Definition 3.6.1 (Block Partition and Concatenation) Let $x = \begin{bmatrix} x^{(1)} \\ x^{(2)} \end{bmatrix}$ with $x^{(1)} \in \mathbb{R}^{d_1}$ and $x^{(2)} \in \mathbb{R}^{d_2}$. A linear map $W \in \mathbb{R}^{k \times (d_1 + d_2)}$ partitions as $W = [W_1 \;\; W_2]$ with $W_i \in \mathbb{R}^{k \times d_i}$, giving

$$Wx = W_1 x^{(1)} + W_2 x^{(2)}.$$

Remark 3.6.1 (Fusion Interpretation) The blocks W_1, W_2 quantify the contribution of each modality. Cross-modal interactions require nonlinear fusion or feature cross-terms; in purely linear fusion they appear only through shared outputs.

3.6.1.2 Block-Diagonal, Block-Sparse, and Shared-Parameter Forms

Block-diagonal structure encodes independent processing: each block acts on its own feature group. Block-sparse structure encodes selective interactions: only certain modality-task pairs are coupled. Shared-parameter forms (e.g., same block repeated) encode parameter tying across tasks or time steps. These structures reduce computation and improve generalization by reducing degrees of freedom.

Definition 3.6.2 (Block-Diagonal and Block-Sparse) A block matrix $A = [A_{ij}]$ is *block-diagonal* if $A_{ij} = 0$ for $i \neq j$. It is *block-sparse* if many blocks are zero according to a specified sparsity pattern.

Remark 3.6.2 (Multi-Task Linear Model) For T tasks with outputs in $\mathbb{R}^T$, one often uses $Y = XW$ with $W \in \mathbb{R}^{d\times T}$. Row-sparsity of W corresponds to selecting a shared subset of features across tasks; block norms encourage this.

3.6.2 *Schur Complements and Elimination*

3.6.2.1 Solving Partitioned Systems and Conditional Structure

Block elimination is the algebraic analogue of "solve for one part, substitute into the other." The Schur complement captures the effective system on remaining variables after eliminating a block. Computationally, it enables efficient solving of structured systems and updates. Conceptually, it isolates how coupling between blocks changes the effective curvature or covariance.

Definition 3.6.3 (Schur Complement) Let

$$M = \begin{bmatrix} A & B \\ C & D \end{bmatrix}$$

with $A \in \mathbb{R}^{p\times p}$ invertible. The *Schur complement of* A *in* M is

$$S_{/A} := D - CA^{-1}B.$$

Similarly, if D is invertible, the Schur complement of D is $S_{/D} := A - BD^{-1}C$.

Proposition 3.6.1 (Block Elimination Formula) *Consider the linear system*

$$\begin{bmatrix} A & B \\ C & D \end{bmatrix} \begin{bmatrix} x \\ y \end{bmatrix} = \begin{bmatrix} u \\ v \end{bmatrix},$$

with A *invertible. Then* y *satisfies*

$$(D - CA^{-1}B)\, y = v - CA^{-1}u,$$

and then $x = A^{-1}(u - By)$.

Remark 3.6.3 (When Schur Complements Help) If A is small or easy to invert and $S_{/A}$ has exploitable structure, this reduces computational cost significantly. Schur complements also appear in constrained optimization (KKT systems) and Gaussian conditioning.

3.6.2.2 Gaussian Conditioning Intuition (Bridge to Probability)

In multivariate Gaussians, conditioning and marginalization correspond to block operations on covariance/precision matrices. The Schur complement appears as the conditional covariance (or precision) of a subset given the other. Thus, block linear algebra is not merely computational; it expresses probabilistic structure. This will be used later when we discuss Gaussian models, linear regression as MAP estimation, and Kalman-like updates.

Remark 3.6.4 (Preview Statement) If a Gaussian vector is partitioned as (X, Y) with covariance blocks, then the conditional covariance of Y given X involves a Schur complement. We will derive the exact formula in the probability/statistics chapters.

3.6.3 Matrix Identities Used in ML

3.6.3.1 Woodbury Identity and Fast Updates

Woodbury is the master identity for updating inverses under low-rank modifications. It explains why ridge regression can be solved efficiently in either primal or dual form depending on whether $n \ll d$ or $d \ll n$, and it underlies online/streaming updates, Gaussian posterior updates, and efficient recomputation of matrix inverses after rank-one changes.

Theorem 3.6.1 (Woodbury Matrix Identity) *Let $A \in \mathbb{R}^{d\times d}$ be invertible, $U \in \mathbb{R}^{d\times k}$, $V \in \mathbb{R}^{k\times d}$, and $C \in \mathbb{R}^{k\times k}$ be invertible. Then*

$$(A + UCV)^{-1} = A^{-1} - A^{-1}U\big(C^{-1} + VA^{-1}U\big)^{-1}VA^{-1}.$$

Remark 3.6.5 (Rank-one Update Special Case) If $k = 1$, $U = u$, $V = v^\top$, $C = 1$, then Woodbury yields the Sherman–Morrison formula:

$$(A + uv^\top)^{-1} = A^{-1} - \frac{A^{-1}uv^\top A^{-1}}{1 + v^\top A^{-1}u},$$

provided $1 + v^\top A^{-1}u \neq 0$.

3.6.3.2 Matrix Determinant Lemma and Log-Determinants

Log-determinants appear in Gaussian likelihoods, information criteria, and Bayesian evidence. The determinant lemma complements Woodbury by providing determinant updates under low-rank modifications. In ML, it enables efficient evaluation of $\log\det(A + UCV)$ without forming the large determinant, crucial for scalable Gaussian processes, Bayesian linear regression, and entropy-like objectives.

Theorem 3.6.2 (Matrix Determinant Lemma) *Let $A \in \mathbb{R}^{d\times d}$ be invertible, $U \in \mathbb{R}^{d\times k}$, $V \in \mathbb{R}^{k\times d}$, and $C \in \mathbb{R}^{k\times k}$. Then*

$$\det(A + UCV) = \det(C^{-1} + VA^{-1}U)\,\det(C)\,\det(A).$$

Equivalently, if $C = I_k$,

$$\det(A + UV) = \det(I_k + VA^{-1}U)\,\det(A).$$

Remark 3.6.6 (Log-det Form) Taking logs gives

$$\log\det(A + UCV) = \log\det(A) + \log\det(C) + \log\det(C^{-1} + VA^{-1}U),$$

often reducing a large log-det to a smaller $k \times k$ log-det.

3.7 Tensor Products, Vectorization, and Reshaping

Beyond basic matrix algebra, modern ML frequently manipulates multiway arrays (tensors), reshapes them, and expresses structured linear maps compactly. Two operator-level tools dominate: the $\mathrm{vec}(\cdot)$ operator (stacking a matrix into a vector) and the Kronecker product $\otimes$. Together they convert bilinear expressions into ordinary matrix–vector products, enabling clean derivations for covariance structures, linear layers, and efficient implementations. We also emphasize outer products as rank-1 tensors that appear naturally in gradients and second moments.

3.7.1 The vec(·) *Operator and Kronecker Products*

3.7.1.1 $\mathrm{vec}(AXB) = (B^\top \otimes A)\mathrm{vec}(X)$

Many expressions in ML are of the form "apply A on the left and B on the right" (e.g., bilinear transforms, covariance updates, whitening, attention mixing). Vectorization turns a matrix unknown X into a vector unknown $\mathrm{vec}(X)$ so that AXB

becomes a standard linear map. The Kronecker product is the matrix representation of this lifted linear map. The identity $\mathrm{vec}(AXB) = (B^\top \otimes A)\mathrm{vec}(X)$ is therefore a universal conversion rule that appears in least squares with matrix-valued unknowns, matrix normal distributions, and linear layers with reshaping (Horn and Johnson 2012).

Definition 3.7.1 (vec(·) **Operator)** For $X \in \mathbb{R}^{m\times n}$, define $\mathrm{vec}(X) \in \mathbb{R}^{mn}$ by stacking columns:

$$\mathrm{vec}(X) := \begin{bmatrix} X_{:1} \\ X_{:2} \\ \vdots \\ X_{:n} \end{bmatrix}.$$

Thus, entry X_{ij} appears at position $i + (j-1)m$ in $\mathrm{vec}(X)$.

Definition 3.7.2 (Kronecker Product) Let $A \in \mathbb{R}^{p\times q}$ and $B \in \mathbb{R}^{r\times s}$. Their *Kronecker product* $A \otimes B \in \mathbb{R}^{(pr)\times(qs)}$ is the block matrix

$$A \otimes B := \begin{bmatrix} a_{11}B & a_{12}B & \cdots & a_{1q}B \\ a_{21}B & a_{22}B & \cdots & a_{2q}B \\ \vdots & \vdots & \ddots & \vdots \\ a_{p1}B & a_{p2}B & \cdots & a_{pq}B \end{bmatrix}.$$

Theorem 3.7.1 (Vec–Kronecker Identity) *Let $A \in \mathbb{R}^{p\times m}$, $X \in \mathbb{R}^{m\times n}$, and $B \in \mathbb{R}^{n\times q}$. Then*

$$\mathrm{vec}(AXB) = (B^\top \otimes A)\,\mathrm{vec}(X).$$

Remark 3.7.1 (Shape Check) $AXB \in \mathbb{R}^{p\times q}$, so $\mathrm{vec}(AXB) \in \mathbb{R}^{pq}$. Also $B^\top \otimes A \in \mathbb{R}^{(pq)\times(mn)}$ and $\mathrm{vec}(X) \in \mathbb{R}^{mn}$, so the product is well-defined.

Proposition 3.7.1 (Basic Kronecker Rules) *For compatible matrices:*

$$(A \otimes B)^\top = A^\top \otimes B^\top, \qquad (A \otimes B)(C \otimes D) = (AC) \otimes (BD),$$

and if A and B are invertible then

$$(A \otimes B)^{-1} = A^{-1} \otimes B^{-1}.$$

3.7.1.2 Applications: Covariance Structure and Linear Layers

Vectorization and Kronecker products arise whenever we model matrix-valued random variables or optimize matrix parameters. In statistics, the covariance of vec(X) often factors as a Kronecker product in matrix-normal models. In deep learning, linear layers can be viewed as matrix multiplications with reshapes; Kronecker identities provide a formal tool to describe these as a single large linear map on a vectorized tensor. This clarifies parameter sharing and structured transforms.

Example 3.7.1 (Covariance with Separable Structure) Let $Z \in \mathbb{R}^{m\times n}$ be a random matrix and assume a separable covariance model

$$\mathrm{Cov}(\mathrm{vec}(Z)) = \Sigma_c \otimes \Sigma_r,$$

where $\Sigma_r \in \mathbb{R}^{m\times m}$ controls row correlations and $\Sigma_c \in \mathbb{R}^{n\times n}$ controls column correlations. This factorization reduces parameters from $(mn)^2$ to $m^2 + n^2$ and is foundational in matrix-normal modeling.

Example 3.7.2 (Linear Layer on a Matrix Input) Let $X \in \mathbb{R}^{m\times n}$ and consider the bilinear transform $Y = AXB$ with learnable A, B. Then by Theorem 3.7.1,

$$\mathrm{vec}(Y) = (B^\top \otimes A)\mathrm{vec}(X),$$

so the bilinear layer is a linear map on the vectorized input with a Kronecker-structured weight matrix. This expresses parameter sharing and can be exploited for speed and memory.

3.7.2 *Tensor (Outer) Products and Rank-1 Structure*

3.7.2.1 Outer Products in Gradients and Second Moments

Outer products $uv^\top$ are the rank-1 building blocks of matrices. They appear naturally in gradients of linear models (e.g., gradient is feature vector times residual), in covariance estimation (second moments), and in attention (queries/keys create outer-product score matrices). Recognizing rank-1 structure allows efficient updates, low-rank approximations, and clear geometric interpretation of how a single example contributes to a parameter update.

Definition 3.7.3 (Outer Product) For $u \in \mathbb{R}^m$ and $v \in \mathbb{R}^n$, the *outer product* is the matrix

$$uv^\top \in \mathbb{R}^{m\times n}, \qquad (uv^\top)_{ij} = u_i v_j.$$

It has rank 1 if $u \neq 0$ and $v \neq 0$.

Example 3.7.3 (Gradient of a Single-Sample Squared Loss) Let $w \in \mathbb{R}^d$ and loss $\ell(w) = \frac{1}{2}(w^\top x - y)^2$ for a single sample (x, y). Then

$$\nabla_w \ell(w) = (w^\top x - y)\, x,$$

which is a scalar residual times the feature vector. For a matrix parameter $W \in \mathbb{R}^{k\times d}$ with loss $\frac{1}{2}\|Wx - y\|_2^2$, the gradient is

$$\nabla_W \ell(W) = (Wx - y)\, x^\top,$$

an outer product of the output residual and the input.

3.7.2.2 Low-Rank Factorizations as Parameterizations

Low-rank factorization $M = UV^\top$ is both a modeling prior and a computational strategy. It reduces parameter count and enforces structured interactions (latent factors). In ML, it underlies matrix completion, collaborative filtering, and low-rank adapters. From an optimization viewpoint, it replaces a hard rank constraint by an unconstrained but nonconvex parameterization.

Definition 3.7.4 (Rank-r Factorization) A matrix $M \in \mathbb{R}^{m\times n}$ has rank at most r if it can be written as $M = UV^\top$ with $U \in \mathbb{R}^{m\times r}$ and $V \in \mathbb{R}^{n\times r}$.

Remark 3.7.2 (Parameter Count) The factorization uses $r(m + n)$ parameters instead of mn. When $r \ll \min\{m, n\}$, this is a dramatic reduction and acts as an inductive bias.

3.8 Idempotents, Projectors, and Pseudoinverses

Not all linear systems have unique solutions. The Moore–Penrose pseudoinverse provides a canonical way to select solutions: it yields least-squares solutions in the overdetermined case and minimum-norm solutions in the underdetermined case (cf. Fig. 3.2). It also provides clean projector formulas and clarifies when using pseudoinverses is numerically safe (and when it is not).

3.8.1 Moore–Penrose Pseudoinverse

3.8.1.1 Overdetermined vs. Underdetermined Systems

When $m > n$, a system $Ax = b$ is typically overdetermined; it may be inconsistent and is solved via least squares. When $m < n$, it is underdetermined; there are

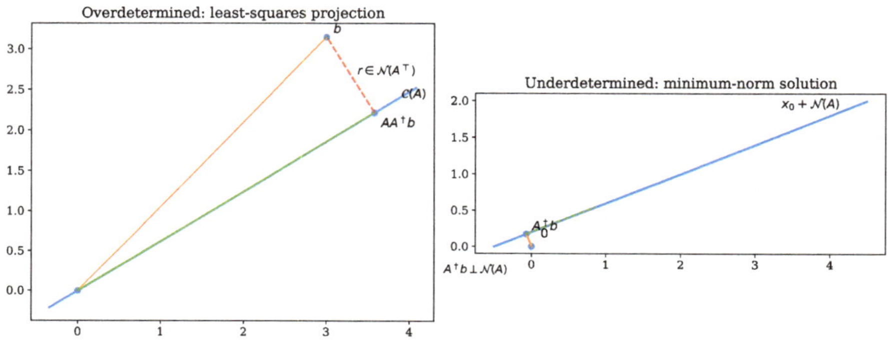

Fig. 3.2 Pseudoinverse geometry. Left: in an overdetermined problem, $A^{\dagger}b$ produces the least-squares fit whose image is the orthogonal projection of b onto the column space. Right: in an underdetermined consistent problem, the solution set is an affine translate of the null space, and $A^{\dagger}b$ selects the minimum-norm point orthogonal to that null space

infinitely many solutions when consistent, and the minimum-norm solution is preferred. The pseudoinverse unifies these cases: $A^{\dagger}b$ is always the minimum-norm least-squares solution.

Definition 3.8.1 (Moore–Penrose Pseudoinverse) For $A \in \mathbb{R}^{m\times n}$, the *Moore–Penrose pseudoinverse* $A^{\dagger} \in \mathbb{R}^{n\times m}$ is the unique matrix satisfying the four Penrose conditions:

$$\text{(i) } AA^{\dagger}A = A, \qquad \text{(ii) } A^{\dagger}AA^{\dagger} = A^{\dagger},$$

$$\text{(iii) } (AA^{\dagger})^{\top} = AA^{\dagger}, \qquad \text{(iv) } (A^{\dagger}A)^{\top} = A^{\dagger}A.$$

Remark 3.8.1 (Existence and Uniqueness) Every matrix has a unique Moore–Penrose pseudoinverse. In practice, it is computed via SVD (introduced in the next chapter).

3.8.1.2 Minimum-Norm Solutions and Geometric Meaning

In underdetermined systems, $Ax = b$ has an affine family of solutions $x = x_0 + v$ with $v \in \mathcal{N}(A)$. Among these, the minimum-norm solution is the one orthogonal to the null space: it lies in the row space $\mathcal{C}(A^{\top})$. In overdetermined least squares, $A^{\dagger}b$ chooses the coefficient vector whose prediction Ax is the projection of b onto the column space and whose coefficient norm is minimal among all least-squares minimizers.

Theorem 3.8.1 (Minimum-Norm Least-Squares Solution) *For any* $A \in \mathbb{R}^{m\times n}$ *and* $b \in \mathbb{R}^m$, *the vector*

$$\widehat{x} := A^\dagger b$$

is a minimizer of $\min_x \|Ax - b\|_2$ *(least squares), and among all minimizers it has minimum Euclidean norm:*

$$\widehat{x} \in \arg\min_x \|Ax - b\|_2 \quad \textit{and} \quad \|\widehat{x}\|_2 = \min\{\|x\|_2 : x \in \arg\min_z \|Az - b\|_2\}.$$

Remark 3.8.2 (Geometric Location) The minimum-norm solution $A^\dagger b$ lies in $\mathcal{C}(A^\top)$ (row space) and is orthogonal to $\mathcal{N}(A)$.

3.8.2 Projectors in Practice

3.8.2.1 Oblique vs. Orthogonal Projections (High-Level Contrast)

A projector P with $P^2 = P$ projects along some complement; only when $P^\top = P$ is the projection orthogonal (closest in ℓ_2). Oblique projectors arise in constrained problems and in certain decomposition methods; they can be useful but are more sensitive to noise because they do not preserve orthogonality. In ML, orthogonal projectors are the default because they connect directly to least squares, PCA, and stable geometry.

Remark 3.8.3 (Two Conditions, Two Geometries) Idempotence $P^2 = P$ says "apply twice equals once" (projection). Symmetry $P^\top = P$ says "projection is orthogonal" (closest-point, Pythagorean decomposition).

3.8.2.2 Using Pseudoinverses Safely (What to Avoid)

The pseudoinverse is conceptually clean but can be numerically dangerous when singular values are tiny: $A^\dagger$ contains reciprocals of singular values, so small singular values amplify noise. Thus, one should avoid explicitly forming $A^\dagger$ for ill-conditioned problems; instead use stable solvers (QR/SVD) and regularization (ridge) or truncated SVD. In ML, this is the practical reason behind ridge regression and early stopping.

Remark 3.8.4 (Instability Mechanism) If $A = U\Sigma V^\top$ (SVD), then $A^\dagger = V\Sigma^\dagger U^\top$ with $\Sigma^\dagger$ containing $1/\sigma_i$ on nonzero singular values. When some σ_i are very small, $1/\sigma_i$ is huge, amplifying perturbations in b.

3.8.3 Least Squares via Pseudoinverse

3.8.3.1 $\widehat{x} = A^{\dagger}b$ and Geometry of Solutions

The pseudoinverse formula unifies the full-rank closed forms: if A has full column rank, $A^{\dagger} = (A^{\top}A)^{-1}A^{\top}$; if A has full row rank, $A^{\dagger} = A^{\top}(AA^{\top})^{-1}$. Geometrically, $AA^{\dagger}$ is the orthogonal projector onto $\mathcal{C}(A)$ and $A^{\dagger}A$ is the orthogonal projector onto $\mathcal{C}(A^{\top})$. Thus, pseudoinverse not only solves least squares but also constructs projectors onto fundamental subspaces.

Proposition 3.8.1 (Full-Rank Formulas) *Let $A \in \mathbb{R}^{m\times n}$.*

- *If* $\operatorname{rank}(A) = n$ *(full column rank), then*

$$A^{\dagger} = (A^{\top}A)^{-1}A^{\top}.$$

- *If* $\operatorname{rank}(A) = m$ *(full row rank), then*

$$A^{\dagger} = A^{\top}(AA^{\top})^{-1}.$$

Proposition 3.8.2 (Projectors from Pseudoinverse) *For any A,*

$$P_{\mathcal{C}(A)} := AA^{\dagger} \text{ is the orthogonal projector onto } \mathcal{C}(A) \subseteq \mathbb{R}^{m},$$

and

$$P_{\mathcal{C}(A^{\top})} := A^{\dagger}A \text{ is the orthogonal projector onto } \mathcal{C}(A^{\top}) \subseteq \mathbb{R}^{n}.$$

3.8.3.2 Numerical Warnings: Instability and Regularization

When A is ill-conditioned, the stable alternative is regularized inversion: replace $A^{\dagger}$ by $(A^{\top}A + \lambda I)^{-1}A^{\top}$ (ridge) or use truncated SVD (discard tiny singular values). These correspond to geometric choices: ridge shrinks all directions, truncated SVD removes near-null directions. Both improve robustness at the cost of bias. Understanding this tradeoff is essential before implementing pseudoinverse-based solvers in ML.

Remark 3.8.5 (Ridge as a Stable Pseudoinverse Surrogate) Define the ridge operator

$$A_{\lambda}^{\dagger} := (A^{\top}A + \lambda I)^{-1}A^{\top}.$$

As $\lambda \downarrow 0$, $A_{\lambda}^{\dagger}$ approaches $A^{\dagger}$ when the limit exists, but for $\lambda > 0$ it avoids division by tiny singular values and is numerically stable.

3.9 Matrix Calculus Primer for Learning (Preview for Optimization)

Training ML models requires gradients. Matrix calculus provides a compact language to compute derivatives of scalar objectives with respect to vectors and matrices, using trace identities and shape-aware rules. This section is a preview: it supplies the most common derivative templates and emphasizes Jacobian shapes and the chain rule, framing backpropagation as repeated Jacobian–vector products (Magnus and Neudecker 2019).

3.9.1 Gradients of Common Scalar Objectives

3.9.1.1 Quadratic Forms, Least Squares, and Regularizers

Quadratic objectives dominate ML: least squares, ridge, Gaussians, and local second-order approximations. They have closed-form gradients and Hessians and reveal the role of symmetry. Keeping shapes explicit prevents mistakes: gradients with respect to vectors are vectors; gradients with respect to matrices are matrices of matching shape.

Proposition 3.9.1 (Basic Gradient Identities (Real Case)) *Let $x \in \mathbb{R}^d$, $A \in \mathbb{R}^{d\times d}$, $b \in \mathbb{R}^d$, and $C \in \mathbb{R}^{m\times d}$. Assume gradients are taken with respect to x.*

1. $\nabla_x(b^\top x) = b$.
2. $\nabla_x\left(\frac{1}{2}\|x\|_2^2\right) = x$.
3. *If A is symmetric,* $\nabla_x\left(\frac{1}{2}x^\top Ax\right) = Ax$. *More generally,* $\nabla_x(x^\top Ax) = (A + A^\top)x$.
4. $\nabla_x\left(\frac{1}{2}\|Cx - y\|_2^2\right) = C^\top(Cx - y)$.

Remark 3.9.1 (Connection to Normal Equations) Setting the gradient in (4) to zero yields $C^\top(Cx - y) = 0$, i.e., the normal equations. Thus, least squares optimality is simultaneously a calculus statement and a projection statement.

3.9.1.2 Trace/Transpose Tricks for Clean Derivations

Trace tricks convert messy sums into compact expressions where cyclicity can be used to place differentials in a convenient position. This is the standard workflow in matrix calculus: write objective as $\mathrm{tr}(\cdot)$, take differential $d(\cdot)$, and match coefficients. Even if you do not memorize derivative tables, you can derive them mechanically with trace rules.

Remark 3.9.2 (Two Common Trace Rewrites) For compatible matrices:

$$\|A\|_F^2 = \operatorname{tr}(A^\top A), \qquad x^\top M x = \operatorname{tr}(M x x^\top).$$

These rewrites are often the first step in gradient derivations with respect to matrix parameters.

3.9.2 Jacobians and Chain Rule

3.9.2.1 Vector-Valued Maps and Jacobian Shapes

In deep learning, layers are vector-valued maps and gradients propagate through Jacobians. Shape discipline is essential: if $f : \mathbb{R}^d \to \mathbb{R}^k$, then $J_f(x) \in \mathbb{R}^{k\times d}$. Confusing left vs. right multiplication is a common error. Keeping Jacobians as linear maps clarifies how to compose them.

Definition 3.9.1 (Jacobian (Recall)) For differentiable $f : \mathbb{R}^d \to \mathbb{R}^k$, the Jacobian at x is $J_f(x) \in \mathbb{R}^{k\times d}$ with $(J_f(x))_{ij} = \partial f_i/\partial x_j$.

Remark 3.9.3 (Jacobian as Best Linear Map) The Jacobian is the linear map such that $f(x + \Delta) \approx f(x) + J_f(x)\Delta$ for small Δ. Thus, Jacobians are the local linear maps of nonlinear models.

3.9.2.2 Backprop as Repeated Jacobian-Vector Products

Backpropagation is essentially the chain rule applied repeatedly, but implemented efficiently by avoiding explicit Jacobian matrices. Instead, one computes Jacobian-vector products (forward-mode) or vector-Jacobian products (reverse-mode). In reverse-mode (standard backprop), gradients propagate by multiplying by transposed Jacobians. This is why transpose rules and shape checks from earlier sections are operationally critical.

Proposition 3.9.2 (Chain Rule (Matrix Form)) *Let $f : \mathbb{R}^d \to \mathbb{R}^k$ and $g : \mathbb{R}^k \to \mathbb{R}^p$ be differentiable. Then $h = g \circ f$ satisfies*

$$J_h(x) = J_g(f(x))\, J_f(x).$$

Remark 3.9.4 (Reverse-Mode Gradient Propagation) If $L : \mathbb{R}^k \to \mathbb{R}$ is a scalar loss and $y = f(x)$, then

$$\nabla_x L(f(x)) = J_f(x)^\top \nabla_y L(y).$$

Thus, gradients flow backward by multiplying by transposed Jacobians.

3.10 Exercises

Unless stated otherwise, all vector spaces are over $\mathbb{R}$, inner products are Euclidean, and $\|\cdot\|$ denotes $\|\cdot\|_2$.

Exercise 3.10.1 (Linearity, Composition, and Shape Discipline) Let $A \in \mathbb{R}^{3\times 2}$, $B \in \mathbb{R}^{2\times 4}$, $C \in \mathbb{R}^{4\times 4}$ be

$$A = \begin{bmatrix} 1 & 2 \\ 0 & -1 \\ 3 & 1 \end{bmatrix}, \quad B = \begin{bmatrix} 1 & 0 & 2 & -1 \\ -2 & 1 & 0 & 3 \end{bmatrix}, \quad C = \begin{bmatrix} 2 & 0 & 1 & 0 \\ 0 & 1 & 0 & 0 \\ 1 & 0 & 3 & -1 \\ 0 & 0 & -1 & 2 \end{bmatrix}.$$

(i) Verify which of the products AB, BA, BC, CB, AC, CA, ABC, $A(BC)$, $(AB)C$ are defined, and state the shape of each defined product. (ii) Compute AB, BC, and $A(BC)$ explicitly and verify associativity by checking $A(BC) = (AB)C$. (iii) Define linear maps $T(x) = Bx$ and $S(y) = Ay$ on appropriate domains; write the matrix of $S \circ T$ and compute $(S \circ T)(x)$ for $x = \begin{bmatrix} 1 \\ -1 \\ 2 \\ 0 \end{bmatrix}$.

Exercise 3.10.2 (Kernel, Image, Rank, and Information Loss) Let $A \in \mathbb{R}^{3\times 4}$ be

$$A = \begin{bmatrix} 1 & 0 & 2 & 1 \\ 0 & 1 & -1 & 2 \\ 1 & 1 & 1 & 3 \end{bmatrix}.$$

(i) Compute a basis for $\mathcal{N}(A)$ by solving $Ax = 0$ and hence determine nullity(A). (ii) Compute rank(A) and verify rank–nullity: rank$(A) + \dim \mathcal{N}(A) = 4$. (iii) Compute a basis for $\mathcal{C}(A)$ (column space) and for $\mathcal{C}(A^\top)$ (row space), and verify the orthogonal decomposition $\mathbb{R}^4 = \mathcal{C}(A^\top) \oplus \mathcal{N}(A)$ by explicitly decomposing $x = \begin{bmatrix} 1 \\ 2 \\ -1 \\ 0 \end{bmatrix}$ as $x = x_r + x_n$ with $x_r \in \mathcal{C}(A^\top)$ and $x_n \in \mathcal{N}(A)$.

Exercise 3.10.3 (Consistency and Affine Solution Sets) Let $A \in \mathbb{R}^{3\times 4}$ be as in Exercise 3.10.2 and consider the two right-hand sides

$$b_1 = \begin{bmatrix} 1 \\ 0 \\ 2 \end{bmatrix}, \quad b_2 = \begin{bmatrix} 1 \\ 0 \\ 0 \end{bmatrix}.$$

(i) Decide whether $Ax = b_1$ and $Ax = b_2$ are consistent by checking whether each b_i lies in $\mathcal{C}(A)$. (ii) For the consistent system(s), find one solution x_0 and then describe the full solution set as $x_0 + \mathcal{N}(A)$. (iii) Among all solutions, find the one with minimum ℓ_2-norm by imposing orthogonality to $\mathcal{N}(A)$ (i.e., the solution lying in $\mathcal{C}(A^\top)$).

Exercise 3.10.4 (Least Squares as Projection and Residual Orthogonality) Let

$$A = \begin{bmatrix} 1 & 0 \\ 1 & 1 \\ 1 & 2 \\ 1 & 3 \end{bmatrix} \in \mathbb{R}^{4\times 2}, \qquad b = \begin{bmatrix} 1 \\ 2 \\ 2 \\ 4 \end{bmatrix} \in \mathbb{R}^4.$$

(i) Solve the least squares problem $\min_{x\in\mathbb{R}^2} \|Ax - b\|_2^2$ by normal equations and compute $\widehat{x}$ exactly. (ii) Compute the fitted vector $\widehat{b} = A\widehat{x}$ and residual $r = b - \widehat{b}$. (iii) Verify $A^\top r = 0$ and interpret this as $r \in \mathcal{N}(A^\top)$. (iv) Compute the projector $P = A(A^\top A)^{-1}A^\top$ and verify $Pb = \widehat{b}$ and $P^2 = P$ and $P^\top = P$.

Exercise 3.10.5 (Ridge/Tikhonov Stabilization and Conditioning) Let

$$A = \begin{bmatrix} 1 & 1 \\ 1 & 1+\varepsilon \\ 1 & 1+2\varepsilon \end{bmatrix} \in \mathbb{R}^{3\times 2}, \qquad b = \begin{bmatrix} 1 \\ 1 \\ 1 \end{bmatrix}, \qquad \varepsilon = 10^{-2}.$$

(i) Compute $A^\top A$ and its eigenvalues (or singular values of A) and estimate $\kappa_2(A)$. (ii) Compute the unregularized least squares solution $\widehat{x}$ via $(A^\top A)^{-1}A^\top b$ (carry at least 5 decimal digits). (iii) Compute the ridge solutions $\widehat{x}_\lambda = (A^\top A + \lambda I)^{-1}A^\top b$ for $\lambda \in \{10^{-4}, 10^{-2}, 10^{-1}\}$. (iv) Compare $\|\widehat{x}_\lambda\|_2$ and $\|A\widehat{x}_\lambda - b\|_2$ across λ, and comment (one paragraph) on the bias–variance/conditioning tradeoff reflected numerically.

Exercise 3.10.6 (Change of Basis and Similarity Invariants) Let

$$A = \begin{bmatrix} 2 & 1 & 0 \\ 0 & 2 & 1 \\ 0 & 0 & 2 \end{bmatrix} \in \mathbb{R}^{3\times 3}, \qquad S = \begin{bmatrix} 1 & 1 & 0 \\ 0 & 1 & 1 \\ 0 & 0 & 1 \end{bmatrix} \in \mathbb{R}^{3\times 3}.$$

(i) Compute $B = S^{-1}AS$ explicitly and verify that A and B are similar. (ii) Compute $\mathrm{tr}(A)$, $\mathrm{tr}(B)$, $\det(A)$, $\det(B)$ and verify invariance. (iii) Compute the eigenvalues of A (and hence of B) and verify that the characteristic polynomials match.

Exercise 3.10.7 (Orthonormal Change of Basis as an Isometry) Let

$$Q = \frac{1}{3}\begin{bmatrix} 2 & -1 & 2 \\ 1 & 2 & -2 \\ 2 & -2 & -1 \end{bmatrix}.$$

(i) Verify that Q is orthogonal by computing $Q^\top Q$. (ii) For $x = \begin{bmatrix} 1 \\ -2 \\ 3 \end{bmatrix}$ and $y = \begin{bmatrix} 2 \\ 0 \\ -1 \end{bmatrix}$, compute $\langle Qx, Qy\rangle$ and $\langle x, y\rangle$ and verify they are equal; also verify $\|Qx\|_2 = \|x\|_2$. (iii) Compute the angle between x and y and between Qx and Qy using cosine similarity and verify equality.

Exercise 3.10.8 (Affine Maps as Homogeneous Linear Maps) Let $f : \mathbb{R}^2 \to \mathbb{R}^2$ and $g : \mathbb{R}^2 \to \mathbb{R}^2$ be affine maps

$$f(x) = A_1x + b_1, \quad A_1 = \begin{bmatrix} 1 & 2 \\ 0 & 1 \end{bmatrix}, \; b_1 = \begin{bmatrix} -1 \\ 3 \end{bmatrix},$$

$$g(x) = A_2x + b_2, \quad A_2 = \begin{bmatrix} 2 & 0 \\ 1 & 1 \end{bmatrix}, \; b_2 = \begin{bmatrix} 0 \\ -2 \end{bmatrix}.$$

(i) Construct homogeneous matrices $\widetilde{A}_1, \widetilde{A}_2$ such that $\widetilde{A}_i \begin{bmatrix} x \\ 1 \end{bmatrix} = \begin{bmatrix} f_i(x) \\ 1 \end{bmatrix}$. (ii) Compute the homogeneous matrix $\widetilde{A} = \widetilde{A}_2\widetilde{A}_1$ and extract the affine parameters (A, b) of $g \circ f$. (iii) Verify your result by directly composing $g(f(x))$ and comparing with (A, b). (iv) Apply $g \circ f$ to $x = \begin{bmatrix} 1 \\ -1 \end{bmatrix}$.

Exercise 3.10.9 (Block Matrices: Multi-View Fusion and Block Elimination) Consider the block system

$$\begin{bmatrix} A & B \\ C & D \end{bmatrix}\begin{bmatrix} x \\ y \end{bmatrix} = \begin{bmatrix} u \\ v \end{bmatrix},$$

where

$$A = \begin{bmatrix} 2 & 1 \\ 1 & 2 \end{bmatrix}, \; B = \begin{bmatrix} 1 & 0 \\ 0 & 1 \end{bmatrix}, \; C = \begin{bmatrix} 1 & -1 \\ 2 & 0 \end{bmatrix}, \; D = \begin{bmatrix} 3 & 1 \\ 1 & 1 \end{bmatrix}, \; u = \begin{bmatrix} 1 \\ 0 \end{bmatrix}, \; v = \begin{bmatrix} 2 \\ -1 \end{bmatrix}.$$

(i) Compute the Schur complement $S_{/A} = D - CA^{-1}B$ and the reduced right-hand side $v - CA^{-1}u$. (ii) Solve for y using $S_{/A}y = v - CA^{-1}u$, then recover

$x = A^{-1}(u - By)$. (iii) Verify your solution by multiplying the full 4×4 block matrix with $\begin{bmatrix} x \\ y \end{bmatrix}$.

Exercise 3.10.10 (Woodbury Identity: Verify Numerically) Let

$$A = \begin{bmatrix} 4 & 1 & 0 \\ 1 & 3 & 1 \\ 0 & 1 & 2 \end{bmatrix}, \quad U = \begin{bmatrix} 1 \\ 0 \\ 2 \end{bmatrix}, \quad V = \begin{bmatrix} 1 & -1 & 0 \end{bmatrix}, \quad C = \begin{bmatrix} 2 \end{bmatrix}.$$

(i) Compute $M = A + UCV$ explicitly and then compute M^{-1} directly (by any exact method). (ii) Compute the Woodbury right-hand side

$$A^{-1} - A^{-1}U\left(C^{-1} + VA^{-1}U\right)^{-1}VA^{-1}$$

and verify it equals M^{-1} exactly (or to numerical tolerance). (iii) Using the matrix determinant lemma, compute $\det(M)$ from $\det(A)$ and the 1×1 term $(C^{-1} + VA^{-1}U)$, and verify against the direct determinant.

Exercise 3.10.11 (Vec and Kronecker Identity By Explicit Computation) Let

$$A = \begin{bmatrix} 1 & 2 \\ 0 & -1 \end{bmatrix}, \quad X = \begin{bmatrix} 2 & -1 & 0 \\ 1 & 3 & 2 \end{bmatrix}, \quad B = \begin{bmatrix} 1 & 0 \\ -1 & 2 \\ 2 & 1 \end{bmatrix}.$$

(i) Compute $Y = AXB$ and then compute $\mathrm{vec}(Y)$ using column-stacking. (ii) Compute $\mathrm{vec}(X)$ and the Kronecker matrix $(B^\top \otimes A)$. (iii) Verify numerically that $\mathrm{vec}(AXB) = (B^\top \otimes A)\mathrm{vec}(X)$.

Exercise 3.10.12 (Outer Products and Rank-1 Updates) Let $u = \begin{bmatrix} 1 \\ -2 \\ 1 \end{bmatrix}$ and $v = \begin{bmatrix} 2 \\ 0 \\ -1 \\ 3 \end{bmatrix}$. (i) Form the outer product $M = uv^\top \in \mathbb{R}^{3\times 4}$ and compute $\mathrm{rank}(M)$. (ii) Compute $M^\top M$ and verify it has rank 1 (by showing two columns are multiples or by computing eigenvalues). (iii) Let $w = \begin{bmatrix} 1 \\ 1 \\ 1 \\ 1 \end{bmatrix}$ and compute Mw and interpret it as a scaled copy of u.

Exercise 3.10.13 (Moore–Penrose pseudoinverse: Over- vs. Under-Determined) Consider the overdetermined matrix $A_o \in \mathbb{R}^{4\times 2}$ and underdetermined matrix $A_u \in$

$\mathbb{R}^{2\times 4}$:

$$A_o = \begin{bmatrix} 1 & 0 \\ 1 & 1 \\ 1 & 2 \\ 1 & 3 \end{bmatrix}, \qquad A_u = \begin{bmatrix} 1 & 0 & 1 & 0 \\ 0 & 1 & 1 & 1 \end{bmatrix}.$$

(i) Compute $A_o^\dagger$ using the full-column-rank formula $(A_o^\top A_o)^{-1}A_o^\top$ and solve $\widehat{x} = A_o^\dagger b$ for $b = \begin{bmatrix} 1 \\ 2 \\ 2 \\ 4 \end{bmatrix}$; verify that this matches the least squares solution in Exercise 3.10.4. (ii) Compute $A_u^\dagger$ using the full-row-rank formula $A_u^\top(A_u A_u^\top)^{-1}$ and compute the minimum-norm solution $\widehat{x} = A_u^\dagger b$ for $b = \begin{bmatrix} 2 \\ -1 \end{bmatrix}$. (iii) For the underdetermined system, describe all solutions to $A_u x = b$ as $\widehat{x} + v$ where $v \in \mathcal{N}(A_u)$ and verify that $\widehat{x} \perp \mathcal{N}(A_u)$.

Exercise 3.10.14 (Projectors from Pseudoinverse and Idempotence) Let $A = \begin{bmatrix} 1 & 0 & 1 \\ 0 & 1 & 1 \end{bmatrix} \in \mathbb{R}^{2\times 3}$ and compute $A^\dagger$ via $A^\top(AA^\top)^{-1}$. (i) Compute $P_1 = AA^\dagger \in \mathbb{R}^{2\times 2}$ and $P_2 = A^\dagger A \in \mathbb{R}^{3\times 3}$. (ii) Verify that $P_1^\top = P_1$, $P_1^2 = P_1$ and similarly for P_2. (iii) For $b = \begin{bmatrix} 1 \\ 2 \end{bmatrix}$, compute $P_1 b$ and interpret it as projection onto $\mathcal{C}(A)$ in $\mathbb{R}^2$.

Exercise 3.10.15 (Matrix Calculus: Gradients by Trace Tricks) Let $C = \begin{bmatrix} 1 & 2 \\ 0 & -1 \\ 3 & 1 \end{bmatrix} \in \mathbb{R}^{3\times 2}$, $y = \begin{bmatrix} 1 \\ 0 \\ -1 \end{bmatrix} \in \mathbb{R}^3$, and consider the objective $f(x) = \frac{1}{2}\|Cx - y\|_2^2 + \frac{\lambda}{2}\|x\|_2^2$ with $\lambda = 3$ and $x \in \mathbb{R}^2$. (i) Rewrite $f(x)$ in trace form using $\|v\|_2^2 = v^\top v$ and $\mathrm{tr}(\cdot)$ identities. (ii) Compute $\nabla f(x)$ and the Hessian $\nabla^2 f(x)$ explicitly. (iii) Solve $\nabla f(x) = 0$ to obtain the unique minimizer $x^\star$ and compute $f(x^\star)$.

Exercise 3.10.16 (Backprop as Jacobian-Transpose Vector Product) Define a two-layer map $f : \mathbb{R}^2 \to \mathbb{R}^2$ by

$$f(x) = W_2\,\sigma(W_1 x + b_1) + b_2,$$

where σ acts elementwise as $\sigma(t) = \tanh(t)$ and

$$W_1 = \begin{bmatrix} 1 & 2 \\ -1 & 1 \\ 2 & 0 \end{bmatrix} \in \mathbb{R}^{3\times 2}, \qquad b_1 = \begin{bmatrix} 0 \\ 1 \\ -1 \end{bmatrix} \in \mathbb{R}^3,$$

$$W_2 = \begin{bmatrix} 1 & 0 & -1 \\ 2 & 1 & 0 \end{bmatrix} \in \mathbb{R}^{2\times 3}, \quad b_2 = \begin{bmatrix} 1 \\ 0 \end{bmatrix} \in \mathbb{R}^2.$$

Let $x = \begin{bmatrix} 1 \\ -1 \end{bmatrix}$ and define the scalar loss $L(x) = \frac{1}{2}\|f(x)\|_2^2$. (i) Compute $z = W_1x + b_1$, $h = \sigma(z)$, and $f(x)$ explicitly. (ii) Compute $\nabla_f \frac{1}{2}\|f\|_2^2 = f$ at $f(x)$ and propagate backward to compute $\nabla_x L(x)$ using $\nabla_x L = J_f(x)^\top f(x)$, without writing the full Jacobian explicitly (use vector-Jacobian products). (iii) Verify your result by computing $J_f(x)$ explicitly via the chain rule and then multiplying by $f(x)$.

Chapter 4
Spectral Theory, SVD, and Principal Components

Spectral methods translate geometry into algebra by decomposing linear operators into *modes*: directions that are preserved (eigenvectors) and scalars that quantify amplification/attenuation (eigenvalues/singular values). In ML, spectra appear whenever we measure variation (covariance/PCA), similarity (kernels), smoothness (graph Laplacians), or curvature (Hessians). This chapter develops the spectral viewpoint rigorously, beginning with eigenvalues/eigenvectors for symmetric matrices and the computational principles that make spectral algorithms practical at scale (Horn and Johnson 2012).

4.1 Why Spectral Thinking Matters in ML

4.1.1 Spectra as "Modes" of Variation

4.1.1.1 Energy, Variance, and Dominant Directions

A quadratic form $x^\top Ax$ measures *energy* of a vector x under an operator A (e.g., covariance, Hessian). Spectral decomposition identifies directions that concentrate this energy. For covariance matrices, eigenvectors are principal directions of variation and eigenvalues quantify variance along those directions. Thus, spectra convert a cloud of points into a small number of dominant axes that often govern downstream performance (compression, denoising, regression stability).

Definition 4.1.1 (Energy of a Direction under a Symmetric Operator) Let $A \in \mathbb{R}^{d\times d}$ be symmetric. The energy of a direction $x \neq 0$ is the scalar

$$\mathcal{E}_A(x) := x^\top Ax.$$

If $A \succeq 0$ this is nonnegative and defines a weighted geometry.

P. Singh, B. Raman, *Foundations of Machine Learning and AI*, Studies in Big Data 199, https://doi.org/10.1007/978-3-032-30336-3_4

Remark 4.1.1 (Variance as Energy) If X is a centered random vector with covariance $\Sigma = \mathbb{E}[XX^\top]$, then for any unit vector u,

$$\mathrm{Var}(u^\top X) = u^\top \Sigma u.$$

Thus, principal variance directions are maximizers of a quadratic form over the unit sphere.

4.1.1.2 Stability, Mixing, and Iterative Algorithms

Spectra govern stability and convergence of iterative procedures. In linear dynamical systems $x_{t+1} = Ax_t$, eigenvalues determine growth/decay and oscillations. In iterative numerical algorithms (power iteration, gradient descent on quadratics), eigenvalues determine contraction factors and rates. In Markov chains and random walks, the spectrum of the transition matrix (or Laplacian) controls mixing time. This is why eigengaps repeatedly appear as the quantitative signature of "fast convergence" and "stable subspaces."

Definition 4.1.2 (Spectral Radius) For $A \in \mathbb{R}^{d\times d}$, the *spectral radius* is

$$\rho(A) := \max\{|\lambda| : \lambda \text{ is an eigenvalue of } A\}.$$

Remark 4.1.2 (Linear Stability Criterion) For the linear iteration $x_{t+1} = Ax_t$, if $\rho(A) < 1$ then $x_t \to 0$ for every initial x_0, whereas if $\rho(A) > 1$ then there exist initial conditions with $\|x_t\|_2 \to \infty$. For symmetric A, $\rho(A) = \|A\|_2$.

4.1.2 Where Eigen/Singular Values Appear

4.1.2.1 Covariance, Kernels, Graph Laplacians, and Hessians

Eigenvalues arise whenever we have a symmetric PSD matrix: covariances, Gram/kernel matrices, and graph Laplacians. Each encodes geometry: covariance encodes spread, kernels encode similarity in feature space, and Laplacians encode smoothness on graphs. Hessians encode local curvature of losses; their eigenvalues determine sharp/flat directions and therefore conditioning and step-size limits. Singular values generalize eigenvalues to non-square/non-symmetric matrices (e.g., data matrices), and connect directly to SVD and PCA.

Remark 4.1.3 (A Single Template) A large fraction of spectral ML can be reduced to analyzing a PSD matrix $M \succeq 0$:

- $M = \widehat{\Sigma} = \frac{1}{n} X_c^\top X_c$ (covariance of centered data);
- $M = K$ with $K_{ij} = k(x_i, x_j)$ (kernel Gram matrix);

- $M = L$ (graph Laplacian or normalized Laplacian);
- $M = \nabla^2 f(w)$ (Hessian of a twice-differentiable loss).

Eigenvectors give modes; eigenvalues give strengths/curvatures.

4.1.2.2 Generalization and Implicit Regularization

Spectra interact with learning not only through computation but also through generalization. Effective dimension, margin, and stability bounds frequently depend on eigenvalues of covariance/kernels. Optimization algorithms can exhibit *implicit regularization*: even without explicit penalties, they prefer solutions aligned with large-variance directions or minimum-norm solutions in overparameterized regimes—phenomena that are most naturally expressed in spectral terms. We treat these ideas as previews; rigorous generalization theory appears later.

Remark 4.1.4 (Preview: Effective Dimension) A common spectral complexity measure is an "effective dimension" such as

$$d_{\mathrm{eff}}(\lambda) := \mathrm{tr}\big(\Sigma(\Sigma + \lambda I)^{-1}\big) = \sum_{i=1}^{d} \frac{\lambda_i(\Sigma)}{\lambda_i(\Sigma) + \lambda},$$

which counts how many eigen-directions are "active" at scale λ. This quantity appears in ridge regression and kernel generalization bounds.

4.2 Eigenvalues, Eigenvectors, and Diagonalization

4.2.1 Spectral Viewpoint

4.2.1.1 Eigen-Geometry: Invariant Directions and Scaling

An eigenvector is a direction that a linear map does not rotate—only scales (and possibly flips sign). Thus, eigenvectors provide a coordinate system in which the operator acts diagonally. In ML, this is the cleanest way to understand how an operator shapes geometry: directions with large eigenvalues are amplified, directions with small eigenvalues are suppressed, and negative eigenvalues indicate sign-flipping curvature (for Hessians, nonconvexity).

Definition 4.2.1 (Eigenvalue and Eigenvector) Let $A \in \mathbb{R}^{d\times d}$. A scalar $\lambda \in \mathbb{R}$ and a nonzero vector $v \in \mathbb{R}^d$ form an *eigenpair* if

$$Av = \lambda v.$$

The set of all eigenvectors associated with λ, together with 0, is the *eigenspace* of λ.

Remark 4.2.1 (Invariant Subspaces) If v is an eigenvector, then $\mathrm{span}\{v\}$ is invariant: $A(\mathrm{span}\{v\}) \subseteq \mathrm{span}\{v\}$. More generally, sums of eigenspaces corresponding to a set of eigenvalues form invariant subspaces.

Example 4.2.1 (Diagonal Action in an Eigenbasis) If A has a basis of eigenvectors $v_1, \dots, v_d$ with eigenvalues $\lambda_1, \dots, \lambda_d$, then any $x = \sum_i \alpha_i v_i$ satisfies

$$Ax = \sum_{i=1}^{d} \alpha_i \lambda_i v_i,$$

so each coordinate is scaled independently. This is the spectral "mode decomposition."

4.2.1.2 Rayleigh Quotient and Variational Characterization

The Rayleigh quotient converts eigen-analysis into optimization on the unit sphere. For symmetric A, maximizing $x^\top Ax$ over $\|x\|_2 = 1$ yields the top eigenvalue, and the maximizer is a top eigenvector. This fact is the formal bridge between linear algebra and PCA: the first principal component is the maximizer of variance (a Rayleigh quotient) under a unit-norm constraint. It also underlies power iteration and many eigenvalue bounds.

Definition 4.2.2 (Rayleigh Quotient) For symmetric $A \in \mathbb{R}^{d\times d}$ and $x \neq 0$, the *Rayleigh quotient* is

$$\mathcal{R}_A(x) := \frac{x^\top Ax}{x^\top x}.$$

Theorem 4.2.1 (Variational Characterization of Extremal Eigenvalues) *Let $A \in \mathbb{R}^{d\times d}$ be symmetric with eigenvalues $\lambda_1 \geq \lambda_2 \geq \cdots \geq \lambda_d$. Then*

$$\lambda_1 = \max_{\|x\|_2=1} x^\top Ax, \qquad \lambda_d = \min_{\|x\|_2=1} x^\top Ax,$$

and the maximizers (resp. minimizers) are exactly the unit eigenvectors associated with λ_1 (resp. λ_d).

Proof Let $A = Q\Lambda Q^\top$ with orthogonal Q and diagonal $\Lambda = \mathrm{diag}(\lambda_1, \dots, \lambda_d)$ (Spectral Theorem, proved next). Write $x = Qz$ with $\|z\|_2 = \|x\|_2 = 1$. Then

$$x^\top Ax = z^\top \Lambda z = \sum_{i=1}^{d} \lambda_i z_i^2, \qquad \sum_{i=1}^{d} z_i^2 = 1.$$

The maximum is achieved by putting all mass on $i = 1$, giving λ_1, and the minimum by $i = d$, giving λ_d. □

Remark 4.2.2 (First Principal Component as a Maximizer) If $\Sigma \succeq 0$ is a covariance matrix, then the direction of maximum variance solves $\max_{\|u\|_2=1} u^\top \Sigma u$, hence is the top eigenvector of Σ.

4.2.2 *Spectral Theorem for Symmetric Matrices*

4.2.2.1 Orthogonal Diagonalization and Geometric Meaning

Symmetric matrices are the perfect class for Euclidean geometry: they admit an orthonormal eigenbasis. Orthogonal diagonalization says a symmetric operator is a rotation/reflection (orthogonal change of basis) followed by axis-aligned scaling (diagonal matrix) followed by the inverse rotation. Hence symmetric matrices act by stretching space along orthogonal axes. This is why covariance matrices define ellipsoids and why quadratic objectives are ellipsoidal.

Theorem 4.2.2 (Spectral Theorem) *If $A \in \mathbb{R}^{d\times d}$ is symmetric, then there exists an orthogonal matrix $Q \in \mathbb{R}^{d\times d}$ and a real diagonal matrix $\Lambda = diag(\lambda_1, \ldots, \lambda_d)$ such that*

$$A = Q\Lambda Q^\top.$$

The columns of Q are orthonormal eigenvectors of A, and λ_i are the corresponding eigenvalues.

Remark 4.2.3 (Geometric Action) For $x \in \mathbb{R}^d$, write $x = Qz$ (coordinates in the eigenbasis). Then

$$Ax = Q\Lambda z,$$

so A scales each eigen-coordinate independently by λ_i. Thus the level sets $\{x : x^\top Ax = c\}$ are ellipsoids aligned with eigenvectors (cf. Fig. 4.1).

4.2.2.2 Positive Definite vs. Semidefinite Matrices

PSD/PD matrices are exactly the symmetric matrices with nonnegative/positive eigenvalues. They define valid energies, covariances, and kernel Gram matrices. In optimization, PSD Hessians correspond to convexity; PD corresponds to strict convexity and uniqueness. Spectrally, PSD means no negative curvature direction; PD means every nonzero direction has strictly positive curvature.

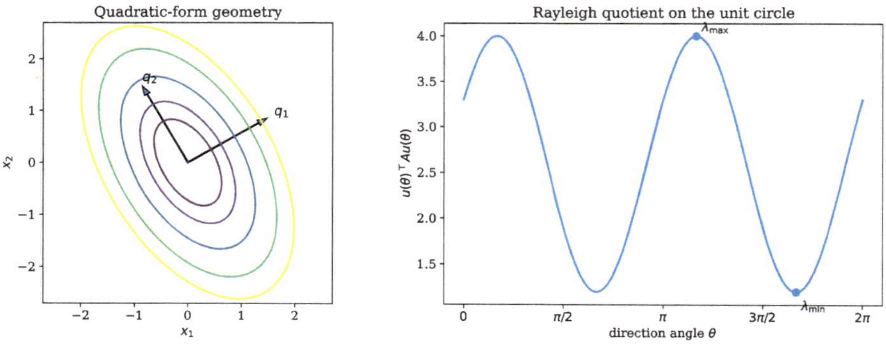

Fig. 4.1 Spectral geometry of a symmetric operator. Left: level sets of a quadratic form are ellipses aligned with eigenvectors. Right: the Rayleigh quotient on the unit circle attains its extrema exactly at the principal eigen-directions

Definition 4.2.3 (PSD and PD Matrices) A symmetric matrix A is *positive semidefinite* (PSD), written $A \succeq 0$, if $x^\top Ax \geq 0$ for all x. It is *positive definite* (PD), written $A \succ 0$, if $x^\top Ax > 0$ for all $x \neq 0$.

Proposition 4.2.1 (PSD/PD via Eigenvalues) *If A is symmetric with eigenvalues $\lambda_1 \geq \cdots \geq \lambda_d$, then*

$$A \succeq 0 \iff \lambda_d \geq 0, \qquad A \succ 0 \iff \lambda_d > 0.$$

Remark 4.2.4 (Covariance and Kernels) Covariance matrices and Gram/kernel matrices are PSD by construction. This is the feasibility condition for "inner-product geometry."

4.2.3 *Eigenvalue Bounds and Intuition*

4.2.3.1 Gershgorin Disks and Quick Localization

Exact eigenvalues are expensive to compute for large matrices, but we often need quick bounds. Gershgorin's theorem provides a fast localization using only row sums: every eigenvalue lies in at least one disk centered at a diagonal entry with radius equal to the sum of off-diagonal magnitudes in that row. This is especially useful for checking PSD heuristics, diagnosing diagonal dominance, and bounding spectra of graph-related matrices.

Theorem 4.2.3 (Gershgorin Disk Theorem) *Let $A \in \mathbb{C}^{d\times d}$. For each $i \in [d]$ define the Gershgorin disk*

$$\mathcal{D}_i := \left\{ z \in \mathbb{C} : \ |z - a_{ii}| \le \sum_{j\neq i} |a_{ij}| \right\}.$$

Then every eigenvalue of A lies in $\bigcup_{i=1}^{d} \mathcal{D}_i$.

Remark 4.2.5 (Diagonal Dominance Heuristic) If A is symmetric and strictly diagonally dominant with positive diagonal (e.g., $a_{ii} > \sum_{j\neq i} |a_{ij}|$), then all Gershgorin disks lie in $(0, \infty)$, strongly suggesting $A \succ 0$ (indeed, such conditions imply PD under mild assumptions).

4.2.3.2 Min–Max Principle (Informal) and Principal Components

The Rayleigh quotient gives the top and bottom eigenvalues. The full spectrum can also be characterized variationally: the k-th eigenvalue is obtained by optimizing over k-dimensional subspaces. This is the min–max principle (Courant–Fischer). In PCA, it states that the best k-dimensional subspace (maximizing explained variance) is spanned by the top k eigenvectors of the covariance matrix. This turns dimensionality reduction into a clean optimization problem.

Theorem 4.2.4 (Courant–Fischer Min–Max Principle) *Let $A \in \mathbb{R}^{d\times d}$ be symmetric with eigenvalues $\lambda_1 \ge \cdots \ge \lambda_d$. For $k \in [d]$,*

$$\lambda_k = \max_{\substack{\mathcal{U}\subseteq\mathbb{R}^d \\ \dim(\mathcal{U})=k}} \ \min_{\substack{x\in\mathcal{U} \\ \|x\|_2=1}} x^\top Ax = \min_{\substack{\mathcal{W}\subseteq\mathbb{R}^d \\ \dim(\mathcal{W})=d-k+1}} \ \max_{\substack{x\in\mathcal{W} \\ \|x\|_2=1}} x^\top Ax.$$

Remark 4.2.6 (PCA Subspace Optimality) If $\Sigma \succeq 0$ is a covariance matrix, then the k-dimensional subspace maximizing captured variance is the span of the top k eigenvectors of Σ, and the captured variance equals $\sum_{i=1}^{k} \lambda_i(\Sigma)$. We will formalize this in the PCA section of this chapter.

4.2.4 Computational Notes

4.2.4.1 Power Iteration Intuition and Convergence Factors

Power iteration is the simplest eigenvector algorithm: repeatedly apply A to a vector and renormalize (Trefethen and Bau 1997). In the eigenbasis, each component is multiplied by λ_i at each step, so the largest-magnitude eigenvalue dominates. The convergence rate depends on the ratio $|\lambda_2/\lambda_1|$ (for symmetric PSD, λ_2/λ_1). This gives an immediate computational meaning to eigengaps: larger gaps imply faster convergence.

Algorithm 1 POWERITERATION (dominant eigenvector, symmetric case)

Require: Symmetric matrix $A \in \mathbb{R}^{d\times d}$, iterations T, initial $x^{(0)} \neq 0$

```
for t = 0, 1, ..., T − 1 do
    y^(t+1) ← A x^(t)
    x^(t+1) ← y^(t+1) / ||y^(t+1)||_2
end for
return x^(T)
```

Proposition 4.2.2 (Convergence Rate (Basic Form)) *Let $A \in \mathbb{R}^{d\times d}$ be symmetric with eigenvalues $\lambda_1 > \lambda_2 \geq \cdots \geq \lambda_d$ and orthonormal eigenvectors q_i. Write $x^{(0)} = \sum_{i=1}^{d} \alpha_i q_i$ with $\alpha_1 \neq 0$. Then the iterates of Algorithm 1 satisfy*

$$dist\big(x^{(t)}, span\{q_1\}\big) = O\left(\left|\frac{\lambda_2}{\lambda_1}\right|^t\right),$$

where distance may be taken as sin *of the angle to q_1 (up to constants depending on α_1).*

Remark 4.2.7 (Practical Note) Power iteration requires only matrix–vector products, so it scales to very large sparse matrices (kernels approximated, graph Laplacians, etc.).

4.2.4.2 Spectral Gaps and Stability of Eigenspaces

Spectral gaps matter not only for convergence speed but also for *stability*: how much eigenvectors/eigenspaces change when the matrix is perturbed (noise, sampling error, numerical error). When the gap between an eigenvalue and the rest is large, the corresponding eigenspace is stable. This is crucial in PCA: the leading eigenvectors estimated from finite samples are meaningful only if the population eigengap is not tiny. Two standard tools express this: Weyl's inequality (eigenvalue perturbations) and Davis–Kahan (eigenspace perturbations).

Theorem 4.2.5 (Weyl's Inequality (Eigenvalue Perturbation Bound)) *Let $A, E \in \mathbb{R}^{d\times d}$ be symmetric, with eigenvalues $\lambda_1(\cdot) \geq \cdots \geq \lambda_d(\cdot)$. Then for each $i \in [d]$,*

$$|\lambda_i(A+E) - \lambda_i(A)| \leq \|E\|_2.$$

Theorem 4.2.6 (Davis–Kahan $\sin\Theta$ Theorem (One-Dimensional Form)) *Let A, E be symmetric and let q be a unit eigenvector of A with eigenvalue λ. Let $\widehat{q}$ be a unit eigenvector of $\widehat{A} := A + E$ with eigenvalue $\widehat{\lambda}$. Assume λ is separated from* *the rest of* $spec(A)$ *by a gap*

$$gap := \min_{\mu \in spec(A),\, \mu \neq \lambda} |\lambda - \mu| > 0.$$

Then the angle θ *between* q *and* $\widehat{q}$ *satisfies*

$$\sin\theta \leq \frac{\|E\|_2}{gap}.$$

Remark 4.2.8 (Interpretation for PCA) If the top eigenvalue of the covariance matrix is well separated from the second, the top principal component direction is stable under perturbations; if the eigengap is tiny, the direction is fragile even if eigenvalues are well estimated.

4.3 Singular Value Decomposition (SVD)

The singular value decomposition is the spectral theorem for *all* real matrices: rectangular, non-symmetric, and rank-deficient. It exposes the intrinsic geometry of a linear map by identifying orthonormal input/output directions (singular vectors) and nonnegative gains (singular values). In ML, SVD is the backbone of PCA, low-rank approximation, conditioning analysis, and efficient numerical algorithms.

4.3.1 SVD as a Universal Factorization

4.3.1.1 Left/Right Singular Vectors and Singular Values

Eigen-decomposition applies to square operators; SVD applies to any matrix $A \in \mathbb{R}^{m\times n}$. It decomposes A into orthogonal changes of basis on input and output spaces plus a diagonal scaling. Right singular vectors describe input directions; left singular vectors describe output directions. Singular values quantify how much A stretches each right-singular direction, and they are invariant under orthogonal basis changes. This is the correct spectral language for data matrices, feature maps, and Jacobians.

Theorem 4.3.1 (Singular Value Decomposition) *For any* $A \in \mathbb{R}^{m\times n}$*, there exist orthogonal matrices* $U \in \mathbb{R}^{m\times m}$*,* $V \in \mathbb{R}^{n\times n}$ *and a diagonal matrix*

$$\Sigma \in \mathbb{R}^{m\times n}, \qquad \Sigma = diag(\sigma_1, \dots, \sigma_r, 0, \dots),$$

with singular values $\sigma_1 \geq \sigma_2 \geq \cdots \geq \sigma_r > 0$ *and* $r = \mathrm{rank}(A)$*, such that*

$$A = U\Sigma V^\top.$$

The columns u_i *of* U *are* left singular vectors, *the columns* v_i *of* V *are* right singular vectors.

Remark 4.3.1 (Outer-Product Form) Writing only the nonzero part,

$$A = \sum_{i=1}^{r} \sigma_i \, u_i v_i^\top,$$

a sum of rank-1 outer products. This expansion makes low-rank truncation transparent.

4.3.1.2 Relationship to $A^\top A$ and $AA^\top$

SVD reduces to symmetric eigenproblems: $A^\top A$ and $AA^\top$ are symmetric PSD, and their eigenvectors are the right/left singular vectors. Their eigenvalues are squared singular values. This explains why singular values measure energy amplification: $\|Ax\|_2^2 = x^\top (A^\top A)x$ and so $A^\top A$ governs how input energy maps to output energy.

Proposition 4.3.1 (Singular Values via Symmetric PSD Matrices) *Let* $A = U\Sigma V^\top$ *be an SVD. Then*

$$A^\top A = V\Sigma^\top \Sigma V^\top, \qquad AA^\top = U\Sigma\Sigma^\top U^\top,$$

and the nonzero eigenvalues of $A^\top A$ *and* $AA^\top$ *are* $\{\sigma_i^2\}_{i=1}^r$. *In particular,* v_i *is an eigenvector of* $A^\top A$ *with eigenvalue* σ_i^2, *and* u_i *is an eigenvector of* $AA^\top$ *with eigenvalue* σ_i^2.

Remark 4.3.2 (Spectral Norm and Frobenius Norm) The SVD yields

$$\|A\|_2 = \sigma_1, \qquad \|A\|_F^2 = \sum_{i=1}^{r} \sigma_i^2.$$

Thus, the top singular value controls worst-case amplification, while the sum of squared singular values is total energy.

4.3.2 *Geometric Action of a Matrix*

4.3.2.1 Rotation–Scaling–Rotation Interpretation

SVD provides a complete geometric description: $V^\top$ rotates/reflections the input coordinates, Σ scales along orthogonal axes, and U rotates/reflections the output. Hence A maps the unit sphere in $\mathbb{R}^n$ to an ellipsoid in $\mathbb{R}^m$ with semi-axis lengths

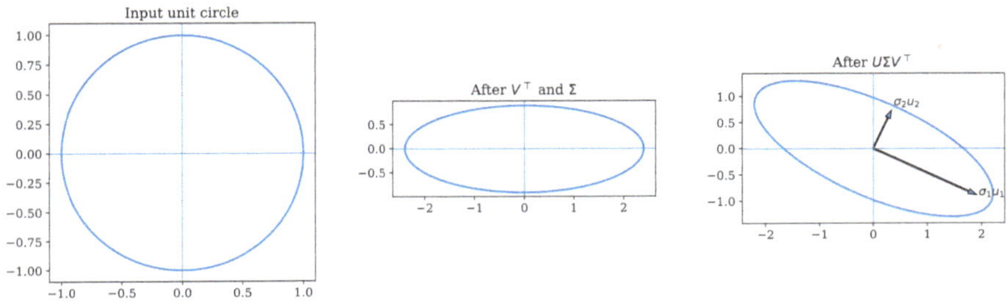

Fig. 4.2 Geometric action of the singular value decomposition. An input sphere is first rotated by $V^\top$, then stretched along orthogonal axes by Σ, and finally rotated by U to produce the output ellipsoid

σ_i. This explains conditioning, anisotropy, and why singular vectors are the principal axes of action.

Proposition 4.3.2 (Ellipsoid Image of the Unit Sphere) *Let $A \in \mathbb{R}^{m\times n}$ have singular values $\sigma_1 \geq \cdots \geq \sigma_r > 0$. Then*

$$\{Ax : \|x\|_2 = 1\}$$

is an ellipsoid (possibly embedded in a lower-dimensional subspace) whose principal semi-axes are σ_i along directions u_i (cf. Fig. 4.2). In particular, $\max_{\|x\|_2=1} \|Ax\|_2 = \sigma_1$ *and* $\min_{\|x\|_2=1,\ x\perp\mathcal{N}(A)} \|Ax\|_2 = \sigma_r$.

4.3.2.2 Condition Number and Distortion Geometry

The condition number compares the maximum and minimum stretching of A on directions not annihilated by A. When $\kappa(A) = \sigma_1/\sigma_r$ is large, A is highly anisotropic: it stretches some directions far more than others. This causes instability in inversion and least squares: small noise in output can correspond to large changes in input along weak directions. In ML, this is the geometric source of multicollinearity issues and slow convergence of gradient methods on quadratic objectives.

Definition 4.3.1 (Spectral Condition Number) If $A \in \mathbb{R}^{m\times n}$ has full column rank ($r = n$), its (spectral) condition number is

$$\kappa_2(A) := \frac{\sigma_{\max}(A)}{\sigma_{\min}(A)} = \frac{\sigma_1}{\sigma_n}.$$

More generally, for rank-r matrices one often uses $\kappa_2(A) := \sigma_1/\sigma_r$ on the nonzero singular spectrum.

Remark 4.3.3 (Distortion of Relative Errors) If A is square and invertible, then $\|A^{-1}\|_2 = 1/\sigma_{\min}(A)$ and $\kappa_2(A) = \|A\|_2\|A^{-1}\|_2$. Large $\kappa_2(A)$ means inversion is sensitive: perturbations are amplified most strongly along the smallest singular directions.

4.3.3 Best Low-Rank Approximation

4.3.3.1 Eckart–Young Theorem and Reconstruction Error

Low-rank approximation is central in compression, denoising, recommender systems, and representation learning. The SVD provides the optimal rank-k approximation by truncating to the top k singular values. The Eckart–Young theorem states that this truncation minimizes error in both spectral and Frobenius norms. This gives a rigorous justification for PCA and for keeping only dominant modes.

Definition 4.3.2 (Truncated SVD) Let $A = U\Sigma V^\top$ with singular values $\sigma_1 \geq \cdots \geq \sigma_r > 0$. For $k < r$, define the rank-k truncation

$$A_k := \sum_{i=1}^{k} \sigma_i u_i v_i^\top = U_k \Sigma_k V_k^\top,$$

where $U_k = [u_1, \ldots, u_k]$, $V_k = [v_1, \ldots, v_k]$, and $\Sigma_k = \mathrm{diag}(\sigma_1, \ldots, \sigma_k)$.

Theorem 4.3.2 (Eckart–Young–Mirsky) *Let $A \in \mathbb{R}^{m\times n}$ with SVD as above and let $k < r = \mathrm{rank}(A)$. Then A_k is a best rank-k approximation:*

$$A_k \in \arg\min_{\mathrm{rank}(B)\leq k} \|A - B\|_2 \quad \textit{and} \quad A_k \in \arg\min_{\mathrm{rank}(B)\leq k} \|A - B\|_F.$$

Moreover,

$$\|A - A_k\|_2 = \sigma_{k+1}, \qquad \|A - A_k\|_F^2 = \sum_{i=k+1}^{r} \sigma_i^2.$$

4.3.3.2 Energy Capture and Truncation Heuristics

The squared singular values quantify energy. For data matrices, they correspond to variance along principal components. Thus, one natural heuristic chooses k so that the energy capture ratio exceeds a threshold, e.g., 95%. Another heuristic looks for an "elbow" in the scree plot (rapid decay then flattening), indicating diminishing returns. These heuristics are not arbitrary; they are grounded in the exact error formulas from Eckart–Young.

Definition 4.3.3 (Energy Capture Ratio) For a matrix A with singular values σ_i, define the fraction of Frobenius energy captured by rank-k truncation as

$$\mathrm{EC}(k) := \frac{\sum_{i=1}^{k} \sigma_i^2}{\sum_{i=1}^{r} \sigma_i^2}.$$

Remark 4.3.4 (Interpretation) $\mathrm{EC}(k)$ measures how much of $\|A\|_F^2$ is retained by A_k. If A is a centered data matrix, this corresponds to explained variance in PCA.

4.3.4 *Perturbation and Sensitivity (High-Level)*

4.3.4.1 Weyl-Type Eigenvalue Stability Intuition

SVD is stable in a spectral sense: small operator-norm perturbations cannot change singular values too much. This mirrors Weyl's inequality for eigenvalues of symmetric matrices. In ML, this matters because data matrices and covariances are noisy estimates; we want to know how singular values and explained variance change under sampling noise or numerical error. The key message: singular values are Lipschitz with respect to $\|\cdot\|_2$ perturbations.

Theorem 4.3.3 (Weyl-Type Bound for Singular Values) *Let $A, E \in \mathbb{R}^{m\times n}$ and let $\sigma_i(\cdot)$ denote singular values in nonincreasing order. Then for each i,*

$$|\sigma_i(A+E) - \sigma_i(A)| \le \|E\|_2.$$

4.3.4.2 Subspace Perturbations and Principal Angles Intuition

While singular values can be stable, singular vectors can rotate substantially if singular values are clustered. Stability of singular subspaces depends on spectral gaps between singular values, analogous to Davis–Kahan for eigenspaces. Principal angles quantify how far two subspaces are from each other; perturbation theorems bound these angles by $\|E\|_2/\text{gap}$. In PCA, this determines whether leading components are reliable.

Remark 4.3.5 (Gap-Controlled Stability) If $\sigma_k(A) \gg \sigma_{k+1}(A)$, then the top-$k$ right singular subspace $\text{span}\{v_1, \dots, v_k\}$ is stable. If $\sigma_k \approx \sigma_{k+1}$, the subspace is fragile: small perturbations can rotate the basis significantly.

4.4 Principal Component Analysis (PCA)

PCA is the canonical method for linear dimensionality reduction (Jolliffe 2002). It identifies an orthonormal basis of directions capturing maximal variance and provides optimal low-rank approximation in squared error. PCA can be derived either from covariance eigen-decomposition or from the SVD of the centered data matrix; these viewpoints are equivalent and complementary.

4.4.1 PCA via Covariance

4.4.1.1 Centering, Covariance Matrices, and Variance Maximization

PCA is defined on centered data. Centering removes translation so that variance reflects spread around the mean. The first principal component is the unit direction maximizing projected variance; subsequent components maximize remaining variance under orthogonality (cf. Fig. 4.3). This is a direct application of Rayleigh quotient and the spectral theorem for the covariance matrix.

Definition 4.4.1 (Centered Data and Sample Covariance) Let $X \in \mathbb{R}^{n\times d}$ have rows $x_i^\top$ and mean $\bar{x}$. Let $X_c = X - \mathbf{1}\bar{x}^\top$ be centered. The (uncorrected) sample covariance matrix is

$$\widehat{\Sigma} := \frac{1}{n} X_c^\top X_c \in \mathbb{R}^{d\times d}.$$

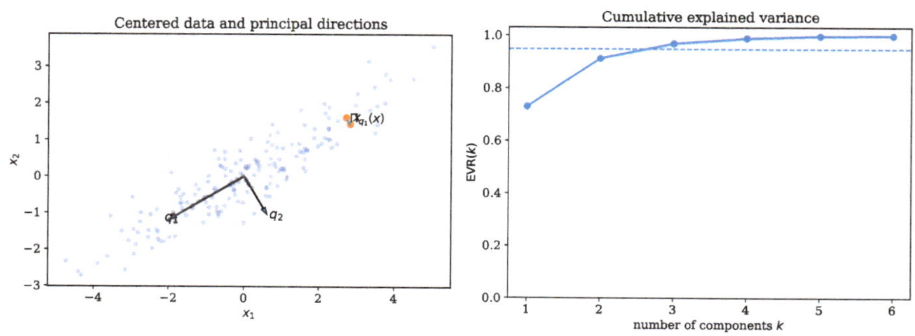

Fig. 4.3 PCA geometry. Left: centered data cloud with first two principal directions and orthogonal projection onto the first component. Right: cumulative explained-variance ratio showing how many components are needed to capture most of the spectral energy

Theorem 4.4.1 (PCA as Variance Maximization) *Let $\widehat{\Sigma}$ be the sample covariance with eigenvalues $\lambda_1 \geq \cdots \geq \lambda_d$ and orthonormal eigenvectors q_i. Then:*

$$q_1 \in \arg \max_{\|u\|_2=1} u^\top \widehat{\Sigma} u,$$

and for $k \geq 2$,

$$q_k \in \arg \max_{\substack{\|u\|_2=1 \\ u \perp q_1,\ldots,q_{k-1}}} u^\top \widehat{\Sigma} u.$$

Moreover, the maximum value at step k equals λ_k.

4.4.1.2 Whitening and Standardized Coordinates

Whitening uses the eigen-decomposition of covariance to map data into coordinates with identity covariance. This makes distances and dot products isotropic and is often a helpful preprocessing step for algorithms sensitive to scaling. In PCA coordinates, whitening is simply scaling by inverse square roots of eigenvalues.

Definition 4.4.2 (PCA Coordinates (Scores) and Whitening) Let $\widehat{\Sigma} = Q\Lambda Q^\top$ with $Q = [q_1, \ldots, q_d]$ orthogonal and $\Lambda = \mathrm{diag}(\lambda_1, \ldots, \lambda_d)$. Define PCA scores (coordinates) for centered data by

$$Z := X_c Q \in \mathbb{R}^{n\times d}.$$

A whitened version is

$$Z_w := X_c Q \Lambda^{-1/2},$$

assuming all $\lambda_i > 0$ (or restricting to positive-eigenvalue subspace). Then $\frac{1}{n} Z_w^\top Z_w = I$.

4.4.2 *PCA via SVD*

4.4.2.1 Data-Matrix SVD and Equivalence to Covariance PCA

Computationally, PCA is often performed via SVD of the centered data matrix, especially when n and d are large. SVD directly yields principal directions and singular values, and avoids forming the covariance explicitly (which may be expensive or ill-conditioned). The equivalence is exact: right singular vectors of X_c are eigenvectors of $\widehat{\Sigma}$, and singular values relate to eigenvalues by σ_i^2/n.

Proposition 4.4.1 (PCA–SVD Equivalence) *Let $X_c \in \mathbb{R}^{n\times d}$ have SVD $X_c = U\Sigma V^\top$ with singular values $\sigma_1 \geq \cdots \geq \sigma_r > 0$. Then*

$$\widehat{\Sigma} = \frac{1}{n}X_c^\top X_c = V\left(\frac{1}{n}\Sigma^\top\Sigma\right)V^\top,$$

so the eigenvalues of $\widehat{\Sigma}$ are $\lambda_i = \sigma_i^2/n$ and the eigenvectors are the right singular vectors v_i. Moreover, the PCA scores satisfy $Z = X_cV = U\Sigma$.

4.4.2.2 Explained Variance, Scree Plots, Choosing *k*

Explained variance is quantified by eigenvalues of covariance (or squared singular values of centered data). A scree plot visualizes eigenvalues in decreasing order; an "elbow" suggests a good truncation dimension. Choosing k by a fixed explained-variance threshold is directly justified by energy capture ratios.

Definition 4.4.3 (Explained Variance Ratio) Let $\lambda_1 \geq \cdots \geq \lambda_d \geq 0$ be eigenvalues of $\widehat{\Sigma}$. The explained variance ratio of the first k components is

$$\mathrm{EVR}(k) := \frac{\sum_{i=1}^k \lambda_i}{\sum_{i=1}^d \lambda_i}.$$

Equivalently, in SVD terms,

$$\mathrm{EVR}(k) = \frac{\sum_{i=1}^k \sigma_i^2}{\sum_{i=1}^r \sigma_i^2}.$$

Remark 4.4.1 (Choosing k) Common heuristics choose the smallest k such that $\mathrm{EVR}(k) \geq \tau$ for $\tau \in [0.9, 0.99]$, or choose k at the scree "elbow." These heuristics balance compression and reconstruction fidelity.

4.4.3 Interpretation and Diagnostics

4.4.3.1 Loadings, Scores, and Feature Influence

PCA outputs both directions (loadings) and coordinates (scores). Loadings quantify how original features contribute to each component, enabling interpretability; scores describe each sample in the low-dimensional coordinate system. Large loading magnitudes indicate features that drive variance, but interpretation requires attention to scaling and centering.

Definition 4.4.4 (Loadings and Scores) Let $Q_k = [q_1, \ldots, q_k] \in \mathbb{R}^{d \times k}$ be the top-k eigenvectors of $\widehat{\Sigma}$. The *scores* are

$$Z_k := X_c Q_k \in \mathbb{R}^{n \times k},$$

and the columns of Q_k are *loadings* (principal directions) in feature space.

Remark 4.4.2 (Reconstruction From Scores) The rank-k reconstruction of centered data is $\widehat{X}_c^{(k)} := Z_k Q_k^\top = X_c Q_k Q_k^\top$. Hence the reconstruction error is the squared Frobenius norm of the residual $X_c(I - Q_k Q_k^\top)$, controlled by discarded eigenvalues/singular values.

4.4.3.2 Outliers, Leverage, and Robust PCA Intuition

Classical PCA is sensitive to outliers: a single extreme point can dominate covariance and rotate principal directions. Leverage scores measure how much a point influences the fitted subspace. Robust PCA aims to decompose data into low-rank structure plus sparse outliers, replacing squared-error objectives by models tolerant to corruption. We treat robust PCA as a preview; the key geometric message is that outliers can distort variance-based directions.

Remark 4.4.3 (Leverage (Preview Definition)) For centered data with top-k left singular vectors $U_k \in \mathbb{R}^{n \times k}$, the i-th leverage score is

$$\ell_i := \|(U_k)_{i:}\|_2^2,$$

quantifying how much sample i participates in the top-k subspace. Large leverage indicates potential outlier/influential points.

4.4.4 *Beyond Classical PCA*

4.4.4.1 Kernel PCA: Nonlinear Feature Spaces

Kernel PCA performs PCA in an implicit feature space using a kernel Gram matrix. It captures nonlinear structure while retaining linear-algebraic solvability. The centered kernel matrix plays the role of a covariance/Gram matrix, and eigenvectors of this matrix yield principal components in feature space.

Remark 4.4.4 (Kernel PCA Sketch) Given kernel matrix K, one centers it as $K_c = HKH$ and performs eigen-decomposition of K_c. Projection of a new point uses kernel evaluations against training points. Full details appear in the kernels chapter.

4.4.4.2 Probabilistic PCA and Latent-Variable Viewpoint

Probabilistic PCA interprets PCA as maximum likelihood in a latent Gaussian factor model. This connects PCA to matrix factorization, EM algorithms, and Bayesian modeling. The spectral solution emerges as the MLE under isotropic noise, revealing PCA as both geometric optimization and probabilistic inference.

Remark 4.4.5 (PPCA Sketch) A basic PPCA model is $x = Wz + \mu + \varepsilon$ with $z \sim \mathcal{N}(0, I_k)$ and $\varepsilon \sim \mathcal{N}(0, \sigma^2 I)$. The MLE for W aligns with the top-k eigenvectors of the sample covariance. We will revisit this in the latent-variable and estimation chapters.

4.5 Subspace Methods and Randomized Spectral Algorithms

Large-scale ML often makes classical decompositions (full eigen-decomposition or full SVD) infeasible. Subspace methods aim to recover only the *dominant subspace* (top-k eigenvectors/singular vectors) using matrix–vector products and a small number of passes over the data. Randomized algorithms further accelerate these tasks by using sketching: random projections that preserve geometry of the relevant subspace. The key theme is that we do not need the entire matrix; we need only its action on a carefully chosen low-dimensional subspace.

4.5.1 Range vs. Null Space Techniques

4.5.1.1 Leverage Scores and Influential Directions

If a matrix has a dominant k-dimensional range, not all rows/columns are equally informative. Leverage scores quantify how much each sample (row) or feature (column) participates in the top-k subspace. They guide importance sampling and explain why uniform sampling can fail in the presence of heavy-tailed or highly nonuniform data. In PCA and regression, leverage scores identify influential observations, connect to outliers, and provide a route to fast sketching.

Definition 4.5.1 (Leverage Scores (Column Space)) Let $X \in \mathbb{R}^{n\times d}$ and let $Q \in \mathbb{R}^{n\times k}$ have orthonormal columns spanning a k-dimensional subspace $\mathcal{U} \subseteq \mathbb{R}^n$ (often the top-k left singular subspace of X). The *(row) leverage scores* are

$$\ell_i := \|e_i^\top Q\|_2^2 = \|Q_{i:}\|_2^2, \qquad i \in [n],$$

where e_i is the i-th standard basis vector.

Remark 4.5.1 (Basic Properties) $0 \le \ell_i \le 1$ and $\sum_{i=1}^n \ell_i = k$ because $\sum_i \ell_i = \mathrm{tr}(QQ^\top) = \mathrm{tr}(Q^\top Q) = k$. Large ℓ_i indicates that row i is strongly represented in the subspace (potentially influential/outlying).

4.5.1.2 Subspace Iteration and Randomized Range Finding

Power iteration extracts a single top eigenvector. Subspace iteration generalizes it to the top-k invariant subspace: apply the matrix repeatedly to a block of vectors and orthonormalize. Randomized range finding replaces the carefully chosen initial block by random test vectors, relying on concentration to ensure the dominant range is captured with high probability. Orthogonalization is essential: it prevents collapse onto the top direction and maintains numerical stability.

Algorithm 2 SUBSPACEITERATION (dominant eigenspace, symmetric case)

Require: Symmetric $A \in \mathbb{R}^{d\times d}$, target rank k, iterations $q \ge 0$, initial $Q^{(0)} \in \mathbb{R}^{d\times k}$ full rank
1: Orthonormalize columns: $Q^{(0)} \leftarrow \mathrm{orth}(Q^{(0)})$
2: **for** $t = 0, 1, \ldots, q-1$ **do**
3: $\quad Y \leftarrow AQ^{(t)}$
4: $\quad Q^{(t+1)} \leftarrow \mathrm{orth}(Y)$ {e.g., QR factorization}
5: **end for**
6: **return** $Q^{(q)}$

Remark 4.5.2 (Randomized Range Finding: The Simplest Idea) Choose $Q^{(0)} = \Omega$ with i.i.d. $\mathcal{N}(0, 1)$ entries (or Rademacher ± 1) and set $Y = A\Omega$. Then $\mathrm{range}(Y)$ approximates the dominant range of A. Power iterations $Y = (AA^\top)^q A\Omega$ amplify spectral gaps and improve accuracy for slowly decaying spectra.

4.5.2 Lanczos/Arnoldi Intuition (Conceptual)

4.5.2.1 Krylov Subspaces as Compressed Representations

Krylov methods build a low-dimensional subspace spanned by $\{x, Ax, A^2x, \ldots\}$. This subspace captures how the matrix acts on an initial vector and often contains good approximations to extremal eigenvectors. Lanczos (symmetric) and Arnoldi (nonsymmetric) compress the operator into a small tridiagonal (Lanczos) or Hessenberg (Arnoldi) matrix whose eigenpairs approximate those of the original. Thus, Krylov subspaces are *compressed operator representations*.

Definition 4.5.2 (Krylov Subspace) For $A \in \mathbb{R}^{d\times d}$ and nonzero $x \in \mathbb{R}^d$, the t-th Krylov subspace is

$$\mathcal{K}_t(A, x) := \mathrm{span}\{x, Ax, A^2x, \ldots, A^{t-1}x\}.$$

Remark 4.5.3 (Why Krylov Works) If A has a dominant eigenvector component in x, then repeated multiplication by A amplifies that component. Unlike power iteration, Krylov methods retain multiple powers, enabling faster convergence and approximation of several eigenpairs simultaneously.

4.5.2.2 When Iterative Eigen-Solvers Win over Decompositions

Full decompositions cost $O(d^3)$ for dense matrices. Iterative methods can be dramatically cheaper when: (i) the matrix is sparse (matrix–vector products are $O(\mathrm{nnz}(A))$), (ii) only the top-k eigenpairs are needed with $k \ll d$, (iii) there is a spectral gap, so convergence is fast. The practical tradeoff is between passes over data (or matvecs) and accuracy, plus sensitivity to numerical issues (loss of orthogonality).

Remark 4.5.4 (Rule of Thumb) Use full SVD/eigendecomposition for moderate d. Use Lanczos/Arnoldi or randomized methods when d is large and you only need a small subspace and can afford only a few passes.

4.5.3 *Randomized SVD (Modern Large-Scale Tool)*

4.5.3.1 Sketching + Orthogonalization Pipeline

Randomized SVD is the modern workhorse for approximate low-rank decompositions (Halko et al. 2011). It uses a random sketch Ω to probe the range of A, forms $Y = A\Omega$, orthonormalizes Y to get Q, then compresses A to a small matrix $B = Q^\top A$ whose SVD is cheap. The only expensive operations are matrix–matrix multiplies involving A, typically implemented as a few passes over data. Orthonormalization ensures numerical stability and good approximation.

Remark 4.5.5 (Why Oversampling Helps) Choosing $k+p$ test vectors (with $p \approx$ 5–20) reduces failure probability and improves capture of the dominant subspace, especially when singular values decay slowly.

4.5.3.2 Accuracy vs. Passes over Data

In large-scale settings, the main cost is reading the data (passes over A or matrix–vector products). Randomized SVD trades additional passes (power steps) for

Algorithm 3 RANDOMIZEDSVD (range sketching)

Require: $A \in \mathbb{R}^{m \times n}$, target rank k, oversampling $p \geq 5$, power steps $q \geq 0$
Draw $\Omega \in \mathbb{R}^{n \times (k+p)}$ with i.i.d. $\mathcal{N}(0, 1)$ entries
$Y \leftarrow A\Omega$
for $t = 1, \ldots, q$ **do**
  $Y \leftarrow A(A^\top Y)$ {power iteration to improve gap}
end for
$Q \leftarrow \text{orth}(Y)$ $\{Q \in \mathbb{R}^{m \times (k+p)}\}$
$B \leftarrow Q^\top A$ $\{B \in \mathbb{R}^{(k+p) \times n}\}$
Compute SVD: $B = \widetilde{U} \Sigma V^\top$
$U \leftarrow Q\widetilde{U}$
return Approximate SVD $A \approx U \Sigma V^\top$

improved accuracy by amplifying spectral gaps: $(AA^\top)^q A$ raises singular values to $(2q + 1)$ powers in effect, separating dominant from tail. Thus, one chooses q small (often 0–2) to balance accuracy and I/O cost.

Remark 4.5.6 (Heuristic) If the spectrum decays fast, $q = 0$ may be sufficient. If the spectrum is flat or clustered, use $q = 1$ or 2 to sharpen separation.

4.6 Spectral Methods in ML Applications

Spectral methods leverage eigenvectors/singular vectors as geometry-aware features. They appear in clustering on graphs, kernel learning, and denoising/compression. This section gives a preview-level but mathematically consistent overview, emphasizing how spectra encode structure.

4.6.1 Spectral Clustering and Graph Laplacians

4.6.1.1 Laplacian Eigenvectors and Partitioning Intuition

Graphs encode pairwise relations. The Laplacian operator measures smoothness of functions on nodes: it penalizes differences across edges. Eigenvectors of the Laplacian provide low-frequency (smooth) modes on the graph. Spectral clustering uses a few smallest-eigenvalue eigenvectors to embed nodes into $\mathbb{R}^k$ where clusters become linearly separable (von Luxburg 2007). The key idea: connectivity structure becomes Euclidean geometry via eigenvectors (cf. Fig. 4.4).

Definition 4.6.1 (Graph Laplacian) For an undirected weighted graph with weight matrix $W \in \mathbb{R}^{n \times n}$ (symmetric, nonnegative) and degree matrix $D = \text{diag}(d_1, \ldots, d_n)$ with $d_i = \sum_j W_{ij}$, the (unnormalized) Laplacian is

$$L := D - W.$$

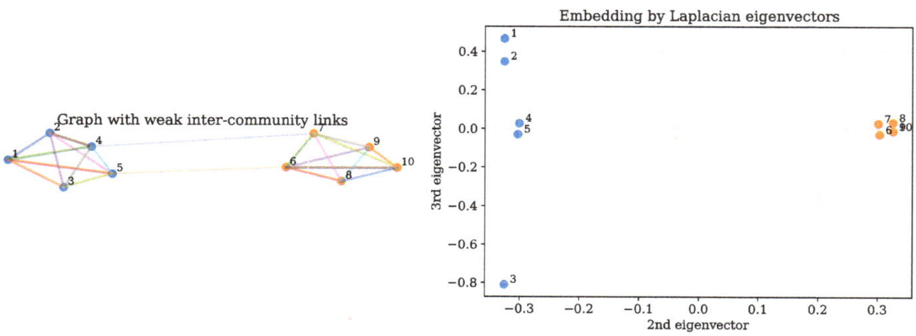

Fig. 4.4 Spectral clustering intuition. Left: a graph with two weakly connected communities. Right: embedding nodes using low-frequency Laplacian eigenvectors separates the communities in Euclidean space, where clustering becomes simple

The symmetric normalized Laplacian is

$$L_{\text{sym}} := I - D^{-1/2} W D^{-1/2},$$

assuming all $d_i > 0$.

Remark 4.6.1 (PSD and Constant Eigenvector) $L \succeq 0$ and $L\mathbf{1} = 0$. The multiplicity of eigenvalue 0 equals the number of connected components.

4.6.1.2 Normalized Cuts and Geometry of Embeddings

Balanced partitioning criteria (like normalized cut) can be relaxed to Rayleigh quotient problems involving L_{sym}, making eigenvectors the natural solution of a relaxed optimization problem. The embedding by eigenvectors transforms the combinatorial clustering problem into geometric clustering (k-means in spectral space).

Remark 4.6.2 (Optimization Preview) Many spectral clustering derivations reduce to minimizing $x^\top L x$ subject to normalization constraints, a Rayleigh quotient problem whose minimizers are eigenvectors corresponding to smallest eigenvalues (excluding the trivial constant mode).

4.6.2 Kernel Methods Through Gram Matrices

4.6.2.1 PSD Kernels, Feature Maps, and Eigenspectra

Kernel methods replace explicit features with a PSD Gram matrix K. Eigendecomposition of K reveals dominant similarity modes in feature space, analogous to PCA. Kernel PCA is PCA performed on K after centering. In learning, the kernel spectrum controls capacity and effective dimension: a rapidly decaying spectrum implies a low effective dimension.

Remark 4.6.3 (Kernel Spectrum as Complexity) If K has eigenvalues $\mu_1 \geq \cdots \geq \mu_n \geq 0$, then quantities like $\sum_i \mu_i/(\mu_i + \lambda)$ measure effective degrees of freedom in kernel ridge regression.

4.6.2.2 Nyström Approximation

Nyström is a randomized low-rank approximation technique for kernel matrices: sample a subset of columns/rows, form a small submatrix, and extend to approximate the full K. This reduces $O(n^2)$ storage and computation. Geometrically, Nyström approximates the dominant eigenspace of the kernel operator using sampled landmark points.

Remark 4.6.4 (Nyström Sketch) If C is the matrix of sampled columns and W the sampled intersection submatrix, a basic approximation is $K \approx CW^{\dagger}C^{\top}$. The quality depends on leverage-score-like quantities for the kernel eigenspace.

4.6.3 Denoising and Compression

4.6.3.1 Low-Rank Denoising and Matrix Thresholding Intuition

If observed data is low-rank signal plus noise, SVD separates dominant singular values (signal) from small ones (noise). Truncation yields optimal low-rank approximation in Frobenius norm, providing a principled denoiser. Soft-thresholding of singular values (nuclear-norm proximal) is a convex relaxation that shrinks rather than hard-truncates, often improving robustness to noise.

Remark 4.6.5 (Two Denoising Rules) Hard thresholding: keep singular values above a cutoff (rank truncation). Soft thresholding: replace σ_i by $\max(\sigma_i - \tau, 0)$ (prox of nuclear norm), producing smoother shrinkage.

4.6.3.2 Anomaly Detection via Reconstruction Error

A low-rank model defines a subspace of "normal" behavior. Points that are poorly reconstructed by the subspace have large residual norms and can be flagged as anomalies. This is a geometric criterion: anomalies lie far from the learned subspace. In practice, one uses PCA reconstruction error $\|x - \Pi_{\mathcal{U}} x\|_2$ or leverage scores as complementary diagnostics.

Remark 4.6.6 (PCA Anomaly Score) With top-k PCA loading matrix Q_k, the reconstruction of centered x is $Q_k Q_k^\top x$ and the anomaly score is $\|x - Q_k Q_k^\top x\|_2$ (or its square).

4.7 Exercises

Unless stated otherwise, all inner products and norms are Euclidean, eigenvalues are ordered in nonincreasing order, and singular values are ordered $\sigma_1 \geq \sigma_2 \geq \cdots \geq 0$.

Exercise 4.7.1 (Rayleigh Quotient and Extremal Eigenvalues) Let

$$A = \begin{bmatrix} 4 & 1 & 0 \\ 1 & 2 & 1 \\ 0 & 1 & 3 \end{bmatrix} \in \mathbb{R}^{3\times 3}.$$

(i) Compute the eigenvalues of A exactly (or to 4 decimals) and verify that $A \succ 0$.
(ii) For the unit vectors $u_1 = \frac{1}{\sqrt{2}}(1, 1, 0)^\top$, $u_2 = \frac{1}{\sqrt{3}}(1, 1, 1)^\top$, and $u_3 = (0, 0, 1)^\top$, compute the Rayleigh quotients $\mathcal{R}_A(u_i) = u_i^\top A u_i$ and compare them with $\lambda_{\max}(A)$ and $\lambda_{\min}(A)$.
(iii) Using Lagrange multipliers, solve $\max_{\|x\|_2=1} x^\top Ax$ and show that any maximizer is an eigenvector associated with $\lambda_{\max}(A)$.

Exercise 4.7.2 (Spectral Theorem and Quadratic-Form Geometry) Let

$$A = \begin{bmatrix} 5 & 2 \\ 2 & 2 \end{bmatrix}.$$

(i) Compute an orthogonal diagonalization $A = Q\Lambda Q^\top$.
(ii) Describe the level set $\mathcal{E} = \{x \in \mathbb{R}^2 : x^\top Ax \leq 1\}$ as an ellipse by giving its principal axes directions and semi-axis lengths.
(iii) Compute $\max_{\|x\|_2=1} x^\top Ax$ and $\min_{\|x\|_2=1} x^\top Ax$ and verify they equal the largest and smallest eigenvalues.

Exercise 4.7.3 (Gershgorin Localization and Quick PSD Check) Consider

$$B = \begin{bmatrix} 6 & -1 & 0 & 1 \\ -1 & 5 & -1 & 0 \\ 0 & -1 & 4 & -1 \\ 1 & 0 & -1 & 3 \end{bmatrix}.$$

(i) Compute all Gershgorin disks and use them to bound the spectrum of B on the real line.

(ii) Using your bounds, argue whether B is likely PD/PSD; then verify by computing the eigenvalues numerically (to 3 decimals) or via leading principal minors.
(iii) If B is PD, compute a Cholesky factorization $B = LL^\top$ (to 3 decimals).

Exercise 4.7.4 (Power Iteration and Eigengap-Controlled Convergence) Let

$$A = \begin{bmatrix} 3 & 1 & 0 \\ 1 & 2 & 0 \\ 0 & 0 & 0.5 \end{bmatrix}, \qquad x^{(0)} = \begin{bmatrix} 1 \\ 1 \\ 1 \end{bmatrix}.$$

(i) Compute the eigenvalues of A and identify $\lambda_1 > \lambda_2 \geq \lambda_3$.
(ii) Perform three steps of power iteration $x^{(t+1)} = Ax^{(t)}/\|Ax^{(t)}\|_2$ and report $x^{(1)}, x^{(2)}, x^{(3)}$.
(iii) Let q_1 be the top eigenvector normalized. Compute $\cos\theta_t = |\langle x^{(t)}, q_1\rangle|$ for $t = 0, 1, 2, 3$ and compare the improvement with the factor $|\lambda_2/\lambda_1|$.

Exercise 4.7.5 (SVD from $A^\top A$ and $AA^\top$) Let

$$A = \begin{bmatrix} 2 & 0 \\ 0 & 1 \\ 2 & 2 \end{bmatrix} \in \mathbb{R}^{3\times 2}.$$

(i) Compute $A^\top A$ and its eigenvalues/eigenvectors; from them obtain the singular values $\sigma_1 \geq \sigma_2$ and right singular vectors v_1, v_2.
(ii) Compute left singular vectors $u_i = \frac{1}{\sigma_i} A v_i$ and verify they are orthonormal.
(iii) Assemble U, Σ, V and verify $A = U\Sigma V^\top$ exactly.

Exercise 4.7.6 (Rotation–Scaling–Rotation Geometry and Ellipsoid Image) Let $A \in \mathbb{R}^{2\times 2}$ be

$$A = \begin{bmatrix} 3 & 1 \\ 0 & 2 \end{bmatrix}.$$

(i) Compute the singular values of A by diagonalizing $A^\top A$.
(ii) Determine the image of the unit circle $\{x : \|x\|_2 = 1\}$ under A as an ellipse and report its principal semi-axis lengths (the singular values) and directions (left singular vectors).
(iii) Compute $\|A\|_2$ and $\|A\|_F$ and verify $\|A\|_2 = \sigma_1$ and $\|A\|_F^2 = \sigma_1^2 + \sigma_2^2$.

Exercise 4.7.7 (Condition Number and Inversion Sensitivity) Let

$$A = \begin{bmatrix} 1 & 1 \\ 1 & 1+\varepsilon \end{bmatrix}, \qquad \varepsilon = 10^{-3}.$$

(i) Compute the singular values of A (to 4 decimals) and hence $\kappa_2(A)$.

(ii) Solve $Ax = b$ for $b = (2, 2+\varepsilon)^\top$ and for $b' = (2, 2)^\top$ and compute $\|x - x'\|_2$.
(iii) Interpret your computation as a numerical demonstration that small changes in b can cause large changes in x when $\kappa_2(A)$ is large.

Exercise 4.7.8 (Eckart–Young Truncation Error (Explicit)) Let a matrix $M \in \mathbb{R}^{3\times 3}$ have SVD singular values $\sigma_1 = 5$, $\sigma_2 = 2$, $\sigma_3 = 1$. Assume its singular vectors are orthonormal and define

$$M = \sum_{i=1}^{3} \sigma_i u_i v_i^\top.$$

(i) Write the rank-1 and rank-2 truncations M_1 and M_2.
(ii) Compute $\|M - M_1\|_2$, $\|M - M_1\|_F$, $\|M - M_2\|_2$, and $\|M - M_2\|_F$ using Theorem 4.3.2.
(iii) Compute the energy capture ratios EC(1) and EC(2) (Definition 4.3.3).

Exercise 4.7.9 (PCA by Covariance: Exact Computation) Let $n = 5$ points in $\mathbb{R}^2$ be

$$x_1 = \begin{bmatrix} 0 \\ 1 \end{bmatrix}, \; x_2 = \begin{bmatrix} 1 \\ 2 \end{bmatrix}, \; x_3 = \begin{bmatrix} 2 \\ 1 \end{bmatrix}, \; x_4 = \begin{bmatrix} 3 \\ 4 \end{bmatrix}, \; x_5 = \begin{bmatrix} 4 \\ 3 \end{bmatrix}.$$

(i) Compute the empirical mean $\bar{x}$ and centered data matrix $X_c \in \mathbb{R}^{5\times 2}$.
(ii) Compute the sample covariance $\widehat{\Sigma} = \frac{1}{n} X_c^\top X_c$.
(iii) Compute eigenvalues/eigenvectors of $\widehat{\Sigma}$ and identify the first principal component direction q_1.
(iv) Compute the PCA scores $z_i = q_1^\top (x_i - \bar{x})$ and the rank-1 reconstructions $\widehat{x_i} = \bar{x} + z_i q_1$.
(v) Compute the total reconstruction SSE $\sum_{i=1}^{n} \|x_i - \widehat{x_i}\|_2^2$ and verify it equals the sum of discarded eigenvalues times n.

Exercise 4.7.10 (PCA via SVD and Equivalence) Using the same dataset as Exercise 4.7.9, form the centered data matrix $X_c \in \mathbb{R}^{5\times 2}$.

(i) Compute the SVD $X_c = U\Sigma V^\top$ (you may compute $X_c^\top X_c$ to find V and Σ).
(ii) Verify that the right singular vectors V match the eigenvectors of $\widehat{\Sigma}$ and that $\lambda_i = \sigma_i^2/n$.
(iii) Compute the scores matrix $Z = X_c V$ and verify $Z = U\Sigma$.

Exercise 4.7.11 (Explained Variance and Choosing k) Let a centered data matrix X_c have singular values

$$\sigma_1 = 10, \; \sigma_2 = 4, \; \sigma_3 = 2, \; \sigma_4 = 1, \; \sigma_5 = 1.$$

(i) Compute EVR(k) for $k = 1, 2, 3, 4, 5$ (Definition 4.4.3).

(ii) Determine the smallest k such that $\mathrm{EVR}(k) \geq 0.90$ and such that $\mathrm{EVR}(k) \geq 0.95$.
(iii) Compute the Frobenius reconstruction error $\|X_c - (X_c)_k\|_F^2$ for $k = 1, 2, 3$ using Eckart–Young.

Exercise 4.7.12 (Weyl Stability: Eigenvalue Perturbations) Let

$$A = \begin{bmatrix} 4 & 1 & 0 \\ 1 & 3 & 1 \\ 0 & 1 & 2 \end{bmatrix}, \qquad E = \begin{bmatrix} 0.1 & 0 & 0 \\ 0 & -0.05 & 0 \\ 0 & 0 & 0.02 \end{bmatrix}, \qquad \widehat{A} = A + E.$$

(i) Compute $\|E\|_2$ exactly (it is diagonal) and use Weyl's inequality (Theorem 4.2.5) to bound $|\lambda_i(\widehat{A}) - \lambda_i(A)|$.
(ii) Compute eigenvalues of A and $\widehat{A}$ numerically (to 4 decimals) and verify the bound.
(iii) Interpret (one paragraph) how eigenvalue stability depends on operator-norm magnitude of noise.

Exercise 4.7.13 (Subspace Perturbation via Principal-Angle Bound (Numerical)) Let

$$A = \begin{bmatrix} 5 & 0 & 0 \\ 0 & 2 & 0 \\ 0 & 0 & 1 \end{bmatrix}, \qquad E = \begin{bmatrix} 0 & 0.2 & 0 \\ 0.2 & 0 & 0 \\ 0 & 0 & 0 \end{bmatrix}, \qquad \widehat{A} = A + E.$$

(i) Compute $\|E\|_2$ and the eigengap between $\lambda_1(A)$ and the rest: gap $= \min\{|\lambda_1 - \lambda_2|, |\lambda_1 - \lambda_3|\}$.
(ii) Compute the top eigenvector q_1 of A and the top eigenvector $\widehat{q}_1$ of $\widehat{A}$ (exactly or to 4 decimals).
(iii) Compute the angle θ between q_1 and $\widehat{q}_1$ and verify the Davis–Kahan bound $\sin\theta \leq \|E\|_2/\mathrm{gap}$ (Theorem 4.2.6).

Exercise 4.7.14 (Randomized Range Finding with a Fixed Sketch Matrix) Let $A \in \mathbb{R}^{4\times 3}$ be

$$A = \begin{bmatrix} 2 & 0 & 0 \\ 0 & 1 & 0 \\ 0 & 0 & 0.5 \\ 2 & 1 & 0 \end{bmatrix}, \qquad \Omega = \begin{bmatrix} 1 & 0 \\ 0 & 1 \\ 1 & 1 \end{bmatrix} \in \mathbb{R}^{3\times 2}.$$

(i) Compute $Y = A\Omega \in \mathbb{R}^{4\times 2}$ and an orthonormal basis $Q = \mathrm{orth}(Y)$ (by Gram–Schmidt or QR).
(ii) Compute the compressed matrix $B = Q^\top A \in \mathbb{R}^{2\times 3}$ and its SVD $B = \widetilde{U}\Sigma V^\top$.
(iii) Form $U = Q\widetilde{U}$ and report the resulting rank-2 approximation $A_2 = U\Sigma V^\top$.
(iv) Compute $\|A - A_2\|_F$ and compare it against the optimal $\|A - (A)_2\|_F$ obtained from the true SVD of A (compute σ_3 and use Eckart–Young).

Exercise 4.7.15 (Spectral Clustering on a Toy Graph) Consider the undirected graph on $n = 6$ nodes with adjacency matrix

$$W = \begin{bmatrix} 0 & 1 & 1 & 0 & 0 & 0 \\ 1 & 0 & 1 & 0 & 0 & 0 \\ 1 & 1 & 0 & 0 & 0 & 0 \\ 0 & 0 & 0 & 0 & 1 & 1 \\ 0 & 0 & 0 & 1 & 0 & 1 \\ 0 & 0 & 0 & 1 & 1 & 0 \end{bmatrix}.$$

(i) Compute the degree matrix D and Laplacian $L = D - W$.
(ii) Compute the eigenvalues and eigenvectors of L and verify that $\lambda = 0$ has multiplicity 2; report a basis for the null space.
(iii) Use the second-smallest eigenvector of L restricted to the nontrivial component(s) (or an eigenvector spanning the second component indicator) to propose a 2-way partition of the nodes.

Exercise 4.7.16 (Kernel Matrix PSD Check and Nyström Approximation (Toy)) Let points in $\mathbb{R}$ be $x_1 = 0$, $x_2 = 1$, $x_3 = 2$, $x_4 = 4$ and define the RBF kernel

$$k(x, z) = \exp\left(-\frac{(x - z)^2}{2\ell^2}\right), \qquad \ell = 1.$$

(i) Form the 4×4 kernel matrix K with $K_{ij} = k(x_i, x_j)$ (to 4 decimals).
(ii) Verify numerically that $K \succeq 0$ by computing its eigenvalues.
(iii) Perform a Nyström approximation using landmark indices $\{1, 3\}$: form $C \in \mathbb{R}^{4\times 2}$ (columns 1 and 3 of K) and $W \in \mathbb{R}^{2\times 2}$ (the 2×2 submatrix on indices $\{1, 3\}$), and compute the approximation $\widetilde{K} = CW^{-1}C^{\top}$.
(iv) Compute the Frobenius error $\|K - \widetilde{K}\|_F$.

Exercise 4.7.17 (Low-Rank Denoising by Singular-Value Thresholding) Let

$$M = \begin{bmatrix} 3 & 0 & 0 \\ 0 & 2 & 0 \\ 0 & 0 & 0 \end{bmatrix}, \qquad N = \begin{bmatrix} 0.2 & -0.1 & 0.0 \\ 0.1 & 0.0 & -0.2 \\ 0.0 & 0.1 & 0.1 \end{bmatrix}, \qquad X = M + N.$$

(i) Compute the singular values of X numerically (to 4 decimals).
(ii) Form the rank-1 and rank-2 truncations X_1, X_2 and compute $\|X - X_k\|_F$ for $k = 1, 2$.
(iii) Perform singular-value soft-thresholding with threshold $\tau = 0.3$, i.e., replace each σ_i by $\max(\sigma_i - \tau, 0)$, and reconstruct $\widetilde{X}$ from the thresholded SVD; compute $\|X - \widetilde{X}\|_F$ and compare with truncation errors.
(iv) Using M as ground truth, compute $\|M - X_k\|_F$ and $\|M - \widetilde{X}\|_F$ to evaluate denoising quality.

Exercise 4.7.18 (Anomaly Detection via PCA Reconstruction Error) Let centered points in $\mathbb{R}^2$ be

$$x_1 = \begin{bmatrix} -2 \\ -2 \end{bmatrix}, \; x_2 = \begin{bmatrix} -1 \\ -1 \end{bmatrix}, \; x_3 = \begin{bmatrix} 0 \\ 0 \end{bmatrix}, \; x_4 = \begin{bmatrix} 1 \\ 1 \end{bmatrix}, \; x_5 = \begin{bmatrix} 2 \\ 2 \end{bmatrix}, \; x_6 = \begin{bmatrix} 0 \\ 5 \end{bmatrix}.$$

(i) Compute the sample covariance $\widehat{\Sigma} = \frac{1}{6}\sum_{i=1}^{6} x_i x_i^\top$.
(ii) Compute the top principal direction q_1 (unit eigenvector of the largest eigenvalue).
(iii) For each i, compute the rank-1 reconstruction $\widehat{x_i} = (q_1 q_1^\top)x_i$ and reconstruction error $e_i = \|x_i - \widehat{x_i}\|_2$.
(iv) Identify the most anomalous point(s) by largest e_i, and justify geometrically in one paragraph.

Chapter 5
Numerical Linear Algebra for Data

Machine learning is implemented on finite-precision hardware and executed at scales where naive linear algebra becomes computationally and numerically fragile. This chapter develops a principled vocabulary for *finite-precision computation*: floating-point models, rounding error, stability (backward vs. forward error), conditioning, and practical diagnostics. The guiding lesson is that *conditioning is a property of the problem*, while *stability is a property of the algorithm*: a stable algorithm cannot rescue an ill-conditioned problem, but an unstable algorithm can fail even on a well-conditioned one (Higham 2002) (cf. Fig. 5.1).

5.1 Why Numerical Linear Algebra Matters for ML

5.1.1 From Exact Math to Finite Precision

5.1.1.1 Floating-Point Arithmetic as a Computational Model

In exact mathematics, $(a+b)+c = a+(b+c)$ and $ab = ba$ always. In floating-point arithmetic, such identities can fail because every operation rounds to a representable number. This makes floating-point arithmetic a *computational model* with its own axioms: (1) numbers are quantized (finite mantissa), (2) operations incur rounding error, (3) extreme magnitudes overflow/underflow, and (4) error can accumulate and amplify through ill-conditioned transformations. In ML, these effects show up as exploding/vanishing gradients, unstable matrix inversions, and silent differences across implementations.

Remark 5.1.1 (What Changes from Real Arithmetic) In floating point, addition is not associative: $(a \oplus b) \oplus c$ may differ from $a \oplus (b \oplus c)$, and subtracting nearly equal numbers can lose many significant digits. Thus, *algebraic equivalence does not imply numerical equivalence.*

P. Singh, B. Raman, *Foundations of Machine Learning and AI*, Studies in Big Data 199, https://doi.org/10.1007/978-3-032-30336-3_5

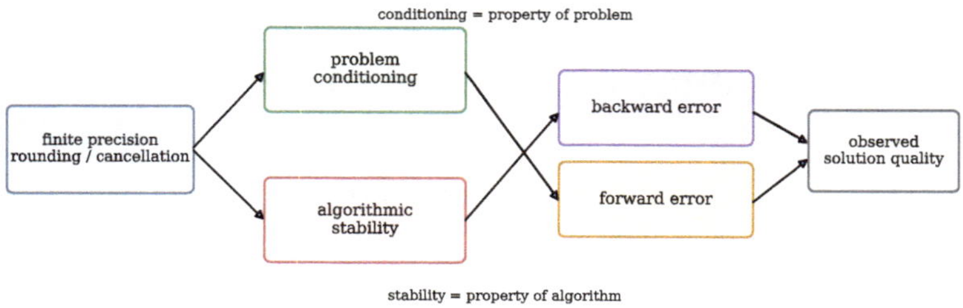

Fig. 5.1 A numerical viewpoint on learning: finite-precision arithmetic injects perturbations, conditioning determines how strongly the problem amplifies them, and algorithmic stability determines whether the computed answer corresponds to a nearby exact problem

5.1.1.2 Why "the Same Formula" Can Behave Differently Numerically

Two mathematically identical expressions can have very different numerical behavior depending on cancellation and amplification. For example, computing a least-squares solution via normal equations $(A^\top A)^{-1}A^\top b$ is mathematically correct, but can be numerically inferior to QR or SVD methods because forming $A^\top A$ squares the condition number. Similarly, computing variance via $\mathbb{E}[X^2] - \mathbb{E}[X]^2$ is algebraically valid but can suffer catastrophic cancellation. The correct workflow is therefore: (i) derive a mathematically correct method, (ii) analyze conditioning of the problem, and (iii) choose a stable algorithm whose error is controlled by that conditioning.

Example 5.1.1 (A Cancellation-Prone Identity) Let $x_1, \dots, x_n$ be large numbers with small spread. The variance identity

$$\frac{1}{n}\sum_{i=1}^{n} x_i^2 - \left(\frac{1}{n}\sum_{i=1}^{n} x_i\right)^2$$

subtracts two close numbers and can lose precision. A more stable computation uses centered values $\frac{1}{n}\sum_{i=1}^{n}(x_i - \bar{x})^2$.

5.1.2 Numerics as a Hidden Bottleneck in Learning

5.1.2.1 Ill-Conditioning, Instability, and Silent Failures

Numerical failures in ML are often *silent*: no explicit error occurs, but outputs drift, gradients become inaccurate, or convergence stalls. The main culprits are (a) ill-conditioning (the problem is sensitive), (b) unstable algorithms (the method amplifies rounding error), and (c) scale mismatch (features/activations vary wildly in magnitude). A small perturbation due to rounding can be magnified by $\kappa(A)$ (condition number), turning benign rounding into a large output error.

Remark 5.1.2 (Two Distinct Notions) Ill-conditioning: small perturbations in input produce large changes in the exact solution. **Instability:** the algorithm produces a result that does not correspond to a nearby exact problem. We can often fix instability by changing the algorithm, but not ill-conditioning without changing the model/data (e.g. regularization).

5.1.2.2 Scalability: Time/Memory Limits at Large n, d

Even stable algorithms may be infeasible at scale. Memory and time costs dominate: storing a dense $n \times d$ matrix costs $O(nd)$ memory; forming $A^\top A$ costs $O(nd^2)$; dense SVD costs roughly $O(\min\{nd^2, dn^2\})$. Thus, large-scale ML relies on structure: sparsity, low rank, sketching, iterative solvers, and minibatching. Numerical linear algebra provides the formal tools to decide which structure-preserving approximation is safe.

Remark 5.1.3 (A Practical Hierarchy) For large dense problems: prefer *matrix–vector* or *matrix–matrix* products over explicit inverses; use QR/SVD for reliability when d is moderate; use iterative/Randomized methods when n, d are huge and only top components are needed.

5.2 Floating-Point Arithmetic and Error Models

5.2.1 Floating-Point Basics

5.2.1.1 Machine Epsilon and Rounding Modes (Conceptual)

Floating-point numbers form a discrete set. The *unit roundoff* (machine precision) u quantifies the maximal relative error from rounding a real number to the nearest representable float. A standard and extremely useful abstraction is the *standard floating-point model*: each basic operation incurs a small relative error bounded by u. Rounding modes (nearest, toward zero, $\pm\infty$) affect bias and error propagation; ML typically uses round-to-nearest.

Definition 5.2.1 (Unit Roundoff/Machine Precision) Let $\mathrm{fl}(\cdot)$ denote rounding to floating point in a fixed format. The *unit roundoff* u is the smallest number such that for all real x in the normal range,

$$\mathrm{fl}(x) = x(1+\delta) \quad \text{for some } \delta \text{ with } |\delta| \le u.$$

Equivalently, u upper-bounds the relative rounding error of a single rounding step.

Proposition 5.2.1 (Standard Model for Basic Operations) *For $\circ \in \{+, -, \times, /\}$ and for floating-point operations without overflow/underflow,*

$$fl(a \circ b) = (a \circ b)\,(1 + \delta), \qquad |\delta| \leq u.$$

Remark 5.2.1 (Accumulated Error Notation) In many error analyses, one uses the shorthand

$$\gamma_k := \frac{ku}{1 - ku},$$

valid when $ku < 1$, to bound the accumulated relative error of k floating-point operations in stable summations/products.

5.2.1.2 Overflow/Underflow and Catastrophic Cancellation

Overflow occurs when a magnitude exceeds the representable maximum; underflow occurs near zero when numbers become denormal/subnormal or round to zero. Both can break the relative error model. Catastrophic cancellation occurs when subtracting nearly equal numbers: most leading digits cancel, leaving a result with very few accurate digits. This is common in naive variance computation, in certain polynomial evaluations, and in ill-conditioned subtraction patterns in optimization.

Definition 5.2.2 (Catastrophic Cancellation (Informal)) Cancellation refers to the loss of significant digits when computing $a - b$ with $a \approx b$, because the exact result is small while the operands are large, so relative error in the result can be enormous.

Example 5.2.1 (Cancellation in Practice) Let $a = 10^8 + 1$ and $b = 10^8$ in a format that cannot represent all integers around 10^8. Even if a and b are individually rounded, computing $\mathrm{fl}(a-b)$ may yield 0 rather than 1 because the "+1" may be lost in representation. This illustrates that cancellation can defeat intuitive arithmetic.

5.2.2 Backward vs. Forward Error

5.2.2.1 Stable Algorithms and Backward Error Viewpoint

Forward error measures how far the computed output is from the true output. Backward error asks a more robust question: is the computed output *exactly correct for a nearby input problem*? Backward stability is the gold standard because it composes well: if each step solves a nearby problem, the whole pipeline is solving a nearby problem. Many classic algorithms (QR-based least squares, Householder transformations) are backward stable, while explicit inversion and naive normal equations may not be.

Definition 5.2.3 (Forward and Backward Error) Let the exact problem be: given data d, compute $x = f(d)$. An algorithm returns $\widehat{x}$.

- *Forward error:* $\|\widehat{x} - x\|$.
- *Backward error:* the smallest perturbation Δd such that $\widehat{x} = f(d + \Delta d)$ (or satisfies the defining equations for perturbed data).

Definition 5.2.4 (Backward Stability (Informal)) An algorithm is *backward stable* if it produces $\widehat{x}$ that is the exact solution to a nearby problem with relative perturbation proportional to unit roundoff:

$$\widehat{x} = f(d + \Delta d), \qquad \frac{\|\Delta d\|}{\|d\|} \leq C u,$$

for a moderate constant C independent of problem size (up to polynomial factors).

Remark 5.2.2 (Why Backward Stability Is the Right Target) If the underlying problem is well-conditioned, a small backward error implies a small forward error. Thus, stability plus conditioning yields correctness.

5.2.2.2 Conditioning vs. Stability: What You Can (and Cannot) Fix

Conditioning quantifies sensitivity of the exact solution to perturbations in data. Stability quantifies whether the algorithm introduces only small perturbations. A stable algorithm ensures backward error is small, but if the problem is ill-conditioned, even tiny backward error can produce a large forward error. Conversely, even a well-conditioned problem can be ruined by an unstable algorithm. Thus, the performance guarantee typically has the form:

$$\frac{\|\widehat{x} - x\|}{\|x\|} \lesssim \kappa(\text{problem}) \times (\text{backward error magnitude}).$$

Remark 5.2.3 (The Core Inequality Template) For many problems, one can show

$$\text{forward error} \lesssim \text{condition number} \times \text{backward error}.$$

Hence, the only robust route to accuracy is: use a stable algorithm and reduce conditioning when possible (via scaling, regularization, feature engineering).

5.2.3 Condition Numbers

5.2.3.1 Conditioning of Linear Systems and Least Squares

For $Ax = b$, the solution depends on A^{-1} when A is nonsingular; if A nearly loses rank, then A^{-1} has large norm and perturbations in b are amplified. For least squares, the relevant matrix is A itself (via its singular values), and forming normal equations squares the condition number, a fundamental reason to prefer QR/SVD over $(A^\top A)^{-1}A^\top$ for numerical robustness.

Definition 5.2.5 (Condition Number of a Nonsingular Matrix) For an invertible matrix $A \in \mathbb{R}^{n\times n}$, the spectral condition number is

$$\kappa_2(A) := \|A\|_2\,\|A^{-1}\|_2 = \frac{\sigma_{\max}(A)}{\sigma_{\min}(A)}.$$

Proposition 5.2.2 (Perturbation Sensitivity of $Ax = b$ (Informal Bound)) *Let A be invertible and consider $(A + \Delta A)(x + \Delta x) = b + \Delta b$ with small perturbations. Under standard smallness assumptions, one has the sensitivity estimate*

$$\frac{\|\Delta x\|_2}{\|x\|_2} \lesssim \kappa_2(A)\left(\frac{\|\Delta A\|_2}{\|A\|_2} + \frac{\|\Delta b\|_2}{\|b\|_2}\right).$$

Thus $\kappa_2(A)$ governs worst-case relative error amplification.

Proposition 5.2.3 (Normal Equations Square Conditioning) *If A has full column rank, then*

$$\kappa_2(A^\top A) = \kappa_2(A)^2.$$

Remark 5.2.4 (Practical Consequence) Solving least squares via normal equations can be dramatically less stable than QR/SVD, because any conditioning issue is squared at the level of $A^\top A$.

5.2.3.2 Practical Diagnostics and Scaling Remedies

Condition numbers are rarely computed exactly at scale, but practitioners can detect conditioning issues via singular value estimates, unusually large coefficients, sensitivity to data resampling, and slow/noisy convergence. Remedies include: (i) feature scaling/standardization, (ii) centering to remove large offsets, (iii) whitening or preconditioning, (iv) regularization (ridge/Tikhonov), and (v) switching to stable solvers (QR/SVD) or iterative methods with preconditioners. The overarching idea is to reshape the geometry so that important directions have comparable scale.

Definition 5.2.6 (Equilibration/Diagonal Scaling (Conceptual)) Given $A \in \mathbb{R}^{m\times n}$, *diagonal scaling* seeks diagonal matrices $D_r \in \mathbb{R}^{m\times m}$, $D_c \in \mathbb{R}^{n\times n}$ with positive diagonals such that the scaled matrix $\widetilde{A} = D_r A D_c$ has more balanced row/column norms. This can improve the conditioning and numerical behavior of solvers.

Remark 5.2.5 (Common ML Scaling Practice) Standardization corresponds to scaling columns so that empirical variances are ≈ 1. In deep learning, normalization layers play a related role: they keep activations in a numerically stable range, reducing conditioning issues in Jacobians and gradients.

Remark 5.2.6 (Solver Choice as a Diagnostic) If normal-equation solutions are unstable but QR/SVD solutions are stable, the issue is conditioning rather than implementation. If both are unstable, the problem itself is ill-conditioned and may require regularization or redesign.

5.3 Conditioning and Stability

This section sharpens the central numerical message for ML: *conditioning* is a property of the problem (data/model geometry), while *stability* is a property of the algorithm (how rounding errors are amplified or controlled). A stable algorithm can be trusted on a well-conditioned problem; an ill-conditioned problem requires changing the formulation (scaling, regularization, reparameterization), not just changing code.

5.3.1 Why Numerical Issues Appear

5.3.1.1 Floating-Point Arithmetic and Roundoff Accumulation

Every floating-point operation introduces a small relative error; long chains of operations can accumulate error. Whether this accumulated error matters depends on cancellation and amplification by subsequent transforms. In linear algebra, repeated summations (dot products), elimination steps, and forming products like $A^\top A$ are common amplification points. In ML, dot products appear everywhere (linear layers, attention), so error accumulation is ubiquitous but usually benign if the problem is well-scaled and algorithms are stable; it becomes harmful when combined with ill-conditioning or extreme dynamic range.

Remark 5.3.1 (Standard Model Reminder) In the normal range and round-to-nearest mode, one uses $\mathrm{fl}(a \circ b) = (a \circ b)(1 + \delta)$ with $|\delta| \le u$. Accumulation over k operations is often bounded by $\gamma_k = \frac{ku}{1-ku}$ when $ku < 1$.

5.3.1.2 Condition Numbers and Sensitivity of Solutions

Condition numbers quantify worst-case sensitivity. For linear systems, $\kappa_2(A)$ measures how much A stretches space anisotropically: directions near the smallest singular vector are weakly constrained and therefore highly sensitive. For least squares, forming normal equations squares the condition number, turning moderate conditioning into severe conditioning. Hence, condition numbers act as *error multipliers*: even backward-stable algorithms can have large forward error when κ is large.

Remark 5.3.2 (Core Template) Forward error $\lesssim$ (conditioning)$\times$(backward error). Thus, stability is necessary, but not sufficient, for accuracy.

5.3.2 Common Failure Modes

5.3.2.1 Ill-Conditioning, Multicollinearity, and Exploding Variance

In regression and many linear readouts, multicollinearity means columns of X are nearly linearly dependent. Geometrically, the column space contains directions with tiny singular values. The consequence is exploding coefficient variance: small perturbations in y produce large changes in $\widehat{w}$, and coefficients become unstable across resamples. Predictions can remain stable (because they depend on projection), creating a dangerous trap: a model may predict well but provide misleading interpretations.

Remark 5.3.3 (A Diagnostic Signature) Large coefficient magnitudes, sensitivity to tiny data perturbations, and large $\kappa(X)$ (or small $\sigma_{\min}(X)$) are classic symptoms.

5.3.2.2 Why "Never Invert a Matrix": Practical Rationale

The advice "never invert a matrix" is a numerical principle: explicitly forming A^{-1} is usually slower, less stable, and less accurate than solving $Ax = b$ by factorization. Inversion introduces unnecessary operations and can square conditioning effects in subsequent multiplications. Moreover, downstream multiplication by A^{-1} can amplify error dramatically along weak directions. Hence, one should compute x via stable solves (LU/QR/Cholesky) or iterative methods, and reuse factorizations when multiple right-hand sides are present.

Remark 5.3.4 (What To Do Instead) To compute $A^{-1}b$, solve $Ax = b$. To compute $A^{-1}B$ for many columns in B, factorize A once and solve triangular systems for each column.

5.3.3 Stabilization Strategies

5.3.3.1 Centering, Scaling, Whitening, and Regularization

Stabilization is geometry engineering: modify coordinates and constraints to reduce anisotropy and sensitivity. Centering removes large offsets that can dominate computations; scaling equalizes feature magnitudes; whitening removes correlations. Regularization (ridge/Tikhonov) shifts eigenvalues away from zero, improving conditioning and reducing variance amplification. These techniques often interact: e.g., standardize then ridge is far more stable than ridge alone on wildly scaled features.

Remark 5.3.5 (Ridge Stabilizes Inversion) Replacing $(X^\top X)^{-1}$ by $(X^\top X + \lambda I)^{-1}$ increases the smallest eigenvalue by at least λ, reducing sensitivity along weak directions.

5.3.3.2 Orthogonalization as a Stability Tool

Orthonormal bases are numerically ideal: orthogonal transformations preserve norms and do not amplify errors. QR factorization is therefore the backbone of stable least squares: it replaces normal equations by orthogonalization. In iterative methods, re-orthogonalization prevents loss of orthogonality and numerical collapse. Thus, orthogonalization is the main stability tool that turns ill-scaled coordinate systems into robust ones.

Remark 5.3.6 (QR vs. Normal Equations) Normal equations use $X^\top X$ and square conditioning; QR uses orthogonal factors and avoids this squaring.

5.4 Direct Methods for Linear Systems

Direct methods solve linear systems by factorizing a matrix into structured factors (triangular/orthogonal) and then performing cheap triangular solves (Golub and Van Loan 2013). Their key advantages are reliability (when implemented with pivoting) and reusability across multiple right-hand sides. Their disadvantages are memory/time costs for large dense matrices and sensitivity to fill-in for sparse matrices.

5.4.1 Gaussian Elimination and LU

5.4.1.1 Pivoting, Growth Factors, and Stability Intuition

Gaussian elimination without pivoting can be catastrophically unstable: small pivots lead to huge multipliers and amplify rounding error. Pivoting swaps rows (or columns) to choose safer pivots, controlling growth in intermediate entries. Partial pivoting (row swaps) is the standard compromise: it is cheap and usually stable in practice, though worst-case growth is possible. The growth factor measures how large intermediate entries become relative to the original matrix; large growth signals potential instability.

Definition 5.4.1 (LU Factorization with Partial Pivoting) For $A \in \mathbb{R}^{n\times n}$, LU with partial pivoting produces a permutation matrix P and factors

$$PA = LU,$$

where L is unit lower triangular and U is upper triangular.

Remark 5.4.1 (Stability Intuition) Pivoting prevents dividing by very small pivots. When pivots are controlled, multipliers in L remain moderate and rounding error is not explosively amplified.

5.4.1.2 Cost Models and Memory Footprints

For dense $n \times n$ matrices, LU factorization costs $\approx \frac{2}{3}n^3$ flops and stores L and U in-place (often within the same array as A), with an additional permutation vector. Triangular solves then cost $O(n^2)$ per right-hand side. In ML, this makes LU feasible for moderate n (e.g., small Newton systems) but infeasible for very large dense problems without exploiting structure.

Remark 5.4.2 (Reuse Is the Point) Factor once ($O(n^3)$), solve many times ($O(n^2)$ each). This is critical in batched inference and in repeated solves with the same operator.

5.4.2 Cholesky for SPD Systems

5.4.2.1 When SPD Arises (Covariances, Normal Equations, Hessians)

Symmetric positive definite (SPD) systems are especially important: covariance matrices, kernel ridge systems, Gauss–Newton approximations, and (in convex problems) Hessians are SPD. Cholesky is the preferred factorization: it is about twice as fast as LU, uses half the storage, and is very stable for SPD matrices. The factorization $A = LL^\top$ reveals geometry directly: it is a matrix square root.

Definition 5.4.2 (Cholesky Factorization) If $A \in \mathbb{R}^{n\times n}$ is symmetric positive definite, then there exists a unique lower triangular L with positive diagonal entries such that

$$A = LL^\top.$$

Remark 5.4.3 (Normal Equations Caution) $X^\top X$ is SPD when X has full column rank, so Cholesky can solve normal equations. But recall: forming $X^\top X$ can square conditioning; QR is often more stable even though Cholesky itself is stable.

5.4.2.2 Failure Modes and Jitter/Regularization Tricks

Cholesky requires SPD. If A is only PSD or is SPD in theory but becomes indefinite numerically (due to rounding or modeling error), Cholesky fails. In ML, this occurs with nearly singular covariances or kernel matrices. A standard fix is *jitter*: add a small multiple of identity, $A \leftarrow A+\lambda I$, making it SPD and improving conditioning. This is exactly ridge/Tikhonov regularization in linear algebra form.

Remark 5.4.4 (Jitter as Numerical Regularization) If Cholesky fails or produces tiny pivots, use $A + \lambda I$ with λ chosen relative to scale (e.g., $\lambda = \epsilon\,\mathrm{tr}(A)/n$). This improves conditioning and stabilizes solves at the cost of biasing the solution slightly.

5.4.3 *Triangular Solves and Reuse*

5.4.3.1 Why Factorizations Are Reusable Across Multiple Right-Hand Sides

Once we have $PA = LU$ or $A = LL^\top$, solving $Ax = b$ reduces to triangular solves, which are cheap and stable. If we need to solve $AX = B$ for many right-hand sides (columns of B), we factorize once and solve each column independently. This pattern appears everywhere: solving for multiple targets in multi-output regression, computing Gaussian posterior means for many observations, and repeatedly solving the same linear operator during iterative optimization.

Remark 5.4.5 (Complexity) After factorization, each solve costs $O(n^2)$ (dense). With r right-hand sides, total solve cost is $O(rn^2)$, amortizing the factorization.

5.4.3.2 Caching and Batching Solves in ML Pipelines

Modern ML systems exploit batching: solve many right-hand sides simultaneously as a matrix solve, using BLAS-optimized kernels. Caching factorizations avoids repeated $O(n^3)$ work when the operator is unchanged (or changes slowly). In second-order methods, one often reuses factorizations across multiple gradient evaluations. In Gaussian processes, kernel matrices are factorized once and reused for prediction at many test points.

Remark 5.4.6 (Implementation Principle) Prefer factorization objects (e.g., storing L or (L, perm)) and call triangular-solve routines rather than forming inverses. This improves performance and numerical robustness.

5.5 Least Squares Computation at Scale

Least squares is the computational core of linear regression, linear probes, ridge regression, and many second-order subroutines in ML. At scale, the bottleneck is not the mathematics but the *numerical pathway*: different algebraically equivalent formulations can have dramatically different stability and cost. This section compares normal equations, QR, and SVD-based solves; then develops regularized least squares as stabilized inversion.

5.5.1 QR and Cholesky in Practice

5.5.1.1 Normal Equations vs. QR: Stability Comparison

The normal equations $(X^\top X)\widehat{w} = X^\top y$ convert least squares into an SPD system, enabling Cholesky. However, forming $X^\top X$ squares the condition number and can amplify rounding error and noise. QR factorization solves least squares without forming $X^\top X$, using orthogonal transformations that preserve norms, and is typically backward stable. Thus QR is the practical default for reliable least squares, while normal equations are used only when conditioning is benign and performance constraints are tight (Trefethen and Bau 1997) (cf. Fig. 5.2).

Definition 5.5.1 (Least Squares Problem) Given $X \in \mathbb{R}^{n \times d}$ and $y \in \mathbb{R}^n$, least squares seeks

$$\widehat{w} \in \arg\min_{w \in \mathbb{R}^d} \|Xw - y\|_2^2.$$

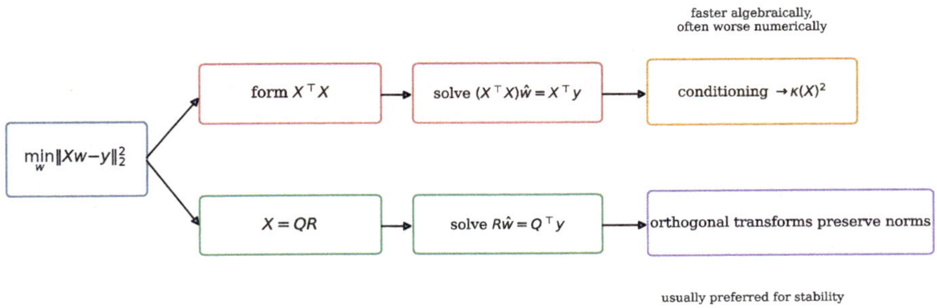

Fig. 5.2 Two computational pathways for least squares. The normal-equation route forms $X^\top X$ and squares conditioning, while the QR route solves the same problem through orthogonalization and preserves Euclidean geometry more faithfully

Proposition 5.5.1 (Normal Equations) *If $\widehat{w}$ minimizes $\|Xw - y\|_2^2$, then it satisfies*

$$X^\top(X\widehat{w} - y) = 0 \quad \Longleftrightarrow \quad (X^\top X)\widehat{w} = X^\top y.$$

If X has full column rank, the minimizer is unique.

Remark 5.5.1 (Conditioning Penalty of Normal Equations) If X has full column rank, then $\kappa_2(X^\top X) = \kappa_2(X)^2$. Thus, even moderate ill-conditioning in X can become severe in $X^\top X$, degrading accuracy.

Definition 5.5.2 (QR Factorization (Thin QR)) Let $X \in \mathbb{R}^{n \times d}$ with $n \geq d$ and full column rank. A (thin) QR factorization is

$$X = QR,$$

where $Q \in \mathbb{R}^{n \times d}$ has orthonormal columns ($Q^\top Q = I_d$) and $R \in \mathbb{R}^{d \times d}$ is upper triangular with positive diagonal.

Proposition 5.5.2 (Least Squares via QR) *If $X = QR$ is a thin QR factorization, then the least squares solution satisfies*

$$\widehat{w} = R^{-1} Q^\top y,$$

i.e., solve the triangular system $R\widehat{w} = Q^\top y$. Moreover, the fitted vector is $\widehat{y} = QQ^\top y$ and the residual is $r = y - \widehat{y}$ with $Q^\top r = 0$.

Remark 5.5.2 (Numerical Advantage) Orthogonal transformations (forming Q) are norm-preserving and do not amplify rounding error. This is the geometric reason QR is typically backward stable.

5.5.1.2 Cholesky for SPD Systems and When It Is Safe

Cholesky is the fastest stable direct solver for SPD matrices. It is safe when the SPD system is formed *without* squaring conditioning (e.g., kernel ridge $K + \lambda I$, covariance matrices with jitter, Gauss–Newton approximations). For normal equations, Cholesky is algebraically valid but can inherit the squared conditioning penalty from forming $X^\top X$. Thus, Cholesky is safe for SPD problems that are intrinsically well-conditioned or explicitly regularized but should be used cautiously as a replacement for QR in ill-conditioned least squares.

Remark 5.5.3 (Safe Cholesky Use-Cases)

- Solve $(K + \lambda I)\alpha = y$ in kernel ridge/Gaussian processes.
- Solve $(A^\top A + \lambda I)w = A^\top y$ in ridge (regularized normal equations).
- Solve $H\Delta = -g$ when H is a well-conditioned SPD Hessian approximation.

5.5.2 SVD-Based Least Squares (Gold Standard)

5.5.2.1 Rank Deficiency and Minimum-Norm Solutions

SVD is the gold standard because it handles rank deficiency gracefully. When X is not full rank, there are infinitely many minimizers; SVD makes the geometry explicit: components in the null space are unconstrained by the data, and the Moore–Penrose solution chooses the minimum-norm minimizer. This is the canonical choice in many theoretical analyses and in stable numerical software.

Proposition 5.5.3 (Least Squares via SVD and Pseudoinverse) *Let $X \in \mathbb{R}^{n\times d}$ have SVD $X = U\Sigma V^\top$ with rank r. Define $\Sigma^\dagger = diag(1/\sigma_1, \ldots, 1/\sigma_r, 0, \ldots)$ and $X^\dagger = V\Sigma^\dagger U^\top$. Then the vector*

$$\widehat{w} := X^\dagger y$$

is the minimum-norm least squares solution, and the fitted values satisfy

$$\widehat{y} = X\widehat{w} = UU^\top y \quad \textit{(projection onto } \mathcal{C}(X)\textit{)}.$$

Remark 5.5.4 (Null-Space Decomposition) Write $y = y_\parallel + y_\perp$ with $y_\parallel \in \mathcal{C}(X)$ and $y_\perp \in \mathcal{N}(X^\top)$. Then $X^\dagger y = X^\dagger y_\parallel$ and $y_\perp$ is the irreducible residual component.

5.5.2.2 Truncated SVD as Regularization

Tiny singular values correspond to directions that are weakly informed by data but heavily amplify noise in $X^\dagger$. Truncated SVD regularization discards these directions, producing a stable approximate inverse. Geometrically, it projects onto the top singular subspace and ignores near-null directions, similar in spirit to ridge but with a hard cutoff. This is a principled denoising mechanism grounded in spectral geometry.

Definition 5.5.3 (Truncated SVD (TSVD) Solution) Let $X = U\Sigma V^\top$ with singular values $\sigma_1 \geq \cdots \geq \sigma_r > 0$. For $k \leq r$, define

$$X_k^\dagger := V_k \Sigma_k^{-1} U_k^\top, \qquad \widehat{w}_{(k)} := X_k^\dagger y.$$

Remark 5.5.5 (TSVD as a Filter) TSVD uses filter factors $f_i = 1/\sigma_i$ for $i \leq k$ and $f_i = 0$ for $i > k$. It eliminates unstable directions entirely.

5.5.3 *Regularized Least Squares*

5.5.3.1 Ridge Regression as Stabilized Solve

Ridge adds an ℓ_2 penalty that makes the normal matrix SPD even when X is rank-deficient. It stabilizes inversion by shifting eigenvalues away from zero. Ridge is therefore both a statistical regularizer (bias–variance tradeoff) and a numerical stabilizer (conditioning improvement). At scale, ridge is often solved by Cholesky on $X^\top X + \lambda I$ or by iterative methods when d is large.

Definition 5.5.4 (Ridge Regression) For $\lambda > 0$, ridge solves

$$\widehat{w}_\lambda \in \arg\min_{w \in \mathbb{R}^d} \|Xw - y\|_2^2 + \lambda \|w\|_2^2.$$

The unique minimizer satisfies

$$(X^\top X + \lambda I)\widehat{w}_\lambda = X^\top y.$$

5.5.3.2 Bias–Variance vs. Conditioning Tradeoff

Ridge trades bias for variance reduction and numerical stability. Spectrally, ridge attenuates directions with small singular values most strongly, exactly the directions that would otherwise amplify noise. Thus, the same mechanism that improves generalization often improves numerical robustness. Choosing λ balances fit quality against stability; cross-validation is the practical choice rule, while spectral analysis explains behavior.

Remark 5.5.6 (Spectral Filtering) In SVD coordinates, ridge uses filter factors $\sigma_i/(\sigma_i^2 + \lambda)$. As λ increases, weak directions are damped more, improving stability but increasing bias.

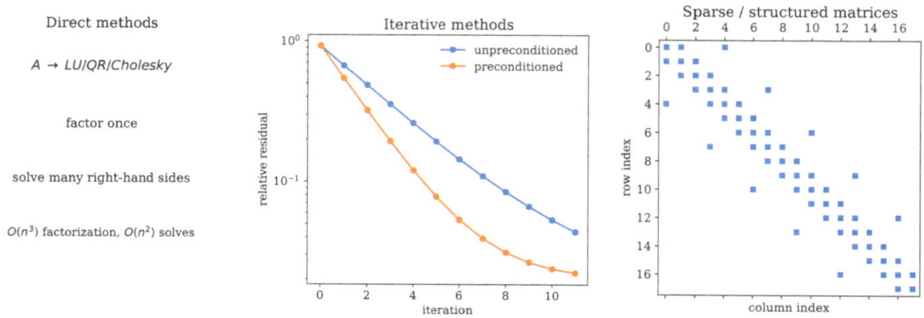

Fig. 5.3 A computational map for large linear problems. Direct methods factorize once and reuse solves; iterative methods use only matrix–vector products and benefit from preconditioning; sparse structure changes both cost and feasibility by replacing dense arithmetic with local interactions over the nonzero pattern

5.6 Iterative Solvers

When n, d are large, direct methods may be too costly in memory or time. Iterative solvers use matrix–vector products and converge to a solution progressively, making them suitable for sparse and structured problems (Saad 2003). Their performance depends strongly on conditioning and therefore on preconditioning (cf. Fig. 5.3).

5.6.1 Large-Scale Linear Systems

5.6.1.1 Gradient Descent for Least Squares: Intuition and Stopping

Least squares has a smooth convex objective with gradient $X^\top(Xw - y)$. Gradient descent is therefore a natural iterative solver that uses only multiplications by X and $X^\top$. Convergence rate depends on the spectrum of $X^\top X$: ill-conditioning leads to slow progress along weak directions. Stopping criteria must reflect both optimization and statistical noise: over-solving may not improve test error.

Proposition 5.6.1 (Gradient and Lipschitz Constant) *Let* $f(w) = \frac{1}{2}\|Xw - y\|_2^2$. *Then*

$$\nabla f(w) = X^\top(Xw - y),$$

and ∇f *is Lipschitz with constant* $L = \|X^\top X\|_2 = \sigma_1(X)^2$*:*

$$\|\nabla f(w) - \nabla f(v)\|_2 \le L\|w - v\|_2.$$

Remark 5.6.1 (Basic Step-Size Rule) A sufficient condition for convergence of gradient descent on f is $0 < \eta < 2/L$. In practice, one may use $\eta \approx 1/L$ or adaptive schemes.

5.6.1.2 Conjugate Gradient as a Krylov Method: Conceptual View

Conjugate gradient (CG) solves SPD linear systems $Hw = g$ using a Krylov subspace method. It can be seen as minimizing a quadratic over expanding Krylov subspaces, achieving finite termination in exact arithmetic in at most d steps, and rapid convergence when eigenvalues are clustered. For least squares and ridge, CG applies to the normal equations (or directly to $X^\top X + \lambda I$) without forming the matrix explicitly, using only X and $X^\top$ products.

Remark 5.6.2 (CG Solves SPD Problems) CG applies to $Hw = g$ when H is SPD. Ridge systems $(X^\top X + \lambda I)w = X^\top y$ are SPD for $\lambda > 0$.

5.6.2 Krylov Methods for Least Squares

5.6.2.1 LSQR/CGNR Intuition and When to Use Them

For rectangular least squares, Krylov methods avoid forming $X^\top X$ explicitly. CGNR applies CG to the normal equations $X^\top X w = X^\top y$ (still squaring conditioning conceptually), while LSQR is a more stable variant related to bidiagonalization (Lanczos on X) and often preferred. The choice depends on sparsity, conditioning, and whether $n \gg d$ or $d \gg n$.

Remark 5.6.3 (Guideline) Use LSQR when solving large sparse least squares reliably. Use CG on $(X^\top X + \lambda I)$ for ridge when matrix–vector products are cheap and a good preconditioner is available.

5.6.2.2 GMRES Idea (Conceptual) for Nonsymmetric Systems

When the system matrix is nonsymmetric or indefinite, CG does not apply. GMRES generalizes Krylov ideas: it builds a Krylov subspace and chooses the iterate that minimizes the residual norm over that subspace. This makes it broadly applicable but memory-intensive (it stores basis vectors); restarted variants trade memory for robustness.

Remark 5.6.4 (GMRES in One Line) Given $Aw = b$, GMRES chooses w_t in $w_0 + \mathcal{K}_t(A, r_0)$ to minimize $\|b - Aw_t\|_2$.

5.6.3 Practical Considerations

5.6.3.1 Preconditioning: Why and How (Overview)

Preconditioning changes the geometry so that the iterative method sees a better-conditioned system. Ideally, a preconditioner $M \approx H$ makes $M^{-1}H$ have clustered eigenvalues. In ML, diagonal preconditioners correspond to feature scaling; more advanced preconditioners use incomplete factorizations or low-rank structure. The same idea appears in optimization as adaptive gradient methods and natural-gradient approximations.

Remark 5.6.5 (Left vs. Right Preconditioning) For $Hw = g$, one may solve $M^{-1}Hw = M^{-1}g$ (left), or $HM^{-1}z = g$ with $w = M^{-1}z$ (right). Either aims to reduce the effective condition number.

5.6.3.2 Stopping Criteria and Numerical Diagnostics

Stopping is part of correctness: one should stop when further progress is below numerical noise or statistical tolerance. Common criteria include relative residual $\|r_t\|/\|r_0\|$, relative step size, or stationarity $\|\nabla f(w_t)\|$. Diagnostics include monitoring condition estimates, detecting stagnation, and checking sensitivity to scaling. In ML, it is often correct to stop early once optimization error falls below estimation error.

Remark 5.6.6 (Two Standard Criteria) For $Hw = g$, stop when $\|g - Hw_t\|_2/\|g\|_2 \le \tau$. For least squares $f(w) = \frac{1}{2}\|Xw - y\|_2^2$, stop when $\|\nabla f(w_t)\|_2/\|\nabla f(w_0)\|_2 \le \tau$ or when validation error stops improving.

5.7 Sparse and Structured Matrices

At modern ML scale, matrices are rarely dense. Sparsity (many zeros) and structure (banded, Toeplitz, low-rank, graph-derived) are the key to feasible computation. This section explains how sparsity changes cost models, why naive elimination can destroy sparsity (fill-in), and how structure can be exploited for large speedups—while also emphasizing robustness checks when structure assumptions are violated.

5.7.1 Sparsity in Real Datasets

5.7.1.1 Sparse Formats and Cost Models (Conceptual)

Storing a dense $n \times d$ matrix costs $O(nd)$ memory; storing a sparse matrix costs $O(\text{nnz})$ where nnz is the number of nonzeros. The dominant operation in iterative solvers and many ML kernels is the matrix–vector product, whose cost scales as $O(\text{nnz})$ for sparse matrices. To benefit, one must use sparse storage formats (CSR/CSC/COO) that avoid explicitly storing zeros and enable cache-friendly traversal of nonzeros. Thus, numerical linear algebra on sparse matrices is as much about data structures and memory access as about arithmetic.

Definition 5.7.1 (Sparsity and nnz) For a matrix $A \in \mathbb{R}^{m\times n}$, let $\text{nnz}(A)$ denote the number of nonzero entries. A is *sparse* if $\text{nnz}(A) \ll mn$.

Remark 5.7.1 (Cost Model for Sparse matvec) A sparse matrix–vector product $y = Ax$ requires $O(\text{nnz}(A))$ multiplications/additions, plus memory traffic. By contrast, dense matvec costs $O(mn)$ operations and memory reads.

5.7.1.2 Graph-Structured Matrices from Networks

Graphs naturally generate sparse matrices: adjacency matrices, Laplacians, incidence matrices, and transition matrices. In ML, these arise in spectral clustering, graph neural networks, Markov models, and manifold learning. The sparsity pattern encodes the graph edges; matrix operations correspond to local message passing. This perspective links graph algorithms and sparse linear algebra: multiplying by a graph matrix is aggregating over neighbors.

Definition 5.7.2 (Adjacency and Laplacian Sparsity) For a graph with n nodes and $|E|$ edges, the adjacency matrix W has $\text{nnz}(W) = O(|E|)$ (up to symmetry), and the Laplacian $L = D - W$ has the same sparsity pattern as W plus diagonal entries. Thus, sparse graphs yield sparse matrices.

5.7.2 *Sparse Factorizations and Fill-In*

5.7.2.1 Why Sparsity Can Be Lost During Elimination

Sparse direct solvers (LU/Cholesky) can suffer from *fill-in*: zeros become nonzero during elimination. Fill-in increases memory and time dramatically and can destroy the advantage of sparsity. Geometrically, elimination couples variables that share neighbors in the elimination graph. Thus, the sparsity pattern is not preserved by naive factorization; controlling fill-in requires careful ordering of elimination.

Remark 5.7.2 (Fill-In as Graph Densification) In Cholesky elimination, eliminating a node connects its neighbors (forming a clique) in the elimination graph, creating new nonzeros in the factor. This is the combinatorial source of fill-in.

5.7.2.2 Ordering Heuristics (Conceptual) and Practical Impacts

Optimal ordering to minimize fill-in is NP-hard, so practice uses heuristics: minimum degree, nested dissection, approximate minimum degree (AMD). The ordering can change runtime by orders of magnitude. In ML systems using sparse direct solvers (e.g., certain graphical models), choosing a good ordering is often the difference between feasible and infeasible computation.

Remark 5.7.3 (Heuristic Examples)

- **(Approx.) minimum degree:** eliminate low-degree nodes first to reduce clique growth.
- **Nested dissection:** recursively separate the graph to localize fill-in.

Both aim to keep intermediate sparsity manageable.

5.7.3 Structure Exploitation

5.7.3.1 Band/Toeplitz/Low-Rank Structure and Speedups

Structure enables algorithms faster than generic $O(n^3)$ factorizations: banded matrices admit $O(nb^2)$ solves (bandwidth b), Toeplitz matrices admit FFT-based methods, and low-rank updates are handled via Woodbury. Many ML problems have approximate structure (e.g., near-Toeplitz covariances in time series, low-rank kernels), and exploiting it can reduce computation dramatically—provided the approximation error is controlled.

Remark 5.7.4 (Three Canonical Speedups)

- **Banded:** cost scales with bandwidth, not matrix size squared.
- **Toeplitz/circulant:** matvec via FFT in $O(n \log n)$.
- **Low-rank:** inverse and log-det updates via Woodbury/determinant lemma.

5.7.3.2 When Structure Breaks: Robustness Checks

Assumed structure can be violated by noise, nonstationarity, or model mismatch. A robust workflow is to (i) quantify approximation error (e.g., residual norm, spectral norm difference), (ii) validate downstream impact (e.g., prediction quality, stability), and (iii) fall back to safer methods when error is too large. In ML, structure exploitation is a controlled approximation: correctness requires monitoring.

Remark 5.7.5 (Two Practical Checks)

- Monitor $\|A - \widetilde{A}\|_2$ or $\|A - \widetilde{A}\|_F$ when using structured approximations $\widetilde{A}$.

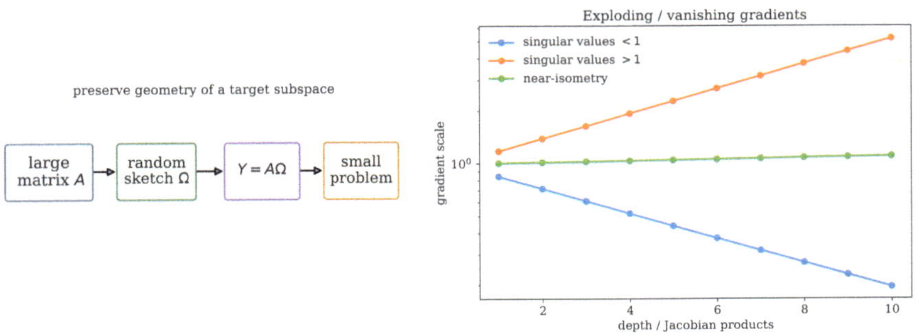

Fig. 5.4 Two modern numerical viewpoints. Left: randomized sketching compresses a large operator to a smaller geometry-preserving problem. Right: repeated Jacobian multiplication in deep learning explains exploding and vanishing gradients through the singular values of the layerwise product

- Verify that key outputs (e.g., residual norms, validation error) are stable under modest changes in approximation parameters.

5.8 Randomized Numerical Linear Algebra (RNLA)

RNLA uses randomness to compress large matrices while approximately preserving the geometry relevant to a task (Halko et al. 2011) (cf. Fig. 5.4). Core tools include subspace embeddings for regression, randomized low-rank approximation for PCA/SVD, and leverage-score sampling for nonuniform data. The main benefit is reduced passes over data and reduced memory footprint, with accuracy controlled by sketch size and oversampling.

5.8.1 Sketching for Regression and PCA

5.8.1.1 Subspace Embeddings and Intuition

A subspace embedding is a random map S that approximately preserves norms of all vectors in a target subspace. For least squares, the target subspace is typically range(X) (or the set of residuals $Xw - y$). If S preserves norms on this subspace, then solving sketched least squares yields a solution close to the original. This is the geometric reason sketching works: preserve distances in the relevant subspace, discard the rest.

Definition 5.8.1 (Subspace Embedding (Informal)) Let $\mathcal{U} \subseteq \mathbb{R}^n$ be a d-dimensional subspace and let $S \in \mathbb{R}^{m \times n}$. We say S is an $(\varepsilon, \mathcal{U})$ subspace embedding if for all $u \in \mathcal{U}$,

$$(1 - \varepsilon)\|u\|_2^2 \leq \|Su\|_2^2 \leq (1 + \varepsilon)\|u\|_2^2.$$

5.8.1.2 Sketched Least Squares: Speed–Accuracy Tradeoffs

Sketched least squares solves $\min_w \|S(Xw - y)\|_2^2$ with $m \ll n$. If S is a subspace embedding for range($[X \;\; y]$), then residual geometry is preserved and the sketched solution has provable relative-error guarantees. Increasing sketch size m improves accuracy but costs more time and memory. In large-scale ML, a small relative error in the solve is often sufficient because statistical noise dominates.

Remark 5.8.1 (Template) Compute $\widetilde{X} = SX$ and $\widetilde{y} = Sy$ and solve $\min_w \|\widetilde{X}w - \widetilde{y}\|_2^2$ by QR/SVD on the smaller problem.

5.8.2 Randomized Low-Rank Approximation

5.8.2.1 Randomized SVD Pipeline (Overview)

Randomized SVD (RSVD) approximates the dominant range by multiplying A with a random test matrix, orthonormalizing, then computing SVD of a small compressed matrix. This yields an approximate low-rank factorization with controllable error and usually requires only a small number of passes over A.

Remark 5.8.2 (RSVD Recap) Draw Ω, form $Y = A\Omega$, orthonormalize $Y \mapsto Q$, compress $B = Q^\top A$, compute SVD of B, lift back: $U = Q\widetilde{U}$. Power iterations improve accuracy when singular values decay slowly.

5.8.2.2 Error Guarantees Intuition and Parameter Choices

RSVD error depends on the tail singular values and the amount of oversampling and power iteration. Oversampling p reduces the probability of missing an important direction; power iterations amplify spectral gaps but add passes. The key qualitative guarantee is: the approximation error is close to optimal (like σ_{k+1} in spectral norm) up to modest factors.

Remark 5.8.3 (Parameter Heuristics) Typical choices: oversampling $p \in [5, 20]$ and power steps $q \in \{0, 1, 2\}$. These often yield near-optimal performance for ML-scale PCA.

5.8.3 Leverage Scores and Sampling

5.8.3.1 Why Leverage Matters for Nonuniform Sampling

Uniform sampling of rows/columns can fail when information is concentrated in a few influential rows/columns. Leverage scores quantify influence relative to a target subspace and provide the correct importance-sampling distribution. They are the spectral analogue of "hard examples" in ML: points with large leverage disproportionately affect the subspace fit.

Remark 5.8.4 (Leverage-Score Sampling Idea) Sample row i with probability proportional to ℓ_i (leverage score) and reweight appropriately; this preserves the subspace with fewer samples.

5.8.3.2 CUR/Nyström Approximations

CUR approximates a matrix by selecting actual columns and rows, improving interpretability (selected features/samples). Nyström approximates PSD kernel matrices using landmark points, reducing $O(n^2)$ memory. Both are sampling-based low-rank approximations whose quality depends on leverage-like quantities.

Remark 5.8.5 (Preview Forms)

$$A \approx CUR \quad \text{(CUR)}, \qquad K \approx CW^{\dagger}C^{\top} \quad \text{(Nyström for PSD } K\text{)}.$$

5.9 Numerics Inside ML Training (Bridges to Optimization)

Training deep models is an optimization problem executed in finite precision. Numerical linear algebra enters through gradient computation, conditioning of Jacobians/Hessians, and implicit regularization phenomena such as early stopping. This section connects numerical stability to core training issues.

5.9.1 Gradient Computations and Stability

5.9.1.1 Exploding/Vanishing Gradients as Conditioning Phenomena

Backpropagation multiplies by Jacobians (often many times). If Jacobians have singular values consistently >1, gradients explode; if <1, gradients vanish. This is a conditioning phenomenon for products of linear maps: the spectrum of the Jacobian product determines stability. Thus, controlling singular values of layers (via normalization, residual connections, careful initialization) is a numerical strategy.

Remark 5.9.1 (One-Line Jacobian Product View) If $h = f_L \circ \cdots \circ f_1$, then $J_h = J_{f_L} \cdots J_{f_1}$. The singular values of this product govern gradient magnitudes.

5.9.1.2 Orthogonal Initialization and Normalization Layers (Geometric View)

Orthogonal matrices have condition number 1 and preserve norms, making them ideal as initial linear maps. Normalization layers rescale activations to keep them in stable ranges and reduce anisotropy. Both can be viewed as geometry engineering to maintain well-conditioned Jacobians.

Remark 5.9.2 (Geometric Moral) Try to keep layerwise transformations close to isometries (or at least not too anisotropic) to preserve signal and gradient flow.

5.9.2 Second-Order Information

5.9.2.1 Hessians, Curvature, and Conditioning

Second-order methods depend on solving linear systems involving Hessians or approximations. Eigenvalues of Hessians determine curvature, step-size limits, and conditioning of Newton steps. In deep learning, exact Hessians are too large, so approximations (diagonal, low-rank, Gauss–Newton/Fisher) are used. The spectral viewpoint is essential: approximations attempt to capture dominant curvature modes efficiently.

Remark 5.9.3 (Curvature Governs Step Sizes) For a quadratic $f(w) = \frac{1}{2}w^\top Hw - g^\top w$, gradient descent convergence depends on $\kappa(H)$. Thus conditioning of curvature is a direct predictor of optimization difficulty.

5.9.2.2 Gauss–Newton and Fisher Approximations (High Level)

Gauss–Newton approximates Hessians by $J^\top J$ in least-squares-like settings, yielding PSD curvature. The Fisher information provides a statistical curvature that often stabilizes learning. Both lead to SPD systems that can be solved by Cholesky/CG with damping, directly tying optimization to numerical linear algebra.

Remark 5.9.4 (Damping as Regularization) Adding λI to curvature matrices (Levenberg–Marquardt/trust-region damping) is the second-order analogue of ridge: it improves SPD-ness and conditioning.

5.9.3 Early Stopping and Numerical Regularization

5.9.3.1 Optimization Tolerance vs. Statistical Precision

It is often pointless (and sometimes harmful) to drive training loss to machine precision when data are noisy. Early stopping acts as an implicit regularizer: it prevents fitting to noise and limits amplification of unstable directions. From a numerical perspective, stopping when residuals/gradients reach the noise floor avoids iterates dominated by rounding and variance.

Remark 5.9.5 (A Practical Principle) Stop when further reduction in training objective does not reduce validation error or when residual/gradient norms plateau at a scale consistent with noise.

5.9.3.2 Diagnostics: Residual Norms, Gradient Norms, and Validation

Numerical diagnostics help detect instability: residual stagnation, non-monotone decrease due to rounding, exploding gradient norms, and sensitivity to batch order or scaling. Tracking both optimization metrics (loss, residual, gradient norm) and statistical metrics (validation) is essential to distinguish numerical failure from statistical limitations.

Remark 5.9.6 (Three Useful Monitors)

$$\|r_t\|_2, \qquad \|\nabla f(w_t)\|_2, \qquad \text{validation loss/error.}$$

Together they triangulate progress, stability, and generalization.

5.10 Exercises

Unless stated otherwise, use Euclidean norms and assume real arithmetic when deriving *exact* answers; then interpret numerical stability/conditioning based on the principles in this chapter.

Exercise 5.10.1 (Floating-Point Rounding and Non-associativity (Explicit Toy Format)) Consider a toy floating-point system $\mathbb{F}$ that rounds to *two significant decimal digits* in base 10 with round-to-nearest (ties to even is not required). Let $\mathrm{fl}(\cdot)$ denote rounding to $\mathbb{F}$.

(i) Compute in $\mathbb{F}$ the values of $(a \oplus b) \oplus c$ and $a \oplus (b \oplus c)$ where

$$a = 9.9 \times 10^1, \qquad b = 1.1, \qquad c = -9.9 \times 10^1,$$

and $\oplus$ denotes floating-point addition ($x \oplus y := \mathrm{fl}(x + y)$); show explicitly that associativity fails.

(ii) Compute $d = \mathrm{fl}(a + c)$ and then $\mathrm{fl}(d + b)$ and compare with $\mathrm{fl}(a \oplus (c \oplus b))$.

(iii) Explain in one paragraph which operation incurred catastrophic cancellation and why (Definition 5.2.2).

Exercise 5.10.2 (Catastrophic Cancellation in Variance: Two Formulas) Let the data be $x_i = 10^6 + i$ for $i = 1, \ldots, 10$.

(i) Compute the exact mean $\bar{x}$ and the exact variance $\frac{1}{10}\sum_{i=1}^{10}(x_i - \bar{x})^2$.

(ii) Compute the exact value of the algebraically equivalent formula

$$\frac{1}{10}\sum_{i=1}^{10} x_i^2 - \bar{x}^2,$$

and verify it equals the variance from (i) in exact arithmetic.

(iii) Now assume rounding to 7 significant decimal digits after each arithmetic operation. Identify which expression is more numerically stable and justify (in one paragraph) using cancellation reasoning; you do *not* need to simulate all rounding steps, but you must point out precisely where the loss of significance arises.

Exercise 5.10.3 (Forward vs. Backward Error for a Scalar Computation) Consider the problem of computing $f(t) = \sqrt{1+t} - 1$ for $t = 10^{-8}$.

(i) Compute $f(t)$ to at least 12 correct decimal digits (using a series expansion or exact arithmetic reasoning).

(ii) Evaluate the naive formula $\sqrt{1+t} - 1$ and the stabilized formula $\frac{t}{\sqrt{1+t}+1}$ in finite precision with $u \approx 10^{-16}$ (double precision intuition) and argue which one has smaller *relative* forward error; provide a clear error-propagation argument, not just a statement.

(iii) Interpret the stabilized formula as a backward-stable reformulation: identify a nearby $\tilde{t}$ such that the computed value equals $f(\tilde{t})$ up to $O(u)$.

Exercise 5.10.4 (Condition Number and Sensitivity of a Linear System) Let

$$A = \begin{bmatrix} 1 & 1 \\ 1 & 1+\varepsilon \end{bmatrix}, \qquad \varepsilon = 10^{-4}, \qquad b = \begin{bmatrix} 2 \\ 2+\varepsilon \end{bmatrix}.$$

(i) Compute $\det(A)$ and A^{-1} exactly in terms of ε.

(ii) Compute $\kappa_2(A)$ by computing the singular values of A (to 4 decimals).

(iii) Let $b' = \begin{bmatrix} 2 \\ 2 \end{bmatrix}$. Solve $Ax = b$ and $Ax = b'$ exactly (in terms of ε) and compute $\|x - x'\|_2$.

(iv) Give a one-paragraph explanation connecting the large sensitivity in (iii) to the condition number from (ii) using Proposition 5.2.2.

Exercise 5.10.5 (Why Normal Equations Can Be Unstable: Conditioning Squared) Let

$$X = \begin{bmatrix} 1 & 1 \\ 1 & 1+\varepsilon \\ 1 & 1+2\varepsilon \end{bmatrix} \in \mathbb{R}^{3\times 2}, \qquad \varepsilon = 10^{-3}, \qquad y = \begin{bmatrix} 1 \\ 1 \\ 1 \end{bmatrix}.$$

(i) Compute $\kappa_2(X)$ by computing singular values of X (to 4 decimals).
(ii) Compute $\kappa_2(X^\top X)$ and verify numerically the identity $\kappa_2(X^\top X) = \kappa_2(X)^2$.
(iii) Compute the least squares solution $\widehat{w}$ using (a) normal equations $(X^\top X)\widehat{w} = X^\top y$ and (b) QR-based method $\widehat{w} = R^{-1}Q^\top y$ with $X = QR$ (thin QR), and compare the two solutions (they should match in exact arithmetic); then explain in one paragraph why method (b) is expected to be more stable in finite precision.

Exercise 5.10.6 (LU Factorization with Partial Pivoting) Let

$$A = \begin{bmatrix} 0 & 2 & 1 \\ 1 & -2 & -1 \\ 2 & 1 & 3 \end{bmatrix}, \qquad b = \begin{bmatrix} 1 \\ 0 \\ 4 \end{bmatrix}.$$

(i) Perform Gaussian elimination *without* pivoting and identify the first pivot failure (division by zero).
(ii) Perform LU with partial pivoting: find a permutation matrix P and factors L, U such that $PA = LU$.
(iii) Solve $Ax = b$ using the factorization via forward and backward substitution, and verify the result by direct multiplication.
(iv) Compute the growth factor $\rho = \frac{\max_{i,j} |u_{ij}|}{\max_{i,j} |a_{ij}|}$ for your U and comment briefly on its implication for stability.

Exercise 5.10.7 (Cholesky Factorization and "jitter") Let

$$K = \begin{bmatrix} 1 & 0.99 & 0.98 \\ 0.99 & 1 & 0.99 \\ 0.98 & 0.99 & 1 \end{bmatrix}.$$

(i) Compute the eigenvalues of K (to 6 decimals) and determine whether $K \succ 0$.
(ii) Attempt a Cholesky factorization $K = LL^\top$; if it fails numerically (or if the smallest eigenvalue is extremely small), set $K_\lambda = K + \lambda I$ with $\lambda = 10^{-3}$ and compute the Cholesky factor L_λ.
(iii) Solve $(K_\lambda)x = y$ for $y = \begin{bmatrix} 1 \\ 0 \\ -1 \end{bmatrix}$ using Cholesky (two triangular solves), and report x.

(iv) Explain in one paragraph why adding jitter improves numerical stability and how it relates to ridge regularization.

Exercise 5.10.8 (Triangular Solves and Reuse Across Multiple Right-Hand Sides) Let

$$L = \begin{bmatrix} 2 & 0 & 0 \\ -1 & 1 & 0 \\ 3 & 2 & 1 \end{bmatrix}, \qquad U = \begin{bmatrix} 1 & -1 & 2 \\ 0 & 2 & 1 \\ 0 & 0 & -1 \end{bmatrix}, \qquad A = LU.$$

Let $B = \begin{bmatrix} 1 & 0 \\ 2 & 1 \\ 0 & 3 \end{bmatrix}$ be two right-hand sides.

(i) Compute A explicitly.
(ii) Solve $AX = B$ by first solving $LY = B$ for Y (forward substitution) and then $UX = Y$ (back substitution), computing both columns of X.
(iii) Count the number of scalar multiply-add operations required for the two solves and compare against the cost of explicitly forming A^{-1} (conceptually) to justify why reuse of factorization is efficient.

Exercise 5.10.9 (QR vs. Cholesky on a Well-Conditioned SPD System) Let $X \in \mathbb{R}^{5\times 3}$ and $y \in \mathbb{R}^5$ be

$$X = \begin{bmatrix} 1 & 0 & 0 \\ 1 & 1 & 0 \\ 1 & 0 & 1 \\ 1 & 1 & 1 \\ 1 & 2 & 1 \end{bmatrix}, \qquad y = \begin{bmatrix} 1 \\ 2 \\ 2 \\ 3 \\ 4 \end{bmatrix}.$$

(i) Compute the normal matrix $G = X^\top X$ and verify that $G \succ 0$.
(ii) Solve $(X^\top X)\widehat{w} = X^\top y$ using Cholesky on G.
(iii) Compute a thin QR factorization $X = QR$ and solve $\widehat{w} = R^{-1}Q^\top y$.
(iv) Verify the two solutions match (up to rounding) and compute $\kappa_2(X)$ and $\kappa_2(G)$; comment on why this example is safe for normal equations.

Exercise 5.10.10 (SVD-Based Least Squares and Rank Deficiency) Let

$$X = \begin{bmatrix} 1 & 2 & 3 \\ 2 & 4 & 6 \\ 1 & 0 & 1 \\ 0 & 1 & 1 \end{bmatrix} \in \mathbb{R}^{4\times 3}, \qquad y = \begin{bmatrix} 1 \\ 2 \\ 0 \\ 1 \end{bmatrix}.$$

(i) Show that X is rank-deficient by exhibiting a nonzero vector v with $Xv = 0$.
(ii) Compute the minimum-norm least squares solution $\widehat{w} = X^\dagger y$ by computing the SVD of X (you may compute $X^\top X$ and diagonalize).

(iii) Compute the fitted vector $\widehat{y} = X\widehat{w}$ and residual norm $\|y - \widehat{y}\|_2$.
(iv) Describe the full set of least squares minimizers as $\widehat{w} + v$ with $v \in \mathcal{N}(X)$ and verify that your $\widehat{w}$ is orthogonal to $\mathcal{N}(X)$.

Exercise 5.10.11 (Truncated SVD as Regularization (Numerical)) Consider the diagonal matrix

$$\Sigma = \operatorname{diag}(10, 1, 0.01) \in \mathbb{R}^{3\times 3}, \quad U = V = I_3, \quad X = U\Sigma V^\top = \Sigma, \quad y = \begin{bmatrix} 1 \\ 1 \\ 1 \end{bmatrix}.$$

(i) Compute the pseudoinverse solution $w = X^\dagger y$ and report $\|w\|_2$.
(ii) Compute the truncated-SVD solution $w_{(k)} = X_k^\dagger y$ for $k = 1, 2$ and report $\|w_{(k)}\|_2$ and residual norms $\|Xw_{(k)} - y\|_2$.
(iii) Explain in one paragraph why the tiny singular value 0.01 causes large coefficient magnitudes and why truncation stabilizes.

Exercise 5.10.12 (Ridge as Stabilized Inversion: Compare with TSVD) Using the same $X = \Sigma$ and y as in Exercise 5.10.11, compute ridge solutions

$$w_\lambda = (X^\top X + \lambda I)^{-1} X^\top y$$

for $\lambda \in \{10^{-6}, 10^{-2}, 1\}$. Report $\|w_\lambda\|_2$ and $\|Xw_\lambda - y\|_2$ and compare against TSVD solutions from Exercise 5.10.11. Then, using the SVD filter-factor viewpoint, identify the effective attenuation factors applied to each singular direction.

Exercise 5.10.13 (Gradient Descent for Least Squares: Step Size and Stopping) Let

$$X = \begin{bmatrix} 2 & 0 \\ 0 & 1 \\ 1 & 1 \end{bmatrix} \in \mathbb{R}^{3\times 2}, \qquad y = \begin{bmatrix} 2 \\ 1 \\ 2 \end{bmatrix}.$$

Define $f(w) = \frac{1}{2}\|Xw - y\|_2^2$.

(i) Compute $L = \|X^\top X\|_2$ by finding eigenvalues of $X^\top X$ and hence give a step size $\eta \in (0, 2/L)$.
(ii) Starting from $w^{(0)} = (0, 0)^\top$, perform 5 iterations of gradient descent $w^{(t+1)} = w^{(t)} - \eta\nabla f(w^{(t)})$ and report $w^{(t)}$ and $\|Xw^{(t)} - y\|_2$ for $t = 1, \ldots, 5$.
(iii) Compute the exact least squares minimizer $\widehat{w}$ by normal equations and compare $\|w^{(5)} - \widehat{w}\|_2$.
(iv) Provide a one-paragraph stopping rule recommendation based on residual norm vs. an assumed noise floor $\sigma = 10^{-2}$ (qualitative, but explicit).

Exercise 5.10.14 (Conjugate Gradient on an SPD System) Let

$$H = \begin{bmatrix} 4 & 1 & 0 \\ 1 & 3 & 1 \\ 0 & 1 & 2 \end{bmatrix} \succ 0, \qquad g = \begin{bmatrix} 1 \\ 2 \\ 0 \end{bmatrix}.$$

(i) Solve $Hw = g$ exactly (by any direct method) to obtain $w^\star$.
(ii) Run two iterations of conjugate gradient starting from $w^{(0)} = 0$ (write out $r^{(0)}, p^{(0)}, \alpha_0$, then $r^{(1)}, p^{(1)}, \alpha_1$) and obtain $w^{(1)}, w^{(2)}$.
(iii) Compute the H-norm errors $\|w^{(t)} - w^\star\|_H := \sqrt{(w^{(t)} - w^\star)^\top H(w^{(t)} - w^\star)}$ for $t = 1, 2$ and comment briefly on the improvement.

Exercise 5.10.15 (Sparse Matrix Formats and matvec Cost) Consider the sparse matrix $A \in \mathbb{R}^{6\times 6}$ with nonzeros

$$A_{11} = 4,\ A_{12} = -1,\ A_{21} = -1,\ A_{22} = 4,\ A_{23} = -1,\ A_{32} = -1,\ A_{33} = 4,$$

$$A_{34} = -1,\ \ldots,\ A_{66} = 4,\ A_{65} = -1,$$

i.e., A is the tridiagonal matrix with 4 on the diagonal and -1 on the super/sub-diagonals.

(i) Write A explicitly and compute $\text{nnz}(A)$.
(ii) Compute Ax for $x = (1, 1, 1, 1, 1, 1)^\top$ using only the nonzero pattern (do not multiply by zeros).
(iii) Give the CSR representation (row pointers, column indices, values) explicitly.
(iv) Compare the arithmetic operation count of dense matvec vs. sparse matvec in this case.

Exercise 5.10.16 (Fill-In in Sparse Cholesky (Small Graph)) Let $A \in \mathbb{R}^{5\times 5}$ be SPD with sparsity pattern (nonzero off-diagonals) corresponding to edges

$$E = \{(1, 2), (2, 3), (3, 4), (4, 5), (2, 4)\},$$

and diagonal entries all $= 10$, off-diagonal entries all $= -1$ for $(i, j) \in E$ (and symmetric).

(i) Write A explicitly.
(ii) Perform one step of Cholesky elimination eliminating variable 2 first and determine which new fill-in edges among $\{1, 3, 4\}$ are created (i.e., which zero entries become nonzero in the Schur complement).
(iii) Repeat eliminating variable 3 first and compare the fill-in pattern. Conclude (one paragraph) why ordering matters.

Exercise 5.10.17 (Toeplitz Structure and Fast matvec (Explicit)) Let $T \in \mathbb{R}^{6\times 6}$ be Toeplitz with first row $(2, -1, 0, 0, 0, 0)$ and first column $(2, -1, 0, 0, 0, 0)^\top$ (so T is the same tridiagonal as in Exercise 5.10.15 but viewed as Toeplitz).

(i) Compute Tx for $x = (1, 2, 3, 4, 5, 6)^\top$.
(ii) Embed T into a 10×10 circulant matrix C (describe the first row of C explicitly) such that the leading 6 components of $C\tilde{x}$ equal Tx for an appropriately padded $\tilde{x}$; this is the standard Toeplitz-to-circulant embedding.
(iii) Explain in one paragraph how FFT-based multiplication would reduce cost asymptotically (no FFT computation required).

Exercise 5.10.18 (Sketched Least Squares with a Fixed Sketch Matrix) Let

$$X = \begin{bmatrix} 1 & 0 \\ 0 & 1 \\ 1 & 1 \\ 2 & 1 \\ 1 & 2 \\ 2 & 2 \end{bmatrix} \in \mathbb{R}^{6\times 2}, \qquad y = \begin{bmatrix} 1 \\ 1 \\ 2 \\ 3 \\ 3 \\ 4 \end{bmatrix}.$$

Let the sketch matrix $S \in \mathbb{R}^{3\times 6}$ be

$$S = \frac{1}{\sqrt{2}} \begin{bmatrix} 1 & 0 & 1 & 0 & 0 & 0 \\ 0 & 1 & 0 & 1 & 0 & 0 \\ 0 & 0 & 0 & 0 & 1 & 1 \end{bmatrix}.$$

(i) Compute $\widetilde{X} = SX$ and $\widetilde{y} = Sy$.
(ii) Solve the sketched least squares problem $\min_w \|\widetilde{X}w - \widetilde{y}\|_2^2$ exactly and report $\widetilde{w}$.
(iii) Solve the full least squares problem $\min_w \|Xw - y\|_2^2$ and report $\widehat{w}$.
(iv) Compute the relative residual increase $\frac{\|X\widetilde{w}-y\|_2 - \|X\widehat{w}-y\|_2}{\|X\widehat{w}-y\|_2}$ and comment briefly on the speed–accuracy tradeoff.

Exercise 5.10.19 (Exploding/Vanishing Gradients via Singular Values (Linear Network)) Consider a depth-3 linear network $f(x) = W_3 W_2 W_1 x$ with

$$W_1 = \begin{bmatrix} 2 & 0 \\ 0 & \frac{1}{2} \end{bmatrix}, \quad W_2 = \begin{bmatrix} 1 & 1 \\ 0 & 1 \end{bmatrix}, \quad W_3 = \begin{bmatrix} 1 & 0 \\ 0 & 2 \end{bmatrix}.$$

(i) Compute $J := W_3 W_2 W_1$ explicitly and compute $\|J\|_2$ and $\sigma_{\min}(J)$ (by diagonalizing $J^\top J$).
(ii) For the scalar loss $L(x) = \frac{1}{2}\|f(x)\|_2^2$, show that $\nabla_x L(x) = J^\top J x$ and compute $\|\nabla_x L(x)\|_2$ for $x = (1, 1)^\top$.
(iii) Interpret (one paragraph) how the singular values of J control gradient magnitudes and relate to exploding/vanishing behavior.

Exercise 5.10.20 (Early Stopping as Numerical Regularization (Toy Ridge Effect)) Let $X = \Sigma = \mathrm{diag}(10, 1, 0.1)$ and $y = (1, 1, 1)^\top$ and consider gradient descent on $f(w) = \frac{1}{2}\|Xw - y\|_2^2$ with step size $\eta = 0.01$ starting from $w^{(0)} = 0$.

(i) Show that the coordinates decouple and satisfy

$$w_i^{(t+1)} = w_i^{(t)} - \eta\sigma_i(\sigma_i w_i^{(t)} - y_i)$$

for $\sigma_1 = 10, \sigma_2 = 1, \sigma_3 = 0.1$.

(ii) Compute $w^{(t)}$ explicitly for $t = 1, 2, 3, 4$ (closed form recursion is fine) and report $\|w^{(t)}\|_2$.

(iii) Compare with the exact least squares solution $w^\star = X^\dagger y$ and explain in one paragraph why early iterations behave like implicit regularization by suppressing the small singular direction.

Chapter 6
Probability Spaces and Random Variables

Learning is fundamentally the problem of making reliable decisions from limited, noisy data. Probability provides the language to (i) represent uncertainty in data and models, (ii) quantify evidence and update beliefs, and (iii) design optimal decision rules under well-specified loss functions. This chapter establishes the foundational measure-theoretic objects (sample spaces, σ-algebras, probability measures), builds the event calculus used throughout generalization and inference, and introduces conditional probability as the algebraic core behind likelihoods and Bayes' rule (Durrett 2019).

6.1 Why Probability Is the Language of Learning

6.1.1 Uncertainty in Data, Models, and Decisions

6.1.1.1 Aleatoric vs. Epistemic Uncertainty (Informal)

Two qualitatively different sources of uncertainty arise in ML. *Aleatoric* uncertainty reflects irreducible randomness in the data generation process (measurement noise, inherent variability of outcomes). *Epistemic* uncertainty reflects lack of knowledge about the data-generating mechanism (finite data, model misspecification); it can, in principle, shrink with more information. Separating these notions clarifies what confidence intervals can achieve, why ensembles and Bayesian posteriors behave differently from noise models, and why "uncertainty estimates" must be interpreted relative to assumptions (cf. Fig. 6.1).

Remark 6.1.1 (Aleatoric vs. Epistemic (Working Definitions)) Informally:

- **Aleatoric:** variability in Y even when X and the true mechanism are fixed (intrinsic noise).

P. Singh, B. Raman, *Foundations of Machine Learning and AI*, Studies in Big Data 199, https://doi.org/10.1007/978-3-032-30336-3_6

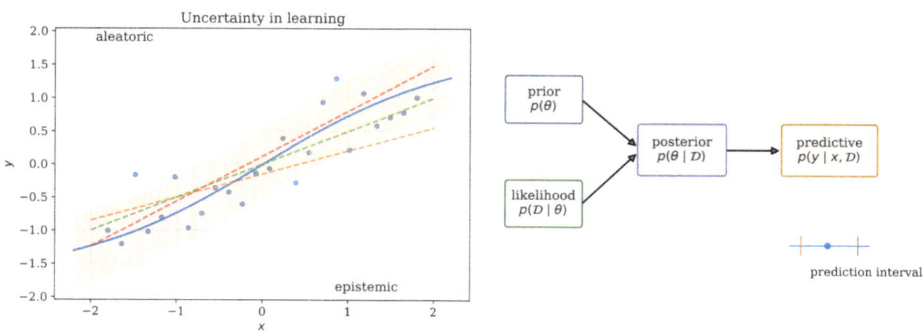

Fig. 6.1 Two probabilistic views of learning. Left: aleatoric uncertainty appears as outcome variability around a fixed input, while epistemic uncertainty appears as multiple plausible predictive rules under limited data. Right: a probabilistic pipeline maps prior assumptions and a likelihood to posterior beliefs and then to predictive distributions and intervals

- **Epistemic:** uncertainty about the mechanism (unknown parameters, unknown function class, limited data).

In a parametric model $p_\theta(y \mid x)$, aleatoric uncertainty is encoded in the conditional distribution itself, while epistemic uncertainty corresponds to uncertainty about θ (or about the model class).

6.1.1.2 Noise Models, Likelihoods, and Prediction Intervals

A learning problem is often specified by a *noise model*—a conditional distribution of outputs given inputs. Training then becomes choosing parameters that make observed data likely (MLE) or likely under a prior (MAP/Bayes). Once a conditional distribution is fitted, we can produce *prediction intervals* that quantify plausible outcomes for a new input. This connects probability directly to decision-making: point prediction minimizes a loss, while interval prediction controls risk under coverage constraints.

Definition 6.1.1 (Conditional Model and Likelihood) Let $\{(x_i, y_i)\}_{i=1}^n$ be data and let $\{p_\theta(y \mid x) : \theta \in \Theta\}$ be a family of conditional distributions. The *likelihood* of θ given data is

$$L(\theta; \mathcal{D}) := \prod_{i=1}^{n} p_\theta(y_i \mid x_i), \qquad \ell(\theta; \mathcal{D}) := \log L(\theta; \mathcal{D}) = \sum_{i=1}^{n} \log p_\theta(y_i \mid x_i).$$

Remark 6.1.2 (Prediction Interval (Conceptual)) Given a fitted conditional distribution $\widehat{p}(y \mid x)$ and confidence level $1-\alpha$, a *prediction interval* is a set $I_\alpha(x) \subseteq \mathcal{Y}$ satisfying (ideally)

$$\mathbb{P}(Y \in I_\alpha(x) \mid X = x) \geq 1 - \alpha.$$

Constructing and validating such intervals is a probabilistic task and will rely on conditioning and concentration later.

6.1.2 *Probability as "Measure of Evidence"*

6.1.2.1 From Frequencies to Beliefs: What ML Uses Operationally

Operationally, ML uses probability in two complementary ways. In classical (frequentist) modeling, probabilities describe long-run frequencies and enable performance guarantees under repeated sampling. In Bayesian modeling, probabilities quantify degrees of belief conditioned on observed data. Even when one is not philosophically Bayesian, the *computational* act of updating beliefs is ubiquitous: posterior-like updates, regularization-as-prior, and uncertainty quantification all follow Bayes-style algebra.

Remark 6.1.3 (Two Operational Roles of Probability)

- **As a data-generating model:** probabilities specify what data you should expect under assumptions; analysis then produces guarantees.
- **As an update rule:** probabilities combine prior information and evidence via Bayes' rule, producing calibrated uncertainty when assumptions are correct.

Both roles require the same core calculus of events and conditioning.

6.1.2.2 Calibration, Scoring Rules, and Probabilistic Predictions

Probabilistic predictions should mean something: predicted probabilities should match empirical frequencies under the model's own notion of uncertainty. *Calibration* formalizes this consistency. *Scoring rules* provide principled objectives for learning probabilities: they assign a numerical score to a probabilistic forecast and the realized outcome. Proper scoring rules reward honest probabilities and connect directly to maximum likelihood (log score) (Gneiting and Raftery 2007). These ideas motivate why we fit likelihoods, why cross-entropy is used, and how to evaluate probabilistic forecasters.

Definition 6.1.2 (Calibration (Binary, Population-Level)) Let $p(X) \in [0, 1]$ be a predicted probability for a binary label $Y \in \{0, 1\}$. We say p is *calibrated* if for all $t \in [0, 1]$ (in an appropriate sense),

$$\mathbb{P}(Y = 1 \mid p(X) = t) = t.$$

Definition 6.1.3 (Scoring Rule and Propriety) A *scoring rule* is a function $S(q, y)$ that assigns a numerical score to a quoted distribution q and an outcome y. It is *proper* if, when the true distribution is p, the expected score is minimized by quoting $q = p$:

$$\mathbb{E}_{Y\sim p}\big[S(q,Y)\big] \ge \mathbb{E}_{Y\sim p}\big[S(p,Y)\big] \qquad \forall q.$$

Example 6.1.1 (Log Score and Cross-Entropy) For discrete outcomes, the log score is $S(q, y) = -\log q(y)$. Minimizing average log score over data yields maximum likelihood / cross-entropy training.

6.2 Probability Axioms and Events

6.2.1 Foundational Objects

6.2.1.1 Sample Spaces and Events as Sets

Probability begins with a set-theoretic description of uncertainty. The sample space Ω contains all possible outcomes; events are subsets of Ω. This matches ML practice: events describe success/failure, threshold exceedance, misclassification, etc. Once events are sets, probability is a set function satisfying axioms; the resulting calculus is what we repeatedly use in generalization, tail bounds, and inference.

Definition 6.2.1 (Sample Space and Event) A *sample space* Ω is a nonempty set whose elements $\omega \in \Omega$ are called outcomes. An *event* is a subset $E \subseteq \Omega$.

Example 6.2.1 (Events from Learning Problems) Let (X, Y) denote random data and $\widehat{h}$ a learned classifier. Typical events include:

$$E_{\text{err}} = \{\widehat{h}(X) \ne Y\}, \qquad E_{\text{large}} = \{\|X\|_2 \ge t\}, \qquad E_{\text{margin}} = \{Y(w^\top X + b) \le \gamma\}.$$

All are subsets of the underlying outcome space.

6.2.1.2 σ-Algebras (Intuition) and Countable Additivity

Not every subset of Ω is always assigned a probability in advanced settings (especially when Ω is uncountable). A σ-algebra specifies which events are measurable (i.e., can be assigned probabilities consistently) and is closed under the operations we need: complements and countable unions. Countable additivity is the core axiom that distinguishes probability measures from finitely additive set functions; it enables limits, convergence results, and all major theorems later (LLN/CLT, concentration).

Definition 6.2.2 (σ-Algebra) A collection $\mathcal{F} \subseteq \mathcal{P}(\Omega)$ is a *σ-algebra* if:

1. $\Omega \in \mathcal{F}$,
2. $E \in \mathcal{F} \Rightarrow E^c \in \mathcal{F}$,

3. if $E_1, E_2, \dots \in \mathcal{F}$, then $\bigcup_{i=1}^{\infty} E_i \in \mathcal{F}$.

Elements of $\mathcal{F}$ are called *measurable events.*

Remark 6.2.1 (Closure Consequences) From the axioms, $\mathcal{F}$ is also closed under countable intersections (by De Morgan) and contains $\emptyset$ (since $\emptyset = \Omega^c$).

Definition 6.2.3 (Probability Measure and Probability Space) A *probability measure* is a function $\mathbb{P} : \mathcal{F} \to [0, 1]$ such that:

1. $\mathbb{P}(\Omega) = 1$,
2. (*Countable additivity*) if $E_1, E_2, \dots \in \mathcal{F}$ are pairwise disjoint, then

$$\mathbb{P}\left(\bigcup_{i=1}^{\infty} E_i\right) = \sum_{i=1}^{\infty} \mathbb{P}(E_i).$$

A *probability space* is a triple $(\Omega, \mathcal{F}, \mathbb{P})$.

6.2.2 *Probability Measures and Basic Identities*

6.2.2.1 Inclusion–Exclusion, Monotonicity, and Continuity of Probability

Once countable additivity is assumed, a rich algebra follows. Inclusion–exclusion corrects overcounting in unions; monotonicity formalizes that larger events have larger probability; and continuity formalizes that probabilities behave well under increasing/decreasing limits of events. These are the technical tools behind almost every probability argument in ML: bounding error events, controlling multiple hypotheses, and passing from finite to infinite approximations.

Proposition 6.2.1 (Basic Consequences) *Let $(\Omega, \mathcal{F}, \mathbb{P})$ be a probability space and let $E, F \in \mathcal{F}$. Then:*

1. $\mathbb{P}(\emptyset) = 0$ *and* $\mathbb{P}(E^c) = 1 - \mathbb{P}(E)$.
2. *(*Monotonicity*) If $E \subseteq F$, then $\mathbb{P}(E) \leq \mathbb{P}(F)$.*
3. *(*Two-set inclusion–exclusion*)*

$$\mathbb{P}(E \cup F) = \mathbb{P}(E) + \mathbb{P}(F) - \mathbb{P}(E \cap F).$$

Proposition 6.2.2 (Continuity from Below and Above) *Let $E_1 \subseteq E_2 \subseteq \cdots$ be an increasing sequence of events and $E = \bigcup_{n=1}^{\infty} E_n$. Then*

$$\mathbb{P}(E) = \lim_{n \to \infty} \mathbb{P}(E_n).$$

If $F_1 \supseteq F_2 \supseteq \cdots$ *is decreasing and* $F = \bigcap_{n=1}^{\infty} F_n$, *then*

$$\mathbb{P}(F) = \lim_{n \to \infty} \mathbb{P}(F_n).$$

Remark 6.2.2 (Why Continuity Matters) Continuity lets us approximate complicated events by simpler ones (finite unions/intersections) and pass to limits. This is essential for convergence theorems and for bounding events via coverings/approximations later.

6.2.2.2 Union Bounds and Why They Appear Everywhere in ML

The union bound (Boole's inequality) is the simplest high-probability tool: it upper-bounds the probability that *any* of many bad events occurs by the sum of their probabilities. This is the workhorse behind multiple testing, confidence bounds with failure probability δ, and uniform generalization bounds (e.g., simultaneously controlling error over many hypotheses). Even when it is loose, it is powerful because it converts a combinatorial "exists" into a sum of manageable terms.

Theorem 6.2.1 (Union Bound) *For any events* $E_1, E_2, \cdots \in \mathcal{F}$,

$$\mathbb{P}\left(\bigcup_{i=1}^{\infty} E_i\right) \leq \sum_{i=1}^{\infty} \mathbb{P}(E_i).$$

Proof Define disjoint events $F_1 = E_1$ and $F_i = E_i \setminus \bigcup_{j=1}^{i-1} E_j$ for $i \geq 2$. Then $\bigcup_i F_i = \bigcup_i E_i$ and $\mathbb{P}(F_i) \leq \mathbb{P}(E_i)$. By countable additivity,

$$\mathbb{P}\left(\bigcup_i E_i\right) = \mathbb{P}\left(\bigcup_i F_i\right) = \sum_i \mathbb{P}(F_i) \leq \sum_i \mathbb{P}(E_i).$$

□

Remark 6.2.3 (A Canonical ML Usage Pattern) To guarantee that *all* of N conditions hold simultaneously with probability at least $1 - \delta$, it suffices to ensure each condition fails with probability at most δ/N and apply the union bound. This "δ/N trick" appears repeatedly in concentration-based generalization arguments.

6.2.3 Conditional Probability

6.2.3.1 Definition and Basic Properties

Conditioning formalizes "restrict to the world where B happened." It is the algebraic mechanism behind likelihoods, posterior updates, and sequential reasoning. In ML, conditional probability appears as $p(y \mid x)$ (noise models), $p(\theta \mid \mathcal{D})$ (posteriors), and $p(\text{error} \mid \text{train})$ (generalization analysis). Mastery of its identities (product rule, chain rule, law of total probability) is essential.

Definition 6.2.4 (Conditional Probability) Let $E, F \in \mathcal{F}$ with $\mathbb{P}(F) > 0$. The *conditional probability* of E given F is

$$\mathbb{P}(E \mid F) := \frac{\mathbb{P}(E \cap F)}{\mathbb{P}(F)}.$$

Proposition 6.2.3 (Product Rule and Chain Rule) *If* $\mathbb{P}(F) > 0$, *then*

$$\mathbb{P}(E \cap F) = \mathbb{P}(E \mid F)\,\mathbb{P}(F) = \mathbb{P}(F \mid E)\,\mathbb{P}(E).$$

More generally, if $E_1, \ldots, E_k$ *have positive probability where needed, then*

$$\mathbb{P}\left(\bigcap_{i=1}^{k} E_i\right) = \mathbb{P}(E_1) \prod_{i=2}^{k} \mathbb{P}\left(E_i \,\middle|\, \bigcap_{j=1}^{i-1} E_j\right).$$

Proposition 6.2.4 (Law of Total Probability) *Let* $\{F_i\}_{i \in I}$ *be a finite or countable partition of* Ω *with* $F_i \in \mathcal{F}$, F_i *disjoint,* $\bigcup_i F_i = \Omega$, *and* $\mathbb{P}(F_i) > 0$. *Then for any event* E,

$$\mathbb{P}(E) = \sum_{i \in I} \mathbb{P}(E \mid F_i)\,\mathbb{P}(F_i).$$

6.2.3.2 Bayes' Rule as Inversion of Conditioning

Bayes' rule expresses an "inversion" principle: it converts $P(E \mid F)$ into $P(F \mid E)$ by reweighting with prior probabilities. In ML this becomes: *posterior* $\propto$ *likelihood* $\times$ *prior*. It is the algebraic foundation of Bayesian inference, naive Bayes classification, and MAP estimation, and it is also a conceptual foundation for many regularization interpretations.

Theorem 6.2.2 (Bayes' Rule (Events)) *Let* $E, F \in \mathcal{F}$ *with* $\mathbb{P}(E) > 0$ *and* $\mathbb{P}(F) > 0$. *Then*

$$\mathbb{P}(E \mid F) = \frac{\mathbb{P}(F \mid E)\,\mathbb{P}(E)}{\mathbb{P}(F)}.$$

More generally, if $\{E_i\}$ *is a partition with* $\mathbb{P}(E_i) > 0$, *then*

$$\mathbb{P}(E_j \mid F) = \frac{\mathbb{P}(F \mid E_j)\,\mathbb{P}(E_j)}{\sum_i \mathbb{P}(F \mid E_i)\,\mathbb{P}(E_i)}.$$

Proof By the product rule, $\mathbb{P}(E \cap F) = \mathbb{P}(F \mid E)\mathbb{P}(E) = \mathbb{P}(E \mid F)\mathbb{P}(F)$. Rearrange to obtain the formula. The partition form follows by applying the law of total probability to $\mathbb{P}(F)$. □

Remark 6.2.4 (Likelihood × Prior/Evidence) In parametric modeling with parameter θ and data $\mathcal{D}$, Bayes' rule yields

$$p(\theta \mid \mathcal{D}) \propto p(\mathcal{D} \mid \theta)\, p(\theta),$$

where the proportionality constant is the evidence $p(\mathcal{D})$. This is the probabilistic counterpart of regularization-as-prior.

6.3 Random Variables and Distributions

A probability space $(\Omega, \mathcal{F}, \mathbb{P})$ describes randomness abstractly. A *random variable* turns that abstraction into a concrete object we can compute with: it is a map from outcomes to numbers/vectors (cf. Fig. 6.2). Distributions (pmf/pdf/cdf) summarize how a random variable behaves, and families of distributions provide a modeling toolkit. This section formalizes random variables as measurable maps, defines distribution functions, lists core distribution families, and discusses how support and tail behavior influence learning.

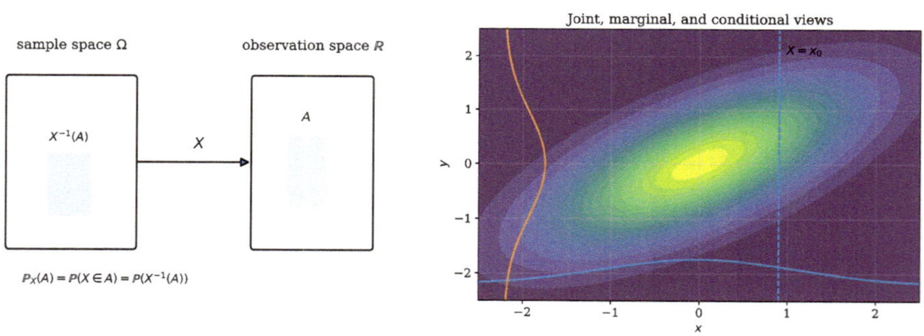

Fig. 6.2 Random variables turn abstract outcomes into observable quantities. Left: a measurable map sends subsets of the observation space back to events in the sample space via preimages. Right: a joint distribution induces marginals by summing or integrating out variables, and conditional structure corresponds to normalized slices of the joint

6.3.1 *Random Variables as Measurable Maps*

6.3.1.1 Real-Valued and Vector-Valued Random Variables

In ML, we model uncertain quantities (labels, features, noise, parameters) as random variables. Measure-theoretically, a random variable is a measurable function, ensuring that events defined by thresholding the variable (e.g., $\{X \le t\}$) are legitimate events in $\mathcal{F}$. Vector-valued random variables bundle multiple coordinates and naturally model feature vectors. Most computations reduce to one of two perspectives: either reason about events via preimages, or reason about expectations by integrating against a distribution.

Definition 6.3.1 (Random Variable (Measurable Map)) Let $(\Omega, \mathcal{F}, \mathbb{P})$ be a probability space and let $(\mathcal{S}, \mathcal{G})$ be a measurable space. A map $X : \Omega \to \mathcal{S}$ is a *random variable* (or random element) if it is measurable:

$$X^{-1}(G) \in \mathcal{F} \quad \forall G \in \mathcal{G}.$$

Definition 6.3.2 (Real-Valued and Vector-Valued Random Variables)

- A *real-valued* random variable is $X : \Omega \to \mathbb{R}$ measurable w.r.t. $(\mathbb{R}, \mathcal{B}(\mathbb{R}))$, where $\mathcal{B}(\mathbb{R})$ is the Borel σ-algebra.
- A *vector-valued* random variable is $X : \Omega \to \mathbb{R}^d$ measurable w.r.t. $(\mathbb{R}^d, \mathcal{B}(\mathbb{R}^d))$.

Remark 6.3.1 (Coordinates Are Random Variables) If $X = (X_1, \dots, X_d) : \Omega \to \mathbb{R}^d$ is measurable, then each coordinate $X_i : \Omega \to \mathbb{R}$ is a real-valued random variable. Conversely, a vector-valued random variable is measurable iff all coordinates are measurable.

6.3.1.2 Events Generated by Random Variables (Preimages)

Events in probability are sets in Ω, but ML statements are typically in data space: "X falls in a region," "loss exceeds ε," "classifier errs." These are all preimages under random variables. This is why preimages are the clean objects: they commute with unions and intersections, so event algebra can be performed in the observable space and pulled back to Ω. This perspective also anticipates measurability requirements: if X is measurable and A is Borel, then $\{X \in A\}$ is a valid event.

Definition 6.3.3 (Events Induced by a Random Variable) Let $X : \Omega \to \mathbb{R}^d$ be a random variable and let $A \subseteq \mathbb{R}^d$ be Borel. The event that X falls in A is

$$\{X \in A\} := X^{-1}(A) = \{\omega \in \Omega : X(\omega) \in A\} \in \mathcal{F}.$$

For real-valued X, typical events are $\{X \le t\}$, $\{|X| \ge t\}$, $\{\|X\|_2 \le r\}$.

Example 6.3.1 (Misclassification as an Event) Let (X, Y) be random data and $\widehat{h}$ a (possibly random) classifier. Then the error event is

$$E_{\text{err}} = \{\widehat{h}(X) \neq Y\} \in \mathcal{F}.$$

Generalization bounds control $\mathbb{P}(E_{\text{err}})$ or its conditional variants.

6.3.2 Distribution Functions

6.3.2.1 PMF/PDF/CDF and How to Move Between Them

A distribution is the pushforward of $\mathbb{P}$ under X. In practice we use three representations: pmf for discrete variables, pdf for absolutely continuous variables, and cdf for general real variables. Moving between them is a computational skill: sums vs. integrals, derivatives of cdfs, and marginalization by summing/integrating out variables. In ML, log-likelihoods depend on pmf/pdf values, while calibration and thresholding often use cdfs.

Definition 6.3.4 (Distribution (Pushforward)) Let $X : \Omega \to (\mathcal{S}, \mathcal{G})$ be measurable. The *distribution* of X is the probability measure $\mathbb{P}_X$ on $(\mathcal{S}, \mathcal{G})$ defined by

$$\mathbb{P}_X(A) := \mathbb{P}(X \in A) = \mathbb{P}(X^{-1}(A)) \qquad \forall A \in \mathcal{G}.$$

Definition 6.3.5 (CDF) For a real-valued random variable X, the *cumulative distribution function* (CDF) is

$$F_X(t) := \mathbb{P}(X \leq t), \qquad t \in \mathbb{R}.$$

Definition 6.3.6 (PMF) If X is discrete with countable support $\mathcal{S} \subseteq \mathbb{R}$, its *probability mass function* (PMF) is

$$p_X(x) := \mathbb{P}(X = x), \qquad x \in \mathcal{S}.$$

Then for any $A \subseteq \mathcal{S}$, $\mathbb{P}(X \in A) = \sum_{x \in A} p_X(x)$.

Definition 6.3.7 (PDF) If X is absolutely continuous w.r.t. Lebesgue measure, it has a *probability density function* (PDF) f_X such that

$$\mathbb{P}(X \in A) = \int_A f_X(x)\, dx$$

for all Borel $A \subseteq \mathbb{R}$, and $F_X(t) = \int_{-\infty}^t f_X(x)\, dx$.

Remark 6.3.2 (From CDF to PDF) If F_X is differentiable at t, then $f_X(t) = F'_X(t)$. Conversely, integrating the PDF yields the CDF.

6.3.2.2 Expectations as Sums/Integrals

Expectation is the main numerical summary of random variables and the foundation of risk minimization. Computationally, expectation is a weighted sum (discrete) or integral (continuous). This preview focuses on correct formulas and their interpretation as averages under a distribution; deeper properties (linearity, conditioning, variance) follow in the next chapter.

Definition 6.3.8 (Expectation (Discrete/Continuous)) Let X be real-valued.

- If X has PMF p_X, then
$$\mathbb{E}[X] = \sum_x x\, p_X(x),$$
whenever the sum converges absolutely.
- If X has PDF f_X, then
$$\mathbb{E}[X] = \int_{\mathbb{R}} x\, f_X(x)\, dx,$$
whenever the integral exists (finite).

More generally, for a measurable function g, $\mathbb{E}[g(X)]$ is defined by replacing x with $g(x)$.

Remark 6.3.3 (Risk as Expectation of Loss) In supervised learning, the population risk is $R(h) = \mathbb{E}[\ell(h(X), Y)]$. Thus, learning is expectation minimization under an unknown distribution.

6.3.3 Common Distribution Families (Core Toolkit)

6.3.3.1 Bernoulli, Binomial, Multinomial, Poisson

Discrete distributions model counts and categorical outcomes, which are ubiquitous: binary labels (Bernoulli), number of successes (Binomial), class counts (Multinomial), event counts in a time window (Poisson) (Ross 2023). They define natural likelihoods for classification and count regression, and their means/variances often yield interpretable parameters.

Definition 6.3.9 (Bernoulli) $X \sim \text{Bernoulli}(p)$ with $p \in [0, 1]$ if $\mathbb{P}(X = 1) = p$ and $\mathbb{P}(X = 0) = 1 - p$.

Definition 6.3.10 (Binomial) $X \sim \text{Binomial}(n, p)$ if

$$\mathbb{P}(X = k) = \binom{n}{k} p^k (1-p)^{n-k}, \qquad k \in \{0, 1, \ldots, n\}.$$

Definition 6.3.11 (Multinomial) Let $p = (p_1, \ldots, p_K) \in \Delta^{K-1}$. A count vector $X = (X_1, \ldots, X_K)$ is Multinomial(n, p) if $\sum_{i=1}^{K} X_i = n$ and

$$\mathbb{P}(X = x) = \frac{n!}{\prod_{i=1}^{K} x_i!} \prod_{i=1}^{K} p_i^{x_i}, \quad x_i \in \mathbb{Z}_{\geq 0}, \ \sum_i x_i = n.$$

Definition 6.3.12 (Poisson) $X \sim \text{Poisson}(\lambda)$ with $\lambda > 0$ if

$$\mathbb{P}(X = k) = e^{-\lambda} \frac{\lambda^k}{k!}, \qquad k \in \mathbb{Z}_{\geq 0}.$$

6.3.3.2 Gaussian, Exponential, Gamma, Beta, Dirichlet

Continuous families model real-valued noise and constrained quantities. Gaussian models light-tailed noise and underlies least squares; Exponential/Gamma model waiting times and positive scales; Beta models probabilities in [0, 1]; Dirichlet models probability vectors on simplices (multiclass probabilities). These are building blocks for Bayesian priors and for likelihoods with support constraints.

Definition 6.3.13 (Gaussian/Normal) $X \sim \mathcal{N}(\mu, \sigma^2)$ has PDF

$$f(x) = \frac{1}{\sqrt{2\pi\sigma^2}} \exp\left(-\frac{(x-\mu)^2}{2\sigma^2}\right).$$

Definition 6.3.14 (Exponential) $X \sim \text{Exponential}(\lambda)$ with $\lambda > 0$ has PDF

$$f(x) = \lambda e^{-\lambda x} \mathbf{1}\{x \geq 0\}.$$

Definition 6.3.15 (Gamma) $X \sim \text{Gamma}(\alpha, \beta)$ with shape $\alpha > 0$ and rate $\beta > 0$ has PDF

$$f(x) = \frac{\beta^\alpha}{\Gamma(\alpha)} x^{\alpha-1} e^{-\beta x} \mathbf{1}\{x \geq 0\}.$$

Definition 6.3.16 (Beta) $X \sim \text{Beta}(\alpha, \beta)$ with $\alpha, \beta > 0$ has PDF

$$f(x) = \frac{1}{B(\alpha, \beta)} x^{\alpha-1}(1-x)^{\beta-1}\mathbf{1}\{0 \le x \le 1\}, \qquad B(\alpha, \beta) = \frac{\Gamma(\alpha)\Gamma(\beta)}{\Gamma(\alpha+\beta)}.$$

Definition 6.3.17 (Dirichlet) Let $\alpha = (\alpha_1, \dots, \alpha_K)$ with $\alpha_i > 0$. A random vector $P \in \Delta^{K-1}$ is Dirichlet(α) if it has density

$$f(p) = \frac{1}{B(\alpha)} \prod_{i=1}^{K} p_i^{\alpha_i - 1} \mathbf{1}\left\{p \in \Delta^{K-1}\right\}, \qquad B(\alpha) = \frac{\prod_{i=1}^{K} \Gamma(\alpha_i)}{\Gamma(\sum_{i=1}^{K} \alpha_i)}.$$

6.3.4 *Distribution Design for Modeling*

6.3.4.1 Support Constraints (Simplex, Positive Reals, Bounded Intervals)

Choosing a distribution begins with respecting the support of the quantity being modeled: probabilities lie on a simplex, rates and scales are positive, proportions lie in [0, 1], counts are nonnegative integers. Support mismatch produces invalid likelihoods and nonsensical predictions. In ML, support constraints also determine which link functions and parameterizations are natural (e.g., softmax for simplex, exp for positivity).

Remark 6.3.4 (Support Dictates Parameterization)

$$\text{Simplex: softmax}, \quad \text{Positive reals: exponential map}, \quad \text{Interval } [0, 1] : \text{logit/sigmoid}.$$

These maps turn unconstrained parameters into constrained random quantities.

6.3.4.2 Heavy Tails vs. Light Tails: Implications for Learning

Tail behavior controls how often extreme events occur and thus how stable empirical averages are. Light-tailed models (e.g., Gaussian) enable strong concentration and small sample complexity for mean estimation, while heavy-tailed data can break naive generalization and require robust estimators. This preview motivates later chapters on concentration inequalities and robust statistics.

Remark 6.3.5 (Preview Intuition) Under heavy tails, rare large outliers can dominate empirical risk and gradients. Robust losses (Huber, ℓ_1-type) and truncation methods can be viewed as geometric tools to limit tail influence.

6.4 Joint, Marginal, Conditional

Most ML models are multivariate: features are vectors and datasets involve multiple random variables. Joint distributions encode all dependence, while marginals and conditionals are obtained by summing/integrating out variables. Factorizations (chain rule, Bayes nets) provide compact representations and motivate model architectures (Grimmett and Stirzaker 2020).

6.4.1 Multivariate Distributions

6.4.1.1 Marginalization, Conditioning, and Factorization

Marginalization is the operation of removing variables you do not observe or do not care about, by summing or integrating them out. Conditioning is the operation of updating the distribution when you learn that an event or variable has a certain value. Factorization describes how a joint distribution can be built from simpler conditional pieces. These operations are the algebraic core of latent-variable models, Bayes classifiers, and probabilistic graphical models.

Definition 6.4.1 (Joint and Marginal (Discrete Case)) Let (X, Y) take values in countable sets. The joint PMF is $p_{X,Y}(x, y) = \mathbb{P}(X = x, Y = y)$. The marginals are

$$p_X(x) = \sum_y p_{X,Y}(x, y), \qquad p_Y(y) = \sum_x p_{X,Y}(x, y).$$

Definition 6.4.2 (Joint and Marginal (Continuous Case)) If (X, Y) has a joint density $f_{X,Y}(x, y)$, then the marginal densities are

$$f_X(x) = \int f_{X,Y}(x, y)\,dy, \qquad f_Y(y) = \int f_{X,Y}(x, y)\,dx,$$

whenever the integrals exist.

Definition 6.4.3 (Conditional Distributions (PMF/PDF)) In the discrete case with $p_Y(y) > 0$,

$$p_{X|Y}(x \mid y) = \frac{p_{X,Y}(x, y)}{p_Y(y)}.$$

In the continuous case (when $f_Y(y) > 0$),

$$f_{X|Y}(x \mid y) = \frac{f_{X,Y}(x, y)}{f_Y(y)}.$$

6.4.1.2 Chain Rule for Probabilities and Densities

The chain rule expresses any joint distribution as a product of conditionals. This is the probability analogue of "factorizing a complex object into modules." In ML it underlies autoregressive models, language modeling, and sequential decision processes. The chain rule also provides a systematic way to write likelihoods for complex models.

Proposition 6.4.1 (Chain Rule (Finite Discrete/Continuous)) *For random variables $X_1, \ldots, X_d$ (discrete or continuous, with the obvious interpretation),*

$$p(x_1, \ldots, x_d) = p(x_1) \prod_{i=2}^{d} p(x_i \mid x_1, \ldots, x_{i-1}).$$

Remark 6.4.1 (Autoregressive Modeling) Autoregressive models choose parametric forms for the conditionals $p(x_i \mid x_{<i})$ and multiply them. This converts joint modeling into a sequence of supervised prediction problems.

6.4.2 *Independence and Dependence*

6.4.2.1 Independence, Conditional Independence, and Examples

Independence is the strongest and most useful simplification: it factorizes a joint into a product of marginals. Conditional independence is more subtle and more realistic: variables become independent after conditioning on a third variable. This is the backbone of graphical models and causal reasoning. In ML, assumptions like "i.i.d. samples" are independence assumptions, and Naive Bayes is a conditional independence model.

Definition 6.4.4 (Independence) Events $E, F \in \mathcal{F}$ are independent if $\mathbb{P}(E \cap F) = \mathbb{P}(E)\mathbb{P}(F)$. Random variables X, Y are independent if for all Borel sets A, B,

$$\mathbb{P}(X \in A, Y \in B) = \mathbb{P}(X \in A)\mathbb{P}(Y \in B).$$

Equivalently (discrete/continuous), $p_{X,Y}(x, y) = p_X(x)p_Y(y)$ or $f_{X,Y}(x, y) = f_X(x)f_Y(y)$.

Definition 6.4.5 (Conditional Independence) X and Y are conditionally independent given Z (written $X \perp Y \mid Z$) if for all measurable A, B,

$$\mathbb{P}(X \in A, Y \in B \mid Z) = \mathbb{P}(X \in A \mid Z)\, \mathbb{P}(Y \in B \mid Z),$$

interpreted as equality of random variables in Z (where defined).

Example 6.4.1 (Naive Bayes Structure) In Naive Bayes, features $(X_1, \ldots, X_d)$ are assumed conditionally independent given class Y:

$$p(x_1, \ldots, x_d \mid y) = \prod_{i=1}^{d} p(x_i \mid y),$$

so

$$p(y \mid x) \propto p(y) \prod_{i=1}^{d} p(x_i \mid y).$$

6.4.2.2 Correlation vs. Independence (What Each Captures)

Correlation measures linear dependence and is blind to many nonlinear relationships. Independence is much stronger: it implies zero correlation (under finite second moments), but not conversely. This distinction is critical in ML: uncorrelated features are not necessarily redundant; nonlinear interactions can remain. Conversely, independence assumptions are often violated even when correlations are small, especially under hidden confounders or mixtures.

Definition 6.4.6 (Covariance and Correlation) For real-valued X, Y with finite second moments,

$$\mathrm{Cov}(X, Y) := \mathbb{E}\big[(X - \mathbb{E}X)(Y - \mathbb{E}Y)\big], \qquad \mathrm{Corr}(X, Y) := \frac{\mathrm{Cov}(X, Y)}{\sqrt{\mathrm{Var}(X)\mathrm{Var}(Y)}}$$

when $\mathrm{Var}(X), \mathrm{Var}(Y) > 0$.

Proposition 6.4.2 (Independence ⇒ Zero Covariance) *If X and Y are independent and have finite second moments, then $Cov(X, Y) = 0$.*

Example 6.4.2 (Zero Correlation Without Independence) Let X be uniform on $\{-1, +1\}$ with $\mathbb{P}(X = 1) = \mathbb{P}(X = -1) = 1/2$, and define $Y = X^2$. Then $Y = 1$ almost surely, so $\mathrm{Cov}(X, Y) = 0$, but X and Y are not independent (since Y is a function of X).

6.4.3 *Bayesian Modeling Patterns*

6.4.3.1 Likelihood–Prior–Posterior Decomposition

Bayesian modeling treats unknown parameters as random variables and updates uncertainty about them after observing data. The posterior combines prior knowledge and evidence from the likelihood. Even if one does not compute full posteriors, the same decomposition motivates MAP estimation and regularization: negative log prior acts like a penalty, and negative log likelihood acts like a loss.

Remark 6.4.2 (Bayes Rule in Density Form) For parameter θ and data $\mathcal{D}$,

$$p(\theta \mid \mathcal{D}) = \frac{p(\mathcal{D} \mid \theta)p(\theta)}{p(\mathcal{D})}, \qquad p(\mathcal{D}) = \int p(\mathcal{D} \mid \theta)p(\theta)\, d\theta.$$

The evidence $p(\mathcal{D})$ normalizes the posterior and plays a key role in model comparison.

6.4.3.2 Latent Variables, Mixtures, and Hierarchical Models

Latent variables explain observed complexity via hidden structure (clusters, topics, factors). Mixture models are the simplest latent-variable models: a hidden label selects a component distribution. Hierarchical models stack latent variables to express multi-level uncertainty (hyperpriors). These patterns will reappear in clustering, EM algorithms, topic models, and Bayesian deep learning.

Remark 6.4.3 (Mixture Model Template) A K-component mixture introduces a latent $Z \in \{1, \ldots, K\}$ with $\mathbb{P}(Z = k) = \pi_k$ and

$$p(x) = \sum_{k=1}^{K} \pi_k\, p(x \mid Z = k).$$

Conditioning yields responsibilities $p(Z = k \mid x)$, central to EM.

6.5 Expectation and Conditional Expectation (Core Operator)

Expectation is the central operator of probability and the computational heart of statistical learning: risk is an expectation of loss, gradients are expectations of stochastic gradients, and uncertainty quantification is expressed via moments. Conditional expectation refines this operator by incorporating information: it is the best predictor given what you know, and it obeys powerful algebraic rules (tower property) that make multi-stage reasoning possible (Williams 1991).

6.5.1 Expectation as an Averaging Operator

6.5.1.1 Linearity and Basic Manipulations

The most important structural fact about expectation is linearity. It allows us to move sums and constants outside, to decompose complicated expressions into manageable parts, and to justify unbiased estimators. Many ML derivations are linearity plus one inequality (Jensen, Cauchy–Schwarz, etc.). Computationally, linearity is what makes empirical averages approximate population expectations.

Definition 6.5.1 (Expectation (General Form)) Let X be an integrable real-valued random variable on $(\Omega, \mathcal{F}, \mathbb{P})$. Its expectation is

$$\mathbb{E}[X] := \int_{\Omega} X(\omega)\, d\mathbb{P}(\omega),$$

whenever the integral exists (finite). If X is discrete with pmf p, then $\mathbb{E}[X] = \sum_x x\, p(x)$; if X has density f, then $\mathbb{E}[X] = \int x\, f(x)\, dx$.

Proposition 6.5.1 (Linearity) *If X, Y are integrable and $a, b \in \mathbb{R}$, then*

$$\mathbb{E}[aX + bY] = a\,\mathbb{E}[X] + b\,\mathbb{E}[Y].$$

More generally, $\mathbb{E}\big[\sum_{i=1}^n a_i X_i\big] = \sum_{i=1}^n a_i \mathbb{E}[X_i]$ for integrable X_i.

Remark 6.5.1 (A Ubiquitous ML Pattern) If $\widehat{\theta} = \frac{1}{n}\sum_{i=1}^n Z_i$ is an empirical average, then $\mathbb{E}[\widehat{\theta}] = \frac{1}{n}\sum_i \mathbb{E}[Z_i]$ by linearity. Thus unbiasedness checks often reduce to computing $\mathbb{E}[Z_i]$ for a single sample.

6.5.1.2 Expectation of Vectors and Matrices (Componentwise)

Most ML objects are vector- or matrix-valued (features, gradients, Jacobians, covariances). Expectation extends componentwise because it is an integral: you integrate each coordinate. This yields clean identities like $\mathbb{E}[AX] = A\mathbb{E}[X]$ for deterministic A and random vector X, which are heavily used in linear models and stochastic optimization.

Definition 6.5.2 (Expectation of Random Vectors/Matrices) Let $X = (X_1, \dots, X_d)^\top$ be an integrable $\mathbb{R}^d$-valued random vector. Define

$$\mathbb{E}[X] := \big(\mathbb{E}[X_1], \dots, \mathbb{E}[X_d]\big)^\top.$$

If $M \in \mathbb{R}^{m\times n}$ is a random matrix with integrable entries, define $\mathbb{E}[M]$ entrywise:

$$(\mathbb{E}[M])_{ij} := \mathbb{E}[M_{ij}].$$

Proposition 6.5.2 (Pulling Out Deterministic Linear Maps) *If A and B are deterministic matrices of compatible sizes and X is an integrable random matrix/vector, then*

$$\mathbb{E}[AX] = A\,\mathbb{E}[X], \qquad \mathbb{E}[XB] = \mathbb{E}[X]\,B, \qquad \mathbb{E}[AXB] = A\,\mathbb{E}[X]\,B.$$

Remark 6.5.2 (Stochastic Gradients) In SGD, one uses random gradients $g(\theta, \xi)$ with $\mathbb{E}[g(\theta, \xi)] = \nabla L(\theta)$. This is precisely an expectation of a vector-valued random variable.

6.5.2 *Conditional Expectation (Intuition and Definition)*

6.5.2.1 $\mathbb{E}[X \mid Y]$ as Optimal Predictor under Squared Loss

Conditional expectation is best understood as prediction with information. Among all functions of Y, the conditional expectation $\mathbb{E}[X \mid Y]$ is the one that minimizes mean squared error (cf. Fig. 6.3). This single fact explains why regression targets are conditional means under squared loss, why Bayes estimators are conditional expectations, and why conditional expectation is a projection operator in L^2 space.

Definition 6.5.3 (Conditional Expectation (Informal Operational Definition)) Let X be integrable and Y be another random variable. A *conditional expectation* of X given Y is a random variable $\mathbb{E}[X \mid Y]$ that is a measurable function of Y and satisfies

$$\mathbb{E}\big[\mathbf{1}\{Y \in B\}\,\mathbb{E}[X \mid Y]\big] = \mathbb{E}\big[\mathbf{1}\{Y \in B\}\,X\big] \qquad \forall \text{ Borel sets } B.$$

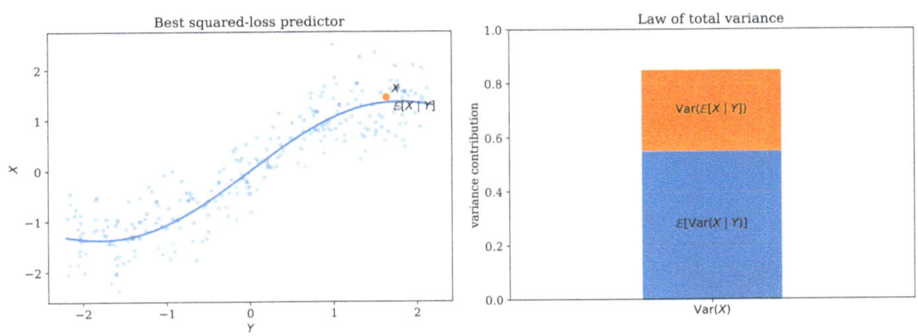

Fig. 6.3 Conditional expectation as an operator. Left: the curve $y \mapsto \mathbb{E}[X \mid Y = y]$ is the optimal squared-loss predictor of X from Y, and the vertical residual is the irreducible conditional noise. Right: the law of total variance decomposes overall variability into expected conditional noise plus explainable variability of the conditional mean

Theorem 6.5.1 (Squared-Loss Optimality) *Assume $X \in L^2$ (i.e., $\mathbb{E}[X^2] < \infty$). Among all measurable functions $g(Y)$ with $\mathbb{E}[g(Y)^2] < \infty$, the minimizer of*

$$\min_g \mathbb{E}\big[(X - g(Y))^2\big]$$

is $g^\star(Y) = \mathbb{E}[X \mid Y]$ (a.s.). Moreover, for any such g,

$$\mathbb{E}\big[(X - g(Y))^2\big] = \mathbb{E}\big[(X - \mathbb{E}[X \mid Y])^2\big] + \mathbb{E}\big[(\mathbb{E}[X \mid Y] - g(Y))^2\big].$$

Remark 6.5.3 (Geometric Interpretation) The decomposition is a Pythagorean theorem in L^2: conditional expectation is the orthogonal projection of X onto the closed subspace of L^2 consisting of functions measurable w.r.t. Y.

6.5.2.2 Tower Property and Law of Total Expectation

The tower property (iterated expectation) is the algebra of multi-stage reasoning: "average over the finer randomness, then over the coarser randomness." In ML, it enables decomposition of risk, variance, and error analysis across latent variables or sampling steps. The law of total expectation is its simplest corollary: $\mathbb{E}[\mathbb{E}[X \mid Y]] = \mathbb{E}[X]$.

Theorem 6.5.2 (Tower Property) *Let X be integrable and let Y, Z be random variables with $\sigma(Y) \subseteq \sigma(Z)$ (i.e., Z is at least as informative as Y). Then*

$$\mathbb{E}\big[\mathbb{E}[X \mid Z] \mid Y\big] = \mathbb{E}[X \mid Y] \qquad a.s.$$

In particular,

$$\mathbb{E}\big[\mathbb{E}[X \mid Y]\big] = \mathbb{E}[X].$$

Example 6.5.1 (Total Expectation under a Finite Partition) If Y takes values in $\{1, \ldots, K\}$ with $\mathbb{P}(Y = k) > 0$, then

$$\mathbb{E}[X] = \sum_{k=1}^{K} \mathbb{E}[X \mid Y = k]\, \mathbb{P}(Y = k).$$

This is the law of total expectation in a discrete form.

6.5.3 *Common Expectations and Moments*

6.5.3.1 Means, Variances, Covariances

The first two moments (mean and variance) summarize location and dispersion. Covariance generalizes variance to pairs and yields covariance matrices for vectors, capturing anisotropy and linear dependence. In ML, covariance matrices define PCA directions, Gaussian likelihoods, and uncertainty ellipsoids; variance decompositions drive bias–variance tradeoffs and generalization heuristics.

Definition 6.5.4 (Mean, Variance, Covariance) For integrable X, define $\mu = \mathbb{E}[X]$. If $\mathbb{E}[X^2] < \infty$, define

$$\mathrm{Var}(X) := \mathbb{E}\big[(X-\mu)^2\big] = \mathbb{E}[X^2] - \mu^2.$$

For X, Y with finite second moments,

$$\mathrm{Cov}(X, Y) := \mathbb{E}\big[(X-\mathbb{E}X)(Y-\mathbb{E}Y)\big] = \mathbb{E}[XY] - \mathbb{E}[X]\mathbb{E}[Y].$$

Definition 6.5.5 (Covariance Matrix) For $X \in \mathbb{R}^d$ with mean $\mu = \mathbb{E}[X]$ and finite second moments, the covariance matrix is

$$\Sigma := \mathbb{E}\big[(X-\mu)(X-\mu)^\top\big] \in \mathbb{R}^{d\times d}.$$

Proposition 6.5.3 (Law of Total Variance) *If $X \in L^2$ and Y is any random variable, then*

$$Var(X) = \mathbb{E}\big[Var(X \mid Y)\big] + Var\big(\mathbb{E}[X \mid Y]\big).$$

Remark 6.5.4 (Learning Interpretation) $\mathbb{E}[\mathrm{Var}(X \mid Y)]$ is irreducible noise after observing Y, while $\mathrm{Var}(\mathbb{E}[X \mid Y])$ is explainable variability due to Y. This mirrors "noise vs. signal explained by features."

6.5.3.2 Moment Generating Functions

MGFs package all moments (when they exist) and convert sums of independent variables into products, making them ideal for tail bounds and concentration. In learning theory, MGF arguments yield Hoeffding/Bernstein-type inequalities and help classify tail behavior (sub-Gaussian, sub-exponential). Here, we record the definition and the key multiplicative property for independence.

Definition 6.5.6 (Moment Generating Function) For a real-valued random variable X, the *moment generating function* is

$$M_X(t) := \mathbb{E}[e^{tX}],$$

for those t where the expectation is finite.

Proposition 6.5.4 (MGF of Sums of Independent Variables) *If X and Y are independent and $M_X(t)$, $M_Y(t)$ are finite, then*

$$M_{X+Y}(t) = M_X(t)\, M_Y(t).$$

More generally, for independent $\{X_i\}_{i=1}^n$, $M_{\sum_i X_i}(t) = \prod_i M_{X_i}(t)$ when finite.

6.6 Transformations of Random Variables

Learning pipelines repeatedly transform random variables: sums of noise, maxima in extreme-value statistics, norms in concentration, and nonlinear feature maps. Transformations are handled cleanly by pushforward distributions: apply a measurable function and obtain a new distribution. For densities, change-of-variables formulas introduce Jacobians, which are central in modern generative modeling (normalizing flows). Finally, sums of independent variables lead to convolutions and motivate central limit phenomena.

6.6.1 Functions of Random Variables

6.6.1.1 Sums, Maxima, Norms, and Indicator Transforms

Most quantities of interest are functions of base random variables: empirical risks (sums), worst-case losses (maxima), regularization terms (norms), and events (indicators). These transforms change the distribution in structured ways. Understanding how expectation and probability behave under such transforms is the basis for risk bounds, concentration, and robust estimation.

Definition 6.6.1 (Function of a Random Variable) If $X : \Omega \to \mathcal{S}$ is a random variable and $g : \mathcal{S} \to \mathbb{R}$ is measurable, then $g(X) : \Omega \to \mathbb{R}$ defined by $g(X)(\omega) = g(X(\omega))$ is a random variable.

Example 6.6.1 (Indicator Transform) For an event $E \in \mathcal{F}$, $\mathbf{1}\{E\}$ is a Bernoulli random variable with $\mathbb{E}[\mathbf{1}\{E\}] = \mathbb{P}(E)$. Thus probabilities can be represented as expectations.

6.6.1.2 Pushforward Distributions (Conceptual)

The pushforward viewpoint is the cleanest formalism: the distribution of $Y = g(X)$ is obtained by mapping sets backward via g^{-1}. This matches the Chap. 1 theme: preimages behave well under set operations, making distribution computation systematic. In practice, this yields CDF methods ($\mathbb{P}(g(X) \leq t)$), change-of-variables for densities, and transformation rules for discrete pmfs.

Definition 6.6.2 (Pushforward Distribution (Again)) If X has distribution $\mathbb{P}_X$ on $(\mathcal{S}, \mathcal{G})$ and $g : \mathcal{S} \to \mathcal{T}$ is measurable, then the distribution of $Y = g(X)$ is the pushforward measure

$$\mathbb{P}_Y = \mathbb{P}_X \circ g^{-1}, \qquad \mathbb{P}_Y(B) = \mathbb{P}_X(g^{-1}(B)).$$

Remark 6.6.1 (CDF Method) For real-valued $Y = g(X)$, $F_Y(t) = \mathbb{P}(Y \leq t) = \mathbb{P}(X \in g^{-1}((-\infty, t]))$. This is often the simplest route to derive distributions of transforms.

6.6.2 *Change of Variables*

6.6.2.1 Jacobian Rule in One and Multiple Dimensions

When variables have densities, transforming coordinates requires correcting by how volumes expand/shrink. In one dimension this correction is $|dx/dy|$; in multiple dimensions it is $|\det J_{g^{-1}}(y)|$. This is the mathematical reason Jacobians appear in density modeling, and it also mirrors linear algebra: determinants measure volume scaling under linear maps.

Theorem 6.6.1 (Change of Variables (One Dimension)) *Let X have density f_X and let $Y = g(X)$ where g is strictly monotone and differentiable with inverse g^{-1} differentiable. Then Y has density*

$$f_Y(y) = f_X(g^{-1}(y)) \left| \frac{d}{dy} g^{-1}(y) \right|.$$

Theorem 6.6.2 (Change of Variables (Multivariate)) *Let $X \in \mathbb{R}^d$ have density f_X and let $Y = g(X)$ where $g : \mathbb{R}^d \to \mathbb{R}^d$ is a bijection, differentiable with differentiable inverse. Then Y has density*

$$f_Y(y) = f_X(g^{-1}(y)) \left|\det J_{g^{-1}}(y)\right|,$$

where $J_{g^{-1}}(y)$ is the Jacobian matrix of g^{-1} at y. Equivalently,

$$f_Y(g(x)) = f_X(x) \left|\det J_g(x)\right|^{-1}.$$

6.6.2.2 Why Jacobians Matter in Normalizing Flows

Normalizing flows build expressive densities by composing invertible maps. Each map contributes a Jacobian determinant term to the log-density, so tractability depends on designing maps with computable $\log|\det J|$ (triangular Jacobians, coupling layers, etc.). Thus, the entire field of flow-based generative modeling is an exercise in Jacobian-aware linear algebra.

Remark 6.6.2 (Log-Density under a Flow (Conceptual)) If $Z \sim p_Z$ and $X = g(Z)$ with invertible g, then

$$\log p_X(x) = \log p_Z(g^{-1}(x)) + \log \left|\det J_{g^{-1}}(x)\right|.$$

This additive form is why log-determinants are central in flows.

6.6.3 *Convolutions and Sums of Independent Variables*

6.6.3.1 Convolution Formula (Conceptual) and Gaussian Closure

Sums of independent random variables are ubiquitous: total noise, aggregated gradients, ensemble averaging. Their distributions are governed by convolution. Gaussians are special because they are closed under addition: sums of independent Gaussians remain Gaussian, with means and variances adding. This is the analytic reason Gaussian noise models are computationally convenient and appear widely in ML.

Definition 6.6.3 (Convolution (Continuous Case)) If X and Y are independent with densities f_X, f_Y, then $S = X + Y$ has density

$$f_S(s) = (f_X * f_Y)(s) = \int_{-\infty}^{\infty} f_X(t)\, f_Y(s-t)\, dt.$$

Proposition 6.6.1 (Gaussian Closure under Addition) *If $X \sim \mathcal{N}(\mu_1, \sigma_1^2)$ and $Y \sim \mathcal{N}(\mu_2, \sigma_2^2)$ are independent, then*

$$X + Y \sim \mathcal{N}(\mu_1 + \mu_2,\ \sigma_1^2 + \sigma_2^2).$$

6.6.3.2 Central Limit Intuition

The CLT explains why Gaussian approximations arise even when underlying noise is not Gaussian: sums of many weakly dependent contributions tend to look normal after centering and scaling. In ML, this motivates Gaussian approximations for aggregated errors, gradient noise, and estimation error. We will state and use CLT rigorously later; here we record the scaling and the qualitative message.

Remark 6.6.3 (CLT Scaling Heuristic) If $X_1, \ldots, X_n$ are i.i.d. with mean μ and variance σ^2, then

$$\frac{1}{\sqrt{n}} \sum_{i=1}^{n} (X_i - \mu)$$

often has a distribution close to $\mathcal{N}(0, \sigma^2)$ for large n. This heuristic underlies many "$1/\sqrt{n}$" error scalings in learning theory.

6.7 Modes of Convergence (Asymptotic Probability)

Asymptotic probability is the language used to formalize "what happens as data grows." Convergence notions (almost sure, in probability, in distribution, in L^p) capture different strengths of approximation. In learning theory, these notions underpin consistency of estimators, stability of empirical risk, and distributional approximations used for uncertainty quantification.

6.7.1 Convergence Notions

6.7.1.1 Almost Sure, in Probability, in Distribution, in L^p

A sequence of random variables $\{X_n\}$ can converge in multiple senses. Almost sure convergence is pathwise and strongest; convergence in probability captures that deviations vanish with high probability; convergence in distribution captures weak convergence of CDFs (useful for CLT); and L^p convergence controls moments and implies convergence in probability by Markov-type arguments. In ML, you choose the notion according to what you need: pointwise guarantee, risk convergence, or asymptotic distribution for confidence intervals.

Definition 6.7.1 (Almost Sure Convergence) $X_n \to X$ *almost surely* (a.s.) if

$$\mathbb{P}\big(\{\omega : \lim_{n\to\infty} X_n(\omega) = X(\omega)\}\big) = 1.$$

Definition 6.7.2 (Convergence in Probability) $X_n \to X$ *in probability* if for every $\varepsilon > 0$,

$$\mathbb{P}\big(|X_n - X| > \varepsilon\big) \to 0 \quad \text{as } n \to \infty.$$

Definition 6.7.3 (Convergence in Distribution) $X_n \Rightarrow X$ *in distribution* if

$$F_{X_n}(t) \to F_X(t) \quad \text{for all continuity points} t \text{of} F_X,$$

where F_{X_n} and F_X are the CDFs.

Definition 6.7.4 (L^p Convergence) For $p \geq 1$, $X_n \to X$ *in* L^p if $\mathbb{E}[|X_n - X|^p] \to 0$ as $n \to \infty$.

Remark 6.7.1 (Vector-Valued Convergence) For $X_n, X \in \mathbb{R}^d$, define convergence in probability or L^p using a norm (typically $\|\cdot\|_2$), e.g., $\mathbb{P}(\|X_n - X\|_2 > \varepsilon) \to 0$. All norms are equivalent in finite dimensions, so the choice affects constants, not the notion.

6.7.1.2 Relationships and Implications

The convergence modes form a hierarchy: a.s. implies in probability, and L^p implies in probability. Convergence in probability implies convergence in distribution, but not conversely. These implications tell you what kind of results you can legitimately derive: for example, $X_n \Rightarrow X$ alone does not control moments, while L^2 does. In learning, guaranteeing risk convergence often requires some integrability (moment) control beyond weak convergence.

Theorem 6.7.1 (Standard Implication Chain) *For random variables X_n, X:*

$$X_n \to X \text{ a.s.} \implies X_n \to X \text{ in probability} \implies X_n \Rightarrow X.$$

Moreover, for any $p \geq 1$,

$$X_n \to X \text{ in } L^p \implies X_n \to X \text{ in probability}.$$

Proof sketch If $X_n \to X$ a.s., then for fixed $\varepsilon > 0$, the indicator $\mathbf{1}\{|X_n - X| > \varepsilon\} \to 0$ a.s.; dominated convergence yields $\mathbb{P}(|X_n - X| > \varepsilon) \to 0$. If $X_n \to X$ in L^p, Markov gives $\mathbb{P}(|X_n - X| > \varepsilon) \leq \mathbb{E}[|X_n - X|^p]/\varepsilon^p \to 0$. Convergence in probability implies convergence in distribution by standard probability theory. □

Remark 6.7.2 (Non-implications Worth Remembering)

- $X_n \Rightarrow X$ does not imply $X_n \to X$ in probability unless the limit X is constant and additional conditions hold.
- $X_n \to X$ in probability does not imply a.s. convergence in general.
- $X_n \Rightarrow X$ does not control $\mathbb{E}[X_n]$ unless uniform integrability conditions are added.

6.7.2 *Why Convergence Matters for Learning*

6.7.2.1 Consistency of Estimators

An estimator $\widehat{\theta}_n$ is *consistent* for θ if it converges to θ as $n \to \infty$. The mode of convergence matters: convergence in probability is the standard notion (weak consistency), while almost sure convergence (strong consistency) is a stronger guarantee. In ML, consistency is a minimum bar: with enough data, the procedure should recover the truth (within the model) or the Bayes-optimal predictor (in nonparametric settings).

Definition 6.7.5 (Consistency) Let $\widehat{\theta}_n$ estimate a parameter θ.

- $\widehat{\theta}_n$ is *weakly consistent* if $\widehat{\theta}_n \to \theta$ in probability.
- $\widehat{\theta}_n$ is *strongly consistent* if $\widehat{\theta}_n \to \theta$ almost surely.

Remark 6.7.3 (Bias/Variance vs. Consistency) Consistency is asymptotic; it does not guarantee small finite-sample error. A biased estimator may still be consistent if its bias vanishes with n, and a low-variance estimator may be inconsistent under misspecification.

6.7.2.2 Empirical Risk and Asymptotic Generalization

Generalization compares empirical quantities (training loss) to population quantities (test loss). As $n \to \infty$, empirical risk should converge to true risk under suitable assumptions. For a fixed hypothesis h, the law of large numbers gives $\widehat{R}_n(h) \to R(h)$. For data-driven $\widehat{h}_n$ chosen by minimizing empirical risk, one needs *uniform* convergence or stability to exchange minimization and limit. This is where union bounds, covering numbers, VC dimension, and concentration enter later.

Definition 6.7.6 (Empirical and Population Risk) Let $(X, Y) \sim \mathcal{D}$ and loss ℓ. For a predictor h, the *population risk* is

$$R(h) := \mathbb{E}[\ell(h(X), Y)].$$

Given i.i.d. data $\{(X_i, Y_i)\}_{i=1}^n$, the *empirical risk* is

$$\widehat{R}_n(h) := \frac{1}{n} \sum_{i=1}^{n} \ell(h(X_i), Y_i).$$

Remark 6.7.4 (Fixed h vs. Learned $\widehat{h}_n$) For fixed h, $\widehat{R}_n(h) \to R(h)$ by LLN under integrability. For $\widehat{h}_n \in \arg\min_{h \in \mathcal{H}} \widehat{R}_n(h)$, one needs additional arguments to show $R(\widehat{h}_n)$ approaches $\inf_{h \in \mathcal{H}} R(h)$.

6.8 Probabilistic Modeling in ML Pipelines

A probabilistic ML pipeline specifies a data-generating process, an observation/noise model (likelihood), and a decision rule. The i.i.d. assumption is a central simplification that enables clean analysis, but it can fail in time series, under covariate shift, and under dependence. Likelihood-based training provides a universal compatibility score between model and data, but misspecification is inevitable in practice. Finally, probabilistic prediction must be evaluated not only by accuracy but also by calibration.

6.8.1 The i.i.d. Assumption and Its Limits

6.8.1.1 i.i.d. Sampling as a Modeling Choice

The i.i.d. assumption states that data are independent and identically distributed from a fixed distribution. It is not a law of nature; it is a modeling choice that makes learning tractable. Under i.i.d., empirical averages concentrate, ERM is analyzable, and likelihood factorizes. In practice, i.i.d. can be approximately valid after careful dataset construction, but violations change the meaning of generalization.

Definition 6.8.1 (i.i.d. Sample) A sequence $\{Z_i\}_{i=1}^n$ is i.i.d. from $\mathcal{D}$ if (i) $Z_i \sim \mathcal{D}$ for all i and (ii) $\{Z_i\}$ are mutually independent. In supervised learning, $Z_i = (X_i, Y_i)$.

Remark 6.8.1 (Factorization Consequence) If $\{Z_i\}$ are i.i.d. with density/pmf p_θ, then the likelihood factorizes: $p_\theta(z_{1:n}) = \prod_{i=1}^n p_\theta(z_i)$. This single algebraic fact underlies most training objectives.

6.8.1.2 Time Series, Covariate Shift, and Dependence

When observations are dependent (time series), or when the test distribution differs from the training distribution (covariate shift), the i.i.d. guarantees can fail. Dependence reduces effective sample size; shift breaks the link between empirical and test risk. This preview flags the right conceptual modifications: mixing conditions for time series and importance weighting for covariate shift.

Remark 6.8.2 (Two Common i.i.d. Failures)

- **Time series:** Z_t depends on past; concentration bounds require mixing assumptions and often slower rates.
- **Covariate shift:** $p_{\text{train}}(x) \neq p_{\text{test}}(x)$ but $p(y \mid x)$ may be stable; reweighting by likelihood ratios can correct bias.

6.8.2 *Data Generating Process and Likelihood*

6.8.2.1 Likelihood as Model–Data Compatibility Score

Likelihood measures how compatible the observed data are with a model. Maximizing likelihood corresponds to minimizing negative log-likelihood, which is a proper scoring rule (log score). Under i.i.d., likelihood decomposes into a sum over samples, yielding scalable objectives. This makes likelihood the universal training principle for probabilistic models, including classifiers (cross-entropy) and generative models.

Definition 6.8.2 (Negative Log-Likelihood (NLL)) Given i.i.d. data $z_{1:n}$ and model p_θ, define

$$\mathrm{NLL}(\theta) := -\sum_{i=1}^{n} \log p_\theta(z_i).$$

In conditional modeling, with $z_i = (x_i, y_i)$ and $p_\theta(y \mid x)$,

$$\mathrm{NLL}(\theta) := -\sum_{i=1}^{n} \log p_\theta(y_i \mid x_i).$$

Remark 6.8.3 (Cross-Entropy View) When p_θ is a classifier producing probabilities over labels, minimizing NLL is exactly cross-entropy minimization.

6.8.2.2 Misspecification: When the Model Is Wrong

No finite model class contains the true data-generating distribution in realistic settings. Under misspecification, MLE converges (under conditions) to the distribution in the model class closest to the truth in KL divergence. This changes interpretation: parameters describe the best approximation within the chosen family, not the true mechanism. In ML, misspecification is the norm, so diagnostics and robustness matter.

Remark 6.8.4 (KL Projection (Informal)) If the true distribution is $p^\star$ and the model class is $\{p_\theta\}$, then under broad conditions, MLE targets $\theta^\star \in \arg\min_\theta \mathrm{KL}(p^\star \| p_\theta)$. Thus likelihood training is a projection in distribution space.

6.8.3 Probabilistic Prediction and Calibration

6.8.3.1 Proper Scoring Rules: Log-Loss and Brier Score

To evaluate probabilistic predictions, we need losses that reward accurate probabilities, not just correct labels. Proper scoring rules ensure honesty: the true probability distribution minimizes expected score. Log-loss is the canonical choice (equivalent to NLL), while Brier score provides a squared-error alternative. These scores link directly to learning objectives and to calibration: a well-calibrated model tends to perform well under proper scores.

Definition 6.8.3 (Log-Loss and Brier Score (Binary)) For $Y \in \{0, 1\}$ and a predicted probability $q \in [0, 1]$ for $Y = 1$:

$$\text{Log-loss:} \quad \ell_{\log}(q, y) := -\big(y \log q + (1 - y) \log(1 - q)\big),$$

$$\text{Brier score:} \quad \ell_{\mathrm{Br}}(q, y) := (q - y)^2.$$

Remark 6.8.5 (Properness) Both log-loss and Brier score are proper scoring rules: the true conditional probability minimizes expected score. We will revisit properness rigorously when discussing decision theory and estimation.

6.8.3.2 Calibration Curves and Reliability Diagrams (Conceptual)

Calibration is about matching predicted probabilities to observed frequencies. Reliability diagrams visualize this by binning predictions and comparing average predicted probability to empirical frequency. This is essential in risk-sensitive ML (medical, finance, autonomous systems) where probability values drive decisions. While calibration is asymptotic by definition, finite-sample diagnostics and recalibration methods (Platt scaling, isotonic regression) are practical tools built on this concept.

Definition 6.8.4 (Reliability Diagram (Binning Definition)) Given predictions $q_i \in [0, 1]$ and labels $y_i \in \{0, 1\}$, choose bins $I_1, \dots, I_B$ partitioning $[0, 1]$. For each bin I_b, compute

$$\widehat{q}_b := \frac{1}{|S_b|} \sum_{i \in S_b} q_i, \qquad \widehat{p}_b := \frac{1}{|S_b|} \sum_{i \in S_b} y_i, \quad S_b := \{i : q_i \in I_b\}.$$

A reliability diagram plots $(\widehat{q}_b, \widehat{p}_b)$; perfect calibration corresponds to the diagonal $\widehat{p}_b = \widehat{q}_b$.

Remark 6.8.6 (Interpretation) Systematic deviations above the diagonal indicate underconfidence; below indicate overconfidence. Calibration error metrics

(ECE/MCE) aggregate these deviations; they are widely used in modern ML evaluation.

6.9 Exercises

Unless stated otherwise, assume all random variables are defined on a common probability space $(\Omega, \mathcal{F}, \mathbb{P})$.

Exercise 6.9.1 (Event Algebra and Inclusion–Exclusion) Let $\Omega = \{1, 2, \ldots, 30\}$ with the uniform probability measure $\mathbb{P}(\{\omega\}) = 1/30$. Define events $A = \{\omega : 2 \mid \omega\}$, $B = \{\omega : 3 \mid \omega\}$, and $C = \{\omega : 5 \mid \omega\}$. Compute $\mathbb{P}(A)$, $\mathbb{P}(B)$, $\mathbb{P}(C)$, $\mathbb{P}(A \cap B)$, $\mathbb{P}(A \cap C)$, $\mathbb{P}(B \cap C)$, and $\mathbb{P}(A \cap B \cap C)$ explicitly; then use inclusion–exclusion to compute $\mathbb{P}(A \cup B \cup C)$ and verify the value by direct counting of $\omega \in \Omega$ satisfying at least one of the divisibility conditions.

Exercise 6.9.2 (Union Bound: Finite and Countable) Let $\Omega = \{1, 2, \ldots, 100\}$ uniform. Define events $E_k = \{\omega : \omega \equiv 0 \pmod{k}\}$ for $k \in \{7, 8, 9, 10\}$.

(i) Compute $\mathbb{P}(\cup_{k=7}^{10} E_k)$ exactly by counting.
(ii) Compute the union bound $\sum_{k=7}^{10} \mathbb{P}(E_k)$ and compare numerically.
(iii) Define additionally $E_k = \{\omega : \omega \equiv 0 \pmod{k}\}$ for $k \in \{11, 12, \ldots, 20\}$ and compute the union bound for $\mathbb{P}(\cup_{k=7}^{20} E_k)$; explain in one paragraph why union bounds become loose as overlaps increase.

Exercise 6.9.3 (Conditional Probability and Bayes' Rule (Events)) An urn contains 4 red balls and 6 blue balls. Two balls are drawn *without replacement*. Let E be the event "the first ball is red" and F the event "exactly one red ball is drawn in total." Compute $\mathbb{P}(E)$, $\mathbb{P}(F)$, $\mathbb{P}(F \mid E)$, $\mathbb{P}(E \mid F)$, and verify Bayes' rule $\mathbb{P}(E \mid F) = \mathbb{P}(F \mid E)\mathbb{P}(E)/\mathbb{P}(F)$ with your computed values.

Exercise 6.9.4 (Random Variables as Measurable Maps via Preimages) Let $\Omega = \{1, 2, \ldots, 12\}$ with uniform probability. Define the random variable $X : \Omega \to \mathbb{R}$ by $X(\omega) = \gcd(\omega, 12)$ and the event $A = \{X \leq 3\}$.

(i) Compute the set $X^{-1}(\{1, 2, 3\})$ explicitly and hence compute $\mathbb{P}(A)$.
(ii) Define $B = \{X \in \{4, 6, 12\}\}$ and compute $\mathbb{P}(B)$.
(iii) Let $g(x) = \mathbf{1}\{x \in \{1, 2, 3\}\}$ and define $Y = g(X)$; compute the distribution (pmf) of Y and verify that $\mathbb{E}[Y] = \mathbb{P}(A)$.

Exercise 6.9.5 (CDF ↔ PDF Computations) Let X have CDF

$$F_X(x) = \begin{cases} 0, & x < 0, \\ 1 - e^{-2x}, & x \geq 0. \end{cases}$$

(i) Identify the distribution family and write the PDF f_X explicitly.
(ii) Compute $\mathbb{P}(0.2 \le X \le 1)$ and $\mathbb{P}(X > 2)$.
(iii) Compute $\mathbb{E}[X]$ and $\mathrm{Var}(X)$ (you may integrate directly or use known formulas, but show the steps).

Exercise 6.9.6 (Discrete Expectations: Binomial and Poisson) Let $N \sim \mathrm{Binomial}(n = 8, p = 0.3)$ and $M \sim \mathrm{Poisson}(\lambda = 2.5)$ be independent.

(i) Compute $\mathbb{P}(N = 2)$ and $\mathbb{P}(N \ge 3)$ exactly.
(ii) Compute $\mathbb{E}[N]$ and $\mathrm{Var}(N)$.
(iii) Compute $\mathbb{P}(M = 0)$ and $\mathbb{P}(M \le 2)$.
(iv) Define $S = N + M$. Without using CLT, compute $\mathbb{E}[S]$ and $\mathrm{Var}(S)$.

Exercise 6.9.7 (Dirichlet and Multinomial Coupling (Simple Numerical Case)) Let $P = (P_1, P_2, P_3) \sim \mathrm{Dirichlet}(\alpha)$ with $\alpha = (2, 1, 3)$. Given P, draw counts $X \mid P \sim \mathrm{Multinomial}(n = 6, P)$.

(i) Compute $\mathbb{E}[P]$ explicitly.
(ii) Compute $\mathbb{P}(X = (2, 1, 3) \mid P)$ as a function of P.
(iii) Compute $\mathbb{E}[X]$ using the tower property $\mathbb{E}[X] = \mathbb{E}[\mathbb{E}[X \mid P]]$ and the fact that $\mathbb{E}[X \mid P] = nP$.
(iv) Compute $\mathbb{P}(X = (2, 1, 3))$ by integrating out P (Dirichlet–Multinomial); you may use the identity

$$\mathbb{P}(X = x) = \frac{n!}{\prod_i x_i!} \cdot \frac{B(\alpha + x)}{B(\alpha)},$$

and compute the numerical value exactly.

Exercise 6.9.8 (Support Constraints and Model Design) For each of the following random quantities, choose a reasonable distribution family from the chapter toolkit and justify (one paragraph each) by explicitly referencing the support and tail behavior: (i) a probability of default $p \in (0, 1)$ for a loan; (ii) a waiting time $T > 0$ until a server responds; (iii) a K-class probability vector $\pi \in \Delta^{K-1}$ with $K = 4$; (iv) a count of rare failures per day. Then, for your chosen families, write down the PDF/PMF (up to normalization constants if needed) with explicit parameters.

Exercise 6.9.9 (Joint, Marginal, Conditional, and Chain Rule (Discrete)) Let (X, Y) be discrete with joint pmf given by the table (rows: $x \in \{0, 1, 2\}$, columns: $y \in \{0, 1\}$)

$p_{X,Y}(x, y) =$

	$y = 0$	$y = 1$
$x = 0$	0.10	0.20
$x = 1$	0.25	0.15
$x = 2$	0.20	0.10

(i) Compute the marginals p_X and p_Y.
(ii) Compute conditionals $p_{X|Y}(\cdot \mid 1)$ and $p_{Y|X}(1 \mid 2)$.
(iii) Verify the chain rule $p(x, y) = p(x)\, p(y \mid x)$ for all pairs (x, y) by explicit multiplication.
(iv) Compute $\mathbb{E}[X]$, $\mathbb{E}[Y]$, and $\mathbb{E}[XY]$.

Exercise 6.9.10 (Independence vs. Correlation) Using the joint pmf in Exercise 6.9.9, determine whether X and Y are independent by checking whether $p_{X,Y}(x, y) = p_X(x)p_Y(y)$ holds for all (x, y). Then compute $\mathrm{Cov}(X, Y) = \mathbb{E}[XY] - \mathbb{E}[X]\mathbb{E}[Y]$ and $\mathrm{Corr}(X, Y)$. Finally, construct a separate example (U, V) on $\Omega = \{-1, +1\}$ with $\mathbb{P}(U = 1) = \mathbb{P}(U = -1) = 1/2$ and $V = U^2$ and verify explicitly that $\mathrm{Cov}(U, V) = 0$ while U and V are not independent.

Exercise 6.9.11 (Conditional Expectation as Best Predictor (Finite Case)) Let (X, Y) be discrete with $Y \in \{0, 1\}$ and

$$\mathbb{P}(Y = 1) = 0.4, \quad \mathbb{P}(Y = 0) = 0.6,$$

and conditional distributions

$$X \mid (Y = 0) \sim \mathrm{Uniform}\{0, 2, 4\}, \qquad X \mid (Y = 1) \sim \mathrm{Uniform}\{1, 3\}.$$

(i) Compute $\mathbb{E}[X \mid Y = 0]$ and $\mathbb{E}[X \mid Y = 1]$ and hence the random variable $\mathbb{E}[X \mid Y]$.
(ii) Compute $\mathbb{E}[X]$ via the law of total expectation.
(iii) Among all predictors of the form $g(Y) = aY+b$, find the minimizer of $\mathbb{E}[(X - g(Y))^2]$ and show it equals $g(Y) = \mathbb{E}[X \mid Y]$.
(iv) Compute $\mathrm{Var}(X)$ and verify the law of total variance $\mathrm{Var}(X) = \mathbb{E}[\mathrm{Var}(X \mid Y)] + \mathrm{Var}(\mathbb{E}[X \mid Y])$ numerically.

Exercise 6.9.12 (MGF and Sum of Independent Variables) Let X_1, X_2, X_3 be independent and each $X_i \sim \mathrm{Bernoulli}(p = 0.2)$.

(i) Compute $M_{X_i}(t) = \mathbb{E}[e^{tX_i}]$ explicitly.
(ii) Use independence to compute the MGF of $S = X_1 + X_2 + X_3$ and identify the distribution of S.
(iii) Compute $\mathbb{P}(S \geq 2)$ exactly and verify it matches the distribution you identified.

Exercise 6.9.13 (Change of Variables (1D) and Density Transformation) Let $X \sim \mathrm{Exponential}(\lambda = 1)$ with density $f_X(x) = e^{-x}\mathbf{1}\{x \geq 0\}$. Define $Y = g(X) = \sqrt{X}$.

(i) Compute the CDF $F_Y(y)$ for $y \geq 0$ using the preimage method.
(ii) Differentiate to obtain the density $f_Y(y)$ and verify it integrates to 1 over $[0, \infty)$.
(iii) Compute $\mathbb{E}[Y]$ by integrating against f_Y (no special-function shortcuts; show the integral clearly).

Exercise 6.9.14 (Multivariate Change of Variables and Jacobians) Let (U, V) be uniform on the unit square $[0, 1]^2$ with joint density $f_{U,V}(u, v) = 1$ on $[0, 1]^2$. Define the transformation

$$X = U, \qquad Y = U + V.$$

(i) Determine the image region $\mathcal{R}$ of (X, Y) in the (x, y)-plane.
(ii) Compute the Jacobian determinant $|\det J_{(u,v)\mapsto(x,y)}|$ and hence derive the joint density $f_{X,Y}(x, y)$ on $\mathcal{R}$.
(iii) Compute the marginal density $f_Y(y)$ for $y \in [0, 2]$ and verify it is triangular (piecewise linear).

Exercise 6.9.15 (Convolution and Gaussian Closure) Let $X \sim \mathcal{N}(1, 2)$ and $Y \sim \mathcal{N}(-1, 3)$ be independent.

(i) Compute the distribution of $S = X + Y$ (mean and variance).
(ii) Compute $\mathbb{P}(S \leq 0)$ in terms of the standard normal CDF Φ (no numeric approximation required).
(iii) Let $T = 2X - Y$. Compute the distribution of T (mean and variance) using linearity of expectation and variance for independent Gaussians.

Exercise 6.9.16 (Modes of Convergence: Verify on Explicit Sequences) Let $U \sim \text{Uniform}(0, 1)$ and define random variables $X_n = U^n$ and $Y_n = \mathbf{1}\{U \leq 1/n\}$.

(i) Show that $X_n \to 0$ almost surely and compute $\mathbb{P}(|X_n - 0| > \varepsilon)$ explicitly as a function of n and ε to verify convergence in probability.
(ii) Compute $\mathbb{E}[X_n]$ and determine whether $X_n \to 0$ in L^1.
(iii) Show that $Y_n \to 0$ almost surely and compute $\mathbb{E}[Y_n]$; determine whether $Y_n \to 0$ in L^1 and in L^2.

Exercise 6.9.17 (Likelihood and Log-Loss (Classification)) Consider binary labels $y_i \in \{0, 1\}$ with predicted probabilities $q_i = \mathbb{P}_\theta(Y = 1 \mid x_i)$. For $n = 8$, let

$$y = (1, 0, 1, 1, 0, 0, 1, 0), \qquad q = (0.9, 0.2, 0.6, 0.7, 0.4, 0.1, 0.8, 0.3).$$

(i) Compute the (conditional) likelihood $L = \prod_{i=1}^{8} q_i^{y_i}(1 - q_i)^{1-y_i}$ and the negative log-likelihood (log-loss)

$$\text{NLL} = -\sum_{i=1}^{8} \big(y_i \log q_i + (1 - y_i)\log(1 - q_i)\big).$$

(ii) Compute the Brier score $\frac{1}{8}\sum_{i=1}^{8}(q_i - y_i)^2$.
(iii) Interpret (one paragraph) which examples contribute most to NLL vs. Brier and why.

Exercise 6.9.18 (Calibration/Reliability Diagram Computation) Using the same (y_i, q_i) as in Exercise 6.9.17, form $B = 4$ bins:

$$I_1 = [0, 0.25), \quad I_2 = [0.25, 0.5), \quad I_3 = [0.5, 0.75), \quad I_4 = [0.75, 1].$$

(i) For each bin I_b, compute $\widehat{q}_b$ (average predicted probability) and $\widehat{p}_b$ (empirical frequency of $y = 1$) as in Definition 6.8.4.
(ii) Compute the expected calibration error (ECE) defined by

$$\mathrm{ECE} = \sum_{b=1}^{4} \frac{|S_b|}{n} |\widehat{p}_b - \widehat{q}_b|.$$

(iii) State whether the model appears overconfident or underconfident in each bin (one sentence per bin).

Chapter 7
Expectation, Moments, and Concentration

Modern learning theory and practice rely on the same quantitative fact: although data are random, suitably averaged quantities become reliable. This reliability is formalized by *concentration of measure*—high-probability bounds showing that empirical quantities (e.g., sample means, empirical risks, gradients) are close to their expectations. The engine behind concentration is a small toolkit of moment identities and inequalities (Markov, Chebyshev, Hoeffding/Bernstein-type), often powered by moment generating functions (Boucheron et al., 2013). This chapter builds that toolkit and shows how it feeds directly into generalization; see Fig. 7.1.

7.1 Why Concentration Is the Engine of Generalization

7.1.1 From Random Data to Reliable Averages

7.1.1.1 Empirical Means as Estimators

The simplest learning object is the empirical mean:

$$\bar{X}_n := \frac{1}{n}\sum_{i=1}^{n} X_i .$$

It already contains the core tension of ML: each X_i is random and noisy, but averaging reduces variability. Most estimators in ML are averages in disguise (risk, gradient estimates, sufficient statistics). Understanding $\bar{X}_n$—its bias, variance, and concentration—is therefore the gateway to understanding why learning works.

P. Singh, B. Raman, *Foundations of Machine Learning and AI*, Studies in Big Data 199, https://doi.org/10.1007/978-3-032-30336-3_7

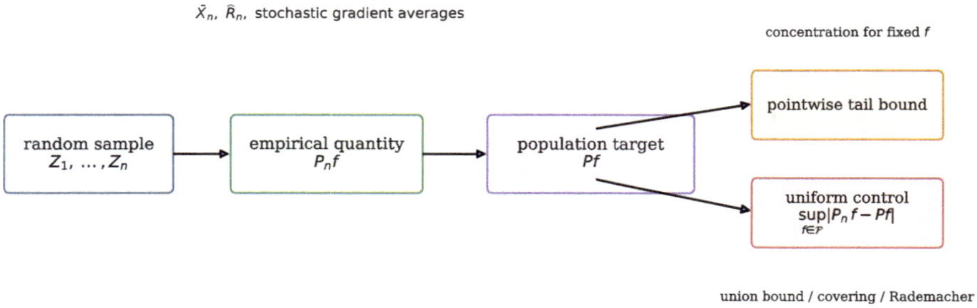

Fig. 7.1 Concentration as the engine of generalization. Random samples produce empirical quantities, concentration controls their deviation from expectations, and a union/complexity step upgrades pointwise concentration to uniform generalization bounds over a hypothesis class

Definition 7.1.1 (Empirical Mean) Let $X_1, \ldots, X_n$ be integrable real-valued random variables. The empirical mean is

$$\bar{X}_n := \frac{1}{n} \sum_{i=1}^{n} X_i.$$

If $X_i \in \mathbb{R}^d$, define $\bar{X}_n$ componentwise.

Proposition 7.1.1 (Unbiasedness and Variance (i.i.d. Case)) *If $X_1, \ldots, X_n$ are i.i.d. with $\mathbb{E}[X_1] = \mu$ and $\mathrm{Var}(X_1) = \sigma^2 < \infty$, then*

$$\mathbb{E}[\bar{X}_n] = \mu, \qquad \mathrm{Var}(\bar{X}_n) = \frac{\sigma^2}{n}.$$

Remark 7.1.1 (The $1/n$ Variance Law) The σ^2/n scaling is the first quantitative form of "averaging reduces noise." Concentration inequalities strengthen this to *high-probability* statements, not just variance.

7.1.1.2 High-Probability Guarantees vs. Expectations

Expectations quantify average behavior but can hide rare catastrophic events. Learning systems often require guarantees that hold with probability at least $1-\delta$: generalization bounds, confidence intervals, safe decision-making, and robust deployment all demand such control. Concentration inequalities transform moment information (variance, boundedness, sub-Gaussian tails) into tail bounds of the form

$$\mathbb{P}\big(|\bar{X}_n - \mu| \geq \varepsilon\big) \leq \delta,$$

which can be inverted to express ε as a function of n and δ.

Remark 7.1.2 (Expectation vs. Tail Control) A statement like $\mathbb{E}|\bar{X}_n - \mu| \le 0.01$ does not exclude the possibility of a 10-unit error occurring with probability 10^{-3}. High-probability bounds control such events directly.

Remark 7.1.3 (The Role of δ) In ML, δ is the acceptable failure probability. Bounds typically scale as $\log(1/\delta)$. Choosing δ small enough to union bound over many hypotheses or time steps is standard practice.

7.1.2 Roadmap: From Moments to Tail Bounds

7.1.2.1 Inequalities as Reusable Templates

Concentration proofs are rarely reinvented from scratch. They are built from reusable templates: Markov $\Rightarrow$ tail bound from nonnegative moments; Chebyshev $\Rightarrow$ tail bound from variance; MGF methods $\Rightarrow$ exponential tails from moment generating functions. Once you master a template, you can apply it across problems: bounding estimation error, bounding gradient noise, controlling maxima via union bounds, etc. This section outlines the templates we will develop.

Remark 7.1.4 (Three Canonical Templates)

1. **Moment/Markov template:** bound $\mathbb{P}(X \ge t)$ via $\mathbb{E}[X]$ for $X \ge 0$.
2. **Variance/Chebyshev template:** bound deviations via $\mathrm{Var}(X)$.
3. **MGF/Chernoff template:** bound tails via $\mathbb{E}[e^{\lambda X}]$ and optimize over λ.

These templates differ in assumptions and sharpness; ML practice chooses based on available information about data.

7.1.2.2 How Bounds Plug into Learning Guarantees

A typical learning guarantee requires controlling the gap between empirical and population quantities:

$$\sup_{h \in \mathcal{H}} \left| \widehat{R}_n(h) - R(h) \right|.$$

Concentration is the tool that turns randomness of the sample into a deterministic bound with high probability. The remaining step is often combinatorial/metric: how large is $\mathcal{H}$, or how complex is it? Union bounds handle finite $\mathcal{H}$; covering numbers and VC/Rademacher methods handle infinite classes. This preview explains why concentration is called the engine: it is the probabilistic part that powers the generalization machine.

Remark 7.1.5 (Finite-Class Prototype) If $|\mathcal{H}| = M$ and for each fixed h we have $\mathbb{P}(|\widehat{R}_n(h) - R(h)| \geq \varepsilon) \leq 2e^{-cn\varepsilon^2}$, then by union bound,

$$\mathbb{P}\Big(\sup_{h\in\mathcal{H}} |\widehat{R}_n(h) - R(h)| \geq \varepsilon \Big) \leq 2Me^{-cn\varepsilon^2}.$$

Solving for ε yields $\varepsilon \asymp \sqrt{\frac{\log M + \log(1/\delta)}{n}}$.

Remark 7.1.6 (Continuous Classes) When $\mathcal{H}$ is infinite, one controls complexity via covering/entropy or Rademacher complexity, then applies concentration on the resulting finite approximation or symmetrized process. We will develop these ideas later.

7.2 Expectation, Variance, Covariance

Expectation and second moments are the computational primitives of probability. They control typical behavior (means), uncertainty (variances), and linear dependence (covariances), and they power concentration inequalities via variance bounds and MGF methods (Durrett, 2019). In ML, these objects appear as population risks, noise models, covariance estimation, PCA, whitening, and the geometry of uncertainty ellipsoids.

7.2.1 Core Operators

7.2.1.1 Linearity of Expectation and Conditioning Tricks

Linearity is the single most reused fact in probability: it allows one to pass expectations through sums and constants without independence assumptions. Conditional expectation adds a second layer of power: by conditioning on a convenient variable, one can simplify computations dramatically. This is the basis of "conditioning tricks" such as the tower property and iterated expectation, used constantly in ML to compute expectations under latent-variable models, dropout-like randomness, and stochastic-gradient estimators.

Proposition 7.2.1 (Linearity (Recall)) *If X, Y are integrable and $a, b \in \mathbb{R}$, then*

$$\mathbb{E}[aX + bY] = a\,\mathbb{E}[X] + b\,\mathbb{E}[Y].$$

Proposition 7.2.2 (Tower Property/Iterated Expectation) *If X is integrable and Y is any random variable, then*

$$\mathbb{E}\big[\mathbb{E}[X \mid Y]\big] = \mathbb{E}[X].$$

More generally, if $\sigma(Y) \subseteq \sigma(Z)$, then

$$\mathbb{E}\big[\mathbb{E}[X \mid Z] \mid Y\big] = \mathbb{E}[X \mid Y] \qquad a.s.$$

Remark 7.2.1 (A Common Trick) To compute $\mathbb{E}[g(X, Y)]$, condition on the variable that makes g simpler:

$$\mathbb{E}[g(X, Y)] = \mathbb{E}\big[\mathbb{E}[g(X, Y) \mid Y]\big].$$

This reduces a joint expectation to an outer expectation of a conditional expectation.

7.2.1.2 Variance Decomposition and Total Variance

Variance measures dispersion around the mean and is the first quantitative control on concentration (via Chebyshev). The law of total variance decomposes uncertainty into *irreducible noise* (expected conditional variance) and *explainable variance* (variance of the conditional mean). This decomposition is the probabilistic analogue of Pythagorean splitting in geometry and is central to bias–variance intuition and regression analysis.

Definition 7.2.1 (Variance) If $X \in L^2$ with $\mu = \mathbb{E}[X]$, then

$$\mathrm{Var}(X) := \mathbb{E}\big[(X - \mu)^2\big] = \mathbb{E}[X^2] - \mu^2.$$

Proposition 7.2.3 (Total Variance) *If $X \in L^2$ and Y is any random variable, then*

$$\mathrm{Var}(X) = \mathbb{E}\big[\mathrm{Var}(X \mid Y)\big] + \mathrm{Var}\big(\mathbb{E}[X \mid Y]\big).$$

Remark 7.2.2 (Interpretation in Learning) $\mathbb{E}[\mathrm{Var}(X \mid Y)]$ is uncertainty that remains even after observing Y (noise), while $\mathrm{Var}(\mathbb{E}[X \mid Y])$ is variability explained by Y (signal). For regression $Y = f(X) + \varepsilon$, this mirrors signal-plus-noise decomposition.

7.2.2 *Covariance Structure*

7.2.2.1 Covariance Matrices and Correlation

Covariance generalizes variance to pairs: it measures whether two quantities co-vary linearly. For vector-valued data, the covariance matrix collects all pairwise covariances and encodes second-order geometry of the distribution. Correlation normalizes covariance to remove scale, making it a pure dependence measure in

$[-1, 1]$ for scalar pairs. In ML, covariance matrices drive PCA, whitening, Gaussian models, and error propagation analyses.

Definition 7.2.2 (Covariance and Correlation) For $X, Y \in L^2$,

$$\mathrm{Cov}(X, Y) := \mathbb{E}\big[(X - \mathbb{E}X)(Y - \mathbb{E}Y)\big] = \mathbb{E}[XY] - \mathbb{E}[X]\mathbb{E}[Y].$$

If $\mathrm{Var}(X), \mathrm{Var}(Y) > 0$, the correlation is

$$\mathrm{Corr}(X, Y) := \frac{\mathrm{Cov}(X, Y)}{\sqrt{\mathrm{Var}(X)\mathrm{Var}(Y)}} \in [-1, 1].$$

Definition 7.2.3 (Covariance Matrix) For $X \in \mathbb{R}^d$ with $\mu = \mathbb{E}[X]$ and finite second moments, the covariance matrix is

$$\Sigma := \mathbb{E}\big[(X - \mu)(X - \mu)^\top\big] \in \mathbb{R}^{d \times d}.$$

Proposition 7.2.4 (Basic Properties) *Let $X \in \mathbb{R}^d$ have covariance Σ.*

1. *Σ is symmetric and positive semidefinite (PSD): $\Sigma = \Sigma^\top$ and $v^\top \Sigma v \geq 0$ for all v.*
2. *$\mathrm{Var}(v^\top X) = v^\top \Sigma v$ for all $v \in \mathbb{R}^d$.*

Remark 7.2.3 (Empirical Covariance (Data Matrix Form)) Given centered samples $x_i - \bar{x}$ stacked as rows in $X_c \in \mathbb{R}^{n \times d}$, the empirical covariance is $\widehat{\Sigma} = \frac{1}{n} X_c^\top X_c$, a Gram/PSD object.

7.2.2.2 Second-Moment Geometry and Ellipsoids

Second moments define geometry. The quadratic form $v^\top \Sigma v$ is the variance along direction v, so Σ shapes an ellipsoid of uncertainty. Eigenvectors give principal directions of variation; eigenvalues give squared axis lengths. This geometric view explains PCA, whitening, and why ill-conditioning corresponds to anisotropy (elongated ellipsoids); cf. Fig. 7.2. It also clarifies how Gaussian likelihoods and Mahalanobis distances arise from covariance.

Definition 7.2.4 (Second-Moment Ellipsoid) Let $\Sigma \succ 0$. The ellipsoid

$$\mathcal{E}_\Sigma(r) := \{x \in \mathbb{R}^d : x^\top \Sigma^{-1} x \leq r^2\}$$

is the set of points within Mahalanobis radius r (centered at 0; translate by μ for center μ).

Remark 7.2.4 (Eigen-Geometry) If $\Sigma = Q \Lambda Q^\top$ with $\Lambda = \mathrm{diag}(\lambda_1, \ldots, \lambda_d)$, then $\mathcal{E}_\Sigma(r)$ has principal axes q_i with semi-axis lengths $r\sqrt{\lambda_i}$. Thus, large eigenvalues correspond to directions of high variance.

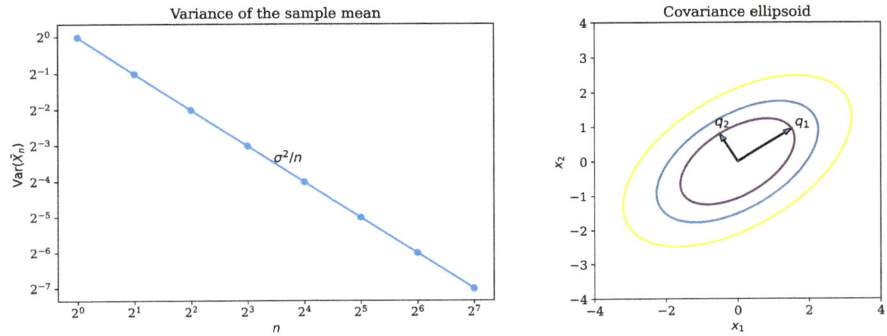

Fig. 7.2 Second-moment geometry. Left: averaging reduces variability of the sample mean at the canonical $1/n$ variance scale. Right: a covariance matrix induces an ellipsoid whose principal axes are the eigenvectors and whose axis lengths are proportional to the square roots of the eigenvalues

7.2.3 Matrix-Valued Moments

7.2.3.1 Quadratic Forms and Covariance as PSD Objects

Many concentration and learning results reduce to controlling quadratic forms $X^\top AX$ and matrix-valued second moments like $\mathbb{E}[XX^\top]$. The key structural fact is PSD-ness: second-moment matrices are PSD by construction. This allows ordering inequalities (Loewner order) and spectral control, which are essential in high-dimensional statistics and in analyzing random matrices (later chapters).

Definition 7.2.5 (Second Moment Matrix and Loewner Order) For $X \in \mathbb{R}^d$ with $\mathbb{E}\|X\|_2^2 < \infty$, define the (uncentered) second moment matrix

$$M_2 := \mathbb{E}[XX^\top].$$

For symmetric matrices A, B, write $A \succeq B$ if $A - B$ is PSD (Loewner order).

Proposition 7.2.5 (PSD Nature of Second Moments) $M_2 = \mathbb{E}[XX^\top]$ *is PSD. More generally, for any fixed PSD matrix* $A \succeq 0$*,*

$$\mathbb{E}[X^\top AX] \geq 0, \qquad \textit{and} \qquad \mathbb{E}[X^\top AX] = \mathrm{tr}\big(A\,\mathbb{E}[XX^\top]\big).$$

Remark 7.2.5 (Directional Variances as Quadratic Forms) Taking $A = vv^\top$ yields $\mathbb{E}[X^\top vv^\top X] = \mathbb{E}[(v^\top X)^2]$, connecting scalar second moments to matrix second moments.

7.2.3.2 Trace Tricks for Second Moments

Trace identities provide a compact calculus for second moments, especially when moving between scalar and matrix views. They allow you to express expectations of quadratic forms as traces, swap cyclic factors to simplify, and connect Frobenius norms to Gram matrices. These tricks become indispensable when deriving covariance identities, computing gradients of quadratic objectives, and analyzing random matrix estimators.

Proposition 7.2.6 (Trace Identities for Second Moments) *Let $X \in \mathbb{R}^d$ have finite second moments and let $A \in \mathbb{R}^{d\times d}$ be deterministic. Then*

$$X^\top AX = \operatorname{tr}(X^\top AX) = \operatorname{tr}(AXX^\top),$$

and therefore

$$\mathbb{E}[X^\top AX] = \operatorname{tr}\big(A\,\mathbb{E}[XX^\top]\big).$$

If X is centered with covariance Σ, then $\mathbb{E}[XX^\top] = \Sigma$ and hence $\mathbb{E}[X^\top AX] = \operatorname{tr}(A\Sigma)$.

Example 7.2.1 (Computing a Quadratic Expectation) Let $X \in \mathbb{R}^2$ be centered with covariance $\Sigma = \begin{bmatrix} 2 & 1 \\ 1 & 3 \end{bmatrix}$ and let $A = \begin{bmatrix} 1 & 0 \\ 0 & 4 \end{bmatrix}$. Then

$$\mathbb{E}[X^\top AX] = \operatorname{tr}(A\Sigma) = \operatorname{tr}\begin{bmatrix} 2 & 1 \\ 4 & 12 \end{bmatrix} = 2 + 12 = 14.$$

7.3 Inequalities

Probability inequalities are reusable templates that convert limited information (nonnegativity, moments, boundedness) into tail bounds and generalization guarantees. They are the “mechanical” part of concentration: once the right inequality is chosen, the rest is algebra and optimization over parameters. This section develops one-shot inequalities (Markov, Chebyshev, Jensen), inequalities for sums (union bounds, max bounds, Cauchy–Schwarz/Hölder), and the standard pipeline by which these inequalities become learning-theoretic risk guarantees.

7.3.1 One-Shot Inequalities

7.3.1.1 Markov and Chebyshev for Tail Bounds

Markov's inequality is the most basic tail bound: it converts a bound on the mean of a nonnegative random variable into a bound on large deviations. Chebyshev refines this by using variance, yielding a $1/t^2$ tail decay for centered variables. These inequalities are often loose compared to exponential bounds, but they require minimal assumptions and are universally applicable. In learning, they provide baseline guarantees and are stepping stones to sharper concentration via mgf methods.

Theorem 7.3.1 (Markov Inequality) *If $X \geq 0$ almost surely and $t > 0$, then*

$$\mathbb{P}(X \geq t) \leq \frac{\mathbb{E}[X]}{t}.$$

Remark 7.3.1 (Moment Version) For $p > 0$ and $X \geq 0$,

$$\mathbb{P}(X \geq t) = \mathbb{P}(X^p \geq t^p) \leq \frac{\mathbb{E}[X^p]}{t^p}.$$

Choosing p is a tuning knob: higher moments can yield sharper bounds when they exist.

Theorem 7.3.2 (Chebyshev Inequality) *If $X \in L^2$ with mean $\mu = \mathbb{E}[X]$ and variance $\sigma^2 = \mathrm{Var}(X)$, then for any $t > 0$,*

$$\mathbb{P}(|X - \mu| \geq t) \leq \frac{\sigma^2}{t^2}.$$

Remark 7.3.2 (Chebyshev from Markov) Apply Markov to the nonnegative random variable $(X - \mu)^2$:

$$\mathbb{P}(|X - \mu| \geq t) = \mathbb{P}((X - \mu)^2 \geq t^2) \leq \frac{\mathbb{E}[(X - \mu)^2]}{t^2} = \frac{\sigma^2}{t^2}.$$

7.3.1.2 Jensen and Convexity-Based Bounds

Jensen's inequality is the bridge between convexity and expectation. It explains why averaging reduces loss for convex losses, why log-likelihood and KL divergence behave nicely, and why $\log \mathbb{E}[\exp(\cdot)]$ upper-bounds expectations (a key move in mgf/Chernoff bounds). In ML, Jensen appears whenever we swap "function of expectation" versus "expectation of function": risk minimization, variational bounds, and ELBO constructions are all Jensen-driven.

Theorem 7.3.3 (Jensen Inequality) *Let $\varphi : \mathbb{R} \to \mathbb{R}$ be convex and let X be integrable. Then*

$$\varphi(\mathbb{E}[X]) \leq \mathbb{E}[\varphi(X)].$$

If φ is concave, the inequality reverses.

Remark 7.3.3 (Two Standard Corollaries)

1. For convex $\|\cdot\|$, $\|\mathbb{E}[Z]\| \leq \mathbb{E}[\|Z\|]$ (norm is convex).
2. For concave log, $\mathbb{E}[\log X] \leq \log \mathbb{E}[X]$ for $X > 0$.

The second is the algebra behind variational lower bounds (ELBO) in latent-variable models.

Example 7.3.1 (Jensen in Risk Minimization) If $\ell(\cdot)$ is convex and $\widehat{\theta}$ is random (e.g., randomized algorithm), then

$$\ell(\mathbb{E}[\widehat{\theta}]) \leq \mathbb{E}[\ell(\widehat{\theta})].$$

Thus, deterministic averaging of parameters can reduce convex loss; this is one reason averaging iterates can stabilize training in convex settings.

7.3.2 Inequalities for Sums

7.3.2.1 Union Bound, Boole's Inequality, and Max Bounds

Bounding a maximum or a "bad event exists" statement is central in learning: uniform generalization over many hypotheses, controlling error across many coordinates, or bounding the maximum of noise terms. The union bound converts a union of events into a sum of probabilities; max bounds convert maxima into unions. These are the standard tools for turning combinatorial complexity (many things can go wrong) into manageable algebra.

Theorem 7.3.4 (Union Bound (Boole's Inequality)) *For events $\{E_i\}_{i \in I}$ (finite or countable),*

$$\mathbb{P}\Big(\bigcup_{i \in I} E_i\Big) \leq \sum_{i \in I} \mathbb{P}(E_i).$$

Proposition 7.3.1 (Max Bound via Union Bound) *Let $X_1, \ldots, X_m$ be real-valued random variables. For any $t \in \mathbb{R}$,*

$$\mathbb{P}\Big(\max_{1 \leq i \leq m} X_i \geq t\Big) = \mathbb{P}\Big(\bigcup_{i=1}^{m} \{X_i \geq t\}\Big) \leq \sum_{i=1}^{m} \mathbb{P}(X_i \geq t).$$

Similarly,

$$\mathbb{P}\Big(\max_{1\le i\le m}|X_i|\ge t\Big)\le\sum_{i=1}^{m}\mathbb{P}(|X_i|\ge t).$$

Remark 7.3.4 (Typical Use with Symmetric Tails) If each X_i has a tail bound $\mathbb{P}(|X_i|\ge t)\le 2e^{-ct^2}$, then

$$\mathbb{P}\Big(\max_{i}|X_i|\ge t\Big)\le 2me^{-ct^2},$$

so $\max_i|X_i|$ is of order $\sqrt{\log m}$ with high probability. This is a standard "$\sqrt{\log m}$" phenomenon in high-dimensional learning.

7.3.2.2 Cauchy–Schwarz/Hölder as Probabilistic Tools

Cauchy–Schwarz and Hölder are not just geometric inequalities; they are probabilistic workhorses. They bound correlations, control expectations of products, and enable decoupling: replace a hard-to-analyze product by a product of norms/moments. This is repeatedly used in concentration proofs, bounding covariances, and analyzing stochastic gradients. In high-level terms: they reduce dependence problems to moment problems.

Theorem 7.3.5 (Cauchy–Schwarz) *If $X,Y\in L^2$, then*

$$|\mathbb{E}[XY]|\le\sqrt{\mathbb{E}[X^2]\,\mathbb{E}[Y^2]}.$$

Theorem 7.3.6 (Hölder) *Let $p,q\in[1,\infty]$ satisfy $\frac{1}{p}+\frac{1}{q}=1$. If $X\in L^p$ and $Y\in L^q$, then*

$$\mathbb{E}[|XY|]\le\big(\mathbb{E}[|X|^p]\big)^{1/p}\big(\mathbb{E}[|Y|^q]\big)^{1/q}.$$

Remark 7.3.5 (Covariance Bound) For $X,Y\in L^2$,

$$|\mathrm{Cov}(X,Y)|=\big|\mathbb{E}[(X-\mathbb{E}X)(Y-\mathbb{E}Y)]\big|\le\sqrt{\mathrm{Var}(X)\mathrm{Var}(Y)}$$

by Cauchy–Schwarz. This is the inequality behind $|\mathrm{Corr}(X,Y)|\le 1$.

7.3.3 How Inequalities Power Learning

7.3.3.1 Bounding Estimation Error via Moments

Learning is estimation. A canonical task is estimating $\mu = \mathbb{E}[X]$ from samples via $\bar{X}_n$. Inequalities transform moment information into error bounds. Variance yields Chebyshev; boundedness plus mgf yields Hoeffding; sub-Gaussianity yields sharp tails. Even when stronger assumptions hold, Chebyshev provides a useful sanity check and a baseline rate.

Proposition 7.3.2 (Chebyshev Bound for Empirical Means) *If $X_1, \ldots, X_n$ are i.i.d. with $\mathbb{E}[X_1] = \mu$ and $\mathrm{Var}(X_1) = \sigma^2 < \infty$, then for any $\varepsilon > 0$,*

$$\mathbb{P}\big(|\bar{X}_n - \mu| \geq \varepsilon\big) \leq \frac{\sigma^2}{n\varepsilon^2}.$$

Equivalently, with probability at least $1 - \delta$,

$$|\bar{X}_n - \mu| \leq \frac{\sigma}{\sqrt{n\delta}}.$$

Remark 7.3.6 (What this Does and Does Not Say) Chebyshev gives a $1/\sqrt{n}$ rate but with a $1/\sqrt{\delta}$ dependence, which is weaker than exponential-tail bounds. Sharper inequalities improve the δ dependence to $\sqrt{\log(1/\delta)/n}$ under stronger assumptions.

7.3.3.2 Risk Bounds: High-Level Pipeline

A generalization bound typically follows a consistent pipeline: (1) express the quantity of interest as an expectation or deviation; (2) bound the deviation for a fixed hypothesis using a one-dimensional inequality (Chebyshev/Hoeffding); (3) extend from one hypothesis to many using a union bound (finite classes) or complexity control (covering/VC/Rademacher); (4) solve for ε in terms of n and δ. This section records the finite-class prototype, which is the cleanest setting to see how inequalities power learning.

Proposition 7.3.3 (Finite-Class Generalization Template) *Let $\mathcal{H}$ be a finite hypothesis class with $|\mathcal{H}| = M$. Assume for each fixed $h \in \mathcal{H}$ and each $\varepsilon > 0$,*

$$\mathbb{P}\big(|\widehat{R}_n(h) - R(h)| \geq \varepsilon\big) \leq \psi(n, \varepsilon),$$

for some tail function ψ decreasing in ε. Then, by union bound,

$$\mathbb{P}\Big(\sup_{h \in \mathcal{H}} |\widehat{R}_n(h) - R(h)| \geq \varepsilon\Big) \leq M\,\psi(n, \varepsilon).$$

In particular, if $\psi(n, \varepsilon) = 2e^{-cn\varepsilon^2}$, *then with probability at least* $1 - \delta$,

$$\sup_{h \in \mathcal{H}} |\widehat{R}_n(h) - R(h)| \leq \sqrt{\frac{1}{cn}\left(\log \frac{2M}{\delta}\right)}.$$

Remark 7.3.7 (What Changes for Infinite Classes) For infinite $\mathcal{H}$, M is replaced by a complexity measure: covering numbers, VC dimension, Rademacher complexity, or algorithmic stability. The probabilistic core (concentration + union-like control) remains the same.

7.4 Limit Theorems

Limit theorems formalize the two most important asymptotic phenomena in learning: (i) empirical averages stabilize (LLN), and (ii) properly normalized fluctuations become approximately Gaussian (CLT). Together they justify why sample means are consistent estimators, why $1/\sqrt{n}$ rates are ubiquitous, and why confidence intervals often use normal approximations (cf. Fig. 7.3). We state the core results, explain what assumptions they need, and preview what changes under dependence or heavy tails.

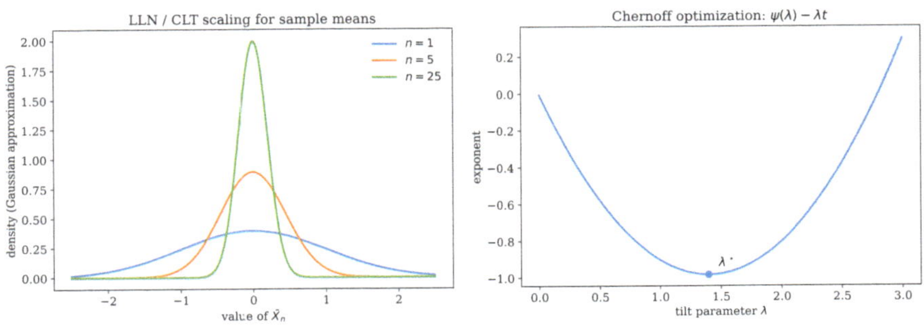

Fig. 7.3 Two asymptotic views of averaging. Left: as n grows, the distribution of the sample mean concentrates around the population mean and its fluctuations shrink at the $1/\sqrt{n}$ scale. Right: Chernoff's method bounds the tail of a random variable by tilting it exponentially and optimizing over the tilt parameter

7.4.1 Law of Large Numbers (LLN)

7.4.1.1 Sample-Mean Consistency and Intuition

The LLN is the mathematical statement behind "averaging reduces randomness." For i.i.d. data with finite mean, the sample mean converges to the population mean. This is the baseline consistency result in statistics and ML: empirical risk approximates true risk, empirical gradients approximate population gradients, and empirical frequencies approximate probabilities. The key is that randomness cancels as n grows, provided no single observation dominates.

Theorem 7.4.1 (Weak Law of Large Numbers (WLLN)) *Let $X_1, X_2, \ldots$ be i.i.d. with $\mathbb{E}[|X_1|] < \infty$ and mean $\mu = \mathbb{E}[X_1]$. Define $\bar{X}_n = \frac{1}{n}\sum_{i=1}^{n} X_i$. Then*

$$\bar{X}_n \to \mu \quad \textit{in probability.}$$

Equivalently, for every $\varepsilon > 0$,

$$\mathbb{P}\big(|\bar{X}_n - \mu| > \varepsilon\big) \to 0.$$

Theorem 7.4.2 (Strong Law of Large Numbers (SLLN)) *Let $X_1, X_2, \ldots$ be i.i.d. with $\mathbb{E}[|X_1|] < \infty$ and mean μ. Then*

$$\bar{X}_n \to \mu \quad \textit{almost surely.}$$

Remark 7.4.1 (Consistency as LLN) The sample mean $\bar{X}_n$ is a consistent estimator of μ (in probability by WLLN, a.s. by SLLN). For bounded losses $\ell(h(X), Y)$, the empirical risk $\widehat{R}_n(h)$ is a sample mean, so LLN provides pointwise consistency: $\widehat{R}_n(h) \to R(h)$ for fixed h.

7.4.1.2 Dependence Caveats: What Breaks

LLN relies on cancellations that can fail under dependence. If samples are correlated, the effective number of independent observations can be much smaller than n. In time series, long-range dependence can prevent convergence or change rates. In ML, this matters for sequential data, reinforcement learning, and distribution shift. The correct fix is not to abandon LLN but to replace independence by structural conditions (mixing, ergodicity) that restore averaging.

Remark 7.4.2 (A Simple Diagnostic: Variance of the Average Under Dependence) If $\{X_i\}$ are stationary with $\mathrm{Cov}(X_i, X_{i+k}) = \gamma(k)$ and finite second moments, then

$$\mathrm{Var}(\bar{X}_n) = \frac{1}{n^2}\sum_{i=1}^{n}\sum_{j=1}^{n}\mathrm{Cov}(X_i, X_j) = \frac{1}{n^2}\Big(n\gamma(0) + 2\sum_{k=1}^{n-1}(n-k)\gamma(k)\Big).$$

If $\sum_{k\geq 1} |\gamma(k)| < \infty$ (short-range dependence), then $\mathrm{Var}(\bar{X}_n) = O(1/n)$ and LLN-like behavior persists. If correlations decay too slowly, the variance may not vanish at $1/n$ and convergence can be slow or fail.

Remark 7.4.3 (What Breaks Operationally) When dependence is strong, treating data as i.i.d. underestimates uncertainty: confidence intervals become too narrow, and generalization bounds can be overly optimistic.

7.4.2 Central Limit Theorem (CLT)

7.4.2.1 Standardization and Asymptotic Normality

LLN says the mean converges; CLT says how it fluctuates around the limit. After centering by μ and scaling by $\sqrt{n}$, the distribution converges to a Gaussian with variance σ^2. This justifies asymptotic confidence intervals and z-tests and explains the ubiquitous $1/\sqrt{n}$ estimation error scaling. In ML, CLT motivates normal approximations for aggregated noise (e.g., gradient noise) and for estimator distributions under large samples.

Theorem 7.4.3 (Central Limit Theorem (Lindeberg–Lévy)) *Let $X_1, X_2, \ldots$ be i.i.d. with mean $\mu = \mathbb{E}[X_1]$ and variance $\sigma^2 = \mathrm{Var}(X_1) \in (0, \infty)$. Then*

$$\sqrt{n}\,\frac{\bar{X}_n - \mu}{\sigma} \Rightarrow \mathcal{N}(0, 1).$$

Equivalently, $\sqrt{n}(\bar{X}_n - \mu) \Rightarrow \mathcal{N}(0, \sigma^2)$.

Remark 7.4.4 (Approximate Confidence Interval (Heuristic)) For large n, $\bar{X}_n$ is approximately normal:

$$\bar{X}_n \approx \mathcal{N}\Big(\mu, \frac{\sigma^2}{n}\Big).$$

Thus a common approximate $(1-\alpha)$ interval is

$$\bar{X}_n \pm z_{1-\alpha/2}\frac{\sigma}{\sqrt{n}},$$

where $z_{1-\alpha/2}$ is a standard normal quantile (in practice, σ is estimated).

7.4.2.2 Delta Method Intuition

Often we care not about $\bar{X}_n$ itself but about a smooth function of it (log, exp, ratio, sigmoid). The delta method says: if $\sqrt{n}(\widehat{\theta}_n - \theta) \Rightarrow \mathcal{N}(0, \tau^2)$ and g is differentiable at θ, then $\sqrt{n}(g(\widehat{\theta}_n) - g(\theta))$ is approximately normal with variance scaled by $(g'(\theta))^2$. Geometrically, this is local linearization (Jacobian) plus CLT.

Remark 7.4.5 (Delta Method (One-Dimensional Statement)) If $\sqrt{n}(\widehat{\theta}_n - \theta) \Rightarrow \mathcal{N}(0, \tau^2)$ and g is differentiable at θ with $g'(\theta) \neq 0$, then

$$\sqrt{n}\big(g(\widehat{\theta}_n) - g(\theta)\big) \Rightarrow \mathcal{N}\big(0,\ (g'(\theta))^2\tau^2\big).$$

This is a CLT through a smooth transformation.

7.4.3 Beyond CLT (Brief Preview)

7.4.3.1 Berry–Esseen Intuition: Rates to Normality (Conceptual)

CLT is asymptotic; Berry–Esseen quantifies how fast the convergence to normality happens. In learning, finite-sample performance matters: if convergence is slow, normal-based intervals can be inaccurate. Berry–Esseen shows the error decays like $O(1/\sqrt{n})$ under a finite third-moment condition, and it identifies what makes normal approximation poor: heavy tails and skewness (large third absolute moment).

Remark 7.4.6 (Berry–Esseen (Informal Rate Statement)) If X_i are i.i.d. with mean 0, variance $\sigma^2 > 0$, and third absolute moment $\rho = \mathbb{E}[|X_1|^3] < \infty$, then

$$\sup_{t\in\mathbb{R}} \left| \mathbb{P}\left(\frac{1}{\sigma\sqrt{n}} \sum_{i=1}^{n} X_i \le t \right) - \Phi(t) \right| \le C \frac{\rho}{\sigma^3\sqrt{n}}$$

for a universal constant C. Thus, skewness/heavy third moment slows convergence.

7.4.3.2 Heavy Tails: When CLT Is Slow or Misleading (Conceptual)

Heavy tails challenge both LLN/CLT-based intuition and concentration. Even when CLT holds, convergence can be slow and finite-sample distributions can be far from normal. When variance is infinite, classical CLT fails and stable laws can arise instead. In ML, heavy tails appear in gradient noise, financial returns, and certain network activations. This motivates robust estimators and tail-sensitive concentration tools (sub-exponential, truncation, median-of-means).

Remark 7.4.7 (Two Failure Modes)

1. **Finite variance but heavy tails:** CLT holds, but Berry–Esseen constant may be large; normal approximation can be poor at practical n.
2. **Infinite variance:** classical CLT fails; sums may converge (after different normalization) to non-Gaussian stable limits.

In either case, "Gaussian intuition" must be used with caution.

7.5 MGFs, CGFs, and Tail Control

Moment generating functions (MGFs) and cumulant generating functions (CGFs) are the main analytic tools for converting moment information into exponential tail bounds. The core idea is simple: apply Markov's inequality to $e^{\lambda X}$, then optimize over λ (Chernoff method). CGFs package this optimization cleanly and connect to large-deviation "rate functions" via Legendre transforms. This section formalizes MGFs/CGFs, highlights existence subtleties, and derives the standard Chernoff bounds that lead to sub-Gaussian and related tail behaviors.

7.5.1 Moment Generating Functions (MGFs)

7.5.1.1 Definitions and Existence Issues

The MGF $M_X(\lambda) = \mathbb{E}[e^{\lambda X}]$ may be finite only on an interval around 0 (or may be infinite everywhere except at 0), especially for heavy-tailed distributions. Many concentration inequalities require finiteness for λ in a neighborhood of 0. In practice, we often work with centered variables $X - \mathbb{E}X$ and bound their MGFs by a quadratic function, capturing "Gaussian-like" tails. Understanding when MGFs exist prevents applying Chernoff bounds in regimes where they are invalid.

Definition 7.5.1 (MGF and log-MGF) For a real-valued random variable X, the *moment generating function* is

$$M_X(\lambda) := \mathbb{E}\big[e^{\lambda X}\big],$$

defined for those $\lambda \in \mathbb{R}$ where the expectation is finite. The *log-MGF* (also called cumulant generating function when it exists) is

$$\psi_X(\lambda) := \log M_X(\lambda),$$

with domain $\{\lambda : M_X(\lambda) < \infty\}$.

Remark 7.5.1 (Existence Caveat) It is possible that $\mathbb{E}[|X|^k] < \infty$ for many k but $M_X(\lambda) = \infty$ for all $\lambda > 0$ (heavy tails). Chernoff bounds rely on $M_X(\lambda)$ being finite for some $\lambda > 0$ (and similarly for negative λ if bounding both tails).

Proposition 7.5.1 (MGF of Sums (Independence)) *If $X_1, \ldots, X_n$ are independent and $M_{X_i}(\lambda)$ are finite, then*

$$M_{\sum_{i=1}^n X_i}(\lambda) = \prod_{i=1}^n M_{X_i}(\lambda), \qquad \psi_{\sum_{i=1}^n X_i}(\lambda) = \sum_{i=1}^n \psi_{X_i}(\lambda).$$

7.5.1.2 Chernoff Bounding Idea (Template)

Chernoff bounding is the exponential upgrade of Markov: to bound $\mathbb{P}(X \geq t)$, apply Markov to $e^{\lambda X}$, which is nonnegative, yielding $\mathbb{P}(X \geq t) \leq e^{-\lambda t}\mathbb{E}[e^{\lambda X}]$. Optimizing over λ yields the best exponential bound available from the MGF. This single template produces Hoeffding, Bernstein, Bennett, and sub-Gaussian tails depending on how we bound the MGF.

Proposition 7.5.2 (Chernoff (Exponential Markov) Bound) *Let X be real-valued and suppose $M_X(\lambda) < \infty$ for some $\lambda > 0$. Then for any $t \in \mathbb{R}$ and any $\lambda > 0$,*

$$\mathbb{P}(X \geq t) \leq e^{-\lambda t}\,\mathbb{E}[e^{\lambda X}] = \exp\big(\psi_X(\lambda) - \lambda t\big).$$

Similarly, if $M_X(-\lambda) < \infty$,

$$\mathbb{P}(X \leq -t) \leq \exp\big(\psi_X(-\lambda) - \lambda t\big).$$

Remark 7.5.2 (Optimization Knob) The bound holds for every λ in the MGF domain. Choosing λ to minimize $\psi_X(\lambda) - \lambda t$ yields the tightest Chernoff bound obtainable by this method.

7.5.2 *Cumulant Generating Functions (CGFs)*

7.5.2.1 Cumulants and Variance Proxies

The CGF $\psi_X(\lambda) = \log \mathbb{E}[e^{\lambda X}]$ linearizes products and turns multiplicative tail bounds into additive ones. Its derivatives at 0 produce cumulants: mean and variance appear as the first two derivatives when they exist. Even beyond exact derivatives, upper bounds on $\psi_X(\lambda)$ act as variance proxies: if $\psi_{X-\mathbb{E}X}(\lambda) \leq \frac{1}{2}\sigma^2\lambda^2$, then σ^2 functions like a variance controlling exponential tails. This is precisely the definition of sub-Gaussianity.

Proposition 7.5.3 (Derivatives of CGF at 0 (When Finite)) *Assume $M_X(\lambda)$ is finite in a neighborhood of 0 so that ψ_X is well-defined and differentiable near 0. Then*

$$\psi_X'(0) = \mathbb{E}[X], \qquad \psi_X''(0) = \mathrm{Var}(X).$$

Definition 7.5.2 (Centered CGF and Variance Proxy) For X with $\mathbb{E}[X]$ finite, define the centered variable $\widetilde{X} := X - \mathbb{E}[X]$ and its CGF $\psi_{\widetilde{X}}(\lambda) = \log \mathbb{E}[e^{\lambda \widetilde{X}}]$

(when finite). An upper bound of the form

$$\psi_{\tilde{X}}(\lambda) \le \frac{\sigma^2\lambda^2}{2} \quad \forall \lambda \in \mathbb{R}$$

treats σ^2 as a *variance proxy* controlling exponential tails.

7.5.2.2 Legendre Transforms and Rate-Function Intuition

Optimizing Chernoff bounds leads naturally to a Legendre transform: the best exponent is the convex conjugate of the CGF. This is the entry point to large deviation theory, where probabilities decay like $\exp(-nI(t))$ for a rate function I. While full large-deviation theory is beyond the scope here, the intuition is valuable: CGF encodes growth of exponential moments, and its Legendre transform encodes tail decay.

Definition 7.5.3 (Convex Conjugate/Legendre Transform) For a convex function $\psi : \mathbb{R} \to \mathbb{R} \cup \{+\infty\}$, its convex conjugate is

$$\psi^*(t) := \sup_{\lambda \in \mathbb{R}} \{\lambda t - \psi(\lambda)\}.$$

Remark 7.5.3 (Chernoff Exponent as a Legendre Transform) From Proposition 7.5.2,

$$\inf_{\lambda>0} \exp(\psi_X(\lambda) - \lambda t) = \exp\big(-\sup_{\lambda>0}\{\lambda t - \psi_X(\lambda)\}\big),$$

so the best exponent is a (one-sided) Legendre transform of ψ_X.

7.5.3 *Bounding Tails via MGFs*

7.5.3.1 Chernoff Bounds for Sums of Independent Variables

Independence makes MGFs multiplicative, so CGFs add. This is why Chernoff bounds become sharp for sums: a sum of n independent terms has CGF equal to n times the single-term CGF (when i.i.d.). The resulting exponent scales linearly with n, yielding exponential concentration in n. This is the mathematical core behind Hoeffding/Bernstein inequalities and thus behind high-probability generalization bounds.

Theorem 7.5.1 (Chernoff Bound for Sums (General Form)) *Let $X_1, \ldots, X_n$ be independent and define $S_n = \sum_{i=1}^n X_i$. Assume $\psi_{X_i}(\lambda)$ is finite for some $\lambda > 0$.*

Then for any $t \in \mathbb{R}$ and any $\lambda > 0$,

$$\mathbb{P}(S_n \geq t) \leq \exp\Big(\sum_{i=1}^{n} \psi_{X_i}(\lambda) - \lambda t\Big).$$

If the X_i are i.i.d. with CGF ψ_X, then

$$\mathbb{P}(S_n \geq t) \leq \exp\big(n\psi_X(\lambda) - \lambda t\big).$$

Remark 7.5.4 (Turning This Into a Usable Bound) To obtain an explicit bound, one must upper-bound $\psi_{X_i}(\lambda)$ by a tractable function (often quadratic for sub-Gaussian variables), then optimize over λ. This is where distributional assumptions enter.

7.5.3.2 From MGF Bounds to Sub-Gaussian Behavior

Sub-Gaussianity is the principle that a random variable has tails no heavier than a Gaussian. It can be defined by an MGF inequality: the centered CGF is bounded by a quadratic. This definition is tailor-made for Chernoff: quadratic CGF yields Gaussian-type tail bounds. Moreover, sums of independent sub-Gaussians remain sub-Gaussian with variance proxies adding, which is exactly the stability property needed for concentration of empirical means and stochastic gradients.

Definition 7.5.4 (Sub-Gaussian (MGF Definition)) A real-valued random variable X with mean $\mu = \mathbb{E}[X]$ is *sub-Gaussian* with parameter σ^2 if for all $\lambda \in \mathbb{R}$,

$$\mathbb{E}\big[e^{\lambda(X-\mu)}\big] \leq \exp\Big(\frac{\sigma^2\lambda^2}{2}\Big).$$

Proposition 7.5.4 (Tail Bound for Sub-Gaussians) *If X is sub-Gaussian with parameter σ^2, then for all $t > 0$,*

$$\mathbb{P}(X - \mu \geq t) \leq \exp\Big(-\frac{t^2}{2\sigma^2}\Big), \qquad \mathbb{P}(|X - \mu| \geq t) \leq 2\exp\Big(-\frac{t^2}{2\sigma^2}\Big).$$

Proof By Chernoff (Proposition 7.5.2),

$$\mathbb{P}(X - \mu \geq t) \leq \inf_{\lambda>0} \exp\Big(\frac{\sigma^2\lambda^2}{2} - \lambda t\Big).$$

The RHS is minimized at $\lambda^\star = t/\sigma^2$, giving $\exp(-t^2/(2\sigma^2))$. Apply the same bound to $-(X - \mu)$ and union bound for two-sided tails. □

Proposition 7.5.5 (Sums of Independent Sub-Gaussians) *Let $X_1, \ldots, X_n$ be independent sub-Gaussian random variables with means μ_i and parameters σ_i^2. Then $S_n = \sum_{i=1}^n X_i$ is sub-Gaussian with mean $\sum_i \mu_i$ and parameter $\sum_i \sigma_i^2$, i.e.,*

$$\mathbb{E}\Big[e^{\lambda(S_n - \mathbb{E}S_n)}\Big] \le \exp\Big(\frac{\lambda^2}{2}\sum_{i=1}^n \sigma_i^2\Big).$$

Remark 7.5.5 (Empirical Mean Concentration) If X_i are i.i.d. sub-Gaussian with parameter σ^2, then $\bar{X}_n$ is sub-Gaussian with parameter σ^2/n, yielding

$$\mathbb{P}(|\bar{X}_n - \mu| \ge t) \le 2\exp\Big(-\frac{nt^2}{2\sigma^2}\Big),$$

the canonical $\sqrt{\log(1/\delta)/n}$ scaling for high-probability mean estimation.

7.6 Sub-Gaussian and Sub-exponential Tails

Tail behavior determines how reliably empirical quantities concentrate (Vershynin, 2018). Two tail classes capture most "well-behaved" regimes in ML: *sub-Gaussian* (Gaussian-like tails) and *sub-exponential* (exponential-like tails). They admit clean, reusable concentration templates (Hoeffding/Bernstein patterns) and can be quantified by Orlicz norms (ψ_2 and ψ_1). This section gives MGF-based definitions, stability under sums, and the conceptual forms of the resulting concentration inequalities used in generalization; see Fig. 7.4.

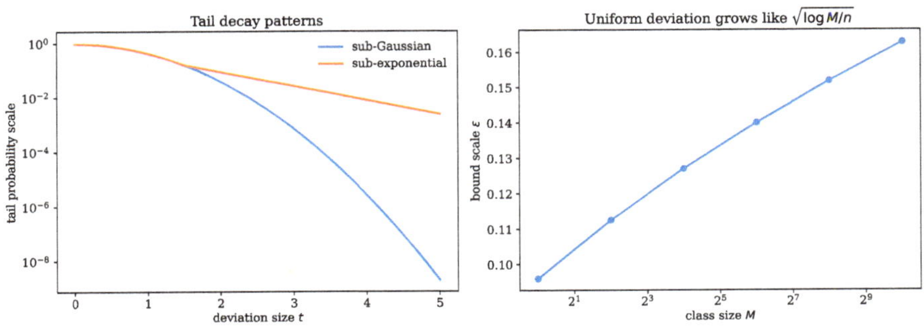

Fig. 7.4 Tail classes and their learning consequences. Left: sub-Gaussian tails decay quadratically in the exponent while sub-exponential tails eventually decay linearly. Right: a finite family of empirical processes is controlled by combining per-function concentration with a union bound, producing the characteristic $\sqrt{(\log M + \log(1/\delta))/n}$ scale

7.6.1 Tail Classes

7.6.1.1 MGF-Based Definitions and Examples

MGF bounds are the most operational definitions because they plug directly into Chernoff. Sub-Gaussianity means the centered log-MGF is bounded by a quadratic for all λ, which implies Gaussian tail decay. Sub-exponentiality relaxes this: the MGF is controlled only on a neighborhood of 0 and yields tails behaving like $\exp(-t)$ beyond a scale. Many ML noises are sub-Gaussian after truncation/normalization, and many heavy-but-not-too-heavy noises are sub-exponential (e.g., products of sub-Gaussians).

Definition 7.6.1 (Sub-Gaussian (MGF Form)) A real-valued random variable X with mean $\mu = \mathbb{E}[X]$ is *sub-Gaussian* with parameter σ^2 if for all $\lambda \in \mathbb{R}$,

$$\mathbb{E}\big[e^{\lambda(X-\mu)}\big] \le \exp\left(\frac{\sigma^2\lambda^2}{2}\right).$$

Definition 7.6.2 (Sub-Exponential (MGF Form)) A real-valued random variable X with mean $\mu = \mathbb{E}[X]$ is *sub-exponential* with parameters (ν^2, b) if for all $|\lambda| < 1/b$,

$$\mathbb{E}\big[e^{\lambda(X-\mu)}\big] \le \exp\left(\frac{\nu^2\lambda^2}{2}\right).$$

Remark 7.6.1 (Interpretation of (ν^2, b)) ν^2 acts like a "variance proxy" for moderate deviations, while b sets the scale beyond which tails behave more like pure exponential. Sub-Gaussian is the special case b arbitrarily small? Operationally, sub-Gaussian implies sub-exponential with some parameters, but the reverse is false.

Example 7.6.1 (Bounded Implies Sub-Gaussian (Conceptual Anchor)) If $X \in [a, b]$ almost surely, then $X - \mathbb{E}X$ is sub-Gaussian with parameter proportional to $(b-a)^2$ (Hoeffding's lemma). This explains why bounded losses yield exponential concentration.

Example 7.6.2 (Gaussian and Exponential Families) A Gaussian $X \sim \mathcal{N}(\mu, \sigma^2)$ is sub-Gaussian with parameter σ^2. An exponential variable is sub-exponential but not sub-Gaussian (tails are exponential, not Gaussian).

7.6.1.2 Sums of Sub-Gaussians and Variance Proxies

The main reason sub-Gaussian variables are beloved in ML is stability under sums: independent sub-Gaussians add their variance proxies. This yields clean concentration for sample means and stochastic gradients. For sub-exponential variables, sums remain sub-exponential with parameters that add in the variance proxy and max in the scale parameter. This is the algebra behind Bernstein-type inequalities.

Proposition 7.6.1 (Sums of Independent Sub-Gaussians) *Let $X_1, \dots, X_n$ be independent sub-Gaussian with means μ_i and parameters σ_i^2. Then $S = \sum_{i=1}^n X_i$ is sub-Gaussian with mean $\sum_i \mu_i$ and parameter $\sum_i \sigma_i^2$:*

$$\mathbb{E}\Big[e^{\lambda(S-\mathbb{E}S)}\Big] \le \exp\Big(\frac{\lambda^2}{2}\sum_{i=1}^n \sigma_i^2\Big) \quad \forall \lambda \in \mathbb{R}.$$

Proposition 7.6.2 (Sums of Independent Sub-Exponentials) *Let $X_1, \dots, X_n$ be independent sub-exponential with parameters (ν_i^2, b_i). Then $S = \sum_{i=1}^n X_i$ is sub-exponential with parameters*

$$\Big(\nu^2,\ b\Big) = \Big(\sum_{i=1}^n \nu_i^2,\ \max_{1\le i\le n} b_i\Big).$$

Remark 7.6.2 (Empirical Mean Scaling) If X_i are i.i.d. sub-Gaussian(σ^2), then $\bar{X}_n$ is sub-Gaussian with parameter σ^2/n. If X_i are i.i.d. sub-exponential(ν^2, b), then $\bar{X}_n$ is sub-exponential with parameters $(\nu^2/n,\ b/n)$.

7.6.2 Concentration Tools for ML

7.6.2.1 Hoeffding/Bernstein-Style Patterns (Conceptual)

Two reusable concentration patterns dominate ML: **Hoeffding-type** (sub-Gaussian) bounds yield purely quadratic exponents e^{-cnt^2}. **Bernstein-type** (sub-exponential) bounds interpolate: quadratic for small deviations and linear for large deviations: $e^{-cn\min(t^2,t)}$ after scaling. The latter reflects that tail decay is Gaussian-like near the mean but only exponential far out. These patterns power confidence intervals, deviation bounds for empirical risks, and analysis of randomized algorithms.

Theorem 7.6.1 (Hoeffding Pattern (Sub-Gaussian Mean, Template)) *If $X_1, \dots, X_n$ are independent, mean μ, and each $X_i - \mu$ is sub-Gaussian with parameter σ^2, then for all $t > 0$,*

$$\mathbb{P}\big(|\bar{X}_n - \mu| \ge t\big) \le 2\exp\Big(-\frac{nt^2}{2\sigma^2}\Big).$$

Theorem 7.6.2 (Bernstein Pattern (Sub-Exponential Mean, Template)) *If $X_1, \dots, X_n$ are independent, mean μ, and each $X_i - \mu$ is sub-exponential with* *parameters* (ν^2, b), *then for all* $t > 0$,

$$\mathbb{P}\big(|\bar{X}_n - \mu| \geq t\big) \leq 2\exp\Big(-c\,n\,\min\Big(\frac{t^2}{\nu^2}, \frac{t}{b}\Big)\Big)$$

for a universal constant $c > 0$.

Remark 7.6.3 (How to Read the "min") For small t (below a scale), the quadratic term dominates: Gaussian-like concentration. For large t, the linear term dominates: exponential-like tails. This is the hallmark of sub-exponential behavior.

7.6.2.2 Uniform Convergence Intuition (Setup for Generalization)

Generalization requires controlling many random quantities simultaneously (over hypotheses, coordinates, time steps). The standard recipe is: (i) get a tail bound for one fixed quantity (Hoeffding/Bernstein), (ii) union bound over a finite set, or (iii) reduce an infinite set to a finite cover plus approximation error. Tail classes matter because they determine whether the per-hypothesis tail is exponential (manageable) or heavy-tailed (hard). This section frames the uniform convergence idea without introducing full VC/Rademacher machinery yet.

Remark 7.6.4 (Finite-Class Uniform Convergence with Sub-Gaussian Tails) If for each fixed $h \in \mathcal{H}$,

$$\mathbb{P}\big(|\widehat{R}_n(h) - R(h)| \geq t\big) \leq 2e^{-cnt^2},$$

then for $|\mathcal{H}| = M$,

$$\mathbb{P}\Big(\sup_{h\in\mathcal{H}} |\widehat{R}_n(h) - R(h)| \geq t\Big) \leq 2Me^{-cnt^2}.$$

Thus $t \asymp \sqrt{(\log M + \log(1/\delta))/n}$ with probability $1 - \delta$.

7.6.3 *Orlicz Norms (Optional but Useful)*

7.6.3.1 ψ_2 and ψ_1 Norms as Tail Measures

Orlicz norms quantify tail behavior in a scale-invariant way. The ψ_2 norm characterizes sub-Gaussian tails and the ψ_1 norm characterizes sub-exponential tails. They allow concise statements like "$\|X\|_{\psi_2} \leq K$" and provide algebraic rules for sums, products, and maxima. In high-dimensional learning, Orlicz norms are often the cleanest language for concentration of random matrices and vectors.

Definition 7.6.3 (Orlicz ψ_α Norms) For $\alpha \geq 1$, define the Orlicz norm

$$\|X\|_{\psi_\alpha} := \inf\left\{c > 0 : \mathbb{E}\big[\exp(|X|^\alpha / c^\alpha)\big] \leq 2\right\}.$$

In particular, $\|X\|_{\psi_2}$ is the sub-Gaussian norm and $\|X\|_{\psi_1}$ is the sub-exponential norm.

Proposition 7.6.3 (Tail Bounds from Orlicz Norms) *If $\|X\|_{\psi_2} \leq K$, then there exists $c > 0$ such that for all $t \geq 0$,*

$$\mathbb{P}(|X| \geq t) \leq 2\exp\Big(-c\,\frac{t^2}{K^2}\Big).$$

If $\|X\|_{\psi_1} \leq K$, then there exists $c > 0$ such that

$$\mathbb{P}(|X| \geq t) \leq 2\exp\Big(-c\,\frac{t}{K}\Big).$$

Remark 7.6.5 (Sub-Gaussian $\Rightarrow$ Sub-Exponential) If $\|X\|_{\psi_2} \leq K$, then $\|X\|_{\psi_1} \leq CK$ for a universal constant C. Thus sub-Gaussian tails are strictly stronger.

7.6.3.2 Rules of Thumb: Converting Between Moments and Tails

A practical skill is translating between moment growth and tail decay. Sub-Gaussian variables have moments growing like $(\sqrt{p})^p$ (equivalently $\|X\|_{L^p} \lesssim K\sqrt{p}$), while sub-exponential variables have moments growing like p^p (equivalently $\|X\|_{L^p} \lesssim Kp$). These rules let you choose whether to work with MGFs, Orlicz norms, or moment bounds depending on what is easiest to estimate.

Remark 7.6.6 (Moment Growth Heuristics) If $\|X\|_{\psi_2} \leq K$, then for all $p \geq 1$,

$$\big(\mathbb{E}|X|^p\big)^{1/p} \lesssim K\sqrt{p}.$$

If $\|X\|_{\psi_1} \leq K$, then for all $p \geq 1$,

$$\big(\mathbb{E}|X|^p\big)^{1/p} \lesssim K\,p.$$

Conversely, if these moment bounds hold for all p with appropriate constants, one can recover sub-Gaussian/sub-exponential tail control.

Remark 7.6.7 (Maxima Rule of Thumb) If $X_1, \ldots, X_m$ are sub-Gaussian with $\|X_i\|_{\psi_2} \leq K$, then $\max_i X_i$ is typically of order $K\sqrt{\log m}$ (up to constants), consistent with union bounds and Orlicz-max inequalities.

7.7 Concentration for Random Vectors and Matrices

Modern ML is high-dimensional: features, gradients, embeddings, and covariances are vector- and matrix-valued. Thus, we need concentration tools beyond scalars. The key themes are: (i) random vectors have norms and pairwise angles that concentrate (near-orthogonality), and (ii) random matrices (especially sample covariance/Gram matrices) have spectra that concentrate, stabilizing linear algebra downstream. This section gives a rigorous preview of the most-used ideas and bounds, with emphasis on what they imply for embeddings, random features, and covariance estimation (Wainwright, 2019).

7.7.1 Vector Concentration Intuition

7.7.1.1 Norm of Random Vectors and Near-Orthogonality

For isotropic random vectors in high dimension, the Euclidean norm is tightly concentrated around its typical scale, and independent vectors are nearly orthogonal. These are consequences of concentration of measure. In ML, they explain why random initialization has predictable norm, why dot products between unrelated embeddings are near 0, and why the geometry of random projections is stable. The cleanest formal model is a sub-Gaussian isotropic vector, for which both norm and inner products admit exponential tail bounds.

Definition 7.7.1 (Isotropic Random Vector) A random vector $X \in \mathbb{R}^d$ is *isotropic* if

$$\mathbb{E}[X] = 0 \qquad \text{and} \qquad \mathbb{E}[XX^\top] = I_d.$$

Equivalently, $\mathbb{E}[\langle X, u\rangle^2] = \|u\|_2^2$ for all $u \in \mathbb{R}^d$.

Remark 7.7.1 (Sub-Gaussian Isotropic Model) A common tractable assumption is that X is isotropic and each one-dimensional projection $\langle X, u\rangle$ is sub-Gaussian with parameter proportional to $\|u\|_2^2$. This captures "Gaussian-like tails in every direction" without requiring X to be Gaussian.

Proposition 7.7.1 (Norm Concentration for Gaussian Vectors (Explicit Case)) *Let $g \sim \mathcal{N}(0, I_d)$. Then $\|g\|_2^2 \sim \chi_d^2$ and for any $t \geq 0$,*

$$\mathbb{P}\Big(\|g\|_2 \geq \sqrt{d} + t\Big) \leq \exp\Big(-\frac{t^2}{2}\Big), \qquad \mathbb{P}\Big(\|g\|_2 \leq \sqrt{d} - t\Big) \leq \exp\Big(-\frac{t^2}{2}\Big).$$

In particular, $\|g\|_2 = \sqrt{d} \pm O_{\mathbb{P}}(1)$.

Proposition 7.7.2 (Near-Orthogonality of Independent Isotropic Vectors (Gaussian Case)) *Let $g, h \sim \mathcal{N}(0, I_d)$ be independent. Then $\langle g, h\rangle$ is mean-zero with variance d, and for any $t \geq 0$,*

$$\mathbb{P}\big(|\langle g, h\rangle| \geq t\big) \leq 2\exp\Big(-\frac{t^2}{2d}\Big).$$

Equivalently, for normalized vectors $\tilde{g} = g/\|g\|_2$ and $\tilde{h} = h/\|h\|_2$, the cosine similarity $\langle \tilde{g}, \tilde{h}\rangle$ is typically of order $1/\sqrt{d}$.

Remark 7.7.2 (Angle Concentration) In high d, most of the mass on the sphere lies near the equator relative to any fixed direction, so independent random directions have angle close to $\pi/2$. This is the geometric meaning of "random vectors are nearly orthogonal."

7.7.1.2 Implications for Random Features and Embeddings

Random features and embeddings rely on concentration: random linear maps preserve norms approximately on finite sets, and random embeddings produce predictable dot-product statistics. This yields practical heuristics: (i) cosine similarities near 0 are expected for unrelated items, (ii) normalized embeddings reduce norm variability and make similarity mostly angular, and (iii) random projections can reduce dimension without destroying geometry. In learning theory, these intuitions become JL-type bounds and random-feature approximation guarantees.

Remark 7.7.3 (Random-Feature Geometry (Conceptual)) If $\phi(x) \in \mathbb{R}^m$ is a random embedding with isotropic/sub-Gaussian coordinates, then $\|\phi(x)\|_2$ concentrates around $\sqrt{m}$ (after appropriate scaling), and dot products $\langle \phi(x), \phi(z)\rangle$ concentrate around their expectation. This is the probabilistic reason random features can approximate kernel inner products.

Remark 7.7.4 (Practical Caution) Concentration is a statement about typical behavior under model assumptions. If embeddings are highly structured (e.g., clustered, anisotropic, heavy-tailed), the "random direction" baseline can fail, and normalization/whitening may be required to restore isotropic behavior.

7.7.2 Matrix Concentration Intuition

7.7.2.1 Spectral Norm Bounds and Stability of Covariance Estimates

Matrices enter ML through sample covariance/Gram matrices, Hessians, Jacobians, and attention matrices. The key stability question is spectral: how close is the random matrix (e.g., empirical covariance) to its expectation in operator norm?

Operator norm control implies stability of inverses, eigenvectors, and projections, which directly governs numerical performance and generalization in linear models and PCA. Matrix concentration results (matrix Bernstein/Chernoff) provide such bounds under sub-Gaussian conditions. Here we give a concrete Gaussian covariance bound as a preview and emphasize the geometric consequences.

Definition 7.7.2 (Spectral Norm) For a matrix $A \in \mathbb{R}^{d\times d}$, the spectral (operator) norm is

$$\|A\|_2 := \sup_{\|x\|_2=1} \|Ax\|_2 = \sqrt{\lambda_{\max}(A^\top A)}.$$

If A is symmetric, $\|A\|_2 = \max\{|\lambda_{\max}(A)|, |\lambda_{\min}(A)|\}$.

Definition 7.7.3 (Empirical Covariance) Let $X_1, \dots, X_n \in \mathbb{R}^d$ be i.i.d. with mean 0. The empirical covariance is

$$\widehat{\Sigma} := \frac{1}{n}\sum_{i=1}^{n} X_i X_i^\top.$$

If $\mathbb{E}[X_i X_i^\top] = \Sigma$, then $\widehat{\Sigma}$ estimates Σ.

Proposition 7.7.3 (Gaussian Covariance Concentration (Preview Statement)) *Let $X_i \sim \mathcal{N}(0, I_d)$ i.i.d. and $\widehat{\Sigma} = \frac{1}{n}\sum_{i=1}^{n} X_i X_i^\top$. Then for any $\delta \in (0, 1)$, with probability at least $1-\delta$,*

$$\|\widehat{\Sigma} - I_d\|_2 \le C\left(\sqrt{\frac{d+\log(1/\delta)}{n}} + \frac{d+\log(1/\delta)}{n}\right)$$

for a universal constant $C > 0$.

Remark 7.7.5 (What this Implies) If $n \gg d$, the RHS is $O\big(\sqrt{d/n}\big)$, so the spectrum of $\widehat{\Sigma}$ lies close to 1:

$$1-\varepsilon \le \lambda_{\min}(\widehat{\Sigma}) \le \lambda_{\max}(\widehat{\Sigma}) \le 1+\varepsilon,$$

meaning the empirical covariance is well-conditioned and invertible. This stability underpins PCA, whitening, and least squares in high dimensions.

Remark 7.7.6 (From Spectral Norm to Inverse Stability) If $\|\widehat{\Sigma} - \Sigma\|_2 \le \varepsilon$ and $\Sigma \succeq \alpha I$, then for $\varepsilon < \alpha$, $\widehat{\Sigma}$ is invertible and

$$\|\widehat{\Sigma}^{-1} - \Sigma^{-1}\|_2 \le \frac{\varepsilon}{\alpha(\alpha-\varepsilon)}.$$

Thus operator-norm concentration yields stable inversion.

7.7.2.2 Why These Bounds Matter in High-Dimensional ML

High-dimensional ML frequently operates near the regime $n \approx d$ or even $n < d$. In these regimes, covariance estimates can be noisy or singular, and matrix concentration quantifies exactly when linear algebra becomes stable: roughly when n dominates effective dimension. These bounds explain why regularization is necessary in high dimensions, why shrinkage estimators outperform naive covariance, and why random features require m large enough to preserve kernel geometry. Conceptually, matrix concentration is the extension of scalar concentration to operator-valued objects, and it powers modern analyses of PCA, random projections, and sketched regression.

Remark 7.7.7 (Two Key ML Consequences)

- **Stable PCA/whitening:** if $\widehat{\Sigma}$ is close to Σ in spectral norm, then eigenvalues/eigenvectors are stable, so principal components estimated from data reflect population structure.
- **Stable regression:** the conditioning of $X^\top X$ (a Gram matrix) controls variance and numerical stability of least squares; matrix concentration gives high-probability control of this conditioning under sub-Gaussian designs.

Remark 7.7.8 (Preview of a Typical "Effective Dimension" Message) Even when d is large, what matters is often an effective dimension such as $\mathrm{tr}(\Sigma)/\|\Sigma\|_2$ or intrinsic rank. Matrix concentration bounds frequently replace d by such quantities, aligning theory with structured-data practice.

7.8 Empirical Processes: From Pointwise to Uniform Bounds

Concentration for a *fixed* function h (pointwise bounds) is only the first step in learning theory. Learning chooses $\widehat{h}$ *based on the same data*, so we need *uniform* control:

$$\sup_{h \in \mathcal{H}} \left| \widehat{R}_n(h) - R(h) \right|.$$

Empirical process theory provides tools to bound such suprema using complexity measures of $\mathcal{H}$ (van der Vaart and Wellner, 1996). This section gives a rigorous preview: the uniform convergence idea, covering-number intuition, and the role of Rademacher complexity as a data-dependent measure that yields generalization bounds via symmetrization and contraction.

7.8.1 Uniform Convergence Idea

7.8.1.1 Supremum over Hypothesis Classes

If h were fixed in advance, we could bound $|\widehat{R}_n(h) - R(h)|$ by a scalar concentration inequality. But learning selects h after seeing the sample, so we must control deviations *simultaneously* over all $h \in \mathcal{H}$. The supremum quantifies the worst-case generalization gap over the class and directly implies performance guarantees for ERM: if the gap is small uniformly, then minimizing empirical risk approximately minimizes true risk.

Definition 7.8.1 (Empirical Process Indexed By a Class) Let $Z_1, \dots, Z_n$ be i.i.d. and let $\mathcal{F}$ be a class of measurable functions $f : \mathcal{Z} \to \mathbb{R}$ with $\mathbb{E}|f(Z)| < \infty$. Define

$$Pf := \mathbb{E}[f(Z)], \qquad P_n f := \frac{1}{n}\sum_{i=1}^{n} f(Z_i).$$

The centered empirical process is $(P_n - P)f$.

Definition 7.8.2 (Uniform Deviation) The uniform deviation of $\mathcal{F}$ is

$$\|P_n - P\|_{\mathcal{F}} := \sup_{f \in \mathcal{F}} |(P_n - P)f|.$$

In supervised learning, one typically takes $f_h(z) = \ell(h(x), y)$ indexed by hypotheses $h \in \mathcal{H}$.

Proposition 7.8.1 (ERM Consequence (Standard Decomposition)) *Let* $\widehat{h} \in \arg\min_{h \in \mathcal{H}} \widehat{R}_n(h)$ *and let* $h^\star \in \arg\min_{h \in \mathcal{H}} R(h)$. *Then*

$$R(\widehat{h}) - R(h^\star) \le 2 \sup_{h \in \mathcal{H}} |\widehat{R}_n(h) - R(h)|.$$

Proof Write

$$R(\widehat{h}) - R(h^\star) = \underbrace{R(\widehat{h}) - \widehat{R}_n(\widehat{h})}_{\le \sup_h (R(h) - \widehat{R}_n(h))} + \underbrace{\widehat{R}_n(\widehat{h}) - \widehat{R}_n(h^\star)}_{\le 0} + \underbrace{\widehat{R}_n(h^\star) - R(h^\star)}_{\le \sup_h (\widehat{R}_n(h) - R(h))}.$$

Combine the two suprema to get the factor 2. □

Remark 7.8.1 (Why this Matters) Proposition 7.8.1 shows that controlling the uniform deviation is enough to control excess risk of ERM. The remaining task is to bound the supremum in terms of n and the complexity of $\mathcal{H}$.

7.8.1.2 Covering Numbers and Complexity (Conceptual)

For infinite hypothesis classes, union bounds over all $h \in \mathcal{H}$ are impossible. Covering numbers replace $\mathcal{H}$ by a finite approximation: an ε-net under an appropriate metric on functions. One then applies concentration on the finite net (union bound) and adds an approximation error term controlled by Lipschitz properties. This converts the geometry/complexity of $\mathcal{H}$ (how many balls needed to cover it) into generalization bounds.

Definition 7.8.3 (ε-Cover and Covering Number) Let $(\mathcal{F}, d)$ be a metric space (e.g., functions with an $L_2(P_n)$ or L_∞ metric). A set $\mathcal{N} \subseteq \mathcal{F}$ is an *ε-cover* of $\mathcal{F}$ if for every $f \in \mathcal{F}$ there exists $g \in \mathcal{N}$ such that $d(f, g) \leq \varepsilon$. The *covering number* is

$$N(\varepsilon, \mathcal{F}, d) := \min\{|\mathcal{N}| : \mathcal{N} \text{ is an } \varepsilon\text{-cover of } \mathcal{F}\}.$$

Remark 7.8.2 (Entropy and the "Log Net Size") The quantity $\log N(\varepsilon, \mathcal{F}, d)$ is called metric entropy. In many bounds, it is the load-bearing term replacing $\log|\mathcal{H}|$ from finite-class analysis.

Remark 7.8.3 (Conceptual Pipeline with Covers) A typical argument structure is:

1. choose an ε-net $\mathcal{N}$ of $\mathcal{F}$ under a data-dependent metric (often $L_2(P_n)$),
2. bound $\max_{g \in \mathcal{N}} |(P_n - P)g|$ via concentration + union bound using $\log|\mathcal{N}|$,
3. transfer the bound from $\mathcal{N}$ to all $\mathcal{F}$ using Lipschitz/smoothness to control $|(P_n - P)(f - g)|$.

This turns the geometry of the class into a uniform probabilistic bound.

7.8.2 Rademacher Complexity (Conceptual Preview)

7.8.2.1 Data-Dependent Complexity Measure

Covering numbers are powerful but can be abstract and distribution-dependent. Rademacher complexity provides a more operational, data-dependent measure of richness: it quantifies how well the function class can correlate with random ± 1 noise on the observed sample. If $\mathcal{F}$ can fit random signs well, it is complex and may overfit; if it cannot, it generalizes better. This measure naturally appears via symmetrization, replacing unknown expectations by an average over random signs.

Definition 7.8.4 (Rademacher Variables) A Rademacher random variable ϵ satisfies $\mathbb{P}(\epsilon = +1) = \mathbb{P}(\epsilon = -1) = 1/2$. Let $\epsilon_1, \ldots, \epsilon_n$ be i.i.d. Rademacher variables independent of the sample $Z_1, \ldots, Z_n$.

Definition 7.8.5 (Empirical Rademacher Complexity) Given a sample $z_{1:n} = (z_1, \ldots, z_n)$ and a function class $\mathcal{F} \subseteq \{f : \mathcal{Z} \to \mathbb{R}\}$, the empirical Rademacher complexity is

$$\widehat{\mathfrak{R}}_n(\mathcal{F}; z_{1:n}) := \mathbb{E}_\epsilon\left[\sup_{f\in\mathcal{F}} \frac{1}{n}\sum_{i=1}^{n} \epsilon_i f(z_i)\right].$$

The (expected) Rademacher complexity is $\mathfrak{R}_n(\mathcal{F}) := \mathbb{E}_Z\big[\widehat{\mathfrak{R}}_n(\mathcal{F}; Z_{1:n})\big]$.

Remark 7.8.4 (Geometric Meaning) The quantity $\sup_{f\in\mathcal{F}}\sum_i \epsilon_i f(z_i)$ measures alignment of the class with pure noise. If this supremum is large, $\mathcal{F}$ can shatter noise patterns, indicating high capacity.

7.8.2.2 How It Yields Generalization Bounds

Rademacher complexity controls uniform deviations by a two-step mechanism: symmetrization bounds $\mathbb{E}\sup_f (P_n - P)f$ by a Rademacher process; then concentration turns expected bounds into high-probability bounds. For Lipschitz losses, contraction inequalities transfer complexity from hypotheses to loss-composed classes. The result is a generalization bound of the form

$$R(h) \le \widehat{R}_n(h) + O(\mathfrak{R}_n(\ell\circ\mathcal{H})) + (\text{confidence term}).$$

This is a key bridge from geometry/complexity to learning guarantees.

Proposition 7.8.2 (Symmetrization (Expected Uniform Deviation, Preview)) *Assume $\mathcal{F}$ is uniformly bounded: $|f(z)| \le b$ for all $f\in\mathcal{F}$ and all z. Then*

$$\mathbb{E}\big[\|P_n - P\|_{\mathcal{F}}\big] \;\le\; 2\,\mathfrak{R}_n(\mathcal{F}).$$

Remark 7.8.5 (High-Probability Upgrade (Conceptual)) With additional concentration arguments (e.g., McDiarmid or Talagrand inequalities), one obtains that with probability at least $1-\delta$,

$$\|P_n - P\|_{\mathcal{F}} \;\lesssim\; 2\,\widehat{\mathfrak{R}}_n(\mathcal{F}; Z_{1:n}) + b\sqrt{\frac{\log(1/\delta)}{n}},$$

up to universal constants. This is the archetypal Rademacher generalization pattern.

Remark 7.8.6 (Contraction Principle) If ϕ is L-Lipschitz and $\phi(0) = 0$, then typically

$$\widehat{\mathfrak{R}}_n(\phi\circ\mathcal{F}) \le L\,\widehat{\mathfrak{R}}_n(\mathcal{F}),$$

which is how one passes from hypothesis complexity to loss complexity.

7.9 Exercises

Throughout, assume all random variables are defined on a common probability space. Unless stated otherwise, logs are natural, and probabilities are with respect to the underlying sample randomness.

Exercise 7.9.1 (Markov and Chebyshev: Explicit Tail Bounds) Let X be a nonnegative random variable with $\mathbb{E}[X] = 3.6$. (i) Use Markov's inequality to bound $\mathbb{P}(X \geq 12)$. (ii) Now assume additionally that $\mathrm{Var}(X) = 9$ and $\mathbb{E}[X] = 3.6$; use Chebyshev to bound $\mathbb{P}(|X - 3.6| \geq 6)$. (iii) Compare the two bounds numerically and explain (one paragraph) which information (mean vs. variance) each inequality uses and why Chebyshev is potentially sharper here.

Exercise 7.9.2 (Moment/Markov Tuning via Powers) Let $X \geq 0$ and suppose $\mathbb{E}[X] = 2$, $\mathbb{E}[X^2] = 10$, and $\mathbb{E}[X^4] = 300$. For $t = 5$, compute the Markov bounds obtained by applying Markov to X, X^2, and X^4:

$$\mathbb{P}(X \geq t) \leq \frac{\mathbb{E}[X]}{t}, \quad \mathbb{P}(X \geq t) \leq \frac{\mathbb{E}[X^2]}{t^2}, \quad \mathbb{P}(X \geq t) \leq \frac{\mathbb{E}[X^4]}{t^4}.$$

Which bound is tightest numerically, and what does this suggest about the usefulness of higher moments in tail control?

Exercise 7.9.3 (Jensen and Log/exp Inequalities) Let X take values $\{1, 2, 4\}$ with probabilities $\mathbb{P}(X = 1) = 0.2$, $\mathbb{P}(X = 2) = 0.5$, $\mathbb{P}(X = 4) = 0.3$. (i) Compute $\mathbb{E}[X]$ and $\mathbb{E}[\log X]$ exactly. (ii) Verify Jensen for concave log: $\mathbb{E}[\log X] \leq \log \mathbb{E}[X]$ numerically. (iii) Define $Y = e^X$ and compute $\log \mathbb{E}[Y]$ and $\mathbb{E}[X]$; verify $\mathbb{E}[X] \leq \log \mathbb{E}[e^X]$ and interpret this inequality as a "soft max" upper bound.

Exercise 7.9.4 (Union Bound and Maxima: Explicit Computation) Let X_1, X_2, X_3, X_4 be real-valued random variables (not necessarily independent) satisfying

$$\mathbb{P}(|X_i| \geq t) \leq 2e^{-t^2/8} \quad \text{for all } t \geq 0, \quad i \in \{1, 2, 3, 4\}.$$

(i) Use the union bound to derive an explicit bound on $\mathbb{P}(\max_{1 \leq i \leq 4} |X_i| \geq t)$. (ii) Find the smallest t (to 2 decimal digits) such that your bound is at most 0.01. (iii) Repeat (ii) for $m = 200$ variables with the same per-coordinate tail bound and compare the resulting t; comment on the $\sqrt{\log m}$ scaling.

Exercise 7.9.5 (Cauchy–Schwarz as a Probabilistic Tool) Let (X, Y) be a pair of random variables with

$$\mathbb{E}[X] = 0, \quad \mathbb{E}[Y] = 0, \quad \mathbb{E}[X^2] = 4, \quad \mathbb{E}[Y^2] = 9, \quad \mathbb{E}[XY] = c.$$

(i) Use Cauchy–Schwarz to bound the possible values of c. (ii) Suppose $c = 5$. Compute $\mathrm{Corr}(X, Y)$ and verify it lies in $[-1, 1]$. (iii) Let $Z = X + 2Y$. Compute $\mathbb{E}[Z]$ and $\mathrm{Var}(Z)$ in terms of c and then specialize to $c = 5$.

Exercise 7.9.6 (Law of Total Variance: Explicit Finite Model) Let $Y \in \{0, 1\}$ with $\mathbb{P}(Y = 1) = 0.3$. Conditional on Y,

$$X \mid (Y = 0) \sim \mathcal{N}(0, 1), \qquad X \mid (Y = 1) \sim \mathcal{N}(2, 4).$$

(i) Compute $\mathbb{E}[X \mid Y]$ and $\mathrm{Var}(X \mid Y)$ as functions of Y. (ii) Compute $\mathbb{E}[X]$ via tower property. (iii) Compute $\mathrm{Var}(X)$ in two ways: directly using $\mathbb{E}[X^2] - \mathbb{E}[X]^2$ and using $\mathrm{Var}(X) = \mathbb{E}[\mathrm{Var}(X \mid Y)] + \mathrm{Var}(\mathbb{E}[X \mid Y])$; verify the two answers match.

Exercise 7.9.7 (MGF Computation and Chernoff Bound (Bernoulli Sum)) Let $X_1, \ldots, X_{50}$ be i.i.d. Bernoulli$(p = 0.2)$ and $S = \sum_{i=1}^{50} X_i$. (i) Compute the MGF $M_{X_1}(\lambda) = \mathbb{E}[e^{\lambda X_1}]$ and hence the MGF of S. (ii) Using Chernoff's method, derive the bound

$$\mathbb{P}(S \geq 15) \leq \inf_{\lambda > 0} \exp\big(\psi_S(\lambda) - 15\lambda\big),$$

where $\psi_S = \log M_S$. Evaluate this bound numerically at $\lambda \in \{0.2, 0.4, 0.6, 0.8\}$ and report the smallest value among these four choices. (iii) Compute the exact probability $\mathbb{P}(S \geq 15)$ by summing the Binomial(50, 0.2) tail (you may leave it as a finite sum), and compare the Chernoff bound to the exact value (qualitatively: loose/tight).

Exercise 7.9.8 (From Sub-Gaussian MGF to Tail Bound) Assume X satisfies $\mathbb{E}[e^{\lambda(X - \mathbb{E}X)}] \leq e^{\sigma^2\lambda^2/2}$ for all $\lambda \in \mathbb{R}$ with $\sigma^2 = 9$. (i) Use Chernoff optimization to show $\mathbb{P}(X - \mathbb{E}X \geq t) \leq \exp(-t^2/18)$. (ii) Compute an explicit t such that $\mathbb{P}(|X - \mathbb{E}X| \geq t) \leq 0.01$. (iii) Let $\bar{X}_n$ be the average of $n = 100$ i.i.d. copies of X; obtain a bound on $\mathbb{P}(|\bar{X}_{100} - \mathbb{E}X| \geq 0.3)$ using the variance-proxy addition rule.

Exercise 7.9.9 (Sub-Exponential Bernstein Pattern: Regime Switch) Assume i.i.d. $X_1, \ldots, X_n$ have mean μ and are sub-exponential with parameters $(\nu^2, b) = (4, 3)$. Using the Bernstein-style pattern

$$\mathbb{P}(|\bar{X}_n - \mu| \geq t) \leq 2\exp\Big(-c\,n \min\Big(\frac{t^2}{\nu^2}, \frac{t}{b}\Big)\Big)$$

with $c = 1/2$ (use this constant), do the following: (i) For $n = 200$, compute the bound for $t \in \{0.1, 0.3, 0.8\}$. (ii) Determine the threshold t_0 at which $\frac{t^2}{\nu^2} = \frac{t}{b}$ and explain in one paragraph why the "min" switches regimes at t_0.

Exercise 7.9.10 (Orlicz Norms and Tail Interpretation) Let X satisfy $\|X\|_{\psi_2} = 2$ in the sense of Definition 7.6.3. (i) Use the tail bound template from Proposition 7.6.3 with constant $c = 1/2$ (use this constant) to give an explicit bound of the form $\mathbb{P}(|X| \geq t) \leq 2\exp(-t^2/8)$. (ii) Using the moment growth heuristic

$\|X\|_{L^p} \lesssim K\sqrt{p}$ with $K = 2$ (treat $\lesssim$ as $\le$ for this exercise), bound $(\mathbb{E}|X|^6)^{1/6}$ numerically. (iii) If instead $\|Y\|_{\psi_1} = 3$, give a bound on $\mathbb{P}(|Y| \ge t)$ of the form $2\exp(-t/6)$ using $c = 1/2$.

Exercise 7.9.11 (Vector Concentration Numerics: Near-Orthogonality) Let $d = 50$ and define $u, v \in \mathbb{R}^{50}$ by

$$u_i = \frac{i}{\sqrt{\sum_{j=1}^{50} j^2}}, \qquad v_i = \frac{(-1)^i}{\sqrt{50}}, \qquad i \in [50].$$

(i) Verify $\|u\|_2 = \|v\|_2 = 1$ and compute $\langle u, v\rangle$ and the angle $\theta = \arccos(\langle u, v\rangle)$. (ii) Compute $\|u - v\|_2$ and verify $\|u - v\|_2^2 = 2 - 2\langle u, v\rangle$. (iii) Compare $|\langle u, v\rangle|$ to the heuristic scale $1/\sqrt{d}$ and comment (one paragraph) on whether the pair appears "nearly orthogonal."

Exercise 7.9.12 (Matrix Concentration Proxy: Stability of Empirical Covariance) Let $X_1, \ldots, X_n \in \mathbb{R}^3$ be the following centered samples (rows), with $n = 6$:

$$X = \begin{bmatrix} 1 & 0 & -1 \\ -1 & 1 & 0 \\ 0 & -1 & 1 \\ 1 & 1 & 0 \\ -1 & 0 & 1 \\ 0 & 0 & 0 \end{bmatrix}.$$

Define $\widehat{\Sigma} = \frac{1}{n} X^\top X$. (i) Compute $\widehat{\Sigma}$ explicitly, its eigenvalues, and its spectral norm $\|\widehat{\Sigma}\|_2$. (ii) Compute $\|\widehat{\Sigma} - I_3\|_2$ (you may compute eigenvalues of $\widehat{\Sigma} - I_3$). (iii) Interpret (one paragraph) what the magnitude of $\|\widehat{\Sigma} - I_3\|_2$ means for the stability of whitening by $\widehat{\Sigma}^{-1/2}$.

Exercise 7.9.13 (Finite-Class Uniform Convergence via Union Bound) Let $\mathcal{H}$ be a finite class with $|\mathcal{H}| = M = 500$ and suppose for each $h \in \mathcal{H}$ the loss is bounded in $[0, 1]$. Assume that for each fixed h we have a Hoeffding-type bound

$$\mathbb{P}\big(|\widehat{R}_n(h) - R(h)| \ge \varepsilon\big) \le 2e^{-2n\varepsilon^2}.$$

(i) Use the union bound to derive a bound on $\mathbb{P}\big(\sup_{h\in\mathcal{H}} |\widehat{R}_n(h) - R(h)| \ge \varepsilon\big)$. (ii) For $n = 2000$ and $\delta = 0.05$, solve for the smallest ε such that this probability is at most δ. (iii) Using Proposition 7.8.1, bound the excess risk of ERM: $R(\widehat{h}) - \inf_{h\in\mathcal{H}} R(h)$ with probability at least $1 - \delta$.

Exercise 7.9.14 (Empirical Rademacher Complexity: Compute Exactly for a Tiny Class) Let the sample be $z_{1:4} = (z_1, z_2, z_3, z_4)$ and consider the finite function class $\mathcal{F} = \{f^{(1)}, f^{(2)}, f^{(3)}\}$ defined by their values on the sample:

$$\left(f^{(1)}(z_i)\right)_{i=1}^{4} = (1, 1, 1, 1), \quad \left(f^{(2)}(z_i)\right)_{i=1}^{4} = (1, -1, 1, -1),$$
$$\left(f^{(3)}(z_i)\right)_{i=1}^{4} = (2, 0, -2, 0).$$

Compute the empirical Rademacher complexity

$$\widehat{\mathfrak{R}}_4(\mathcal{F}; z_{1:4}) = \mathbb{E}_\epsilon \left[\sup_{f \in \mathcal{F}} \frac{1}{4} \sum_{i=1}^{4} \epsilon_i f(z_i) \right],$$

where ϵ_i are i.i.d. Rademacher. Do this by enumerating all 2^4 sign vectors $\epsilon \in \{\pm 1\}^4$, computing the supremum for each, and averaging.

Exercise 7.9.15 (Covering-Number Intuition: Explicit ε-net for an Interval Class) Let $\mathcal{F} = \{f_\theta : \theta \in [0, 1]\}$ where $f_\theta(x) = \mathbf{1}\{x \leq \theta\}$ on $\mathcal{Z} = [0, 1]$. Consider the metric $d_\infty(f, g) = \sup_{x \in [0,1]} |f(x) - g(x)|$. (i) Show that $d_\infty(f_\theta, f_{\theta'}) \in \{0, 1\}$ for $\theta \neq \theta'$ and hence determine $N(\varepsilon, \mathcal{F}, d_\infty)$ for $\varepsilon < 1$ and for $\varepsilon \geq 1$. (ii) Now consider instead the empirical L_1 metric on the sample points $x_{1:5} = (0.1, 0.2, 0.4, 0.8, 0.9)$:

$$d_{1,n}(f, g) = \frac{1}{5} \sum_{i=1}^{5} |f(x_i) - g(x_i)|.$$

Compute $N(\varepsilon, \mathcal{F}, d_{1,n})$ exactly for $\varepsilon \in \{0, 0.2, 0.4\}$ by constructing a smallest ε-cover.

Chapter 8
Statistical Estimation: MLE, MAP, and Bayes

At the core of ML lies a simple computational task: *estimate unknown quantities from data*. Sometimes the unknown is a finite-dimensional parameter vector; sometimes it is an entire function (a predictor); sometimes it is a latent structure (clusters, topics, states). Regardless of form, estimation provides the bridge from observations to predictions, and it supplies the language for uncertainty quantification and principled decision-making. This chapter develops three tightly connected estimation paradigms: maximum likelihood (MLE), maximum a posteriori (MAP), and Bayesian inference (Casella and Berger 2002); see the overall estimation workflow in Fig. 8.1.

8.1 Why Estimation Is the Heart of ML

8.1.1 From Data to Parameters to Predictions

8.1.1.1 Likelihood-Based Learning as a Unifying Principle

A remarkable fraction of ML can be written as *choosing a model that makes observed data probable*. This is likelihood-based learning: specify a probabilistic model for the data (or conditional outputs given inputs), then choose parameters that maximize likelihood (equivalently minimize negative log-likelihood). This principle unifies regression (Gaussian noise $\Rightarrow$ least squares), classification (categorical likelihood $\Rightarrow$ cross-entropy), density estimation, language modeling, and generative modeling. Even when one starts from an optimization viewpoint, many standard losses are exactly negative log-likelihoods under an assumed noise model.

Definition 8.1.1 (Statistical Model and Parameter) A *statistical model* is a family of probability distributions $\{P_\theta : \theta \in \Theta\}$ on an observation space $(\mathcal{Z}, \mathcal{G})$.

P. Singh, B. Raman, *Foundations of Machine Learning and AI*, Studies in Big Data 199, https://doi.org/10.1007/978-3-032-30336-3_8

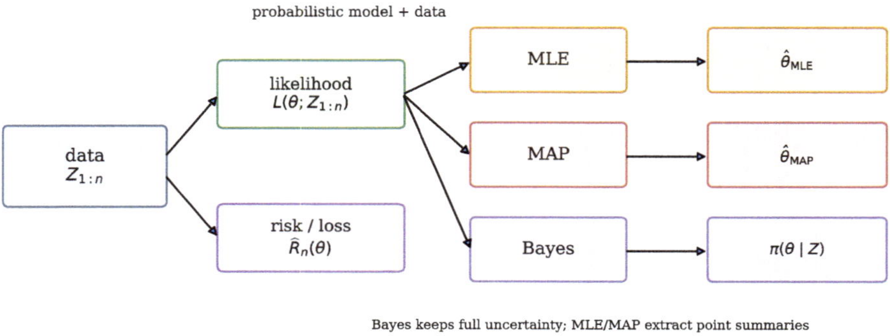

Fig. 8.1 A unifying view of estimation in ML. Data induce a likelihood and an empirical risk; MLE chooses the parameter maximizing likelihood, MAP adds prior information as regularization, and Bayesian inference retains the full posterior and propagates it to predictive uncertainty and decisions

Here $\theta \in \Theta$ is the *parameter* (unknown in practice). Given data $Z_1, \ldots, Z_n$ (often i.i.d. from some distribution), estimation produces $\widehat{\theta} = \widehat{\theta}(Z_{1:n})$.

Definition 8.1.2 (Likelihood and Log-Likelihood) Assume $Z_1, \ldots, Z_n$ are i.i.d. with density/pmf p_θ under P_θ. The *likelihood* of θ given the sample is

$$L(\theta; Z_{1:n}) := \prod_{i=1}^{n} p_\theta(Z_i), \qquad \ell(\theta; Z_{1:n}) := \log L(\theta; Z_{1:n}) = \sum_{i=1}^{n} \log p_\theta(Z_i).$$

Definition 8.1.3 (Maximum Likelihood Estimator (MLE)) Any maximizer of the likelihood is a *maximum likelihood estimator*:

$$\widehat{\theta}_{\text{MLE}} \in \arg\max_{\theta\in\Theta} L(\theta; Z_{1:n}) \quad \Longleftrightarrow \quad \widehat{\theta}_{\text{MLE}} \in \arg\max_{\theta\in\Theta} \ell(\theta; Z_{1:n}).$$

Equivalently, it minimizes the negative log-likelihood (NLL) $-\ell(\theta; Z_{1:n})$.

Remark 8.1.1 (Conditional Likelihood: Supervised Learning) In supervised learning with $Z_i = (X_i, Y_i)$, one often models $p_\theta(y \mid x)$. Then (conditional) log-likelihood is

$$\ell(\theta; (x_{1:n}, y_{1:n})) = \sum_{i=1}^{n} \log p_\theta(y_i \mid x_i),$$

and minimizing NLL is exactly cross-entropy training for classification models.

Example 8.1.1 (Gaussian Noise ⇒ Least Squares) Let $Y \mid X = x$ satisfy $Y = w^\top x + b + \varepsilon$ with $\varepsilon \sim \mathcal{N}(0, \sigma^2)$ independent of X. Then

$$-\log p_{w,b}(y \mid x) = \frac{1}{2\sigma^2}(y - (w^\top x + b))^2 + \text{const},$$

so MLE of (w, b) is equivalent to minimizing squared loss $\sum_i (y_i - (w^\top x_i + b))^2$.

8.1.1.2 Risk Minimization vs. Probabilistic Estimation

There are two equivalent lenses for much of ML. The *optimization* lens chooses a predictor by minimizing an expected loss (risk). The *probabilistic* lens chooses parameters by maximizing likelihood under an assumed data model. These coincide when the loss is a proper scoring rule derived from a likelihood (e.g., log-loss). When they differ, the distinction is instructive: risk minimization can be model-free (any loss), whereas likelihood-based estimation imposes a probabilistic structure that enables uncertainty quantification and principled prediction intervals.

Definition 8.1.4 (Loss and Population Risk) Let $(X, Y) \sim \mathcal{D}$ and let h be a predictor. A *loss* is $\ell : \widehat{\mathcal{Y}} \times \mathcal{Y} \to [0, \infty)$. The *population risk* is

$$R(h) := \mathbb{E}\big[\ell(h(X), Y)\big].$$

Given data $\{(X_i, Y_i)\}_{i=1}^n$, the empirical risk is

$$\widehat{R}_n(h) := \frac{1}{n}\sum_{i=1}^{n} \ell(h(X_i), Y_i).$$

Proposition 8.1.1 (NLL Is Empirical Risk under Log-Loss) *Let $p_\theta(y \mid x)$ be a conditional model and define the loss*

$$\ell_\theta(x, y) := -\log p_\theta(y \mid x).$$

Then minimizing negative conditional log-likelihood equals empirical risk minimization for ℓ_θ:

$$\arg\min_\theta \left\{ -\sum_{i=1}^{n} \log p_\theta(y_i \mid x_i) \right\} = \arg\min_\theta \sum_{i=1}^{n} \ell_\theta(x_i, y_i).$$

Remark 8.1.2 (What the Probabilistic Lens Adds) Risk minimization alone does not specify uncertainty about predictions. A probabilistic model $p_\theta(y \mid x)$ gives a full predictive distribution, enabling calibrated probabilities, prediction intervals, and decision rules under asymmetric costs.

8.1.2 Frequentist and Bayesian Lenses

8.1.2.1 Parameters as Fixed vs. Random

The frequentist lens treats parameters as fixed but unknown constants; randomness comes only from the data. An estimator $\widehat{\theta}$ is random because it depends on random data, and its quality is evaluated by its sampling distribution, bias/variance, and confidence sets. The Bayesian lens treats parameters as random variables with a prior distribution capturing uncertainty; data update this uncertainty via the posterior. Importantly, both lenses often yield similar point estimates but differ in how they quantify uncertainty and how they incorporate prior knowledge.

Definition 8.1.5 (Frequentist Viewpoint) A parameter $\theta \in \Theta$ is a fixed (nonrandom) element. Data $Z_{1:n}$ are random with distribution P_θ. An estimator $\widehat{\theta} = \widehat{\theta}(Z_{1:n})$ is evaluated via its distribution under P_θ (e.g., $\mathbb{E}_\theta[\widehat{\theta}]$, $\mathrm{Var}_\theta(\widehat{\theta})$).

Definition 8.1.6 (Bayesian Viewpoint) A parameter Θ is a random variable with prior density/pmf $\pi(\theta)$. Given data $Z_{1:n}$ with likelihood $p_\theta(Z_{1:n})$, the posterior is

$$\pi(\theta \mid Z_{1:n}) = \frac{p_\theta(Z_{1:n})\,\pi(\theta)}{\int_\Theta p_\vartheta(Z_{1:n})\,\pi(\vartheta)\,d\vartheta},$$

whenever the denominator is finite and nonzero.

Remark 8.1.3 (Same Algebra, Different Interpretation) The identity "posterior $\propto$ likelihood $\times$ prior" is an algebraic consequence of Bayes' rule. The difference between frequentist and Bayesian approaches lies in interpretation of θ (fixed vs. random) and in what uncertainty means (sampling uncertainty vs. posterior uncertainty).

8.1.2.2 Uncertainty Quantification and Decision-Making

Estimation is not just about point values; it is about making decisions under uncertainty. Frequentist uncertainty is expressed via confidence intervals/regions with coverage guarantees under repeated sampling. Bayesian uncertainty is expressed via posterior distributions and credible sets. Decision theory unifies both: choose an action (estimate, classification decision, etc.) to minimize expected loss. Under Bayesian modeling, the optimal action is obtained by minimizing posterior expected loss; under frequentist modeling, optimality can be posed as minimax or risk-based criteria.

Definition 8.1.7 (Decision Rule and Bayes Action) Let A be an action space (e.g., estimates in $\mathbb{R}^d$) and let $L : A \times \Theta \to [0, \infty)$ be a loss. Given data Z, a decision rule outputs $a(Z) \in A$. In the Bayesian setting, the *Bayes action* minimizes posterior expected loss:

$$a^\star(Z) \in \arg\min_{a\in A} \mathbb{E}\big[L(a,\Theta) \mid Z\big] = \arg\min_{a\in A} \int_\Theta L(a,\theta)\,\pi(\theta \mid Z)\,d\theta.$$

Remark 8.1.4 (Posterior Mean and MAP as Special Cases)

- If $L(a,\theta) = \|a-\theta\|_2^2$, then $a^\star(Z) = \mathbb{E}[\Theta \mid Z]$ (posterior mean).
- If $L(a,\theta) = \mathbf{1}\{a \neq \theta\}$ on a discrete Θ, then $a^\star(Z)$ is a posterior mode (MAP).

Thus, "best estimate" depends on the loss, not only on the posterior.

Definition 8.1.8 (Confidence Interval vs. Credible Interval (One-Dimensional)) A (frequentist) $(1-\alpha)$ confidence interval is a random interval $C(Z)$ such that for all θ,

$$\mathbb{P}_\theta\big(\theta \in C(Z)\big) \geq 1-\alpha.$$

A (Bayesian) $(1-\alpha)$ credible interval is a set $B(Z)$ satisfying

$$\mathbb{P}\big(\Theta \in B(Z) \mid Z\big) \geq 1-\alpha.$$

They answer different questions (coverage under repeated sampling vs. posterior probability given observed data).

Remark 8.1.5 (ML Relevance) In deployment, decisions often depend on uncertainty (medical diagnosis, risk scoring, safety constraints). Probability models enable calibrated uncertainty, while concentration and asymptotics enable guarantees. The next sections (MLE, MAP, Bayes) develop how these estimates are computed and how their uncertainty is quantified.

8.2 Point Estimators and Properties

A *point estimator* compresses data $Z_{1:n}$ into a single value $\widehat{\theta}$ intended to approximate an unknown target θ. To compare estimators rigorously we need quality criteria (bias, variance, MSE, consistency) and performance limits (Cramér–Rao type) (Bickel and Doksum 2015). This section develops these criteria, shows how common estimators are constructed, and previews sufficiency and exponential families as a principled data-compression viewpoint that later explains why MLE/MAP often have simple forms.

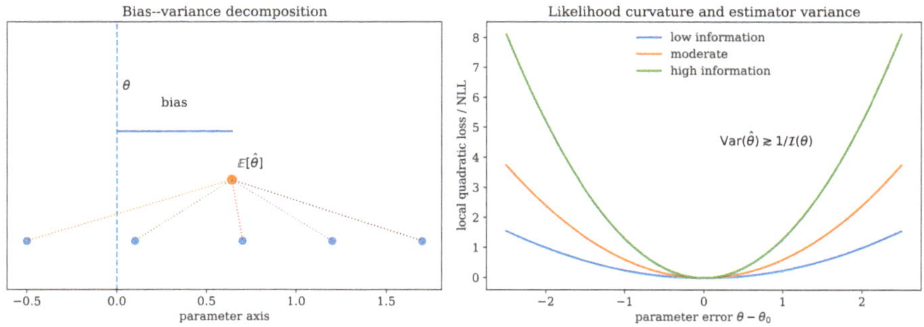

Fig. 8.2 Two views of estimator quality. Left: the mean squared error decomposes into variance plus squared bias. Right: Fisher information controls the minimum attainable variance of unbiased estimators through the Cramér–Rao perspective; sharper likelihood curvature means smaller achievable uncertainty

8.2.1 Quality Criteria

8.2.1.1 Bias, Variance, MSE, and Consistency

An estimator is a random variable, so it has a distribution under the data-generating process. Bias measures systematic error, variance measures sensitivity to sampling noise, and MSE combines both into a single quadratic risk (cf. Fig. 8.2). Consistency is asymptotic: does the estimator converge to the target as $n \to \infty$? In ML, these criteria explain why some estimators are stable but biased (regularized), why others are unbiased but high-variance (ill-conditioned), and why asymptotic correctness does not guarantee finite-sample performance.

Definition 8.2.1 (Bias, Variance, MSE) Let $\widehat{\theta}$ estimate a scalar parameter $\theta \in \mathbb{R}$. Assuming required moments exist, define

$$\mathrm{Bias}(\widehat{\theta};\theta) := \mathbb{E}_\theta[\widehat{\theta}] - \theta, \qquad \mathrm{Var}_\theta(\widehat{\theta}) := \mathbb{E}_\theta\big[(\widehat{\theta} - \mathbb{E}_\theta\widehat{\theta})^2\big],$$

$$\mathrm{MSE}_\theta(\widehat{\theta}) := \mathbb{E}_\theta\big[(\widehat{\theta} - \theta)^2\big].$$

Proposition 8.2.1 (Bias–Variance Decomposition) *For any estimator $\widehat{\theta}$ with finite second moment,*

$$MSE_\theta(\widehat{\theta}) = Var_\theta(\widehat{\theta}) + Bias(\widehat{\theta};\theta)^2.$$

Definition 8.2.2 (Consistency) A sequence of estimators $\widehat{\theta}_n$ is

- *Consistent* for θ if $\widehat{\theta}_n \to \theta$ in probability under P_θ;
- *Strongly consistent* if $\widehat{\theta}_n \to \theta$ almost surely.

Remark 8.2.1 (Consistency Is Not a Finite-Sample Guarantee) Consistency only describes the limit as $n \to \infty$. Two consistent estimators can have very different finite-sample MSE, and an inconsistent estimator can still be useful if it trades bias for much lower variance.

8.2.1.2 Efficiency Intuition and Cramér–Rao Viewpoint

Efficiency asks: among (approximately) unbiased estimators, who has the smallest possible variance? The Cramér–Rao bound (CRB) provides a fundamental lower bound on the variance of any unbiased estimator under regularity conditions. It is derived from Cauchy–Schwarz applied to the score function and reveals the central role of Fisher information: information grows with sample size and shrinks the best achievable variance like $1/n$. In ML, CRB motivates the curvature/Hessian viewpoint and clarifies why well-specified likelihood models yield statistically efficient estimators.

Definition 8.2.3 (Score and Fisher Information (Scalar Parameter)) Let Z have density/pmf $p_\theta(z)$ with respect to a base measure, and assume differentiability in θ. The *score* is

$$s_\theta(Z) := \frac{\partial}{\partial\theta} \log p_\theta(Z).$$

The *Fisher information* is

$$\mathcal{I}(\theta) := \mathbb{E}_\theta\big[s_\theta(Z)^2\big],$$

whenever finite. For i.i.d. samples $Z_{1:n}$, $\mathcal{I}_n(\theta) = n\mathcal{I}(\theta)$.

Theorem 8.2.1 (Cramér–Rao Lower Bound (Scalar, Unbiased Case)) *Assume regularity conditions under which $\mathbb{E}_\theta[s_\theta(Z)] = 0$ and differentiation under the integral is valid. If $\widehat{\theta}$ is an unbiased estimator of θ based on n i.i.d. samples, then*

$$Var_\theta(\widehat{\theta}) \geq \frac{1}{\mathcal{I}_n(\theta)} = \frac{1}{n\mathcal{I}(\theta)}.$$

Remark 8.2.2 (Efficiency (Informal)) An unbiased estimator that attains the CRB (or asymptotically attains it) is called efficient (or asymptotically efficient). Many MLEs are asymptotically efficient under correct specification.

Remark 8.2.3 (Biased Estimators and Tradeoffs) CRB is stated for unbiased estimators. Allowing bias can reduce MSE substantially (e.g., ridge regression). A generalized CRB exists for biased estimators, but the main ML lesson is: bias can buy variance reduction.

8.2.2 *From Data to Estimators*

8.2.2.1 Estimating Means, Variances, and Proportions

The most common estimation tasks in ML are estimating means (average loss/gradient), variances (uncertainty/noise scale), and proportions (class probabilities). These are building blocks for risk estimation, calibration, and probabilistic modeling. They also showcase the basic workflow: write the quantity as an expectation under a distribution, then replace expectation by an empirical average.

Definition 8.2.4 (Sample Mean and Sample Proportion) Let $X_1, \ldots, X_n$ be i.i.d. with mean μ. The sample mean is $\bar{X}_n = \frac{1}{n}\sum_{i=1}^n X_i$. If $X_i \in \{0, 1\}$ with $\mathbb{P}(X_i = 1) = p$, then $\bar{X}_n$ is the sample proportion estimator of p.

Definition 8.2.5 (Sample Variance) Let $X_1, \ldots, X_n$ be i.i.d. with mean μ and variance σ^2 and define $\bar{X}_n$. Two common estimators of σ^2 are

$$\widehat{\sigma}_n^2 := \frac{1}{n}\sum_{i=1}^n (X_i - \bar{X}_n)^2, \qquad s_n^2 := \frac{1}{n-1}\sum_{i=1}^n (X_i - \bar{X}_n)^2.$$

The second is unbiased for σ^2 under finite second moments.

Remark 8.2.4 (Why $n - 1$ Appears) Replacing μ by $\bar{X}_n$ "uses up" one degree of freedom; the factor $(n - 1)$ corrects this bias in the variance estimator.

8.2.2.2 Plug-In Principle and Invariance Heuristics

Often we want to estimate a function of a parameter, $g(\theta)$, rather than θ itself. The plug-in principle says: estimate θ by $\widehat{\theta}$, then estimate $g(\theta)$ by $g(\widehat{\theta})$. This is ubiquitous in ML: estimate mean/variance then standardize; estimate class probabilities then compute entropy; estimate covariance then compute Mahalanobis distance. Invariance heuristics formalize when such transformations behave well: for MLE, invariance often holds exactly ($g(\widehat{\theta}_{\mathrm{MLE}})$ is MLE of $g(\theta)$ when g is one-to-one).

Definition 8.2.6 (Plug-In Estimator) Let $\widehat{\theta}$ estimate θ and let g be a measurable function. The *plug-in estimator* of $g(\theta)$ is $g(\widehat{\theta})$.

Remark 8.2.5 (Bias Propagation) Even if $\widehat{\theta}$ is unbiased, $g(\widehat{\theta})$ need not be unbiased unless g is affine. This is a direct consequence of Jensen's inequality when g is convex/concave.

Remark 8.2.6 (Delta-Method Preview) If $\sqrt{n}(\widehat{\theta} - \theta)$ is asymptotically normal, then $g(\widehat{\theta})$ inherits asymptotic normality with variance scaled by $g'(\theta)^2$. Thus plug-in estimators often have computable asymptotic uncertainty.

8.2.3 *Sufficiency and Exponential Families*

8.2.3.1 Sufficient Statistics and Data Compression Intuition

Sufficiency formalizes when data can be compressed without losing information about a parameter. A statistic $T(Z_{1:n})$ is sufficient if the conditional distribution of the full data given T does not depend on θ. In ML terms: if T is sufficient, then all likelihood-based inference about θ can be done using T alone. This viewpoint explains why counts, sums, and sample means repeatedly appear: they are sufficient in many standard models.

Definition 8.2.7 (Sufficient Statistic) A statistic $T(Z_{1:n})$ is *sufficient* for θ in the family $\{P_\theta\}$ if the conditional distribution of $Z_{1:n}$ given T does not depend on θ.

Remark 8.2.7 (Factorization Criterion (Preview Statement)) A standard way to prove sufficiency is the Neyman–Fisher factorization theorem: if the likelihood can be written as $L(\theta; z_{1:n}) = g_\theta(T(z_{1:n}))\, h(z_{1:n})$, then T is sufficient. We will use this later in exponential family models.

8.2.3.2 Exponential Family Form and Its Benefits

Exponential families include many core ML distributions (Bernoulli, categorical, Gaussian with known variance, Poisson, Gamma). They admit a canonical likelihood form involving a low-dimensional sufficient statistic $T(z)$. This yields computational advantages: convex log-likelihoods, clean gradients, conjugate priors, and interpretable moments. They are also the natural foundation for generalized linear models.

Definition 8.2.8 (Exponential Family (Canonical Form)) A distribution family $\{p_\theta(z)\}$ is an exponential family if it can be written as

$$p_\theta(z) = h(z) \exp\Big(\langle \eta(\theta), T(z) \rangle - A(\theta)\Big),$$

where $T(z)$ is the sufficient statistic, $\eta(\theta)$ is the natural parameter, $A(\theta)$ is the log-partition function, and $h(z)$ is the base measure term.

Remark 8.2.8 (Why $A(\theta)$ Matters) $A(\theta)$ ensures normalization and encodes moments: its derivatives (with respect to natural parameters) yield expectations and covariances of $T(Z)$. This is why exponential families are analytically tractable.

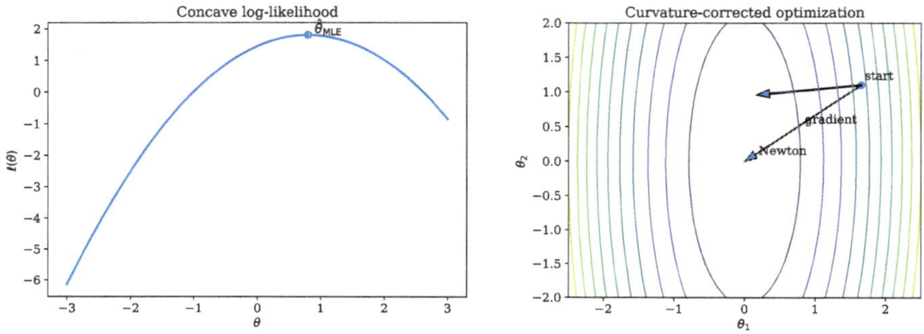

Fig. 8.3 Likelihood geometry and optimization. Left: a concave log-likelihood has a unique global maximizer with curvature governed by observed/Fisher information. Right: gradient ascent follows the local slope, while Newton's method rescales by curvature and moves more directly toward the optimum in anisotropic landscapes

8.3 Maximum Likelihood Estimation (MLE)

Maximum likelihood is a universal estimation principle: choose parameters that make the observed data most probable under the assumed model. For i.i.d. data it yields an additive objective (sum of log-likelihoods), enabling scalable optimization. MLE connects directly to generalization via proper scoring rules (log-loss), and to uncertainty via curvature (Fisher information). This section derives MLE through score equations, illustrates exponential-family templates, and explains computation and curvature (Bishop 2006); the optimization and curvature viewpoint is summarized in Fig. 8.3.

8.3.1 Deriving MLE

8.3.1.1 Likelihood, Log-Likelihood, and Score Equations

The likelihood $L(\theta)$ is a function of θ with the data fixed. Maximizing likelihood is equivalent to maximizing log-likelihood because log is monotone. The score equation $\nabla_\theta \ell(\theta) = 0$ gives candidate optima and is the basis of Newton/IRLS and asymptotic theory. In ML, many objectives are negative log-likelihoods; the score is the gradient used by optimization algorithms, and its Jacobian/Hessian captures curvature.

Definition 8.3.1 (Likelihood and Log-Likelihood (i.i.d.)) Let $Z_1, \ldots, Z_n$ be i.i.d. with density/pmf $p_\theta(z)$ for $\theta \in \Theta \subseteq \mathbb{R}^d$. Define

$$L(\theta; Z_{1:n}) = \prod_{i=1}^{n} p_\theta(Z_i), \qquad \ell(\theta; Z_{1:n}) = \log L(\theta; Z_{1:n}) = \sum_{i=1}^{n} \log p_\theta(Z_i).$$

An MLE is any

$$\widehat{\theta}_{\text{MLE}} \in \arg\max_{\theta \in \Theta} \ell(\theta; Z_{1:n}).$$

Definition 8.3.2 (Score and Score Equations) Assume $\ell(\theta)$ is differentiable. The *score* is

$$U(\theta) := \nabla_\theta \ell(\theta; Z_{1:n}) \in \mathbb{R}^d.$$

Any interior maximizer $\widehat{\theta}$ satisfies the *score equations*

$$U(\widehat{\theta}) = 0.$$

Remark 8.3.1 (Constrained MLE) If Θ imposes constraints (e.g., $\sigma^2 > 0$, simplex constraints), then maximizers may occur on the boundary and the score equation must be combined with constraint conditions (KKT).

Example 8.3.1 (Bernoulli MLE (One Paragraph Computation)) Let $Z_1, \ldots, Z_n$ be i.i.d. Bernoulli(p) with $p \in (0, 1)$. Then

$$\ell(p) = \sum_{i=1}^{n} \big(Z_i \log p + (1 - Z_i) \log(1 - p)\big).$$

Differentiating,

$$\frac{d\ell}{dp} = \sum_{i=1}^{n} \left(\frac{Z_i}{p} - \frac{1 - Z_i}{1 - p} \right) = \frac{\sum_i Z_i}{p} - \frac{n - \sum_i Z_i}{1 - p}.$$

Setting to zero yields $\widehat{p} = \frac{1}{n} \sum_{i=1}^{n} Z_i$ (sample proportion).

8.3.1.2 Exponential Families: Templates and Examples

Exponential families provide a plug-and-play template: log-likelihood becomes a linear function of sufficient statistics minus a convex log-partition. This yields concise score equations and often closed-form MLEs (or convex objectives). In ML, this is why Bernoulli/logistic, categorical/softmax, Poisson regression, and Gaussian regression fit naturally in one framework.

Definition 8.3.3 (Exponential Family Likelihood Structure) Suppose

$$p_\theta(z) = h(z) \exp\big(\langle \eta(\theta), T(z) \rangle - A(\theta)\big).$$

For i.i.d. data, the log-likelihood is

$$\ell(\theta) = \sum_{i=1}^{n} \log h(Z_i) + \left\langle \eta(\theta), \sum_{i=1}^{n} T(Z_i) \right\rangle - nA(\theta).$$

Thus the data enter only through the aggregated sufficient statistic $\sum_i T(Z_i)$.

Example 8.3.2 (Gaussian Mean with Known Variance) Let $Z_i \sim \mathcal{N}(\mu, \sigma^2)$ i.i.d. with known σ^2. Then

$$\ell(\mu) = -\frac{n}{2} \log(2\pi\sigma^2) - \frac{1}{2\sigma^2} \sum_{i=1}^{n} (Z_i - \mu)^2.$$

Differentiating and setting to zero yields $\widehat{\mu} = \bar{Z}_n = \frac{1}{n} \sum_i Z_i$.

Example 8.3.3 (Poisson Rate) Let $Z_i \sim \text{Poisson}(\lambda)$ i.i.d. with $\lambda > 0$. The log-likelihood is

$$\ell(\lambda) = \sum_{i=1}^{n} \big(Z_i \log \lambda - \lambda - \log(Z_i!) \big).$$

The score equation $\frac{d\ell}{d\lambda} = 0$ gives $\sum_i \frac{Z_i}{\lambda} - n = 0$, hence $\widehat{\lambda} = \bar{Z}_n$.

8.3.2 Properties and Computation

8.3.2.1 Invariance and Asymptotic Normality: Informal Roadmap

MLE has two flagship properties: *invariance* (transforming the parameter transforms the MLE) and *asymptotic normality* (the estimator is approximately Gaussian with covariance given by inverse Fisher information). These results require regularity conditions (identifiability, differentiability, nondegenerate information). In ML, invariance justifies plug-in estimates of functions of parameters, and asymptotic normality motivates Wald-type confidence intervals and curvature-based uncertainty estimates.

Proposition 8.3.1 (Invariance of MLE) *Let $\widehat{\theta}$ be an MLE for $\theta \in \Theta$, and let $g : \Theta \to \Gamma$ be a one-to-one transformation. Then the MLE for $\gamma = g(\theta)$ is $\widehat{\gamma} = g(\widehat{\theta})$.*

Remark 8.3.2 (Asymptotic Normality (Roadmap Statement)) Under standard regularity conditions, for true parameter θ_0,

$$\sqrt{n}\big(\widehat{\theta}_{\text{MLE}} - \theta_0\big) \Rightarrow \mathcal{N}\big(0, \ \mathcal{I}(\theta_0)^{-1}\big),$$

where $\mathcal{I}(\theta_0)$ is Fisher information for one sample. Thus $\widehat{\theta}_{\mathrm{MLE}}$ has asymptotic covariance $\mathcal{I}(\theta_0)^{-1}/n$. We will connect this to curvature in the next subsection.

8.3.2.2 Numerical MLE: Gradient/Hessian-Based Solvers

Closed-form MLEs are the exception; most require numerical optimization. Because log-likelihood is typically smooth (and often concave for exponential families), gradient-based methods are natural. Newton's method uses the Hessian (second derivative) for fast local convergence; quasi-Newton methods approximate it. In GLMs, iteratively reweighted least squares (IRLS) is Newton's method in disguise. The practical message: the score gives the gradient; the Hessian/information gives curvature and conditioning.

Definition 8.3.4 (Observed Hessian of the Log-Likelihood) Assume $\ell(\theta)$ is twice differentiable. Define the observed Hessian

$$H(\theta) := \nabla_\theta^2 \ell(\theta; Z_{1:n}) \in \mathbb{R}^{d\times d}.$$

For maximization, one often works with the negative Hessian of the negative log-likelihood, which is typically PSD near optimum in well-specified models.

Remark 8.3.3 (Newton Step (Maximization Form)) A Newton update for maximizing ℓ is

$$\theta^{(t+1)} = \theta^{(t)} - H(\theta^{(t)})^{-1}\, U(\theta^{(t)}),$$

where $U = \nabla\ell$. When maximizing concave ℓ, $-H$ is PSD and the step is well-defined locally when H is nonsingular.

Remark 8.3.4 (Gradient Ascent) A first-order update is

$$\theta^{(t+1)} = \theta^{(t)} + \eta_t\, U(\theta^{(t)}),$$

with step-size η_t. Conditioning of the objective controls how sensitive convergence is to η_t.

8.3.3 *Curvature and Information*

8.3.3.1 Observed vs. Expected Information (Fisher Information)

Information measures curvature of the log-likelihood around the truth. Observed information is the negative Hessian at the observed data; expected (Fisher) information is its expectation under the model. Both quantify how sharply peaked the likelihood is: sharper curvature means parameters are more identifiable and estimators have lower variance. In ML, this is the bridge between optimization geometry (Hessian conditioning) and statistical uncertainty (estimator variance).

Definition 8.3.5 (Observed Information and Fisher Information) Let $\ell(\theta; Z_{1:n})$ be the log-likelihood. The *observed information* is

$$\mathcal{J}_n(\theta) := -\nabla_\theta^2 \ell(\theta; Z_{1:n}).$$

For a single sample Z, define the Fisher information matrix

$$\mathcal{I}(\theta) := \mathbb{E}_\theta\Big[\nabla_\theta \log p_\theta(Z)\ \nabla_\theta \log p_\theta(Z)^\top\Big],$$

assuming finiteness. For i.i.d. samples, $\mathcal{I}_n(\theta) = n\,\mathcal{I}(\theta)$.

Remark 8.3.5 (Equivalent Fisher Forms (under Regularity)) Under conditions allowing interchange of differentiation and integration,

$$\mathcal{I}(\theta) = -\mathbb{E}_\theta\Big[\nabla_\theta^2 \log p_\theta(Z)\Big],$$

and hence $\mathbb{E}_\theta[\mathcal{J}_n(\theta)] = n\,\mathcal{I}(\theta)$.

8.3.3.2 Why Curvature Controls Uncertainty and Optimization

Curvature is a common currency for statistics and optimization. Statistically, asymptotic covariance of MLE is $\mathcal{I}(\theta_0)^{-1}/n$: flatter likelihood (smaller information) means higher uncertainty. Computationally, curvature determines conditioning: if the Hessian has eigenvalues with large ratio, gradient methods zig-zag and converge slowly; Newton/quasi-Newton exploit curvature to accelerate. Thus, the same eigenvalues that govern confidence ellipsoids also govern optimization difficulty.

Remark 8.3.6 (Wald-Type Confidence Ellipsoid) If $\widehat{\theta}$ is an MLE and $\mathcal{J}_n(\widehat{\theta})$ is nonsingular, then an approximate $(1-\alpha)$ confidence ellipsoid is

$$\Big\{\theta : (\theta - \widehat{\theta})^\top \mathcal{J}_n(\widehat{\theta})\,(\theta - \widehat{\theta}) \le \chi^2_{d,\,1-\alpha}\Big\},$$

where $\chi^2_{d,\,1-\alpha}$ is a chi-square quantile. This is an ellipsoid shaped by the observed information.

Remark 8.3.7 (Condition Number and Optimization Speed) If the negative log-likelihood is locally quadratic with Hessian $H \succeq 0$, then the condition number $\kappa(H) = \lambda_{\max}(H)/\lambda_{\min}(H)$ controls gradient method convergence. Large κ means strong anisotropy: narrow valleys, slow progress unless preconditioning or second-order methods are used.

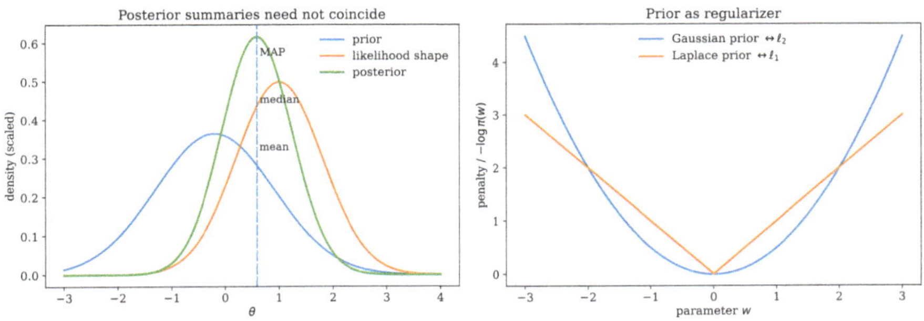

Fig. 8.4 Bayesian and regularized estimation viewed together. Left: the posterior combines likelihood and prior, and different posterior summaries (mean, median, MAP) need not coincide. Right: priors induce regularizers: a Gaussian prior yields smooth shrinkage, while a Laplace prior promotes sparsity through sharper corners

8.4 MAP and Bayesian Estimation

Bayesian estimation treats unknown parameters as random variables and updates uncertainty using Bayes' rule. This yields a full posterior distribution, from which one can extract point estimates (posterior mean, MAP, median), credible intervals, and posterior predictive distributions (Gelman et al. 2014). MAP estimation also reveals a deep connection between priors and regularization: negative log priors become penalty terms (cf. Fig. 8.4). This section develops these ideas with careful contrasts and computational templates.

8.4.1 Bayes Rule as Update

8.4.1.1 Priors, Likelihood, Posterior; Conjugacy Patterns

Bayes' rule converts prior beliefs into posterior beliefs by reweighting with the likelihood. Conjugate priors make this update algebraic: posterior belongs to the same family as the prior with updated hyperparameters. This is valuable in ML for fast online updates, smoothing of probabilities, and interpretable regularization. Even when conjugacy does not hold, the same structure motivates approximate inference (Laplace, variational, MCMC).

Definition 8.4.1 (Prior, Likelihood, Posterior) Let Θ be a parameter with prior density/pmf $\pi(\theta)$ and let data $Z_{1:n}$ have likelihood $p_\theta(Z_{1:n})$. The posterior is

$$\pi(\theta \mid Z_{1:n}) = \frac{p_\theta(Z_{1:n})\, \pi(\theta)}{\int_\Theta p_\vartheta(Z_{1:n})\, \pi(\vartheta)\, d\vartheta},$$

whenever the denominator is finite and nonzero. The denominator is the *evidence* (marginal likelihood) $p(Z_{1:n})$.

Remark 8.4.1 (Log Form) Taking logs yields

$$\log \pi(\theta \mid Z_{1:n}) = \log p_\theta(Z_{1:n}) + \log \pi(\theta) - \log p(Z_{1:n}),$$

so posterior maximization (MAP) is likelihood maximization plus a prior term.

Definition 8.4.2 (Conjugate Prior (Pattern Statement)) A prior family $\{\pi(\theta; \alpha)\}$ is *conjugate* to a likelihood family $\{p_\theta(\cdot)\}$ if the posterior $\pi(\theta \mid Z_{1:n})$ belongs to the same prior family with updated hyperparameters:

$$\pi(\theta \mid Z_{1:n}) = \pi(\theta; \alpha'),$$

where α' is a deterministic function of α and the sufficient statistics of the data.

Example 8.4.1 (Beta–Bernoulli Conjugacy) Let $Z_i \sim \text{Bernoulli}(p)$ i.i.d. and prior $p \sim \text{Beta}(\alpha, \beta)$. With $s = \sum_{i=1}^n Z_i$, the posterior is

$$p \mid Z_{1:n} \sim \text{Beta}(\alpha + s,\ \beta + n - s).$$

Example 8.4.2 (Dirichlet–Multinomial Conjugacy) Let counts $X \mid \pi \sim \text{Multinomial}(n, \pi)$ and prior $\pi \sim \text{Dirichlet}(\alpha)$. Then $\pi \mid X \sim \text{Dirichlet}(\alpha + X)$. This is the algebra behind additive smoothing in multiclass probabilities.

8.4.1.2 Posterior Mean vs. MAP vs. Median

A posterior distribution encodes uncertainty; a point estimate extracts a single representative. Which point is "best" depends on the loss function: posterior mean minimizes squared error, posterior median minimizes absolute error, and MAP maximizes posterior density (equivalently minimizes 0–1 loss on a discrete parameter space, or is a mode for continuous). In ML, MAP is computationally convenient (optimization), while posterior mean is often better under squared loss and yields smoother estimates. Understanding these distinctions prevents misinterpretation of MAP as "the Bayesian answer" in all cases.

Definition 8.4.3 (Posterior Mean, Median, MAP) Let Θ be a real-valued parameter with posterior distribution $\pi(\theta \mid Z)$.

- **Posterior mean:** $\widehat{\theta}_{\text{PM}} := \mathbb{E}[\Theta \mid Z]$ (if finite).
- **Posterior median:** any $\widehat{\theta}_{\text{Med}}$ satisfying $\mathbb{P}(\Theta \leq \widehat{\theta}_{\text{Med}} \mid Z) \geq \frac{1}{2}$ and $\mathbb{P}(\Theta \geq \widehat{\theta}_{\text{Med}} \mid Z) \geq \frac{1}{2}$.
- **MAP:** any maximizer

$$\widehat{\theta}_{\text{MAP}} \in \arg\max_\theta\ \pi(\theta \mid Z) \quad \Longleftrightarrow \quad \widehat{\theta}_{\text{MAP}} \in \arg\max_\theta\ \log p_\theta(Z) + \log \pi(\theta).$$

Proposition 8.4.1 (Bayes Actions under Common Losses) *Let $L(a, \theta)$ be a loss and consider minimizing posterior expected loss $\mathbb{E}[L(a, \Theta) \mid Z]$ over actions a.*

1. *If $L(a, \theta) = (a - \theta)^2$, then $a^\star = \mathbb{E}[\Theta \mid Z]$.*
2. *If $L(a, \theta) = |a - \theta|$, then any posterior median is optimal.*
3. *If Θ is discrete and $L(a, \theta) = \mathbf{1}\{a \neq \theta\}$, then any posterior mode (MAP) is optimal.*

Remark 8.4.2 (Continuous Parameters and MAP) For continuous θ, MAP depends on the parameterization (change of variables changes the density by a Jacobian factor). Thus MAP is not invariant under reparameterization unless one uses an invariant prior or works with appropriate measures. Posterior means/medians are generally more stable under smooth reparameterizations (via delta-method behavior).

8.4.2 *Credible Intervals and Posterior Uncertainty*

8.4.2.1 Credible vs. Confidence Intervals: Careful Contrast

Credible intervals and confidence intervals can look similar numerically, but they answer different questions. A Bayesian credible interval is a *posterior probability statement* conditioned on the observed data. A frequentist confidence interval is a *coverage guarantee* under repeated sampling for each fixed parameter. In well-specified regular models with "matching priors," they may approximately coincide asymptotically, but they can differ substantially, especially under misspecification or strong priors. ML practitioners must be explicit which interpretation is intended.

Definition 8.4.4 (Credible Interval) A (Bayesian) $(1 - \alpha)$ credible interval for a real parameter Θ is any random interval $C(Z)$ such that

$$\mathbb{P}(\Theta \in C(Z) \mid Z) \geq 1 - \alpha.$$

A common choice is the equal-tailed interval:

$$C_{\mathrm{ET}}(Z) = \left[\, F_{\Theta|Z}^{-1}(\alpha/2),\ F_{\Theta|Z}^{-1}(1 - \alpha/2)\,\right],$$

where $F_{\Theta|Z}$ is the posterior CDF.

Definition 8.4.5 (Confidence Interval (Contrast)) A (frequentist) $(1 - \alpha)$ confidence interval is a random interval $I(Z)$ such that for all θ,

$$\mathbb{P}_\theta(\theta \in I(Z)) \geq 1 - \alpha.$$

Remark 8.4.3 (Key Contrast in One Line) Credible: probability over Θ given observed Z. Confidence: probability over Z given fixed θ. They are not the same probability space.

8.4.2.2 Posterior Predictive Distributions

Point estimates ignore parameter uncertainty. Bayesian prediction integrates it out. The posterior predictive distribution averages the likelihood over the posterior, producing better-calibrated uncertainty when the model is correct and providing a principled way to compute prediction intervals. In ML, posterior predictives underpin Bayesian linear regression, Gaussian processes, and Bayesian neural networks. Even in approximate forms, this "integrate over parameters" idea is central.

Definition 8.4.6 (Posterior Predictive) Let Θ have posterior $\pi(\theta \mid Z_{1:n})$ and suppose Z_{new} has likelihood $p_\theta(z_{\text{new}})$ conditionally independent given θ. The posterior predictive distribution is

$$p(z_{\text{new}} \mid Z_{1:n}) = \int p_\theta(z_{\text{new}})\, \pi(\theta \mid Z_{1:n})\, d\theta.$$

In conditional modeling, for a new input x, this becomes

$$p(y \mid x, Z_{1:n}) = \int p_\theta(y \mid x)\, \pi(\theta \mid Z_{1:n})\, d\theta.$$

Remark 8.4.4 (Prediction Intervals) A $(1-\alpha)$ Bayesian prediction interval for Y_{new} given x can be constructed from quantiles of $p(y \mid x, Z_{1:n})$ (e.g., equal-tailed).

8.4.3 Priors as Regularizers

8.4.3.1 MAP $\leftrightarrow$ Regularization Correspondence

MAP estimation converts Bayesian inference into optimization: maximize posterior is maximize log-likelihood plus log-prior. Thus, negative log-prior is a regularizer. This correspondence explains many regularization choices: Gaussian priors yield ℓ_2 penalties (ridge); Laplace priors yield ℓ_1 penalties (lasso); group Laplace-like priors yield group sparsity. This is a deep unification: priors encode geometric preferences, and regularization encodes those preferences in optimization language.

Proposition 8.4.2 (MAP as Penalized MLE) *Let $\pi(\theta)$ be a prior density and let $\ell(\theta)$ be the log-likelihood for data $Z_{1:n}$. Then any MAP estimator satisfies*

$$\widehat{\theta}_{MAP} \in \arg\max_{\theta}\{\ell(\theta) + \log \pi(\theta)\} \quad \Longleftrightarrow \quad \widehat{\theta}_{MAP} \in \arg\min_{\theta}\{-\ell(\theta) + \Omega(\theta)\},$$

where the penalty is $\Omega(\theta) := -\log \pi(\theta)$ (up to an additive constant).

Example 8.4.3 (Gaussian Prior $\Rightarrow$ Ridge) For linear regression with Gaussian noise, $-\ell(w)$ is proportional to $\sum_i (y_i - w^\top x_i)^2$. If the prior is $w \sim \mathcal{N}(0, \tau^2 I)$, then $-\log \pi(w) = \frac{1}{2\tau^2}\|w\|_2^2 + \text{const}$, so MAP solves ridge regression.

Example 8.4.4 (Laplace Prior $\Rightarrow$ Lasso) If $\pi(w) \propto \exp(-\lambda\|w\|_1)$ (independent Laplace components), then $-\log \pi(w) = \lambda\|w\|_1 + \text{const}$, so MAP yields an ℓ_1-penalized objective (lasso-type).

8.4.3.2 Choosing Priors: Scale, Sparsity, and Invariances (Conceptual)

Prior choice encodes assumptions and invariances: scale priors control typical magnitude of parameters; sparsity priors prefer many zeros; invariance priors respect symmetries (rotation invariance suggests spherical Gaussian priors, permutation invariance suggests exchangeable priors). Poorly chosen priors can dominate likelihood or break reparameterization stability. In ML practice, priors are often chosen for their regularization effect and computational tractability (conjugacy or convex penalties), but principled choices align with domain knowledge and symmetry.

Remark 8.4.5 (Three Guiding Principles for Prior Design)

1. **Support:** $\pi(\theta)$ must place mass on feasible parameter values (positivity, simplex constraints, etc.).
2. **Scale:** hyperparameters should reflect plausible magnitudes (e.g., τ in Gaussian priors), often tied to feature scaling.
3. **Symmetry/invariance:** if the model has a symmetry, priors that respect it avoid introducing arbitrary coordinate dependence.

Remark 8.4.6 (MAP Is not Invariant under Reparameterization) If $\phi = g(\theta)$ is a smooth bijection, then densities transform by a Jacobian factor, so the MAP in ϕ-space need not map back to the MAP in θ-space unless the prior is chosen carefully. This is one reason posterior summaries (means/intervals) can be preferable when invariance matters.

8.5 Risk and Decision Theory

Estimation is ultimately in service of a decision: a point prediction, a class label, an interval, or an action. Decision theory provides the unifying framework: specify a loss function that quantifies the cost of decisions, then choose the decision rule that minimizes expected loss (*risk*) (Berger 1985). This perspective clarifies why different estimators (mean, median, MAP) are optimal under different losses, why classification has a Bayes-optimal decision boundary, and how robustness considerations motivate alternative losses under misspecification or heavy tails.

8.5.1 Loss Functions and Decisions

8.5.1.1 Squared, Absolute, 0–1 Loss, and Bayes Actions

A loss function formalizes what "good" means. Under squared loss, large errors are penalized quadratically and the optimal predictor is a mean. Under absolute loss, the optimal predictor is a median, reflecting robustness to outliers. Under 0–1 loss, the optimal decision is the most probable class (mode). These are not heuristics but direct consequences of minimizing conditional expected loss. In ML, these losses connect to regression, robust regression, and classification, respectively.

Definition 8.5.1 (Decision Rule and Risk) Let (X, Y) be random variables and let $\mathcal{A}$ be an action space (predictions/decisions). A decision rule is a measurable map $d : \mathcal{X} \to \mathcal{A}$. Given a loss $L : \mathcal{A} \times \mathcal{Y} \to [0, \infty)$, the (population) risk is

$$R(d) := \mathbb{E}\big[L(d(X), Y)\big].$$

The conditional risk at x is

$$R(d; x) := \mathbb{E}\big[L(d(x), Y) \mid X = x\big].$$

A Bayes rule minimizes conditional risk pointwise in x.

Definition 8.5.2 (Bayes Action) Fix a distribution $P(Y \mid X = x)$ and a loss $L(a, y)$. A *Bayes action* at x is any

$$a^\star(x) \in \arg\min_{a \in \mathcal{A}} \mathbb{E}[L(a, Y) \mid X = x].$$

A Bayes decision rule is $d^\star(x) = a^\star(x)$.

Example 8.5.1 (Three Canonical Losses) For regression with $a \in \mathbb{R}$:

$$L_2(a, y) = (a - y)^2, \qquad L_1(a, y) = |a - y|.$$

For classification with labels $y \in \{1, \ldots, K\}$ and decisions $a \in \{1, \ldots, K\}$:

$$L_{0\text{–}1}(a, y) = \mathbf{1}\{a \neq y\}.$$

8.5.1.2 Proper Scoring Rules: High-Level Motivation

When models output probabilities, we need losses that evaluate a *distributional forecast*, not just a point label. Proper scoring rules ensure that quoting the true conditional distribution minimizes expected score. Log-loss (cross-entropy) and Brier score are standard examples. In ML, training with log-loss corresponds to maximum likelihood, and properness provides the conceptual justification: honest probabilities are optimal under the score.

Definition 8.5.3 (Scoring Rule (Discrete Outcome)) Let Y take values in $\{1, \ldots, K\}$. A scoring rule assigns a loss to a quoted distribution $q \in \Delta^{K-1}$ and realized outcome y:

$$S(q, y) \in \mathbb{R}.$$

It is *proper* if for the true distribution p,

$$\mathbb{E}_{Y \sim p}[S(q, Y)] \geq \mathbb{E}_{Y \sim p}[S(p, Y)] \qquad \forall q.$$

Remark 8.5.1 (Why This Matters for ML) If a learning algorithm minimizes empirical proper score, then (under enough data and correct specification) it is encouraged to produce calibrated probabilities rather than merely correct hard labels.

8.5.2 Optimal Predictors Under Common Losses

8.5.2.1 Mean/Median/Mode as Bayes Estimators

The mean/median/mode triad is the canonical illustration that "best estimate depends on loss." These results are derived by minimizing conditional expected loss at each x. They also connect to Bayesian estimation: posterior mean/median/MAP are Bayes estimators under squared/absolute/0–1 losses when Y is the parameter of interest. In ML regression, the conditional mean $\mathbb{E}[Y \mid X = x]$ is the Bayes regressor for squared loss.

Theorem 8.5.1 (Bayes Estimators for Scalar Regression) *Fix x and let $F_{Y|X=x}$ be the conditional distribution.*

1. *Under squared loss, any Bayes action is the conditional mean:*

$$a_2^\star(x) = \mathbb{E}[Y \mid X = x].$$

2. *Under absolute loss, any Bayes action is a conditional median:*

$$a_1^\star(x) \in \mathit{Median}(Y \mid X = x).$$

3. *If Y is discrete and loss is 0–1 on values, any Bayes action is a conditional mode:*

$$a_{0-1}^\star(x) \in \arg\max_y \mathbb{P}(Y = y \mid X = x).$$

Remark 8.5.2 (Robustness Implication) The median is robust to outliers: moving a small fraction of probability mass to extreme values can change the mean drastically, but changes the median only when it crosses the 50% threshold. This is why absolute-loss and quantile regression are robust.

8.5.2.2 Classification Risk and Bayes Classifier

In classification, the conditional label distribution $p(y \mid x)$ determines the Bayes-optimal decision rule. For 0–1 loss, the Bayes classifier picks the most probable class (argmax). Its risk is the Bayes risk, a fundamental lower bound on achievable classification error. This result is the conceptual foundation of discriminative modeling: learning aims to approximate $p(y \mid x)$ well enough so that argmax decisions are near-optimal.

Definition 8.5.4 (Bayes Classifier and Bayes Risk) For K-class classification with 0–1 loss, the Bayes classifier is

$$h^\star(x) \in \arg\max_{y \in \{1,\ldots,K\}} \mathbb{P}(Y = y \mid X = x).$$

Its risk $R(h^\star)$ is the *Bayes risk*, the minimal possible risk under 0–1 loss.

Remark 8.5.3 (Binary Case: Threshold Rule) For $Y \in \{0, 1\}$, the Bayes rule is a threshold: predict 1 iff $\mathbb{P}(Y = 1 \mid X = x) \geq 1/2$. More general thresholds arise under asymmetric costs.

8.5.3 Model Misspecification and Robustness

8.5.3.1 When MLE/MAP Target the "Closest" Model (Conceptual)

When the assumed model family is wrong (misspecified), MLE/MAP do not recover the truth but instead recover the best approximation within the model class. For

MLE, the target is the KL-projection: the parameter minimizing $\mathrm{KL}(p^\star \| p_\theta)$. For MAP, the target balances KL fit with the prior penalty. This explains why strong priors can dominate in low-data regimes, and why likelihood-based methods can be fragile when tails are misspecified.

Remark 8.5.4 (KL Projection Viewpoint (Informal)) If $p^\star$ is the true distribution and $\{p_\theta\}$ is the model class, then under broad conditions

$$\widehat{\theta}_{\mathrm{MLE}} \to \theta^\star \in \arg\min_\theta \mathrm{KL}(p^\star \| p_\theta).$$

Thus MLE is a projection in distribution space, not necessarily truth recovery.

8.5.3.2 Robust Losses and Heavy-Tailed Likelihoods

Heavy tails and outliers can destroy squared-loss and Gaussian-likelihood assumptions. Robust estimation replaces Gaussian-like losses with losses that grow more slowly (Huber, absolute loss, Tukey), or replaces Gaussian noise models with heavy-tailed likelihoods (Laplace, Student-t). Geometrically, these methods limit the influence of extreme residuals; probabilistically, they correspond to likelihoods with heavier tails and weaker moment assumptions. This preview motivates robust ML under distributional uncertainty.

Remark 8.5.5 (Robust Regression Templates)

- **Absolute loss:** $L(a, y) = |a - y|$ corresponds to Laplace-like noise, robust to outliers.
- **Huber loss:** quadratic near 0, linear in the tails; balances efficiency and robustness.
- **Student-t likelihood:** heavy-tailed noise model, reduces sensitivity to extreme residuals.

We will develop these more formally in later chapters on optimization and robust learning.

8.6 Likelihood Geometry and Optimization (Bridge)

Likelihood-based learning is simultaneously a statistical problem (which model best explains data?) and an optimization problem (how do we find the parameter values efficiently and reliably?). The geometry of the likelihood—its curvature, conditioning, and possible multimodality—controls both the statistical uncertainty of estimators and the computational difficulty of training. This section bridges these perspectives: convexity vs. nonconvexity, second-order structure (Newton/IRLS), a preview of Fisher/natural-gradient geometry, and the geometry of regularization paths and hyperparameter selection.

8.6.1 Convex vs. Nonconvex Likelihoods

8.6.1.1 Log-Concavity and Global Optima

For optimization, the best-case scenario is a concave log-likelihood (equivalently, a convex negative log-likelihood). Then any local maximum is global, and gradient-based methods have strong convergence guarantees. Many classical models used in ML are in this regime: exponential families under canonical links yield convex NLLs. Log-concavity is also statistically favorable: it yields unimodal posteriors (under log-concave priors) and stable inference.

Definition 8.6.1 (Log-Concavity/Convex NLL) A likelihood $L(\theta)$ is *log-concave* if $\ell(\theta) = \log L(\theta)$ is concave on Θ. Equivalently, the negative log-likelihood $f(\theta) := -\ell(\theta)$ is convex.

Proposition 8.6.1 (Global Optimality under Log-Concavity) *If ℓ is concave on a convex parameter set Θ, then any stationary point $\widehat{\theta}$ of ℓ in the interior of Θ (i.e., $\nabla \ell(\widehat{\theta}) = 0$) is a global maximizer of ℓ over Θ. Equivalently, if $f = -\ell$ is convex, any point with $\nabla f(\widehat{\theta}) = 0$ is a global minimizer of f.*

Remark 8.6.1 (Examples in ML) Logistic regression (with ℓ_2 regularization) has convex NLL; Poisson regression under canonical link is convex; Gaussian least squares is convex. These models admit reliable optimization and clean statistical theory.

8.6.1.2 Multimodality, Identifiability, and Local Minima

Nonconvex likelihoods arise when models are latent-variable, overparameterized, or exhibit symmetries. Then the objective can be multimodal (many local maxima), and optimization becomes sensitive to initialization. Moreover, identifiability issues (different parameters implying the same distribution) create flat directions or ridges, complicating both estimation and uncertainty quantification. In ML, mixture models and deep networks exemplify this: local minima and saddle points are part of the landscape.

Definition 8.6.2 (Identifiability (Conceptual)) A parametric family $\{P_\theta : \theta \in \Theta\}$ is *identifiable* if

$$P_{\theta_1} = P_{\theta_2} \implies \theta_1 = \theta_2.$$

If not, distinct parameters produce the same distribution, leading to non-unique optima and flat directions.

Remark 8.6.2 (Multimodality from Latent Variables) Mixture models are typically nonconvex and have label-switching symmetry: permuting component labels yields the same likelihood. Thus the likelihood has multiple equivalent modes, and naive curvature-based uncertainty can be misleading without handling symmetry.

Remark 8.6.3 (Local Minima vs. Saddle Points) In high-dimensional nonconvex problems, saddle points are common. Second-order methods distinguish them via negative curvature directions. This is one motivation for curvature-aware optimization.

8.6.2 Second-Order Structure

8.6.2.1 Newton/IRLS as Curvature-Corrected Optimization

First-order methods use gradients but ignore curvature, so they can zig-zag in ill-conditioned landscapes. Newton's method rescales the gradient by the inverse Hessian, yielding steps that are invariant to affine reparameterization and converge rapidly near optimum when the Hessian is well-behaved. In generalized linear models, Newton updates take a special form: iteratively reweighted least squares (IRLS), which reduces each iteration to a weighted least squares solve. This connects likelihood optimization back to linear algebra and projection geometry.

Definition 8.6.3 (Newton Step for Minimizing NLL) Let $f(\theta) = -\ell(\theta)$ be the negative log-likelihood and assume f is twice differentiable. The Newton update is

$$\theta^{(t+1)} = \theta^{(t)} - \left(\nabla^2 f(\theta^{(t)})\right)^{-1} \nabla f(\theta^{(t)}),$$

whenever the Hessian is invertible.

Remark 8.6.4 (Why Newton Is "curvature-corrected") If locally $f(\theta) \approx f(\theta_0) + \frac{1}{2}(\theta - \theta_0)^\top H(\theta - \theta_0)$, then Newton steps solve the quadratic approximation exactly in one iteration. When H is ill-conditioned, Newton automatically rescales directions by curvature, unlike vanilla gradient descent.

Remark 8.6.5 (IRLS (Conceptual)) For many GLMs, $\nabla^2 f(\theta) = X^\top W(\theta) X$ for a weight matrix $W(\theta)$, so the Newton step solves a weighted least squares system. Hence each iteration is linear algebra plus an updated weight rule.

8.6.2.2 Natural Gradient and Fisher Geometry

Gradient descent depends on the chosen parameterization because it measures step size in Euclidean parameter space. But probability models live on a statistical manifold where the natural local metric is the Fisher information. The natural gradient rescales the gradient by the inverse Fisher information, producing updates that are invariant under smooth reparameterizations and often better conditioned for probabilistic models. This is the geometric link: curvature as information, optimization as geodesic-steepest descent.

Definition 8.6.4 (Fisher Information Metric (Recall)) For a model $p_\theta(z)$, the Fisher information matrix is

$$\mathcal{I}(\theta) = \mathbb{E}_\theta\Big[\nabla_\theta \log p_\theta(Z)\, \nabla_\theta \log p_\theta(Z)^\top\Big].$$

It defines a Riemannian metric on parameter space (the Fisher metric).

Definition 8.6.5 (Natural Gradient) The natural-gradient direction for minimizing $f(\theta)$ is

$$\widetilde{\nabla} f(\theta) := \mathcal{I}(\theta)^{-1} \nabla f(\theta),$$

and a natural-gradient update is $\theta^{(t+1)} = \theta^{(t)} - \eta_t\, \widetilde{\nabla} f(\theta^{(t)})$.

Remark 8.6.6 (Observed Hessian vs. Fisher) For well-specified models, the observed information $-\nabla^2 \ell(\theta)$ and the Fisher information $\mathcal{I}(\theta)$ often coincide in expectation and can be close near the optimum. Natural gradient can be seen as a stabilized second-order method using expected curvature rather than observed curvature.

8.6.3 Regularization Paths and Hyperparameters

8.6.3.1 Tuning λ and Prior Scales

Regularization introduces hyperparameters (e.g., λ) controlling the strength of the prior/penalty. As λ varies, the solution traces a *regularization path*. Small λ fits data more closely but can overfit or become ill-conditioned; large λ stabilizes and shrinks estimates but increases bias. In Bayesian terms, λ corresponds to prior scale (variance, sparsity rate), and its choice encodes how strongly we trust the prior relative to the data.

Definition 8.6.6 (Penalized Likelihood/MAP Objective) Given NLL $f(\theta) = -\ell(\theta)$ and penalty $\Omega(\theta)$, consider

$$\widehat{\theta}_\lambda \in \arg\min_\theta \big\{ f(\theta) + \lambda\, \Omega(\theta) \big\}.$$

In Bayesian MAP, $\Omega(\theta) = -\log \pi(\theta)$ (up to constants) and λ corresponds to prior scale.

Remark 8.6.7 (Path Intuition) The mapping $\lambda \mapsto \widehat{\theta}_\lambda$ often changes smoothly in convex problems and may change abruptly for nonsmooth penalties (e.g., ℓ_1 causing sparsity pattern changes). Understanding the path explains model selection behavior.

8.6.3.2 Cross-Validation and Evidence

Hyperparameters are chosen to generalize, not merely to fit training data. Cross-validation (CV) chooses λ by approximating test performance from data splits. Bayesian evidence chooses hyperparameters by maximizing marginal likelihood $p(\mathcal{D})$, automatically balancing fit and complexity (Occam's razor). These are two sides of the same coin: CV is predictive and frequentist; evidence is probabilistic and Bayesian. Both depend on the same core quantities: likelihood, complexity, and generalization error.

Remark 8.6.8 (Cross-Validation (Conceptual)) K-fold CV selects λ minimizing average validation loss across folds:

$$\widehat{\lambda} \in \arg\min_{\lambda} \frac{1}{K} \sum_{k=1}^{K} \widehat{R}_{\text{val}}^{(k)}(\widehat{\theta}_{\lambda}^{(-k)}),$$

where $\widehat{\theta}_{\lambda}^{(-k)}$ is trained on data excluding fold k.

Remark 8.6.9 (Evidence/Marginal Likelihood (Conceptual)) In Bayesian modeling,

$$p(\mathcal{D} \mid \lambda) = \int p(\mathcal{D} \mid \theta)\, \pi(\theta \mid \lambda)\, d\theta$$

is the evidence as a function of hyperparameter λ (prior scale). Maximizing $p(\mathcal{D} \mid \lambda)$ selects λ by trading off fit and prior complexity.

8.7 Practical Estimation Workflows in ML

Estimation in ML is not only a mathematical procedure; it is a workflow: data are split to approximate out-of-sample performance, hyperparameters are tuned, uncertainty is assessed, and diagnostics guard against misspecification and numerical artifacts. This section provides a rigorous conceptual protocol for train/validation/test, previews uncertainty estimation methods (bootstrap vs. Bayesian), and lists sanity checks that detect overfitting, non-identifiability, and ill-conditioning.

8.7.1 Train/Validation/Test as Estimation Protocol

8.7.1.1 Generalization as Out-of-Sample Estimation

A trained model is an estimator of a prediction rule. Its quality must be evaluated on data not used to fit it, otherwise the estimate is biased optimistically (selection on noise). The train/validation/test protocol is therefore an estimation procedure: training estimates parameters, validation estimates hyperparameter performance, and testing estimates final generalization. When done correctly, test performance is an approximately unbiased estimator of out-of-sample risk. When done incorrectly (leakage, repeated tuning on test), the test estimate becomes biased and underestimates uncertainty.

Definition 8.7.1 (Split Protocol (Conceptual)) Given data $\mathcal{D}$, split into disjoint subsets

$$\mathcal{D} = \mathcal{D}_{\text{train}} \,\dot{\cup}\, \mathcal{D}_{\text{val}} \,\dot{\cup}\, \mathcal{D}_{\text{test}}.$$

Training produces $\widehat{\theta}(\lambda)$ from $\mathcal{D}_{\text{train}}$ for each hyperparameter λ. Validation selects $\widehat{\lambda}$ using $\mathcal{D}_{\text{val}}$. Testing reports $\widehat{R}_{\text{test}}(\widehat{\theta}(\widehat{\lambda}))$ on $\mathcal{D}_{\text{test}}$.

Remark 8.7.1 (Out-of-Sample Risk as an Estimation Target) The test loss is an estimator of the population risk $R(\widehat{h})$ of the *entire pipeline* (training algorithm + tuning), not just of the parametric family. Thus, the split protocol estimates performance of a stochastic procedure.

8.7.1.2 Overfitting as Variance Explosion

Overfitting is fundamentally a variance phenomenon: the estimator adapts too strongly to sample noise, so its out-of-sample error increases. In linear regression, this is visible as high variance in coefficients when $X^{\top}X$ is ill-conditioned. In complex models, it manifests as large sensitivity of predictions to small data perturbations or training randomness. Regularization, early stopping, and model selection are variance-control mechanisms. This viewpoint aligns overfitting with estimation error in the bias–variance decomposition.

Remark 8.7.2 (Bias–Variance Lens (Conceptual)) Increasing model capacity typically decreases approximation bias but increases estimation variance. Overfitting corresponds to choosing capacity where variance dominates and test error increases. Regularization shifts the operating point toward lower variance at the cost of higher bias.

Remark 8.7.3 (Practical Symptom) A canonical symptom is a widening gap between training loss (decreasing) and validation/test loss (increasing or stagnant). This is an empirical indicator that variance is dominating.

8.7.2 *Uncertainty Estimation*

8.7.2.1 Bootstrap vs. Bayesian Credible Intervals (Conceptual Comparison)

Uncertainty estimates answer: "how much would my estimator change if I repeated the data collection?" Bootstrap approximates this sampling variability by resampling the dataset and refitting; it is frequentist and model-agnostic. Bayesian credible intervals quantify uncertainty under a probabilistic prior and likelihood; they are model-based and can be sharper when the model is well specified. Both are valuable: bootstrap is robust to mild misspecification but computationally expensive; Bayesian intervals are principled under correct models but can be sensitive to priors and misspecification.

Remark 8.7.4 (Bootstrap Workflow (Conceptual)) Given data $\mathcal{D} = \{z_i\}_{i=1}^n$ and estimator $\widehat{\theta} = \widehat{\theta}(\mathcal{D})$:

1. For $b = 1, \dots, B$, draw a bootstrap sample $\mathcal{D}^{(b)}$ by sampling n points with replacement from $\mathcal{D}$.
2. Compute $\widehat{\theta}^{(b)} = \widehat{\theta}(\mathcal{D}^{(b)})$.
3. Use the empirical distribution of $\{\widehat{\theta}^{(b)}\}$ to form standard errors and confidence intervals.

Remark 8.7.5 (Bayesian Credible Interval Workflow (Conceptual)) Given prior $\pi(\theta)$ and likelihood $p_\theta(\mathcal{D})$, compute/post-approximate the posterior $\pi(\theta \mid \mathcal{D})$ and take posterior quantiles to obtain a credible interval. Posterior predictive uncertainty integrates both noise and parameter uncertainty.

8.7.2.2 Calibration of Predictive Probabilities

A probabilistic classifier outputs $q(x) \approx \mathbb{P}(Y = 1 \mid X = x)$. Calibration checks whether these probabilities correspond to observed frequencies. Even highly accurate classifiers can be poorly calibrated (overconfident or underconfident), which matters in decision-making under uncertainty. Calibration methods (Platt scaling, isotonic regression, temperature scaling) can be viewed as secondary estimation problems: fit a mapping that corrects probability outputs on a validation set without changing ranking too much.

Remark 8.7.6 (Calibration as a Post-processing Estimator) Given predicted probabilities q_i and labels y_i on a calibration set, one estimates a monotone map g so that $g(q)$ is calibrated. This is an estimation problem on $[0, 1]$ with constraints (monotonicity, smoothness) chosen to control variance.

8.7.3 Diagnostics and Sanity Checks

8.7.3.1 Residuals, Likelihood Checks, and Posterior Predictive Checks

Diagnostics test whether the fitted model is compatible with data beyond a single scalar score. In regression, residuals reveal misspecification (heteroscedasticity, nonlinearity, outliers). In likelihood-based modeling, one checks whether fitted distributions reproduce key statistics of the data. In Bayesian modeling, posterior predictive checks simulate new data from the fitted posterior predictive and compare summaries to observed data. These are not optional: they are the empirical guardrails that prevent false confidence in a wrong model.

Remark 8.7.7 (Residual Diagnostics (Regression)) Given fitted values $\widehat{y}_i$ and residuals $r_i = y_i - \widehat{y}_i$, common checks include:

- plotting r_i vs. $\widehat{y}_i$ (detect nonlinearity/heteroscedasticity),
- checking heavy tails/outliers (large $|r_i|$),
- checking correlation patterns (non-i.i.d. errors).

Remark 8.7.8 (Posterior Predictive Checks (Bayesian)) Sample $\theta^{(s)} \sim \pi(\theta \mid \mathcal{D})$ and simulate replicated data $\mathcal{D}^{(s)} \sim p_{\theta^{(s)}}(\cdot)$. Compare statistics $T(\mathcal{D}^{(s)})$ to $T(\mathcal{D})$ (histograms, moments, tail frequencies). Systematic mismatch indicates misspecification.

8.7.3.2 Detecting Non-identifiability and Numerical Pathologies

Non-identifiability means different parameter values produce essentially the same predictions/likelihood. It causes flat likelihood ridges, unstable estimates, and meaningless uncertainty quantification. Numerical pathologies (ill-conditioning, near-singularity) amplify noise and cause solvers to behave erratically. In ML practice, these issues appear as wildly varying parameter estimates across runs, exploding standard errors, or sensitivity to tiny perturbations. Geometry provides the detection tools: inspect rank/condition numbers, Hessian eigenvalues, and parameter correlations; apply regularization or reparameterization to restore identifiability and stability.

Remark 8.7.9 (Common Numerical Red Flags)

- **Ill-conditioned design:** $\kappa(X^\top X) \gg 1$ in regression; coefficients unstable.
- **Nearly singular information:** observed information $\mathcal{J}_n(\widehat{\theta})$ has tiny eigenvalues; standard errors explode.
- **Symmetry modes:** multiple equivalent optima (e.g., mixture label switching); MAP depends on initialization.
- **Optimization instability:** large steps, divergence, NaNs; often due to scale mismatch or poor curvature.

Remark 8.7.10 (Practical Remedies) Center/scale features, use orthonormal or whitened coordinates, add ridge-type regularization, constrain parameters (e.g., ordered mixture means), and prefer stable solvers (QR/SVD) over explicit inverses.

8.8 Exercises

Unless stated otherwise, assume i.i.d. sampling and use natural logarithms.

Exercise 8.8.1 (Bias–Variance–MSE and Consistency Check) Let $X_1, \dots, X_n$ be i.i.d. with $\mathbb{E}[X_i] = \mu$ and $\mathrm{Var}(X_i) = \sigma^2$, and consider the estimator

$$\widehat{\mu}_c := c\,\bar{X}_n, \qquad \bar{X}_n = \frac{1}{n}\sum_{i=1}^{n} X_i, \quad c \in \mathbb{R}.$$

(i) Compute $\mathrm{Bias}(\widehat{\mu}_c; \mu)$, $\mathrm{Var}(\widehat{\mu}_c)$, and $\mathrm{MSE}(\widehat{\mu}_c)$ in closed form.
(ii) Find the value of c that minimizes $\mathrm{MSE}(\widehat{\mu}_c)$ and express the minimum MSE.
(iii) Determine for which values of c the estimator $\widehat{\mu}_c$ is (weakly) consistent for μ as $n \to \infty$.

Exercise 8.8.2 (Cramér–Rao Bound for Bernoulli) Let $Z_1, \dots, Z_n$ be i.i.d. $\mathrm{Bernoulli}(p)$ with $p \in (0, 1)$.

(i) Compute the log-likelihood $\ell(p)$ and score $s(p) = \partial\ell/\partial p$.
(ii) Compute the Fisher information for one sample, $\mathcal{I}(p)$, and hence $\mathcal{I}_n(p)$.
(iii) State the Cramér–Rao lower bound for any unbiased estimator of p.
(iv) For the sample proportion $\widehat{p} = \bar{Z}_n$, compute $\mathrm{Var}(\widehat{p})$ and verify whether it attains the CRB.

Exercise 8.8.3 (MLE for Gaussian Mean and Variance (with Full Computation)) Let $z_1, \dots, z_6$ be observed data:

$$z = (2.1,\ 1.7,\ 3.4,\ 2.8,\ 1.9,\ 2.6).$$

Assume $Z_i \overset{\text{iid}}{\sim} \mathcal{N}(\mu, \sigma^2)$ with both $\mu \in \mathbb{R}$ and $\sigma^2 > 0$ unknown.

(i) Derive the MLE $\widehat{\mu}$ and $\widehat{\sigma}^2_{\text{MLE}}$ by differentiating the log-likelihood (show the score equations).
(ii) Compute numerical values of $\widehat{\mu}$ and $\widehat{\sigma}^2_{\text{MLE}}$ for the given data.
(iii) Compute the unbiased sample variance $s^2 = \frac{1}{n-1}\sum_{i=1}^{n}(z_i - \widehat{\mu})^2$ and compare it to $\widehat{\sigma}^2_{\text{MLE}}$.

Exercise 8.8.4 (Invariance of MLE under Reparameterization) Let $Z_1, \ldots, Z_n$ be i.i.d. Exponential(λ) with unknown rate $\lambda > 0$. Suppose you observe $n = 5$ samples

$$z = (0.4,\ 1.2,\ 0.7,\ 0.1,\ 0.6).$$

(i) Derive the MLE $\widehat{\lambda}$.
(ii) Let $\theta = 1/\lambda$ be the mean parameter. Use the invariance property to obtain the MLE $\widehat{\theta}$.
(iii) Compute $\widehat{\lambda}$ and $\widehat{\theta}$ numerically for the given z.
(iv) Verify by direct optimization of the likelihood written in terms of θ that the maximizer equals $\widehat{\theta}$.

Exercise 8.8.5 (Observed vs. Expected Information (Numerical Comparison)) Let $Z_1, \ldots, Z_n$ be i.i.d. Poisson(λ) with unknown $\lambda > 0$ and observed data

$$z = (1, 0, 3, 2, 1, 4, 0, 2) \quad (n = 8).$$

(i) Compute the MLE $\widehat{\lambda}$.
(ii) Compute the observed information $\mathcal{J}_n(\lambda) = -\frac{d^2}{d\lambda^2}\ell(\lambda)$ and evaluate it at $\widehat{\lambda}$.
(iii) Compute the Fisher information $\mathcal{I}_n(\lambda)$ and evaluate it at $\widehat{\lambda}$.
(iv) Compare the two numerically and give one paragraph interpreting why they are close (or not) for this dataset.

Exercise 8.8.6 (MAP for Bernoulli with Beta Prior; Posterior Mean vs. MAP vs. Median) Let $Z_1, \ldots, Z_{20}$ be Bernoulli trials with $\sum_{i=1}^{20} Z_i = 7$ successes. Assume the prior $p \sim \text{Beta}(\alpha, \beta)$ with $\alpha = 2$ and $\beta = 3$.

(i) Compute the posterior distribution of p.
(ii) Compute the posterior mean $\mathbb{E}[p \mid Z]$.
(iii) Compute the MAP estimator $\widehat{p}_{\text{MAP}}$ (assume $\alpha + s > 1$ and $\beta + n - s > 1$ and use the Beta mode formula).
(iv) Compute the equal-tailed 90% credible interval endpoints as posterior quantiles of $\text{Beta}(\alpha + s, \beta + n - s)$ (*you may leave your answer in terms of Beta inverse-CDF notation* $F^{-1}_{\text{Beta}(a,b)}(\cdot)$).
(v) State which loss (squared/absolute/0–1 on a discretized grid) each of mean/median/MAP corresponds to.

Exercise 8.8.7 (MAP $\leftrightarrow$ Regularization in Linear Regression) Consider the linear model $y = Xw + \varepsilon$ with $\varepsilon \sim \mathcal{N}(0, \sigma^2 I_n)$ and prior $w \sim \mathcal{N}(0, \tau^2 I_d)$. Let

$$X = \begin{bmatrix} 1 & 0 \\ 1 & 1 \\ 1 & 2 \\ 1 & 3 \end{bmatrix}, \quad y = \begin{bmatrix} 1 \\ 2 \\ 2 \\ 4 \end{bmatrix}, \quad \sigma^2 = 1, \quad \tau^2 = \frac{1}{2}.$$

(i) Derive the MAP estimator w_{MAP} by minimizing the negative log posterior and show it equals ridge:

$$w_{\mathrm{MAP}} = (X^\top X + \lambda I)^{-1} X^\top y, \quad \lambda = \sigma^2/\tau^2.$$

(ii) Compute λ numerically and then compute w_{MAP} exactly.
(iii) Compute the unregularized least squares solution (if it exists uniquely) and compare it to w_{MAP}.

Exercise 8.8.8 (Posterior Predictive in Conjugate Gaussian Model) Let $Z_1, \ldots, Z_n \mid \mu \overset{\text{iid}}{\sim} \mathcal{N}(\mu, \sigma^2)$ with known $\sigma^2 = 1$ and prior $\mu \sim \mathcal{N}(\mu_0, \tau_0^2)$. Take $\mu_0 = 0$, $\tau_0^2 = 4$, and observed data $z = (2.1, 1.7, 3.4, 2.8, 1.9, 2.6)$ (so $n = 6$).

(i) Compute the posterior distribution $\mu \mid z$ (mean and variance).
(ii) Compute the posterior predictive distribution for a new observation $Z_{\text{new}} \mid z$ (mean and variance).
(iii) Compute a 95% Bayesian prediction interval for Z_{new} in the form $\widehat{m} \pm 1.96\widehat{s}$.

Exercise 8.8.9 (Decision Theory: Bayes Estimator under Asymmetric Absolute Loss) Let $Y \mid X = x$ have a continuous conditional CDF $F_{Y|x}$. Consider the asymmetric absolute loss

$$L_\alpha(a, y) = \alpha(y - a)\mathbf{1}\{y \geq a\} + (1 - \alpha)(a - y)\mathbf{1}\{y < a\}, \qquad \alpha \in (0, 1).$$

Show that any Bayes action $a^\star(x)$ minimizing $\mathbb{E}[L_\alpha(a, Y) \mid X = x]$ is an α-quantile of $Y \mid X = x$, i.e.,

$$F_{Y|x}(a^\star(x)) \geq \alpha \ \text{ and } \ 1 - F_{Y|x}(a^\star(x)-) \geq 1 - \alpha.$$

(You may assume differentiability of the conditional risk in a and justify the first-order condition.)

Exercise 8.8.10 (Bayes Classifier and Bayes Risk (Finite Computation)) Let $Y \in \{0, 1\}$ and $X \in \{a, b, c\}$ with joint probabilities:

$$\mathbb{P}(X = a, Y = 1) = 0.12, \ \mathbb{P}(X = a, Y = 0) = 0.18;$$

$$\mathbb{P}(X = b, Y = 1) = 0.08, \ \mathbb{P}(X = b, Y = 0) = 0.22;$$

$$\mathbb{P}(X = c, Y = 1) = 0.20, \ \mathbb{P}(X = c, Y = 0) = 0.20.$$

(i) Compute $\eta(x) = \mathbb{P}(Y = 1 \mid X = x)$ for $x \in \{a, b, c\}$.
(ii) Write the Bayes classifier under 0–1 loss and compute its risk (Bayes risk).
(iii) Consider a classifier $\widehat{h}$ that predicts 1 for $x \in \{a, b\}$ and 0 for $x = c$; compute its risk and compare to Bayes risk.

Exercise 8.8.11 (Proper Scoring Rules: Log-Loss vs. Brier on a Dataset) For $n = 10$, let binary labels and predicted probabilities be

$$y = (1, 0, 1, 1, 0, 0, 1, 0, 1, 0),$$
$$q = (0.9, 0.1, 0.6, 0.7, 0.4, 0.2, 0.8, 0.3, 0.55, 0.05).$$

(i) Compute the average log-loss

$$\frac{1}{n}\sum_{i=1}^{n} -\big(y_i \log q_i + (1 - y_i)\log(1 - q_i)\big)$$

and the average Brier score $\frac{1}{n}\sum_{i=1}^{n}(q_i - y_i)^2$.

(ii) Identify which examples contribute disproportionately to log-loss and explain (one paragraph) why log-loss is more sensitive to extreme miscalibration than Brier.

Exercise 8.8.12 (Likelihood Geometry: Concavity Check and Newton Step) Consider logistic regression with a single parameter $\theta \in \mathbb{R}$ and data (x_i, y_i) for $i \in [6]$:

$$x = (-2, -1, 0, 1, 2, 3), \qquad y = (0, 0, 0, 1, 1, 1).$$

Model: $\mathbb{P}_\theta(Y = 1 \mid x) = \sigma(\theta x)$ where $\sigma(t) = 1/(1 + e^{-t})$.

(i) Write the log-likelihood $\ell(\theta) = \sum_{i=1}^{6}\big(y_i \log\sigma(\theta x_i) + (1 - y_i)\log(1 - \sigma(\theta x_i))\big)$.

(ii) Compute the score $\ell'(\theta)$ and Hessian $\ell''(\theta)$.

(iii) Show that ℓ is concave by proving $\ell''(\theta) \le 0$ for all θ.

(iv) Starting from $\theta^{(0)} = 0$, compute one Newton step $\theta^{(1)} = \theta^{(0)} - \ell'(\theta^{(0)})/\ell''(\theta^{(0)})$ numerically.

Exercise 8.8.13 (Cross-Validation as Hyperparameter Estimation) You fit ridge regression $\widehat{w}_\lambda = (X^\top X + \lambda I)^{-1}X^\top y$ with

$$X = \begin{bmatrix} 1 & 0 \\ 1 & 1 \\ 1 & 2 \\ 1 & 3 \\ 1 & 4 \\ 1 & 5 \end{bmatrix}, \qquad y = \begin{bmatrix} 0 \\ 1 \\ 1 \\ 2 \\ 2 \\ 3 \end{bmatrix}.$$

Use a threefold CV with folds $\{1, 2\}$, $\{3, 4\}$, $\{5, 6\}$ (indices of rows). For each $\lambda \in \{0,\ 0.5,\ 2\}$: train on two folds (4 points), evaluate mean squared error on the held-out fold (2 points), and average across the three held-out folds. Compute the CV score for each λ and select the minimizer. (You must show each fold's fitted $\widehat{w}_\lambda^{(-k)}$ and validation error explicitly.)

Exercise 8.8.14 (Detecting Non-identifiability and Numerical Pathologies) Consider the design matrix

$$X = \begin{bmatrix} 1 & 1 & 2 \\ 1 & 2 & 4 \\ 1 & 3 & 6 \\ 1 & 4 & 8 \end{bmatrix} \in \mathbb{R}^{4\times 3}$$

and response $y = \begin{bmatrix} 1 \\ 2 \\ 3 \\ 4 \end{bmatrix}$.

(i) Compute rank(X) and exhibit a nonzero vector $v \in \mathcal{N}(X)$.
(ii) Show that if w is any solution to least squares, then $w + tv$ yields the same fitted values for all $t \in \mathbb{R}$.
(iii) Compute the minimum-norm least squares solution $w^\dagger = X^\dagger y$ (use pseudoinverse formula or a stable approach), and explain (one paragraph) why regularization would be recommended here.

Exercise 8.8.15 (Bootstrap vs. Bayesian: Uncertainty for a Mean) Suppose you observe $n = 8$ data points

$$z = (1.2,\ 0.7,\ 1.5,\ 0.9,\ 1.1,\ 0.8,\ 1.4,\ 1.0).$$

Assume a Gaussian model with known variance $\sigma^2 = 0.04$ and prior $\mu \sim \mathcal{N}(0, 1)$.

(i) Compute the posterior $\mu \mid z$ and a 95% credible interval for μ.
(ii) Compute the plug-in normal 95% CI for μ using $\bar{z} \pm 1.96\,(\sigma/\sqrt{n})$.
(iii) Perform a *manual bootstrap with* $B = 6$ *resamples* using the following resample index-multisets (each of size 8):

$$(1, 1, 2, 3, 4, 5, 6, 7),\ (2, 2, 3, 4, 5, 6, 7, 8),\ (1, 3, 3, 4, 4, 6, 7, 8),$$

$$(1, 2, 4, 4, 5, 5, 7, 8),\ (1, 2, 2, 2, 6, 6, 7, 8),\ (3, 4, 5, 6, 6, 7, 7, 8),$$

compute the six bootstrap means, their sample standard deviation, and a bootstrap normal CI $\bar{z} \pm 1.96\,\widehat{\text{se}}_{\text{boot}}$.
(iv) Compare the three intervals (one paragraph) and discuss which assumptions differ.

Chapter 9
Hypothesis Testing and Resampling

Hypothesis testing is the classical formalism for making *binary decisions* under uncertainty: accept or reject a claim, ship or do not ship a model change, declare a feature effective or ineffective. In ML, testing appears in model comparison (is model A truly better than model B or is it noise?), online experiments (A/B tests), and diagnostics (is a coefficient significant? is calibration improved?). This chapter develops hypothesis testing as decision theory with controlled error rates, clarifies the meaning of p-values, and motivates resampling methods as practical inference tools when analytic distributions are unavailable.

9.1 Why Hypothesis Testing Appears in ML

9.1.1 Testing as a Decision Problem Under Uncertainty

9.1.1.1 From Scientific Hypotheses to Model Comparison

A hypothesis test formalizes a decision between two competing explanations of the data: a null model H_0 and an alternative H_1. In ML, the "hypotheses" are often about performance gaps, distribution shifts, or parameter constraints: e.g., H_0: the mean improvement in accuracy is 0; H_1: it is positive. Testing is therefore an instance of decision theory: choose an action (reject/not reject) based on a statistic computed from data, while controlling the probability of false discoveries (Type I error) and quantifying detection capability (power). This perspective is essential because ML evaluation is noisy: metrics fluctuate across samples, seeds, and time (Casella and Berger 2002) (cf. Fig. 9.1).

Definition 9.1.1 (Hypotheses and a Test) Let Z denote observed data with distribution in a family $\{P_\theta : \theta \in \Theta\}$. A *null hypothesis* is a subset $\Theta_0 \subseteq \Theta$ and an *alternative* is $\Theta_1 \subseteq \Theta$ (often $\Theta_1 = \Theta \setminus \Theta_0$). A (non-randomized) *test* is a

P. Singh, B. Raman, *Foundations of Machine Learning and AI*, Studies in Big Data 199, https://doi.org/10.1007/978-3-032-30336-3_9

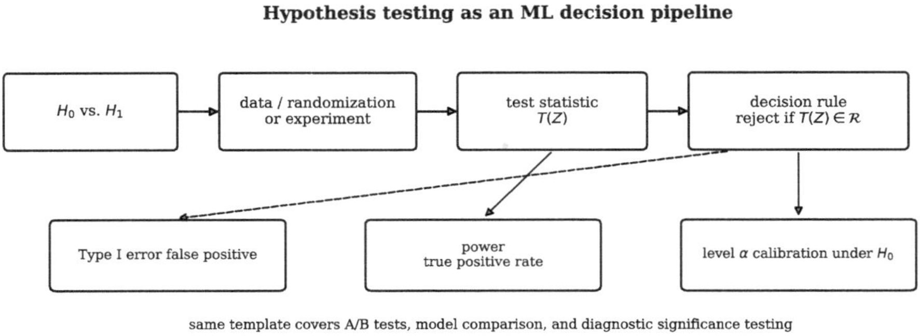

Fig. 9.1 Hypothesis testing as an ML decision pipeline: define null and alternative, randomize or sample data, compute a test statistic, and decide by comparing to a calibrated rejection rule with controlled Type I error

measurable function $\varphi : \mathcal{Z} \to \{0, 1\}$, where $\varphi(Z) = 1$ means "reject H_0" and $\varphi(Z) = 0$ means "do not reject H_0."

Definition 9.1.2 (Type I Error, Size, Type II Error, Power) For a test φ:

- *Type I error* at $\theta \in \Theta_0$ is $\mathbb{P}_\theta(\varphi(Z) = 1)$ (rejecting H_0 when H_0 is true). The *size* of the test is $\sup_{\theta \in \Theta_0} \mathbb{P}_\theta(\varphi(Z) = 1)$.
- *Type II error* at $\theta \in \Theta_1$ is $\mathbb{P}_\theta(\varphi(Z) = 0)$. The *power* at $\theta \in \Theta_1$ is $1 - \mathbb{P}_\theta(\varphi(Z) = 0) = \mathbb{P}_\theta(\varphi(Z) = 1)$.

A test has *significance level* α if its size is at most α.

Remark 9.1.1 (Model Comparison as a Composite Hypothesis Test) In ML, comparing two models often induces a composite null: $H_0 : \Delta \leq 0$ vs. $H_1 : \Delta > 0$ for an unknown performance gap Δ. Composite hypotheses require care: size must be controlled uniformly over all distributions in Θ_0.

9.1.1.2 A/B Testing, Online Experiments, and ML Evaluation

A/B testing is hypothesis testing in production: deploy variant A and B, collect outcomes, and decide whether to ship B. The null is typically "no improvement"; the alternative is "improvement" in a business metric (CTR, retention) or ML metric (conversion uplift). Two issues make ML A/B tests subtle: dependence (users generate multiple interactions) and sequential monitoring (optional stopping). A rigorous test specification forces you to define the unit of analysis, randomization scheme, estimand (mean difference), and stopping rule.

Definition 9.1.3 (A/B Test Estimand (Difference in Means)) Let $Y^{(A)}$ and $Y^{(B)}$ denote outcomes under variants A and B (e.g., per-user conversion). A common estimand is the mean difference

$$\Delta := \mathbb{E}[Y^{(B)}] - \mathbb{E}[Y^{(A)}].$$

The hypotheses are $H_0 : \Delta \leq 0$ vs. $H_1 : \Delta > 0$ (one-sided) or $H_0 : \Delta = 0$ vs. $H_1 : \Delta \neq 0$ (two-sided).

Remark 9.1.2 (Unit of Randomization Matters) If assignment is per-user, the data are approximately i.i.d. at the user level, not at the event level. Treating events as independent inflates the effective sample size and produces overly small p-values.

Remark 9.1.3 (Offline Model Evaluation as Repeated Testing) Repeatedly trying many model variants and reporting the best one is a form of multiple testing/selection. Without correction (or a proper holdout protocol), claimed improvements can be artifacts of noise.

9.2 p-Values, Evidence, and Common Misunderstandings

9.2.1 What a p-Value Is (and Is Not)

9.2.1.1 What a p-Value Is (and Is Not)

A p-value is a property of a *test statistic* under a *specified null distribution*. It measures how extreme the observed statistic is *assuming H_0 is true*. It is not the probability that H_0 is true, and it is not a measure of effect size. In ML, p-values are often used to judge whether an observed improvement could plausibly be due to randomness. To use them correctly, you must specify: the null model, the statistic, the sampling/assignment mechanism, and the tail definition.

Definition 9.2.1 (Test Statistic and p-Value) Let $T(Z)$ be a real-valued test statistic such that larger values are more evidence against H_0. For a simple null $H_0 : \theta = \theta_0$, define the (one-sided) p-value as

$$p(Z) := \mathbb{P}_{\theta_0}\big(T(Z') \geq T(Z)\big),$$

where Z' is an independent replicate drawn from P_{θ_0}. For two-sided tests, one uses an appropriate two-sided extremeness definition (e.g., $|T|$).

Proposition 9.2.1 (Validity/Super-Uniformity Under the Null) *If $p(Z)$ is defined as in Definition 9.2.1 and T has a continuous distribution under H_0, then under H_0 the random variable $p(Z)$ is uniform on* $[0, 1]$. *In general (including discreteness), it is* super-uniform*:*

$$\mathbb{P}_{\theta_0}(p(Z) \leq \alpha) \leq \alpha \qquad \forall \alpha \in [0, 1].$$

Remark 9.2.1 (Operational Meaning) A level-α test that rejects when $p(Z) \leq \alpha$ controls Type I error at α for the null distribution used to compute p.

Remark 9.2.2 (What a p-Value is Not)

- Not $\mathbb{P}(H_0 \text{ true} \mid Z)$ (that is a Bayesian posterior probability and requires a prior).
- Not an effect size measure (a tiny effect can yield tiny p-value with large n).
- Not a guarantee of practical significance (shipping decisions must consider magnitude, cost, and uncertainty).

9.2.1.2 Practical Interpretation Pitfalls in ML Practice

ML workflows create systematic opportunities for invalid p-values: repeated peeking at results, hyperparameter search with selective reporting, non-i.i.d. data (time series, repeated users), and metric hacking (choosing the metric after looking). These issues effectively change the null distribution of the statistic, destroying Type I error control. A rigorous workflow makes the test plan *pre-specified* (metric, sample size or stopping rule, and analysis unit), or uses sequential-testing corrections and multiple-testing control.

Remark 9.2.3 (Common Pitfalls (with the Underlying Failure Mode))

1. **Optional stopping / peeking:** repeatedly checking p-values and stopping when significant inflates Type I error unless using a sequentially valid method.
2. **Multiple comparisons:** testing many metrics, segments, or model variants without correction increases false discoveries (family-wise or FDR inflation).
3. **Non-independence:** correlated samples (same user multiple times, time dependence) reduce effective sample size; naive tests understate uncertainty.
4. **Post-selection inference:** choosing the best model/hyperparameters on the same data and then reporting a p-value as if the model were fixed is invalid.

Remark 9.2.4 (Practitioner-Grade Safeguards)

- Define the primary metric and analysis plan before running the experiment.
- Use a clean holdout (or nested CV) for model selection vs. final reporting.
- Aggregate to the correct i.i.d. unit (e.g., user-level) or use clustered/robust standard errors.
- If monitoring continuously, use sequential tests (alpha-spending, SPRT, always-valid p-values).
- If testing many hypotheses, control FWER (Bonferroni) or FDR (Benjamini–Hochberg) (Benjamini and Hochberg 1995).

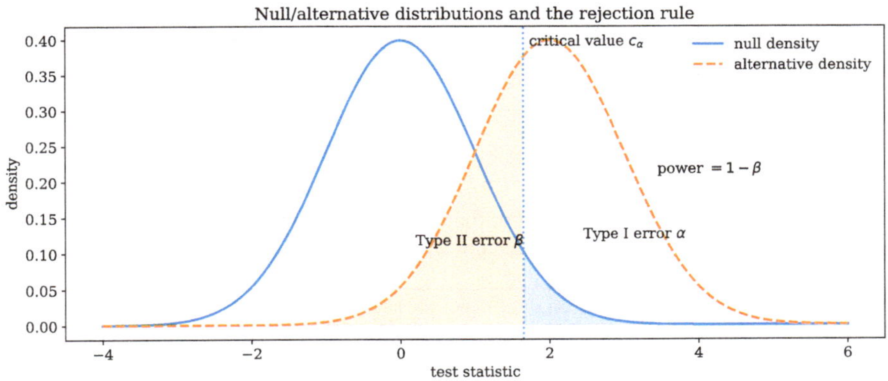

Fig. 9.2 Neyman–Pearson geometry for a one-sided test. The rejection threshold is calibrated under the null to control Type I error at level α; under the alternative, the same threshold determines Type II error and power

9.3 Neyman–Pearson Framework and Classical Tests

The Neyman–Pearson (NP) framework formalizes hypothesis testing as a *decision rule* with controlled error probabilities (cf. Fig. 9.2). The central design choice is the significance level α (Type I error control), after which we seek tests with high power (low Type II error) against meaningful alternatives. Likelihood ratio tests (LRTs) provide a principled and broadly applicable construction, and the classical Z/t/χ^2/F tests are special cases or close relatives under common parametric assumptions.

9.3.1 Testing Pipeline

9.3.1.1 Null/Alternative, Test Statistics, Rejection Regions

A test is specified by (i) a null model class Θ_0, (ii) an alternative class Θ_1, (iii) a statistic that extracts evidence from data, and (iv) a rejection region where we reject H_0. The key principle is *calibration under the null*: choose the rejection region so that false rejections occur with probability at most α for all distributions in Θ_0 (size control). Once α is fixed, the remaining freedom should be used to increase power against relevant alternatives.

Definition 9.3.1 (Hypotheses and Test) Let Z denote observed data with distribution P_θ for $\theta \in \Theta$. A null hypothesis is $\Theta_0 \subseteq \Theta$ and an alternative is $\Theta_1 \subseteq \Theta$ (often $\Theta_1 = \Theta \setminus \Theta_0$). A (non-randomized) test is a measurable map $\varphi : \mathcal{Z} \to \{0, 1\}$ where $\varphi(Z) = 1$ means "reject H_0." The *rejection region* is $\mathcal{R} := \{z : \varphi(z) = 1\}$.

Definition 9.3.2 (Test Statistic and Critical Region) A *test statistic* is a real-valued function $T : \mathcal{Z} \to \mathbb{R}$ such that "large" (or "small") values indicate evidence against H_0. A common test form is $\varphi(z) = \mathbf{1}\{T(z) \geq c\}$ (upper-tail), where c is a critical value chosen to control Type I error.

Remark 9.3.1 (Simple vs. Composite Hypotheses) If $\Theta_0 = \{\theta_0\}$ is a single point, the null is *simple*; otherwise it is *composite*. Composite nulls require uniform Type I control over all $\theta \in \Theta_0$, which is the main technical difficulty in many ML testing problems (model selection, multiple segments, distribution shift).

9.3.1.2 Type I/II Errors, Power, and Effect Sizes

Error control is the contract of hypothesis testing. Type I error is the false positive rate under the null; we cap it at α. Type II error and power depend on the true parameter under the alternative; tests are compared via their *power function*. Effect size separates statistical significance from practical relevance: with large n, tiny effects can become significant, so one should report both a p-value and an effect size (with uncertainty).

Definition 9.3.3 (Size and Power Function) Let φ be a test. Its *Type I error* at $\theta \in \Theta_0$ is $\mathbb{P}_\theta(\varphi(Z) = 1)$. The *size* is

$$\text{size}(\varphi) := \sup_{\theta \in \Theta_0} \mathbb{P}_\theta(\varphi(Z) = 1).$$

The *power function* is $\beta_\varphi(\theta) := \mathbb{P}_\theta(\varphi(Z) = 1)$ for $\theta \in \Theta$. For $\theta \in \Theta_1$, $1 - \beta_\varphi(\theta)$ is the Type II error.

Remark 9.3.2 (Effect Size vs. p-Value) A p-value measures extremeness under H_0; it does not measure the magnitude of the effect. In ML reporting, include a point estimate of the effect (e.g., mean difference) plus a standard error/interval, and interpret significance in the context of practical thresholds.

Example 9.3.1 (Standardized Mean Difference (Cohen's d, Two-Sample)) If group means are $\bar{x}_A, \bar{x}_B$ with sample variances s_A^2, s_B^2 and sizes n_A, n_B, a common standardized effect size is

$$d := \frac{\bar{x}_B - \bar{x}_A}{s_p}, \qquad s_p^2 := \frac{(n_A - 1)s_A^2 + (n_B - 1)s_B^2}{n_A + n_B - 2},$$

when equal-variance modeling is plausible. Under heteroscedasticity, one may use alternatives (e.g., Glass's Δ, or report the raw difference).

9.3.2 *Likelihood Ratio Tests (LRT)*

9.3.2.1 Neyman–Pearson Lemma for Simple Hypotheses (Conceptual)

When both H_0 and H_1 are simple, there is an optimal test in a precise sense: for a fixed size α, the most powerful test rejects for large likelihood ratio $p_1(z)/p_0(z)$. This is the Neyman–Pearson lemma, the foundational result that justifies likelihood ratio statistics. In ML terms, it says: under a fully specified null and alternative, the likelihood ratio is the best evidence aggregator.

Theorem 9.3.1 (Neyman–Pearson Lemma) *Let $H_0 : \theta = \theta_0$ and $H_1 : \theta = \theta_1$ be simple hypotheses with densities/pmfs $p_0(z)$ and $p_1(z)$. Fix $\alpha \in (0, 1)$. Among all tests φ with $\mathbb{P}_{\theta_0}(\varphi(Z) = 1) \le \alpha$, a most powerful test is of the form*

$$\varphi^\star(z) = \mathbf{1}\left\{\frac{p_1(z)}{p_0(z)} \ge c\right\},$$

with threshold c chosen so that $\mathbb{P}_{\theta_0}\left(\frac{p_1(Z)}{p_0(Z)} \ge c\right) = \alpha$ (or $\le \alpha$ with randomization on the boundary when needed).

Remark 9.3.3 (Likelihood Ratio as Sufficient Evidence Aggregator) The NP lemma formalizes that, for simple hypotheses, no other statistic can uniformly outperform the likelihood ratio in power at fixed size. This is why LRTs are the canonical starting point for test design (Lehmann and Romano 2005).

9.3.2.2 Generalized Likelihood Ratio Tests and Asymptotics

In practice, hypotheses are composite (unknown nuisance parameters). The generalized likelihood ratio test (GLRT) compares the best likelihood under the null to the best likelihood overall. Under regularity conditions and large samples, the distribution of $-2 \log \Lambda$ is approximately χ^2 with degrees of freedom equal to the difference in parameter dimensions (Wilks' phenomenon). This yields a broadly reusable asymptotic testing recipe for nested models, widely used in ML for model comparison and diagnostics.

Definition 9.3.4 (Generalized Likelihood Ratio) Let $L(\theta; z)$ be the likelihood. Define

$$\Lambda(z) := \frac{\sup_{\theta \in \Theta_0} L(\theta; z)}{\sup_{\theta \in \Theta} L(\theta; z)} \in [0, 1].$$

A GLRT rejects H_0 for small $\Lambda(z)$ (equivalently, large $-2 \log \Lambda(z)$).

Remark 9.3.4 (Nested-Model Form) If $\Theta_0 \subseteq \Theta$ and $\widehat{\theta_0} \in \arg\max_{\Theta_0} \ell(\theta)$, $\widehat{\theta} \in \arg\max_{\Theta} \ell(\theta)$, then

$$-2\log \Lambda = 2\big(\ell(\widehat{\theta}) - \ell(\widehat{\theta_0})\big),$$

i.e., twice the log-likelihood gap between the unrestricted and restricted fits.

Theorem 9.3.2 (Wilks' Theorem (Asymptotic, Preview Statement)) *Assume regularity conditions and that the null is true with θ_0 in the interior of Θ_0. If Θ_0 is defined by k independent smooth constraints (equivalently,* $\dim(\Theta) - \dim(\Theta_0) = k$*), then*

$$-2\log \Lambda(Z_{1:n}) \Rightarrow \chi_k^2 \quad \text{as } n \to \infty.$$

Remark 9.3.5 (When Wilks Can Mislead) Wilks can fail under boundary constraints, non-identifiability, mixture models (label switching), or small samples. In such cases, resampling/permutation/bootstrap methods often provide more reliable calibration.

9.3.3 Canonical Tests

9.3.3.1 Z/t/χ^2/F Tests: When Each Applies

Classical tests are specialized LRT-like procedures under normal or large-sample approximations. Z-tests assume known variance (or rely on asymptotic normality); t-tests replace unknown variance by a sample estimate; χ^2 tests arise from squared-normal structures (variance tests, goodness-of-fit); F tests arise from ratios of quadratic forms (variance ratios, nested linear-model comparisons). In ML practice, these appear in A/B testing, coefficient testing in regression, and ANOVA-style comparisons.

Definition 9.3.5 (One-Sample Z Test (Known Variance)) Let $X_1, \ldots, X_n \overset{iid}{\sim} \mathcal{N}(\mu, \sigma^2)$ with *known* σ^2. Test $H_0 : \mu = \mu_0$ vs. $H_1 : \mu \neq \mu_0$ using

$$Z := \frac{\bar{X}_n - \mu_0}{\sigma/\sqrt{n}}.$$

Reject for $|Z| \geq z_{1-\alpha/2}$ (two-sided), where $z_{1-\alpha/2}$ is a standard normal quantile.

Definition 9.3.6 (One-Sample t Test (Unknown Variance)) Let $X_1, \ldots, X_n \overset{iid}{\sim} \mathcal{N}(\mu, \sigma^2)$ with unknown σ^2. Let $s^2 = \frac{1}{n-1}\sum_{i=1}^{n}(X_i - \bar{X}_n)^2$. Test $H_0 : \mu = \mu_0$ using

$$T := \frac{\bar{X}_n - \mu_0}{s/\sqrt{n}},$$

which follows a t_{n-1} distribution under H_0 (exactly in the Gaussian model).

Definition 9.3.7 (χ^2 Variance Test (Gaussian)) Let $X_1, \ldots, X_n \overset{iid}{\sim} \mathcal{N}(\mu, \sigma^2)$. Under $H_0 : \sigma^2 = \sigma_0^2$, the statistic

$$Q := \frac{(n-1)s^2}{\sigma_0^2}$$

has a χ^2_{n-1} distribution. This yields two-sided variance tests by comparing Q to χ^2 quantiles.

Definition 9.3.8 (F Test: Ratio of Variances (Gaussian)) Let $X_1, \ldots, X_{n_1} \sim \mathcal{N}(\mu_1, \sigma_1^2)$ and $Y_1, \ldots, Y_{n_2} \sim \mathcal{N}(\mu_2, \sigma_2^2)$ independent. Under $H_0 : \sigma_1^2 = \sigma_2^2$, the ratio

$$F := \frac{s_X^2}{s_Y^2}$$

has an $F_{n_1-1,\, n_2-1}$ distribution, where s_X^2, s_Y^2 are sample variances.

Remark 9.3.6 (F Test for Nested Linear Models (ANOVA Form)) In linear regression with Gaussian errors, comparing a restricted model (fewer parameters) to a full model yields an F statistic based on the reduction in residual sum of squares (RSS) per added degree of freedom. This is an LRT/Wilks-type consequence specialized to Gaussian linear models.

9.3.3.2 Assumptions, Robustness, and Diagnostics

Classical tests come with assumptions (normality, independence, equal variances, correct model specification). Violations can invalidate p-values and inflate Type I error. Robust alternatives (Welch's t-test, permutation tests, bootstrap CIs, sandwich standard errors) mitigate these issues. Diagnostics are therefore part of responsible testing: check distributional shape, dependence, and heteroscedasticity; use transformations or robust methods when needed.

Remark 9.3.7 (Assumption Checklist) Before applying Z/t/χ^2/F tests, check:

- **i.i.d. unit:** are observations independent at the analysis unit (e.g., user-level in A/B tests)?
- **Distributional shape:** is a Gaussian approximation plausible (via CLT or residual diagnostics)?
- **Homoscedasticity:** are variances comparable across groups (especially for two-sample t-tests)?
- **Model fit:** for regression-based tests, are residuals roughly unstructured (no obvious nonlinearity/heteroscedasticity)?

Remark 9.3.8 (Robust Replacements (Common in ML))

- **Welch's t-test:** replaces equal-variance t-test when variances differ.
- **Permutation tests:** provide exact/approximate null calibration under exchangeability.
- **Bootstrap:** provides uncertainty for complex statistics without closed-form sampling distributions.
- **Robust/sandwich SEs:** mitigate heteroscedasticity in regression-based inference.

These tools are especially relevant in ML where metrics can be non-normal and data can be dependent.

9.4 Confidence Intervals and Duality with Tests

Confidence intervals (CIs) are the estimation counterpart of hypothesis tests: tests answer "reject or not?" while intervals answer "which parameter values remain plausible?" A fundamental duality connects them: many CIs can be obtained by *inverting* a family of level-α tests. This section formalizes coverage, clarifies interpretation, contrasts common CI constructions (Wald/score/likelihood), and previews bootstrap intervals (percentile, basic, and BCa intuition).

9.4.1 Confidence Intervals as Inverted Tests

9.4.1.1 Coverage and Interpretation

A CI is a random set computed from data; its defining property is frequentist coverage under repeated sampling: for each fixed true parameter, the CI contains that parameter with probability at least $1 - \alpha$. This is a statement about the randomness of the data, not about probability of the parameter. The inversion view clarifies why a CI is meaningful: it collects all null values that would not be rejected by a test at level α. In ML evaluation, CIs quantify uncertainty of metrics (accuracy uplift, loss reduction) and help distinguish practical vs. noise-level improvements.

Definition 9.4.1 (Confidence Interval and Coverage) Let $\{P_\theta : \theta \in \Theta\}$ be a model and let Z denote data. A (two-sided) $(1 - \alpha)$ *confidence interval* for θ is a measurable map $C : \mathcal{Z} \to 2^\Theta$ such that for all $\theta \in \Theta$,

$$\mathbb{P}_\theta\big(\theta \in C(Z)\big) \geq 1 - \alpha.$$

If equality holds for all θ, the interval has exact coverage.

Remark 9.4.1 (Interpretation Pitfall) The statement $\mathbb{P}_\theta(\theta \in C(Z)) \geq 1-\alpha$ does *not* mean $\mathbb{P}(\theta \in C(Z) \mid Z) \geq 1 - \alpha$. The latter is Bayesian and requires a prior.

Theorem 9.4.1 (Inversion Duality (Tests $\leftrightarrow$ CIs)) *Suppose for each $\theta_0 \in \Theta$ we have a level-α test $\varphi_{\theta_0}(Z) \in \{0, 1\}$ of $H_0 : \theta = \theta_0$ vs. $H_1 : \theta \neq \theta_0$ (or another alternative), satisfying*

$$\sup_{\theta:\theta=\theta_0} \mathbb{P}_\theta\big(\varphi_{\theta_0}(Z) = 1\big) \leq \alpha.$$

Define the set-valued estimator

$$C(Z) := \{\theta_0 \in \Theta : \varphi_{\theta_0}(Z) = 0\} \quad \textit{(do not reject } \theta_0\textit{)}.$$

Then $C(Z)$ is a $(1 - \alpha)$ confidence set for θ.

Proof Fix true θ. Then

$$\mathbb{P}_\theta(\theta \notin C(Z)) = \mathbb{P}_\theta(\varphi_\theta(Z) = 1) \leq \alpha,$$

so $\mathbb{P}_\theta(\theta \in C(Z)) \geq 1 - \alpha$. □

9.4.1.2 Wald vs. Score vs. Likelihood intervals (Conceptual)

There are three canonical CI constructions in likelihood-based models. Wald intervals use asymptotic normality of an estimator and an estimated standard error (fast, but can misbehave near boundaries or under skewness). Score intervals invert a score test and often have better finite-sample behavior. Likelihood (or likelihood-ratio) intervals invert the LRT and are invariant to reparameterization; they often behave well in small samples but require solving an optimization/root-finding problem. Understanding these options is essential in ML when probabilities lie near $0/1$, counts are small, or models are constrained.

Remark 9.4.2 (Wald Interval (Template)) If $\widehat{\theta}$ is approximately normal with standard error $\widehat{\mathrm{se}}(\widehat{\theta})$, a two-sided Wald interval is

$$\widehat{\theta} \pm z_{1-\alpha/2}\, \widehat{\mathrm{se}}(\widehat{\theta}),$$

where $z_{1-\alpha/2}$ is the standard normal quantile. Wald is simple but may under-cover for skewed distributions or boundary parameters.

Remark 9.4.3 (Score Interval (Template)) Score intervals invert a score test: they are the set of θ_0 values for which the score statistic is not too extreme. For binomial proportions, score intervals (Wilson) often outperform Wald near 0 or 1.

Remark 9.4.4 (Likelihood-Ratio Interval (Template)) Likelihood intervals invert the LRT:

$$C_{\mathrm{LR}}(Z) = \left\{\theta : 2(\ell(\widehat{\theta}) - \ell(\theta)) \le \chi^2_{1,\,1-\alpha}\right\}.$$

This is invariant under smooth reparameterizations and often has good coverage properties.

9.4.2 Bootstrap Confidence Intervals

9.4.2.1 Percentile and Basic Intervals

Bootstrap constructs CIs by approximating the sampling distribution of an estimator via resampling from the empirical distribution (cf. Fig. 9.3). Percentile intervals take quantiles of the bootstrap estimator distribution; basic intervals correct bias by centering around the original estimate. These methods are widely used in ML for metrics and estimators whose analytic sampling distributions are messy. Their validity depends on i.i.d. sampling at the correct unit and on the bootstrap being a good approximation for the estimator at hand.

Definition 9.4.2 (Bootstrap Resamples and Bootstrap Replicates) Given data $\mathcal{D} = \{z_1, \ldots, z_n\}$ and an estimator $\widehat{\theta} = \widehat{\theta}(\mathcal{D})$, bootstrap draws $\mathcal{D}^{(b)}$ by sampling n

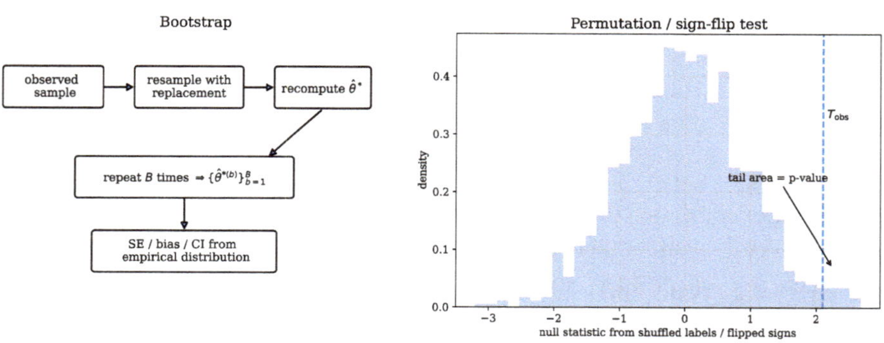

Fig. 9.3 Two resampling workflows. Left: bootstrap resamples the independent units with replacement to estimate variability or confidence intervals. Right: permutation or sign-flip tests generate a null distribution by exploiting exchangeability under the null

points with replacement from $\mathcal{D}$, and computes

$$\widehat{\theta}^{(b)} := \widehat{\theta}(\mathcal{D}^{(b)}), \qquad b = 1, \ldots, B.$$

Let $\widehat{F}^*$ denote the empirical CDF of $\{\widehat{\theta}^{(b)}\}_{b=1}^{B}$.

Definition 9.4.3 (Percentile Bootstrap CI) A two-sided $(1-\alpha)$ percentile interval is

$$C_{\text{perc}} = \left[\, (\widehat{F}^*)^{-1}(\alpha/2),\; (\widehat{F}^*)^{-1}(1-\alpha/2) \,\right],$$

i.e., the $(\alpha/2)$ and $(1-\alpha/2)$ quantiles of the bootstrap replicates.

Definition 9.4.4 (Basic Bootstrap CI) Let q_a denote the a-quantile of the bootstrap replicates $\{\widehat{\theta}^{(b)}\}$. A two-sided basic interval is

$$C_{\text{basic}} = \left[\, 2\widehat{\theta} - q_{1-\alpha/2},\; 2\widehat{\theta} - q_{\alpha/2} \,\right].$$

This reflects the bootstrap distribution around $\widehat{\theta}$ to correct for bias.

Remark 9.4.5 (When Percentile Can Misbehave) Percentile intervals can have poor coverage when the estimator is biased or when the sampling distribution is skewed. Basic and BCa attempt to address these issues.

9.4.2.2 BCa Intuition (Conceptual)

BCa (bias-corrected and accelerated) intervals adjust percentile cutoffs using two parameters: a bias correction z_0 (how far the estimator is from the median of its bootstrap distribution) and an acceleration a (how the standard error changes with the parameter, related to skewness via jackknife influence). BCa often improves coverage for skewed distributions and nonlinear estimators, at the cost of additional computation. In ML, BCa can be valuable for metrics with skewed sampling distributions (AUC, F1, rare-event rates).

Remark 9.4.6 (BCa (Conceptual Form)) BCa replaces the nominal quantiles $\alpha/2$ and $1-\alpha/2$ by adjusted levels

$$\alpha_1 = \Phi\!\left(z_0 + \frac{z_0 + z_{\alpha/2}}{1 - a(z_0 + z_{\alpha/2})}\right), \qquad \alpha_2 = \Phi\!\left(z_0 + \frac{z_0 + z_{1-\alpha/2}}{1 - a(z_0 + z_{1-\alpha/2})}\right),$$

and then uses the bootstrap quantiles at α_1, α_2. Here $z_u = \Phi^{-1}(u)$. We will not derive BCa fully here; the key message is that it corrects both bias and skewness/curvature effects.

Remark 9.4.7 (Practical Caution) Bootstrap assumes resampling approximates the sampling distribution. If data are dependent (time series, clustered users), one must use block/bootstrap-at-unit methods; naive bootstrap can severely understate uncertainty.

9.5 Multiple Testing

Multiple testing is unavoidable in modern ML: we screen many features, compare many models/hyperparameters, monitor many metrics, and often iterate repeatedly until something "works." Each test carries a chance of a false positive; with many tests, false positives become likely unless we correct. This section formalizes error-rate control (FWER and FDR), presents the Bonferroni family and the Benjamini–Hochberg (BH) procedure, and discusses dependence and reporting via q-values.

9.5.1 Why Multiple Testing Is Unavoidable in ML

9.5.1.1 Feature Screening, Hyperparameter Sweeps, and Leaderboard Effects

In ML workflows, we routinely run dozens to thousands of comparisons: feature selection (each feature tested for association), hyperparameter sweeps (each configuration implicitly tested), ablation studies, and repeated evaluation on public leaderboards. Even if each p-value is valid marginally, the probability of at least one false discovery grows quickly with the number of tests. This is not a niche statistical issue: it is the mathematical explanation for why repeated tuning on a fixed benchmark produces illusory progress and why proper holdout protocols are essential.

Remark 9.5.1 (A Simple Back-of-the-Envelope) If m independent tests are performed at level α, then the probability of at least one false positive is $1-(1-\alpha)^m \approx m\alpha$ for small α. For $m = 100$ and $\alpha = 0.05$, this is about 0.994. Thus "some significant result" is almost guaranteed without correction.

9.5.1.2 Selective Reporting and "p-Hacking" Analogues

P-hacking in ML appears as selective reporting: choosing the best run, best seed, best metric, best subset, best stopping time, or best hyperparameter setting, and then reporting a p-value as if the choice were fixed in advance. This invalidates nominal Type I error because selection changes the null distribution. The correct response is procedural (pre-registration, clean holdouts) and statistical (multiple-testing corrections, selective inference tools).

Remark 9.5.2 (Selection Changes the Null) If you choose the maximum of many noisy improvements and then test it as if it were a single pre-specified hypothesis, you implicitly test a *different* null (the null distribution of the maximum), which has much heavier tails.

9.5.2 Error-Rate Control

9.5.2.1 FWER and Bonferroni Family

FWER (family-wise error rate) controls the probability of making *any* false rejection among a family of hypotheses. It is stringent and appropriate when even one false discovery is costly (safety-critical claims, high-stakes scientific findings). Bonferroni is the simplest FWER control method: test each hypothesis at level α/m. Its strength is universality (works under arbitrary dependence); its weakness is conservativeness when m is large.

Definition 9.5.1 (FWER) Consider m null hypotheses $H_{0,1}, \ldots, H_{0,m}$ and a multiple-testing procedure that outputs rejections. Let V be the number of false rejections (true nulls rejected). The *family-wise error rate* is

$$\text{FWER} := \mathbb{P}(V \geq 1).$$

Theorem 9.5.1 (Bonferroni Correction) *Let $p_1, \ldots, p_m$ be valid p-values for hypotheses $H_{0,1}, \ldots, H_{0,m}$ (each super-uniform under its null). Reject $H_{0,i}$ if $p_i \leq \alpha/m$. Then FWER $\leq \alpha$ (under arbitrary dependence).*

Proof Let $E_i = \{p_i \leq \alpha/m\}$ for true nulls. Then $\mathbb{P}(E_i) \leq \alpha/m$. By union bound,

$$\text{FWER} = \mathbb{P}\Big(\bigcup_{\text{true } H_{0,i}} E_i \Big) \leq \sum_{\text{true } H_{0,i}} \mathbb{P}(E_i) \leq m \cdot \frac{\alpha}{m} = \alpha.$$

□

Remark 9.5.3 (Holm Step-Down (Bonferroni Improvement)) Holm's procedure is a uniformly more powerful step-down variant that still controls FWER under arbitrary dependence: sort p-values and compare to progressively less stringent thresholds $\alpha/(m - k + 1)$.

9.5.2.2 FDR and Benjamini–Hochberg Procedure

When m is large (feature screening, genomics-like ML settings), controlling FWER is too strict. FDR (false discovery rate) controls the *expected proportion* of false rejections among all rejections. This is often the right balance: allow some false positives but keep their proportion small. The Benjamini–Hochberg (BH) procedure is the canonical FDR controller and is widely used due to simplicity and good power.

Definition 9.5.2 (FDR) Let R be the total number of rejections and V the number of false rejections. Define the false discovery proportion (FDP) as

$$\mathrm{FDP} := \begin{cases} \frac{V}{R}, & R > 0, \\ 0, & R = 0. \end{cases}$$

The *false discovery rate* is $\mathrm{FDR} := \mathbb{E}[\mathrm{FDP}]$.

Theorem 9.5.2 (Benjamini–Hochberg (BH) Procedure) *Given p-values $p_1, \ldots, p_m$, sort them as $p_{(1)} \leq \cdots \leq p_{(m)}$. Let*

$$k^\star := \max\left\{k \in [m] : \; p_{(k)} \leq \frac{k}{m} q\right\},$$

with the convention that $k^\star = 0$ if the set is empty. Reject all hypotheses with p-values $p_i \leq p_{(k^\star)}$. Under independence (and under certain positive dependence conditions), BH controls FDR at level q:

$$FDR \leq q.$$

Remark 9.5.4 (What BH Does Geometrically) BH sets a data-dependent cutoff that becomes less stringent as k grows: larger families of rejections require stronger evidence only up to the line $p_{(k)} \leq (k/m)q$. This adaptivity is why BH has much higher power than Bonferroni when many signals exist.

9.5.3 Dependence and Practical Corrections

9.5.3.1 Positive Dependence and When BH Still Works (Conceptual)

In ML, tests are rarely independent: features are correlated, model variants share data and initialization, and metrics co-move. Dependence can invalidate naive FDR control. However, BH remains valid under a broad class of positive dependence conditions (often summarized as PRDS-type conditions). When dependence is arbitrary or unknown, one can use more conservative corrections (Benjamini–Yekutieli) or rely on resampling/permutation-based calibration.

Remark 9.5.5 (Benjamini–Yekutieli (BY) Correction (Dependence-Safe)) A dependence-robust variant replaces q by q/c_m where $c_m = \sum_{i=1}^{m} \frac{1}{i} \approx \log m + \gamma$. Equivalently, BH thresholds become

$$p_{(k)} \leq \frac{k}{m} \frac{q}{c_m}.$$

This controls FDR under arbitrary dependence but can be conservative for large m.

Remark 9.5.6 (Practical ML Guidance) If tests are strongly correlated (e.g., overlapping hyperparameter sweeps), it is often better to: (i) use a clean holdout for final claims, (ii) treat the sweep as model selection (not as multiple independent tests), and (iii) report uncertainty via bootstrap or repeated-split evaluation rather than relying solely on p-values.

9.5.3.2 q-Values and Reporting Standards

A q-value is the FDR analogue of a p-value: it assigns to each hypothesis the minimum FDR level at which it would be called significant. Reporting q-values (rather than only p-values) communicates multiple-testing adjustment directly. In ML papers, good reporting includes: the family of tests, the correction method (Bonferroni/Holm/BH/BY), the chosen level, and the number of tests. Without this, significance claims are not interpretable.

Definition 9.5.3 (q-Value (BH-Style Interpretation)) Given a multiple-testing procedure controlling FDR, the *q-value* of hypothesis i is

$$q_i := \inf\{q \in (0, 1] : H_{0,i} \text{ would be rejected when controlling FDR at level } q\}.$$

Thus, q_i is the smallest FDR target level under which hypothesis i becomes significant.

Remark 9.5.7 (Reporting Standards (Minimal Checklist)) When reporting multiple-testing results, include:

- number of hypotheses m and how they were generated (feature screening, sweep size, etc.),
- which error rate was controlled (FWER or FDR) and at what level (α or q),
- correction method (Bonferroni/Holm/BH/BY) and any dependence assumptions,
- both effect sizes and adjusted significance (q-values or adjusted p-values), not only "significant/non-significant."

9.6 Resampling Methods

When analytic sampling distributions are unavailable or unreliable, resampling provides practical inference. Bootstrap approximates the sampling distribution of an estimator by resampling from the empirical distribution. Permutation (randomization) tests generate an exact or near-exact null distribution under exchangeability. Jackknife and cross-validation are closely related to resampling: they probe sensitivity to individual data points and provide nearly unbiased risk estimates in certain regimes. This section develops the core mechanics and the assumptions that make resampling valid.

9.6.1 *Bootstrap*

9.6.1.1 Bootstrap Distributions and Standard Errors

The bootstrap constructs an empirical approximation to the sampling distribution of a statistic $\widehat{\theta} = T(\mathcal{D})$. By repeatedly resampling the dataset (with replacement) and recomputing the statistic, we obtain bootstrap replicates that mimic the variability of $\widehat{\theta}$ under repeated sampling. The bootstrap standard error is the sample standard deviation of these replicates. In ML, bootstrap is widely used for uncertainty of metrics (AUC, accuracy, log-loss), for model parameter uncertainty (when refitting is feasible), and for calibration curves.

Definition 9.6.1 (Bootstrap Replicates) Let $\mathcal{D} = \{z_1, \ldots, z_n\}$ be an i.i.d. sample and let $\widehat{\theta} = T(\mathcal{D})$ be a statistic. For $b = 1, \ldots, B$, draw a bootstrap dataset $\mathcal{D}^{(b)}$ by sampling n points with replacement from $\mathcal{D}$, and define

$$\widehat{\theta}^{(b)} := T(\mathcal{D}^{(b)}).$$

The empirical distribution of $\{\widehat{\theta}^{(b)}\}_{b=1}^{B}$ is the *bootstrap distribution.*

Definition 9.6.2 (Bootstrap Standard Error) The bootstrap estimate of standard error is

$$\widehat{\text{se}}_{\text{boot}} := \sqrt{\frac{1}{B-1}\sum_{b=1}^{B}\left(\widehat{\theta}^{(b)} - \overline{\widehat{\theta}}^{*}\right)^2}, \qquad \overline{\widehat{\theta}}^{*} := \frac{1}{B}\sum_{b=1}^{B}\widehat{\theta}^{(b)}.$$

Remark 9.6.1 (I.i.d. at the Correct Unit) Bootstrap validity depends on resampling the correct independent units (users, sessions, sequences). If observations are clustered or time-dependent, naive bootstrap underestimates uncertainty; one must use block/bootstrap-at-cluster variants.

9.6.1.2 Bias Correction and Variance Estimation

Bootstrap can estimate not only variance but also bias: compare the mean of bootstrap replicates to the original estimate. This yields a bias estimate and a bias-corrected estimator. Bootstrap also supports multiple CI constructions (percentile, basic, BCa). In ML pipelines, bias correction is relevant when the estimator is nonlinear (e.g., AUC, F1, entropy) and when n is modest, where plug-in estimates can be systematically biased.

Definition 9.6.3 (Bootstrap Bias Estimate and Correction) The bootstrap bias estimate is

$$\widehat{\text{bias}}_{\text{boot}} := \overline{\widehat{\theta}}^{*} - \widehat{\theta}.$$

A simple bias-corrected estimator is

$$\widehat{\theta}_{\mathrm{bc}} := \widehat{\theta} - \widehat{\mathrm{bias}}_{\mathrm{boot}} = 2\widehat{\theta} - \overline{\widehat{\theta}}^{*}.$$

Remark 9.6.2 (Variance Estimation) The bootstrap variance estimate is $(\widehat{\mathrm{se}}_{\mathrm{boot}})^2$. For smooth statistics, this often approximates the true sampling variance well; for nonsmooth statistics (medians, maxima), special care or alternative methods may be needed.

9.6.2 *Permutation Tests*

9.6.2.1 Exchangeability Assumptions

Permutation tests generate a null distribution by rearranging labels or assignments in ways that should not matter under H_0. The key assumption is exchangeability: under the null, the joint distribution is invariant to the permutation applied. This yields exact finite-sample Type I control without normality assumptions, making permutation tests attractive in ML evaluation. However, exchangeability can fail under dependence, covariate shift, or unequal variances; in such cases, the permutation null is wrong.

Definition 9.6.4 (Exchangeability) A sequence $(Z_1, \ldots, Z_n)$ is *exchangeable* if for every permutation π of $[n]$,

$$(Z_1, \ldots, Z_n) \overset{d}{=} (Z_{\pi(1)}, \ldots, Z_{\pi(n)}).$$

Permutation tests rely on an exchangeability statement under H_0 for the transformed data used in the test.

Remark 9.6.3 (Randomization in A/B Tests) In a properly randomized A/B test with fixed outcomes and random assignment, exchangeability of assignments under H_0 is justified by design, which makes randomization tests particularly compelling for online experiments.

9.6.2.2 Randomization Tests for Model Comparison

In ML, model comparison often uses paired evaluations: for each data point (or fold/seed), compute a difference in metric between two models. Under the null of no performance difference, the signs of paired differences can be randomized (sign-flip test) or labels can be permuted, yielding a null distribution for the mean difference (or another statistic). This avoids reliance on t-test normality, especially when differences are skewed. The test is exact when the randomization symmetry holds.

Definition 9.6.5 (Permutation Test (Generic)) Let $T(Z)$ be a test statistic and let $\mathcal{G}$ be a set of transformations (permutations/sign flips) such that under H_0,

$$Z \overset{d}{=} g(Z) \quad \forall g \in \mathcal{G}.$$

Given observed z, define the permutation p-value

$$p_{\text{perm}}(z) := \frac{1}{|\mathcal{G}|} \sum_{g \in \mathcal{G}} \mathbf{1}\{T(g(z)) \geq T(z)\}$$

(for an upper-tail test; modify appropriately for two-sided). In practice, approximate by sampling g uniformly at random when $|\mathcal{G}|$ is large.

Example 9.6.1 (Paired Sign-Flip Test for Two Models) Let paired differences be $d_i = m_i^{(B)} - m_i^{(A)}$ for $i \in [n]$ (e.g., per-fold metric differences). Under the null that the distribution of d_i is symmetric about 0, sign flips $d_i \mapsto s_i d_i$ with $s_i \in \{-1, +1\}$ preserve the joint distribution. Using statistic $T(d) = \sqrt{n}\, \bar{d}/s_d$ (or simply $\bar{d}$), one obtains an exact/approximate p-value by random sign flips.

Remark 9.6.4 (What is Being Tested) Permutation tests typically test a null of exchangeability (no association) rather than a specific parametric form. Thus they are robust, but only when the invariance group truly reflects the null.

9.6.3 Jackknife and Cross-Validation Connections

9.6.3.1 Leave-One-Out Intuition

The jackknife probes estimator sensitivity by systematically leaving out each observation. It yields a simple estimate of bias and variance for smooth statistics and connects to influence functions. Leave-one-out cross-validation (LOOCV) uses the same idea for predictive performance: train on $n - 1$ points, test on the held-out point, repeat. The conceptual link is that both are "leave-one-out" perturbations of the dataset.

Definition 9.6.6 (Jackknife Replicates) Let $\widehat{\theta} = T(\mathcal{D})$ with $\mathcal{D} = \{z_1, \ldots, z_n\}$. Define the leave-one-out dataset $\mathcal{D}_{(-i)} = \mathcal{D} \setminus \{z_i\}$ and jackknife replicate

$$\widehat{\theta}_{(-i)} := T(\mathcal{D}_{(-i)}), \qquad i \in [n].$$

Definition 9.6.7 (Jackknife Variance Estimate (Classical Form)) Define $\overline{\widehat{\theta}}_{(-\cdot)} = \frac{1}{n} \sum_{i=1}^{n} \widehat{\theta}_{(-i)}$. The jackknife variance estimate is

$$\widehat{\text{Var}}_{\text{jack}}(\widehat{\theta}) := \frac{n-1}{n} \sum_{i=1}^{n} \left(\widehat{\theta}_{(-i)} - \overline{\widehat{\theta}}_{(-\cdot)}\right)^2.$$

Remark 9.6.5 (LOOCV Risk Estimate (Connection)) If $\widehat{h}_{(-i)}$ is trained on $\mathcal{D}_{(-i)}$, then LOOCV estimates risk by

$$\widehat{R}_{\mathrm{LOO}} := \frac{1}{n} \sum_{i=1}^{n} \ell\big(\widehat{h}_{(-i)}(x_i), y_i\big).$$

This shares the same leave-one-out structure as jackknife replicates.

9.6.3.2 Influence and Sensitivity Viewpoint (Conceptual)

Jackknife is a discrete approximation to an influence function: how much does the estimator change when we perturb the data distribution at one point? In ML, influence/sensitivity analyses diagnose outliers, dataset poisoning, and instability. For smooth M-estimators, one can approximate leave-one-out updates using a single Hessian inverse (Newton step), linking influence to curvature. This is the same geometry: curvature controls stability.

Remark 9.6.6 (Influence-Function Intuition) If $\widehat{\theta}$ solves an estimating equation $\sum_{i=1}^{n} \psi(z_i, \theta) = 0$ and the Jacobian is well-conditioned, then removing one point changes $\widehat{\theta}$ by approximately

$$\widehat{\theta}_{(-i)} - \widehat{\theta} \approx \frac{1}{n} H^{-1} \psi(z_i, \widehat{\theta}),$$

where H is a Hessian/Jacobian-like matrix. Thus, large influence corresponds to large ψ values and/or ill-conditioned curvature.

9.7 Model Comparison and Evaluation Protocols

Model comparison in ML is hypothesis testing under repeated evaluation and selection. A rigorous protocol must respect: (i) the *unit of analysis* (paired comparisons on the same examples/users), (ii) the *metric* being compared (accuracy, AUC, log-loss), and (iii) the *dependence structure* (time series, clustered users). This section presents principled paired-testing templates, permutation tests for performance gaps, metric-aware caveats, and dependence-aware evaluation via blocked resampling and careful backtesting.

9.7.1 Comparing Predictors

9.7.1.1 Paired Tests on Losses (Paired t-Test, Wilcoxon; Conceptual)

Most comparisons should be *paired*: evaluate both models on the same test instances and compare per-instance losses. This removes nuisance variability from sampling difficulty and isolates the model difference. A paired t-test assumes approximate normality of the mean loss difference (often justified by CLT when n is large), while the Wilcoxon signed-rank test is a nonparametric alternative relying on symmetry assumptions and ranks. In ML papers, paired tests are appropriate when you have per-instance (or per-fold) differences and want to test whether the mean (or median) difference is nonzero.

Definition 9.7.1 (Paired Loss Differences) Given test examples $\{z_i\}_{i=1}^n$ and two predictors h_A, h_B, define per-example losses $\ell_{A,i} = \ell(h_A; z_i)$ and $\ell_{B,i} = \ell(h_B; z_i)$ and the paired differences

$$d_i := \ell_{B,i} - \ell_{A,i}, \qquad i \in [n].$$

Then the mean difference is $\bar{d} = \frac{1}{n}\sum_{i=1}^n d_i$. A negative $\bar{d}$ indicates B has lower loss on average (better under the loss).

Remark 9.7.1 (Paired t-Test Template (Conceptual)) Assuming d_i are i.i.d. with finite variance and approximately normal mean via CLT, test $H_0 : \mathbb{E}[d_i] = 0$ using

$$T = \frac{\bar{d}}{s_d/\sqrt{n}}, \qquad s_d^2 = \frac{1}{n-1}\sum_{i=1}^{n}(d_i - \bar{d})^2,$$

and compare T to a t_{n-1} distribution (exact under normality of d_i).

Remark 9.7.2 (Wilcoxon Signed-Rank (Conceptual)) Wilcoxon tests whether the distribution of d_i is symmetric about 0 (and whether its median is 0), using ranks of $|d_i|$ with signs. It is robust to heavy tails and outliers compared to the t-test, but still assumes i.i.d. pairing.

9.7.1.2 Permutation Tests for Performance Gaps

Permutation/randomization tests provide a distribution-free way to assess performance gaps under a null symmetry: under H_0 of no difference, swapping model labels within each pair (or flipping signs of d_i) leaves the joint distribution unchanged. This yields an exact test conditional on the observed magnitudes when the symmetry holds. Permutation tests are especially useful when losses are skewed or bounded and when the t approximation is dubious.

Definition 9.7.2 (Paired Permutation (Sign-Flip) Test) Given paired differences $d_1, \dots, d_n$ and a statistic $T(d)$ (e.g., $T(d) = \bar{d}$), generate random signs $\epsilon_i \in \{-1, +1\}$ i.i.d. and compute $T(\epsilon \odot d)$, where $(\epsilon \odot d)_i = \epsilon_i d_i$. The permutation p-value for an upper-tail test is

$$p_{\text{perm}} = \mathbb{P}_\epsilon\big(T(\epsilon \odot d) \geq T(d)\big),$$

estimated by Monte Carlo if needed.

Remark 9.7.3 (Null Symmetry Requirement) The sign-flip test is valid if under H_0 the joint distribution of $(d_1, \dots, d_n)$ is symmetric: $(d_1, \dots, d_n) \stackrel{d}{=} (\epsilon_1 d_1, \dots, \epsilon_n d_n)$ for independent Rademacher ϵ_i. This is implied by exchangeability of model labels within each pair.

9.7.2 Metrics-Aware Testing

9.7.2.1 Accuracy vs. AUC vs. Log-Loss: What Can Be Tested

Different metrics correspond to different random variables and thus different test statistics and assumptions. Accuracy is a mean of Bernoulli indicators (correct/incorrect), so it admits binomial-type tests and paired tests such as McNemar for paired classifications. Log-loss is a mean of per-example negative log probabilities, naturally suited to paired difference tests. AUC is a U-statistic based on pairwise ranking comparisons and has a more complex sampling distribution; naive i.i.d. assumptions on per-example contributions are not directly valid. Thus, "test the metric difference" requires matching the test to the metric's mathematical structure.

Remark 9.7.4 (Metric Structures)

- **Accuracy:** $\widehat{\text{Acc}} = \frac{1}{n}\sum_{i=1}^{n} \mathbf{1}\{\widehat{y}_i = y_i\}$ (mean of Bernoullis).
- **Log-loss:** $\widehat{\text{LL}} = \frac{1}{n}\sum_{i=1}^{n} -\log q_i(y_i)$ (mean of real losses).
- **AUC:** $\widehat{\text{AUC}} = \frac{1}{n_+ n_-}\sum_{i\in\mathcal{P}}\sum_{j\in\mathcal{N}} \mathbf{1}\{s_i > s_j\}$ (U-statistic over positive–negative pairs).

Each implies a different variance structure and, consequently, different inference tools.

Remark 9.7.5 (What Can be Tested in Practice)

- Accuracy differences: use paired tests (e.g., McNemar) or paired bootstrap on per-example correctness.
- Log-loss differences: paired t-test / Wilcoxon on per-example loss differences, or paired permutation/sign-flip.
- AUC differences: DeLong-type variance methods or bootstrap/permutation that respect pairwise structure.

9.7.2.2 Confidence Intervals for AUC and Calibration Error

AUC and calibration metrics are nonlinear functionals of the sample and often have skewed sampling distributions. Bootstrap is a practical default: resample data (at the correct unit), recompute the metric, and take quantiles for a CI (Efron and Tibshirani 1993). For AUC, one must preserve the pairing structure between positives and negatives; for calibration error (ECE), binning introduces discretization noise, so repeated resampling helps quantify variability. This preview emphasizes protocol correctness: resample at the unit that is i.i.d. (often user-level), not event-level.

Remark 9.7.6 (Bootstrap CI for AUC (Protocol Sketch)) Given a labeled dataset with scores, draw bootstrap resamples (with replacement) of the labeled pairs (s_i, y_i), compute AUC on each resample, and take percentile or BCa quantiles. If data are clustered by user, resample users (cluster bootstrap) and include all their events.

Remark 9.7.7 (Bootstrap CI for Calibration Error (Protocol Sketch)) For ECE computed from bins $\{I_b\}$, bootstrap resample the prediction-label pairs, recompute ECE for each resample, and take quantiles. Because ECE depends on binning, report bins used and consider sensitivity analysis to bin choice.

9.7.3 Time-Series and Dependent Data

9.7.3.1 Blocked/Bootstrap for Dependence (Conceptual)

Dependence violates the i.i.d. assumptions behind classical tests and naive bootstrap. In time series, observations close in time are correlated; resampling individual points destroys dependence structure and underestimates uncertainty. Blocked bootstrap (or moving-block bootstrap) resamples contiguous blocks to preserve short-range dependence. Similarly, evaluation should use time-respecting splits (train on past, test on future), and tests should be performed on block-aggregated statistics to approximate independence.

Remark 9.7.8 (Moving Block Bootstrap (Conceptual)) Choose a block length L and construct blocks $(Z_t, \ldots, Z_{t+L-1})$. Sample blocks with replacement and concatenate to length n to form a bootstrap series. This preserves local dependence up to scale L.

Remark 9.7.9 (Effective Sample Size Intuition) Under dependence, variance of sample averages decays slower than $1/n$. One can think in terms of an effective sample size $n_{\text{eff}} < n$. Using i.i.d. standard errors implicitly assumes $n_{\text{eff}} = n$ and will be overly optimistic.

9.7.3.2 Backtesting Pitfalls and Leakage

Backtesting is evaluation on historical time series data. Leakage is the primary risk: using future information in feature engineering, normalization, hyperparameter tuning, or target construction. Leakage produces inflated performance estimates and invalid inference. A rigorous backtest protocol enforces causal ordering (only past influences present), uses walk-forward validation, and separates model selection from final evaluation. Statistical tests must then respect dependence and multiple testing (many strategies tried).

Remark 9.7.10 (Common Leakage Patterns)

- **Look-ahead features:** using future prices/labels indirectly (rolling windows computed with future points).
- **Global normalization:** scaling using full-series mean/variance including test period.
- **Hyperparameter selection on test:** repeated tuning using the same backtest window.
- **Survivorship bias:** evaluating only on assets that remain in the dataset.

Remark 9.7.11 (Walk-Forward Evaluation (Conceptual)) Split time into successive windows: train on $[1, t]$, validate on $(t, t + \Delta]$, test on the next interval, then roll forward. This mimics deployment and reduces leakage while providing multiple evaluation blocks for uncertainty estimation.

9.8 Model Validation

Model validation is the estimation protocol that turns training performance into a reliable estimate of out-of-sample performance. Because ML pipelines include model selection (architecture, features, hyperparameters, stopping time), validation is itself a statistical procedure subject to bias and variance. This section presents splits and cross-validation (including nested CV), time-series validation under dependence, leakage patterns and safeguards, and the interpretation of hyperparameter selection as multiple comparisons.

9.8.1 Splits and Cross-Validation

9.8.1.1 Train/Validation/Test and Nested CV

A single train/test split is the simplest protocol but yields a high-variance estimate of generalization. Adding a validation set supports hyperparameter tuning but introduces selection bias if test data influence decisions. Cross-validation reduces variance by averaging over folds; however, using CV both to tune and to report performance can still be optimistically biased. Nested CV resolves this by separating the roles: the outer loop estimates generalization, the inner loop selects hyperparameters (Arlot and Celisse 2010). This is the gold-standard protocol when data are limited and model selection is substantial.

Definition 9.8.1 (Basic Split Protocol) Partition the dataset into disjoint sets:

$$\mathcal{D} = \mathcal{D}_{\text{train}} \dot{\cup} \mathcal{D}_{\text{val}} \dot{\cup} \mathcal{D}_{\text{test}}.$$

Train on $\mathcal{D}_{\text{train}}$, select hyperparameters (and early stopping) using $\mathcal{D}_{\text{val}}$, and report final performance on $\mathcal{D}_{\text{test}}$ exactly once.

Definition 9.8.2 (K-Fold Cross-Validation) Partition indices $[n]$ into K disjoint folds $F_1, \ldots, F_K$ of (approximately) equal size. For each k, train on $\mathcal{D} \setminus F_k$ and evaluate on F_k. The CV estimate of risk is

$$\widehat{R}_{\text{CV}} := \frac{1}{K} \sum_{k=1}^{K} \widehat{R}_{\text{val}}^{(k)},$$

where $\widehat{R}_{\text{val}}^{(k)}$ is the average loss on fold F_k from the model trained without F_k.

Definition 9.8.3 (Nested Cross-Validation) Nested CV uses an *outer* CV loop to estimate generalization and an *inner* CV loop (within each outer training set) to select hyperparameters. Concretely: for each outer fold F_k,

1. run inner CV on $\mathcal{D} \setminus F_k$ to select $\widehat{\lambda}^{(-k)}$,
2. retrain on $\mathcal{D} \setminus F_k$ with $\widehat{\lambda}^{(-k)}$,
3. evaluate on F_k to obtain an outer-fold test loss.

Average outer losses to obtain an (approximately) unbiased estimate of the *full selection procedure*'s risk.

Remark 9.8.1 (What Nested CV Estimates) Nested CV estimates the generalization performance of the entire pipeline "train + tune," not just a fixed hyperparameter model. This is exactly what you want when reporting final performance.

9.8.1.2 Time-Series Splits and Non-i.i.d. Considerations

Standard CV assumes i.i.d. samples and breaks when data are temporally dependent or when the distribution drifts. Randomly shuffling time series introduces leakage (future information in training), yielding inflated performance. Time-series validation must preserve temporal order: train on the past, validate/test on the future. Additionally, dependence reduces effective sample size, so uncertainty estimates require blocked resampling or block-based aggregation.

Definition 9.8.4 (Time-Series Split (Walk-Forward)) For ordered data $\{Z_t\}_{t=1}^T$, choose cut times $t_1 < t_2 < \cdots$ and evaluate by training on $Z_{1:t_k}$ and testing on $Z_{t_k+1:t_{k+1}}$, then rolling forward. This preserves causality and approximates deployment.

Remark 9.8.2 (Non-i.i.d. Implications) When data are dependent, naive standard errors and p-values based on i.i.d. assumptions typically understate uncertainty. Validation protocols should use blocks (or higher-level units) that are closer to independent.

9.8.2 Leakage and Safeguards

9.8.2.1 Data Leakage Patterns (Preprocessing, Target Leakage)

Leakage is any information path from test (or future) data into training or model selection. It invalidates evaluation by creating an optimistic bias. Leakage often arises from preprocessing done before splitting (global normalization), feature engineering that uses future outcomes, or inadvertently including target information in features (target leakage). The fix is procedural: split first, then fit preprocessing only on training data, then apply to validation/test; and enforce causal construction in time-series settings.

Remark 9.8.3 (Common Leakage Patterns)

- **Preprocessing leakage:** standardizing/whitening using statistics computed on the full dataset (train+test).
- **Target leakage:** features contain information derived from Y (or future Y) directly or indirectly.
- **Temporal leakage:** using future observations when constructing features (rolling windows that peek ahead).
- **Group leakage:** same entity (user/patient) appears in both train and test when the unit of independence is entity-level.

Remark 9.8.4 (Safeguard Principle) **Fit on train only.** Any operation that learns from data (scalers, PCA, feature selection, imputation, calibration) must be fitted using $\mathcal{D}_{\text{train}}$ only, then applied to validation/test. This includes label-dependent selection steps.

9.8.2.2 Reproducibility and Robust Evaluation Protocols

Validation is an estimator and thus has variability. Reproducibility requires controlling randomness: fixed seeds, documented splits, and repeated runs. Robust protocols use multiple seeds, multiple splits, and report distribution of outcomes rather than a single number. For comparisons, use paired evaluation whenever possible (same splits/seeds across models) to reduce variance. In ML, robust reporting also includes confidence intervals or standard errors for key metrics.

Remark 9.8.5 (Minimum Reporting Checklist)

- exact split protocol (random split, CV folds, nested CV, time-series walk-forward),
- number of repetitions/seeds and how randomness was controlled,
- metric definitions and any preprocessing/calibration steps,
- paired vs. unpaired comparisons and statistical test/CI method if used,
- leakage safeguards implemented.

9.8.3 Hyperparameter Selection as Multiple Comparisons

9.8.3.1 Why Validation Overfits and How Nested CV Helps

Hyperparameter tuning is multiple testing in disguise: each tried configuration is a hypothesis about performance. Selecting the best validation score is selecting the maximum of many noisy estimates, which is optimistically biased. This "validation overfitting" becomes worse as the search space grows. Nested CV reduces this bias by using an outer holdout to evaluate the selection procedure, not the selected configuration on the same data.

Remark 9.8.6 (Selection Bias as a Max Effect) If $\widehat{R}_{\text{val}}(\lambda)$ are noisy estimates, then $\min_\lambda \widehat{R}_{\text{val}}(\lambda)$ is biased downward (compared to $\min_\lambda R(\lambda)$). The bias increases with the number of tried λ values and with noise variance.

Remark 9.8.7 (Nested CV as Correction) Nested CV approximates the distribution of the selected model's test risk by repeating the selection process on multiple outer splits, thereby estimating the generalization of the selection rule itself.

9.8.3.2 Early Stopping and Selection Bias (Conceptual)

Early stopping chooses a training time/epoch based on validation performance, so it is also a model-selection procedure. Repeatedly evaluating validation loss across epochs and picking the best epoch inflates optimism: you are selecting the minimum of many noisy measurements. This is particularly subtle because it occurs within every training run. Mitigations include using a patience rule, smoothing validation curves, using a separate calibration set for stopping decisions, or treating early stopping as part of the pipeline whose performance must be evaluated by outer validation (nested protocols).

Remark 9.8.8 (Early Stopping is Hyperparameter Tuning) The stopping time $t^\star$ is a hyperparameter selected using validation data. Thus, final performance claims should be evaluated as "train with early stopping" as a single algorithm, ideally with an outer test set untouched by stopping decisions.

Remark 9.8.9 (Practical Mitigation) Use *patience* (require improvement sustained over several epochs), reduce validation frequency (fewer looks), and always report the protocol (how $t^\star$ was chosen) to make the selection procedure auditable.

9.9 Sequential Testing and Online Experiments

Online experiments are rarely analyzed once at a fixed sample size. In practice, teams *peek* at results repeatedly and stop early if an effect looks significant or if harm is suspected. Classical fixed-horizon p-values do *not* remain valid under optional stopping; repeated looks inflate false positives. Sequential testing provides designs and decision rules that control Type I error under continuous monitoring. This section previews the core issues (peeking problem), and two foundational remedies: SPRT-style likelihood ratio boundaries and alpha-spending/group-sequential designs.

9.9.1 Why Sequential Matters

9.9.1.1 Peeking Problem and Inflated False Positives

If you repeatedly compute a fixed-horizon p-value after each new batch of data and stop when it drops below α, the overall probability of a false rejection can be far larger than α. Intuitively, even under H_0, random fluctuations will eventually cross a fixed threshold if you look often enough. This is the same multiple-testing phenomenon, but over time rather than across hypotheses. In ML A/B testing, this shows up as "we checked daily and shipped when p<0.05"—a recipe for false discoveries.

Remark 9.9.1 (A Simple Conceptual Model) Let p_t be the p-value computed at time t using all data up to t. Under H_0, each p_t is (approximately) uniform if computed once at a fixed t, but the event $\{\exists t : p_t \le \alpha\}$ can have probability much larger than α when monitoring many times. Thus, optional stopping breaks naive Type I calibration unless the procedure is designed to be sequentially valid.

9.9.1.2 Stopping Rules and Decision Boundaries (Conceptual)

Sequential testing replaces a single critical value with *decision boundaries* over time. At each look, the statistic is compared to a boundary; crossing triggers stopping and a decision. Good designs stop early when evidence is strong (either for benefit or harm), while controlling Type I error. Mathematically, a sequential test is a stopping time plus a terminal decision rule, and the key requirement is: *the probability of ever crossing the rejection boundary under H_0 is at most α.*

Definition 9.9.1 (Sequential Test (Conceptual)) A sequential test consists of:

- a filtration $\{\mathcal{F}_t\}$ representing information up to time t,
- a stopping time τ (the first time a boundary is crossed),
- a terminal decision $\varphi \in \{0, 1\}$ measurable w.r.t. $\mathcal{F}_\tau$.

It has level α if $\sup_{\theta \in \Theta_0} \mathbb{P}_\theta(\varphi = 1) \leq \alpha$.

Remark 9.9.2 (Two-Sided Monitoring) In online experiments, one often wants early stopping both for positive effects (ship) and for negative effects (abort). This corresponds to two boundaries: an upper boundary for rejecting H_0 in favor of H_1 and a lower boundary for accepting H_0 or declaring harm.

9.9.2 *Basic Sequential Ideas*

9.9.2.1 SPRT Intuition (Conceptual)

The Sequential Probability Ratio Test (SPRT) is the sequential analogue of Neyman–Pearson for *simple* hypotheses. It monitors the cumulative likelihood ratio between H_1 and H_0 and stops when the ratio becomes very large (favoring H_1) or very small (favoring H_0). The likelihood ratio is a martingale under H_0 (after normalization), which enables error control via optional stopping theorems. In ML A/B tests, SPRT-like logic yields early decisions with controlled false positives when the hypotheses are well specified.

Definition 9.9.2 (Likelihood Ratio Process) For i.i.d. observations $Z_1, Z_2, \ldots$ with densities/pmfs p_0 under H_0 and p_1 under H_1 (simple hypotheses), define the cumulative likelihood ratio after t samples as

$$\Lambda_t := \prod_{i=1}^{t} \frac{p_1(Z_i)}{p_0(Z_i)}, \qquad \log \Lambda_t = \sum_{i=1}^{t} \log \frac{p_1(Z_i)}{p_0(Z_i)}.$$

Remark 9.9.3 (SPRT Stopping Rule (Template)) Choose thresholds $A > 1$ and $B \in (0, 1)$. Stop at

$$\tau = \inf\{t : \Lambda_t \geq A \text{ or } \Lambda_t \leq B\},$$

and decide H_1 if $\Lambda_\tau \geq A$, else decide H_0 if $\Lambda_\tau \leq B$. Thresholds are chosen to target Type I and Type II error levels (approximately) via $A \approx (1 - \beta)/\alpha$ and $B \approx \beta/(1 - \alpha)$.

Remark 9.9.4 (Why SPRT Can Stop Early) If the true effect is large, log-likelihood ratios accumulate quickly and cross a boundary at small t. If the effect is near zero, boundaries may not be crossed quickly, reflecting inherent uncertainty.

9.9.2.2 Alpha Spending and Group Sequential Designs (Conceptual)

Many practical experiments are monitored at a finite number of planned looks (e.g., daily for 2 weeks). Group sequential designs allocate the total Type I error budget α across looks using an *alpha-spending* function. At each look k, the critical threshold is adjusted so that the cumulative Type I spent so far is controlled. This preserves overall Type I error while allowing interim analyses. Common spending rules (O'Brien–Fleming, Pocock) trade off early vs. late stopping aggressiveness.

Definition 9.9.3 (Alpha Spending (Conceptual)) Let planned analysis times be $1 \leq t_1 < \cdots < t_K$ and let $\alpha \in (0, 1)$ be the total Type I budget. An *alpha-spending plan* specifies numbers $0 \leq \alpha_1 \leq \cdots \leq \alpha_K = \alpha$, where α_k is the cumulative Type I error allowed up to look k. Critical boundaries at look k are chosen so that

$$\mathbb{P}_{H_0}(\text{reject by look } k) \leq \alpha_k,$$

and hence $\mathbb{P}_{H_0}(\text{reject by look } K) \leq \alpha$.

Remark 9.9.5 (Group Sequential Design (Template)) At each look k, compute a standardized statistic T_k (often approximately normal under H_0). Choose critical values c_k such that rejecting when $T_k \geq c_k$ at the first crossing controls the cumulative Type I spend. O'Brien–Fleming spends very little α early (very strict early thresholds) and more later; Pocock spends more evenly (more permissive early stopping, slightly stricter later).

Remark 9.9.6 (Practical ML Takeaway) If you will peek, plan the peeks. Pre-specify the number/timing of looks and use a sequentially valid method. Otherwise, fixed-horizon p-values and classical CIs are not trustworthy under repeated monitoring.

9.10 Exercises

Unless stated otherwise, use natural logarithms, treat samples as i.i.d. at the stated unit, and compute all p-values as exact or to four decimal places.

Exercise 9.10.1 (Type I/II Errors and Power for a Fully Specified Test) Let $X \sim \mathcal{N}(\mu, 1)$ and consider testing $H_0 : \mu = 0$ vs. $H_1 : \mu = 1$ using the test $\varphi(x) = \mathbf{1}\{x \geq c\}$ for a threshold $c \in \mathbb{R}$. (i) Choose c so that the test has size $\alpha = 0.05$ (i.e., $\mathbb{P}_{\mu=0}(X \geq c) = 0.05$) and compute that c numerically in terms of the standard normal quantile. (ii) Compute the Type II error $\beta = \mathbb{P}_{\mu=1}(X < c)$ and the power $1 - \beta$. (iii) Repeat (ii) for $\mu = 0.5$ (a smaller effect size) and comment (one paragraph) on how power depends on effect size.

Exercise 9.10.2 (Neyman–Pearson Likelihood Ratio Test (Simple Hypotheses)) Let $Z_1, \ldots, Z_n$ be i.i.d. with $Z_i \sim \mathcal{N}(\mu, 1)$ and consider simple hypotheses $H_0 : \mu = 0$ vs. $H_1 : \mu = 1$. (i) Derive the likelihood ratio $\Lambda(z) = \frac{p_1(z)}{p_0(z)}$ and show that $\log \Lambda(z)$ is an affine function of $\sum_{i=1}^{n} z_i$. (ii) Conclude from the Neyman–Pearson lemma that the most powerful level-α test rejects for large $\bar{Z}_n$. (iii) For $n = 9$ and observed data

$$z = (-0.2,\ 0.4,\ 1.3,\ 0.7,\ -0.1,\ 0.9,\ 1.1,\ 0.0,\ 0.8),$$

compute $\bar{z}$ and perform the level-$\alpha = 0.05$ NP test (compute the rejection threshold explicitly).

Exercise 9.10.3 (GLRT and Wilks: Nested Gaussian Mean Test) Let $Z_1, \ldots, Z_{12}$ be i.i.d. $\mathcal{N}(\mu, \sigma^2)$ with both μ and σ^2 unknown. Test $H_0 : \mu = 0$ vs. $H_1 : \mu \neq 0$ using a generalized likelihood ratio test. Given observed data

$$z = (0.9,\ 1.1,\ 0.2,\ -0.1,\ 0.5,\ 1.4,\ 0.7,\ 0.3,\ 0.6,\ -0.4,\ 0.8,\ 0.0),$$

(i) Compute the unrestricted MLEs $(\widehat{\mu}, \widehat{\sigma}^2)$ and the restricted MLE under H_0, $\widehat{\sigma}_0^2$. (ii) Compute the GLRT statistic $-2 \log \Lambda = 2(\ell(\widehat{\mu}, \widehat{\sigma}^2) - \ell(0, \widehat{\sigma}_0^2))$ explicitly. (iii) Using Wilks' theorem (1 degree of freedom), compute the approximate p-value as $\mathbb{P}(\chi_1^2 \geq -2 \log \Lambda)$.

Exercise 9.10.4 (Z-Test and t-Test on the Same Sample) You observe $n = 10$ measurements:

$$x = (2.4,\ 2.1,\ 1.9,\ 2.8,\ 2.5,\ 2.0,\ 2.2,\ 2.7,\ 2.3,\ 2.6).$$

Test $H_0 : \mu = 2.0$ vs. $H_1 : \mu \neq 2.0$. (i) Compute $\bar{x}$ and s^2. (ii) Perform a two-sided Z-test assuming known variance $\sigma^2 = 0.25$ (i.e., $\sigma = 0.5$); report the Z statistic and p-value. (iii) Perform a two-sided t-test with unknown variance; report the t statistic, degrees of freedom, and p-value. (iv) Compare the conclusions at $\alpha = 0.05$ and explain (one paragraph) why the p-values differ.

Exercise 9.10.5 (χ^2 Variance Test and F Variance-Ratio Test) Group A has $n_A = 8$ observations and Group B has $n_B = 10$ observations:

$$x^{(A)} = (4.1,\ 3.9,\ 4.5,\ 4.0,\ 4.2,\ 3.8,\ 4.4,\ 4.1),$$

$$x^{(B)} = (3.2,\ 3.1,\ 3.7,\ 3.3,\ 3.5,\ 3.0,\ 3.6,\ 3.4,\ 3.2,\ 3.8).$$

Assume Gaussianity within each group. (i) Compute s_A^2 and test $H_0 : \sigma_A^2 = 0.04$ using the χ^2 statistic $Q = (n_A - 1)s_A^2/0.04$ (two-sided, $\alpha = 0.05$). (ii) Test H_0 :

$\sigma_A^2 = \sigma_B^2$ vs. $H_1 : \sigma_A^2 \neq \sigma_B^2$ using the F statistic $F = s_A^2/s_B^2$ (two-sided, $\alpha = 0.05$). (iii) State clearly which assumptions are needed for the exact null distributions and what would break if data were heavy-tailed.

Exercise 9.10.6 (p-Values: Compute and Interpret Correctly) A test statistic under H_0 has distribution $T \sim \chi_3^2$. You observe $t_{\text{obs}} = 9.50$. (i) Compute the p-value $p = \mathbb{P}(\chi_3^2 \geq 9.50)$. (ii) State (one paragraph) two correct interpretations of this p-value and two incorrect interpretations commonly seen in ML reports. (iii) If the same test is performed $m = 20$ times and the smallest p-value observed is 0.03, compute the Bonferroni-adjusted threshold at $\alpha = 0.05$ and state whether 0.03 is significant after correction.

Exercise 9.10.7 (Bonferroni, Holm, and BH on the Same p-Values) You test $m = 8$ hypotheses and obtain p-values

$$p = (0.001,\ 0.007,\ 0.012,\ 0.021,\ 0.034,\ 0.041,\ 0.08,\ 0.20).$$

(i) Under FWER control at $\alpha = 0.05$, determine which hypotheses are rejected by Bonferroni. (ii) Apply Holm's step-down procedure at $\alpha = 0.05$ and list the rejected hypotheses. (iii) Under FDR control at $q = 0.10$, apply Benjamini–Hochberg and list the rejected hypotheses. (iv) Compute BH-style q-values for each hypothesis (report in increasing order of p-values) using the standard monotone adjustment.

Exercise 9.10.8 (Paired Model Comparison: Paired t-Test vs. Sign-Flip Permutation) Two models A and B are evaluated on the same $n = 12$ folds, producing paired *log-loss* values. Let differences be $d_i = \ell_{B,i} - \ell_{A,i}$ (negative favors B), with

$$d = (-0.020,\ -0.015,\ -0.005,\ 0.000,\ -0.010,\ -0.008,\ 0.004,\ -0.012,$$
$$-0.006,\ 0.002,\ -0.014,\ -0.003).$$

(i) Compute $\bar{d}$, s_d, and the paired t-statistic $T = \bar{d}/(s_d/\sqrt{n})$; report the two-sided p-value using t_{n-1}. (ii) Perform a *Monte Carlo sign-flip permutation test* with $B = 12$ sign patterns given below (each row is a sign vector in $\{\pm 1\}^{12}$):

$$(+,+,+,+,+,+,+,+,+,+,+,+),\ (-,+,+,+,+,+,+,+,+,+,+,+),$$
$$(+,-,+,+,+,+,+,+,+,+,+,+),$$

$$(+,+,-,+,+,+,+,+,+,+,+,+),\ (+,+,+,-,+,+,+,+,+,+,+,+),$$
$$(+,+,+,+,-,+,+,+,+,+,+,+),$$

$$(-,-,+,+,+,+,+,+,+,+,+,+),\ (-,+,-,+,+,+,+,+,+,+,+,+),$$
$$(-,+,+,-,+,+,+,+,+,+,+,+),$$

$$(-,+,+,+,-,+,+,+,+,+,+,+),\ (-,-,-,+,+,+,+,+,+,+,+,+),$$
$$(-,-,-,-,+,+,+,+,+,+,+,+).$$

Using $T(d) = \bar{d}$ as the statistic, compute the permutation p-value for the one-sided alternative "B better" (i.e., $\bar{d} < 0$) by comparing $T(\epsilon \odot d)$ to $T(d)$ under these sign patterns. (iii) Comment (one paragraph) on when the permutation test is more trustworthy than the t-test.

Exercise 9.10.9 (Bootstrap Standard Error and CI for a Metric) You evaluate a model on $n = 10$ examples and record per-example losses

$$\ell = (0.30,\ 0.25,\ 0.40,\ 0.10,\ 0.15,\ 0.22,\ 0.18,\ 0.35,\ 0.28,\ 0.12).$$

Let $\widehat{\theta} = \bar{\ell}$. Use the following $B = 8$ bootstrap index-multisets (each of size 10) to compute bootstrap replicates of $\bar{\ell}$:

$$(1,2,3,4,5,6,7,8,9,10),\ (1,1,2,3,4,5,6,7,8,9),\ (2,3,4,5,6,7,8,9,10,10),$$
$$(1,3,3,4,4,6,7,8,9,10),\ (1,2,4,4,5,5,7,8,9,10),\ (2,2,2,3,6,6,7,8,9,10),$$
$$(1,5,5,5,6,7,7,8,9,10),\ (3,4,5,6,6,7,7,8,9,10).$$

(i) Compute the eight bootstrap means $\widehat{\theta}^{(b)}$. (ii) Compute the bootstrap standard error $\widehat{\text{se}}_{\text{boot}}$. (iii) Compute a 90% percentile bootstrap CI and a 90% basic bootstrap CI for θ using the bootstrap replicate quantiles.

Exercise 9.10.10 (Permutation Test for A/B Conversion Uplift (Randomization Test)) In an A/B test, $n = 12$ users are independently randomized to A or B with equal probability. You observe binary conversions $Y_i \in \{0, 1\}$ and the assignment vector $G_i \in \{A, B\}$:

$$(G, Y) = \big((A,1),(A,0),(A,0),(A,1),(A,0),(A,0),(B,1),(B,1),(B,0),$$
$$(B,1),(B,0),(B,1)\big).$$

Let the test statistic be the difference in conversion rates $T = \bar{Y}_B - \bar{Y}_A$. (i) Compute T_{obs}. (ii) Under the sharp null of no treatment effect, the outcomes are fixed and only assignments are random. Compute the exact randomization distribution of T by enumerating all assignments with 6 users in B (balanced designs), or (if you prefer) compute it for the following 10 alternative balanced assignments (given by the set of indices assigned to B):

$$\{1,2,3,4,5,6\}, \{1,2,3,4,5,7\}, \{1,2,3,4,7,8\}, \{1,2,3,7,8,9\}, \{1,2,7,8,9,10\},$$
$$\{1,7,8,9,10,11\}, \{7,8,9,10,11,12\}, \{2,4,6,7,9,12\}, \{1,3,5,8,10,12\},$$
$$\{2,3,4,8,11,12\}.$$

Compute T for each assignment and estimate a one-sided p-value for "B better" as the fraction with $T \geq T_{\text{obs}}$. (iii) State precisely the exchangeability/randomization assumption that justifies this test.

Exercise 9.10.11 (Jackknife Variance and Influence of a Single Point) You observe $n = 7$ values

$$x = (1.0,\ 1.2,\ 0.9,\ 1.1,\ 3.8,\ 1.0,\ 1.1).$$

Let $\widehat{\theta} = \bar{x}$. (i) Compute the n jackknife replicates $\widehat{\theta}_{(-i)}$ (leave one out). (ii) Compute the jackknife variance estimate $\widehat{\text{Var}}_{\text{jack}}(\widehat{\theta}) = \frac{n-1}{n} \sum_{i=1}^{n} (\widehat{\theta}_{(-i)} - \overline{\widehat{\theta}}_{(-\cdot)})^2$. (iii) Identify which observation has the largest influence on $\widehat{\theta}$ in the jackknife sense and explain (one paragraph) why.

Exercise 9.10.12 (CI–Test Duality: Invert a Z-Test) Let $X_1, \ldots, X_n \overset{iid}{\sim} \mathcal{N}(\mu, \sigma^2)$ with known $\sigma = 2$ and $n = 25$. You observe $\bar{x} = 1.2$. (i) Write the level-$\alpha = 0.05$ two-sided Z-test for $H_0 : \mu = \mu_0$ in terms of the rejection region for μ_0. (ii) Invert this test to derive the corresponding 95% confidence interval for μ and compute it numerically. (iii) Verify explicitly that μ_0 is rejected at level 0.05 iff μ_0 lies outside your derived interval.

Exercise 9.10.13 (Sequential Peeking Inflation (Toy Calculation)) Suppose you (incorrectly) perform two independent looks at a level-$\alpha = 0.05$ test under H_0 and reject if either look rejects. Assuming independence between the two looks' p-values under H_0 (toy assumption), compute the actual Type I error:

$$\mathbb{P}(\text{reject at look 1 or look 2}).$$

Then generalize to K independent looks and compute the Type I error as a function of K. Finally, solve for the per-look level α' such that $K = 10$ looks yield overall Type I error 0.05 under independence.

Exercise 9.10.14 (Alpha-Spending Toy Design (Discrete Allocation)) You plan $K = 4$ interim looks with total Type I budget $\alpha = 0.05$ and decide to use a simple alpha-spending plan:

$$\alpha_1 = 0.005, \quad \alpha_2 = 0.015, \quad \alpha_3 = 0.030, \quad \alpha_4 = 0.050,$$

where α_k is the cumulative spend by look k. Assume (toy) that the test statistic at each look is standard normal under H_0 and independent across looks. (i) Compute the per-look incremental alpha spends $\Delta_k = \alpha_k - \alpha_{k-1}$ (with $\alpha_0 = 0$). (ii) For each look k, compute a one-sided critical value c_k such that $\mathbb{P}(Z \geq c_k) = \Delta_k$ for $Z \sim \mathcal{N}(0, 1)$. (iii) Interpret in one paragraph how the critical values change over time and why early looks are stricter.

Chapter 10
Linear Models: Regression and Classification

Linear models occupy a unique position in ML: they are simultaneously (i) strong baselines that often remain competitive with careful feature engineering, (ii) mathematically transparent objects whose behavior can be characterized exactly, and (iii) the local building blocks of far more complex systems (deep networks, kernel methods, and probabilistic models). They connect directly to the geometry developed earlier (subspaces, projections, Gram matrices) and to the estimation principles developed later (MLE/MAP/Bayes under noise models). This chapter treats linear models as the canonical "first principles" of prediction: if we cannot reason about linear models, we cannot reliably reason about the rest of ML.

10.1 Why Linear Models Remain Foundational

10.1.1 Linear Models as Baselines and Building Blocks

10.1.1.1 Interpretability and Fast Training

A linear predictor decomposes prediction into additive contributions from features: each coefficient has a direct meaning (marginal effect under scaling conventions), enabling interpretability, debugging, and scientific insight. Computationally, linear models are fast: convex objectives, closed-form solutions in least squares, and highly optimized solvers (QR, coordinate descent). This makes them ideal baselines and reliable components in larger pipelines (e.g., linear probes for representation quality, calibration layers, and last-layer fine-tuning) (Hastie et al. 2009) (cf. Fig. 10.1).

P. Singh, B. Raman, *Foundations of Machine Learning and AI*, Studies in Big Data 199, https://doi.org/10.1007/978-3-032-30336-3_10

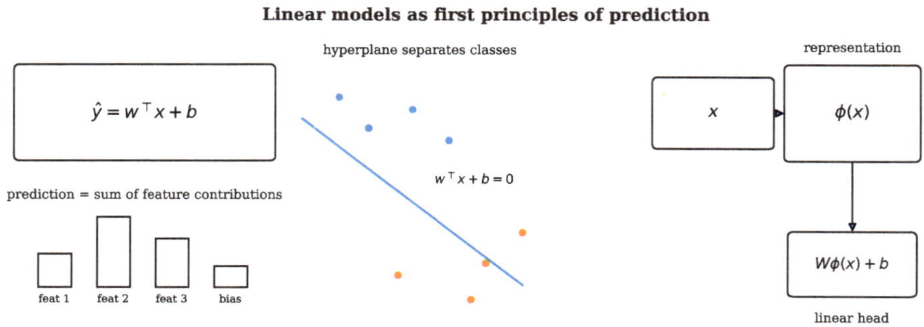

Fig. 10.1 Why linear models remain foundational: additive feature contributions for interpretation, geometric decision boundaries in input space, and linear heads operating on learned representations in modern pipelines

Definition 10.1.1 (Linear Regression Model (Prediction Form)) Let $x \in \mathbb{R}^d$ be a feature vector. A linear regression predictor is

$$\widehat{y} = w^\top x + b, \qquad w \in \mathbb{R}^d,\ b \in \mathbb{R}.$$

For a dataset $\{(x_i, y_i)\}_{i=1}^n$, with $X \in \mathbb{R}^{n\times d}$ (rows $x_i^\top$) and $y \in \mathbb{R}^n$, the prediction vector is $\widehat{y} = Xw + b\mathbf{1}$.

Definition 10.1.2 (Linear Classification Model (Score Form)) For binary classification with labels $Y \in \{-1, +1\}$, a linear score is

$$s(x) = w^\top x + b.$$

A hard classifier is $h(x) = \text{sign}(s(x))$. In probabilistic classification, one typically maps scores to probabilities (e.g., via a sigmoid or softmax), discussed later.

Remark 10.1.1 (Coordinate Conventions) Throughout, vectors are columns, $X \in \mathbb{R}^{n\times d}$ stores samples as rows, and $w^\top x$ is a scalar. If an intercept is included, one can augment features by $\tilde{x} = \begin{bmatrix} x \\ 1 \end{bmatrix}$ and parameters by $\tilde{w} = \begin{bmatrix} w \\ b \end{bmatrix}$ so that $\widehat{y} = \tilde{w}^\top \tilde{x}$.

10.1.1.2 Linear Layers Inside Deep Networks

Even when a model is nonlinear end-to-end, it is assembled from affine (linear + bias) pieces. Fully connected layers are affine maps; convolutions are structured linear maps; attention includes linear projections. Moreover, many deep models terminate in a linear readout on top of a learned representation $\phi(x)$:

$$\widehat{y} = W\,\phi(x) + b.$$

This makes linear models the interface between representation learning (geometry) and decision-making (prediction). It also explains why linear-probe performance is a standard diagnostic: it isolates representation quality from nonlinear head capacity.

Definition 10.1.3 (Affine Layer) An affine (dense) layer is a map $f : \mathbb{R}^m \to \mathbb{R}^k$ of the form

$$f(z) = Wz + b, \qquad W \in \mathbb{R}^{k\times m},\ b \in \mathbb{R}^k.$$

Example 10.1.1 (Deep Representation + linear head) Let $\phi : \mathcal{X} \to \mathbb{R}^m$ be a nonlinear network producing embeddings. A linear head for K-class classification is

$$s(x) = W\phi(x) + b \in \mathbb{R}^K, \qquad \widehat{y} = \arg\max_{k\in[K]} s_k(x).$$

Training only (W, b) with ϕ frozen is a convex optimization problem and measures the linear separability of $\phi(\mathcal{X})$.

Remark 10.1.2 (Why Linear Modules Are Ubiquitous) Linear/affine modules are stable under composition, easy to differentiate, and can be optimized reliably. Thus they are used extensively even in nonlinear architectures, where the nonlinearity is introduced by activation functions and gating.

10.1.2 From Geometry to Statistics

10.1.2.1 Projections, Residuals, and Subspace Viewpoints

The geometry of linear regression is the geometry of orthogonal projection. With $X \in \mathbb{R}^{n\times d}$, predictions Xw lie in the column space $\mathcal{C}(X) \subseteq \mathbb{R}^n$. Fitting by least squares chooses the closest point in $\mathcal{C}(X)$ to y, and the residual is orthogonal to $\mathcal{C}(X)$. This viewpoint immediately explains: (i) why fitted values are unique even when coefficients are not (rank deficiency), (ii) why $X^\top r = 0$ (residual uncorrelated with features), and (iii) why squared-error decomposes by Pythagoras.

Definition 10.1.4 (Fitted Values and Residual) Given $X \in \mathbb{R}^{n\times d}$ and $y \in \mathbb{R}^n$, for any w define

$$\widehat{y}(w) := Xw, \qquad r(w) := y - \widehat{y}(w).$$

In least squares, $\widehat{w} \in \arg\min_w \|r(w)\|_2^2$, and we write $\widehat{y} := \widehat{y}(\widehat{w})$ and $r := r(\widehat{w})$.

Proposition 10.1.1 (Normal Equations and Residual Orthogonality) *Any least-squares minimizer $\widehat{w}$ satisfies*

$$X^\top(X\widehat{w} - y) = 0,$$

equivalently $X^\top r = 0$. Hence $r \in \mathcal{N}(X^\top) = \mathcal{C}(X)^\perp$ and $\widehat{y} \in \mathcal{C}(X)$.

Remark 10.1.3 (Hat Matrix) If X has full column rank, the orthogonal projector onto $\mathcal{C}(X)$ is

$$P = X(X^\top X)^{-1}X^\top,$$

and $\widehat{y} = Py$, $r = (I - P)y$. This ties regression to the projection operators developed earlier.

10.1.2.2 Noise Models and Likelihood Interpretations

A geometric fitting rule becomes a statistical estimator once we specify how data are generated. Under a Gaussian noise model, least squares is maximum likelihood; under other noise models, different losses arise (e.g., Laplace noise $\Rightarrow$ absolute loss). For classification, mapping linear scores to probabilities yields logistic/softmax likelihoods and cross-entropy loss. This probabilistic lens provides: (i) principled objectives, (ii) uncertainty quantification via curvature (Fisher information), and (iii) coherent probabilistic predictions (calibration). We will develop these connections rigorously in the estimation chapters.

Remark 10.1.4 (Gaussian Noise $\Rightarrow$ Squared Loss) If $Y \mid X = x$ satisfies $Y = w^\top x + b + \varepsilon$ with $\varepsilon \sim \mathcal{N}(0, \sigma^2)$, then maximizing conditional likelihood is equivalent to minimizing $\sum_i (y_i - (w^\top x_i + b))^2$ (up to constants).

Remark 10.1.5 (Logistic Likelihood $\Rightarrow$ Cross-Entropy) For binary $Y \in \{0, 1\}$, if $\mathbb{P}(Y = 1 \mid X = x) = \sigma(w^\top x + b)$ with sigmoid σ, then negative log-likelihood is the logistic (cross-entropy) loss. Thus linear classification is naturally a likelihood-based estimation problem.

10.2 Regression Setup and Loss Functions

Regression models a real-valued response as a function of input features. Linear regression assumes the predictor is affine in the features, and learning amounts to choosing coefficients to minimize a loss. This section fixes notation (design matrix, targets, hypothesis class), distinguishes empirical from population risk, and introduces loss functions used in regression (squared, absolute, Huber), including weighted least squares for heteroscedastic noise. Finally, we emphasize feature engineering as geometry: basis expansions and interactions change the ambient space where linear methods operate.

10.2.1 Problem Formulation

10.2.1.1 Design Matrix, Targets, and Hypothesis Class

A regression dataset is an input matrix plus a target vector. The design matrix encodes geometry of inputs: its column space determines what the linear model can represent. The hypothesis class is the set of all affine functions of the features; when features are expanded, the hypothesis class changes even if the model remains linear in the expanded coordinates. This makes notation and shape discipline essential for avoiding conceptual errors.

Definition 10.2.1 (Regression Dataset and Design Matrix) Let $\{(x_i, y_i)\}_{i=1}^n$ be data with $x_i \in \mathbb{R}^d$ and $y_i \in \mathbb{R}$. Define the design matrix $X \in \mathbb{R}^{n\times d}$ by $X_{i:} = x_i^\top$ and the target vector $y \in \mathbb{R}^n$ by y_i. An affine predictor is $x \mapsto w^\top x + b$, with $w \in \mathbb{R}^d$ and $b \in \mathbb{R}$.

Definition 10.2.2 (Augmented Design for Intercept) Define $\widetilde{x}_i = \begin{bmatrix} x_i \\ 1 \end{bmatrix} \in \mathbb{R}^{d+1}$, $\widetilde{w} = \begin{bmatrix} w \\ b \end{bmatrix} \in \mathbb{R}^{d+1}$, and $\widetilde{X} \in \mathbb{R}^{n\times(d+1)}$ with rows $\widetilde{x}_i^\top$. Then the affine model becomes linear: $\widehat{y} = \widetilde{X}\widetilde{w}$.

Definition 10.2.3 (Linear Hypothesis Class) The linear regression hypothesis class (with intercept absorbed) is

$$\mathcal{H} := \left\{ h_{\widetilde{w}} : \mathbb{R}^{d+1} \to \mathbb{R},\ h_{\widetilde{w}}(\widetilde{x}) = \widetilde{w}^\top \widetilde{x},\ \widetilde{w} \in \mathbb{R}^{d+1} \right\}.$$

10.2.1.2 Empirical Risk and Population Risk

Learning is risk minimization: empirical risk uses observed data, population risk uses the true data distribution. In regression, empirical risk is an average loss over samples; population risk is an expectation of the same loss under (X, Y). This distinction is crucial: minimizing empirical risk can overfit, while minimizing population risk defines the ideal predictor. Concentration results later explain when empirical minimization approximates population minimization.

Definition 10.2.4 (Empirical and Population Risk) Let $\ell : \mathbb{R} \times \mathbb{R} \to [0, \infty)$ be a loss. For a predictor h, define empirical risk

$$\widehat{R}_n(h) := \frac{1}{n} \sum_{i=1}^{n} \ell(h(x_i), y_i),$$

and population risk (with (X, Y) distributed as the data-generating process)

$$R(h) := \mathbb{E}\big[\ell(h(X), Y)\big].$$

Remark 10.2.1 (Estimation vs. Approximation) If the true regression function is $f^{\star}(x) = \mathbb{E}[Y \mid X = x]$, then even the best linear predictor may have nonzero approximation error if $f^{\star}$ is not affine in the chosen features. Feature engineering changes this approximation error.

10.2.2 Loss Functions for Regression

10.2.2.1 Squared Loss, Absolute Loss, Huber Loss (Robustness)

The squared loss is analytically convenient: it yields closed-form solutions and orthogonal projection geometry, and corresponds to Gaussian noise. However, it is sensitive to outliers because residuals are squared. Absolute loss corresponds to Laplace-like noise and is more robust, but yields nonsmooth optimization. The Huber loss interpolates: quadratic near zero (efficiency) and linear in the tails (robustness). This triad is the standard regression loss toolkit and illustrates how probabilistic assumptions and optimization properties align.

Definition 10.2.5 (Residual) For a prediction $\widehat{y}$ and target y, the residual is $r = \widehat{y} - y$. Losses below are typically functions of r.

Definition 10.2.6 (Squared Loss)

$$\ell_2(\widehat{y}, y) := \frac{1}{2}(\widehat{y} - y)^2 = \frac{1}{2}r^2.$$

Definition 10.2.7 (Absolute Loss)

$$\ell_1(\widehat{y}, y) := |\widehat{y} - y| = |r|.$$

Definition 10.2.8 (Huber Loss) Fix $\delta > 0$. The Huber loss is

$$\ell_{\text{Huber},\delta}(\widehat{y}, y) := \begin{cases} \frac{1}{2}r^2, & |r| \le \delta, \\ \delta\big(|r| - \frac{\delta}{2}\big), & |r| > \delta. \end{cases}$$

Remark 10.2.2 (Robustness Intuition) Squared loss has unbounded influence for large residuals; absolute and Huber losses cap the growth (linear tails), making them robust to outliers and heavy-tailed noise.

10.2.2.2 Weighted Least Squares and Heteroscedasticity

When noise variance depends on x (heteroscedasticity), treating all residuals equally is inefficient. Weighted least squares (WLS) assigns higher weight to more reliable observations and lower weight to noisier ones. Geometrically, WLS changes the inner product on $\mathbb{R}^n$ by weighting residual coordinates, yielding a different projection. Statistically, WLS corresponds to MLE under Gaussian noise with non-constant variance.

Definition 10.2.9 (Weighted Least Squares) Given weights $w_i > 0$, define the WLS objective

$$\min_{\beta} \sum_{i=1}^{n} w_i \, (\beta^\top \widetilde{x}_i - y_i)^2.$$

Let $W = \mathrm{diag}(w_1, \dots, w_n)$. In matrix form:

$$\min_{\widetilde{w}} \ \|W^{1/2}(\widetilde{X}\widetilde{w} - y)\|_2^2.$$

Remark 10.2.3 (Heteroscedastic Gaussian Noise) If $Y_i = \widetilde{w}^\top \widetilde{x}_i + \varepsilon_i$ with $\varepsilon_i \sim \mathcal{N}(0, \sigma_i^2)$ independent, then MLE corresponds to WLS with $w_i = 1/\sigma_i^2$.

10.2.3 Feature Engineering as Geometry

10.2.3.1 Polynomial Features and Basis Expansions

A linear model is only as expressive as its feature map. Basis expansions (polynomials, splines, Fourier features) embed inputs into a higher-dimensional space where linear prediction approximates nonlinear functions. Geometrically, this changes the subspace $\mathcal{C}(X)$ of representable prediction vectors. Computationally, it increases dimension and can introduce severe collinearity, motivating regularization.

Definition 10.2.10 (Feature Map and Linear Model in Feature Space) Let $\phi : \mathbb{R}^d \to \mathbb{R}^p$ be a feature map. A *linear model in feature space* is

$$\widehat{y} = \theta^\top \phi(x), \qquad \theta \in \mathbb{R}^p.$$

With $\Phi \in \mathbb{R}^{n \times p}$ defined by $\Phi_{i:} = \phi(x_i)^\top$, training is linear regression in Φ.

Example 10.2.1 (Univariate Polynomial Features) For scalar input $x \in \mathbb{R}$ and degree k, define

$$\phi_k(x) = (1, x, x^2, \ldots, x^k)^\top \in \mathbb{R}^{k+1}.$$

Then $\theta^\top \phi_k(x)$ is a polynomial of degree at most k.

10.2.3.2 Interactions and High-Dimensional Design Matrices

Interactions capture multiplicative effects: $x_i x_j$ terms allow the model to respond to feature co-occurrence. In high-dimensional settings, including all interactions can blow up feature dimension from d to $O(d^2)$, making design matrices huge and ill-conditioned. This is where sparsity assumptions, regularization, and feature selection become essential. Geometrically, interactions rotate and warp the feature space; statistically, they increase variance unless constrained.

Definition 10.2.11 (Pairwise Interaction Expansion) For $x \in \mathbb{R}^d$, define the interaction feature map that includes all pairwise products:

$$\phi_{\text{int}}(x) := \left(1,\ x_1, \ldots, x_d,\ \{x_i x_j\}_{1 \le i \le j \le d}\right)^\top.$$

Its dimension is $1 + d + \frac{d(d+1)}{2}$.

Remark 10.2.4 (High-Dimensional Caution) When $p \gg n$, the least squares problem is underdetermined and coefficients are not identifiable without additional structure (e.g., ridge/lasso). This is a geometric consequence of $\text{rank}(\Phi) \le n$ and a statistical consequence of variance blow-up.

10.3 Ordinary Least Squares (OLS)

Ordinary least squares (OLS) is the canonical regression method: choose the linear predictor that minimizes squared error. It is simultaneously (i) a geometric projection onto a column space, (ii) an MLE under Gaussian noise, and (iii) a computational linear algebra problem where conditioning determines numerical reliability. This section derives OLS via normal equations and projections, states the roadmap of statistical guarantees (Gauss–Markov, variance formulas, coefficient intervals), and discusses stable computation (QR/SVD) and rank-deficiency issues (Seber and Lee 2003) (cf. Fig. 10.2).

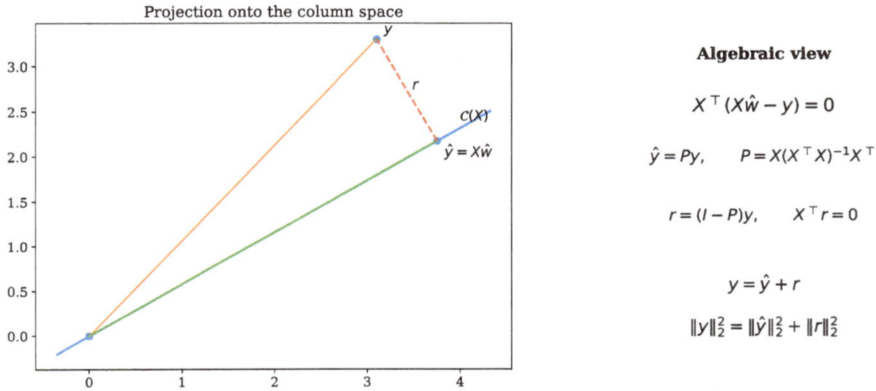

Fig. 10.2 OLS as orthogonal projection. The response vector is decomposed into a fitted component in the column space of the design matrix and a residual orthogonal to that space

10.3.1 Derivation and Geometry

10.3.1.1 Normal Equations and Projection Interpretation

The OLS objective is a convex quadratic in w. Differentiating yields normal equations. Geometrically, the fitted vector $X\widehat{w}$ is the orthogonal projection of y onto the column space $\mathcal{C}(X)$ in $\mathbb{R}^n$. This explains uniqueness and invariances: fitted values are unique even if coefficients are not (rank deficiency), and OLS is the "closest-point" approximation in Euclidean geometry.

Definition 10.3.1 (OLS Objective) Given $X \in \mathbb{R}^{n\times d}$ and $y \in \mathbb{R}^n$, OLS solves

$$\widehat{w} \in \arg\min_{w\in\mathbb{R}^d} \|Xw - y\|_2^2.$$

(Intercepts can be handled by augmenting X with a column of ones.)

Proposition 10.3.1 (Normal Equations) *Any minimizer $\widehat{w}$ satisfies*

$$X^\top(X\widehat{w} - y) = 0 \quad \Longleftrightarrow \quad (X^\top X)\widehat{w} = X^\top y.$$

If X has full column rank, then $X^\top X$ is invertible and the minimizer is unique:

$$\widehat{w} = (X^\top X)^{-1}X^\top y.$$

Remark 10.3.1 (Projection Form of Fitted Values) If X has full column rank, define the hat matrix

$$P := X(X^\top X)^{-1}X^\top \in \mathbb{R}^{n\times n}.$$

Then $\widehat{y} = X\widehat{w} = Py$ is the orthogonal projection of y onto $\mathcal{C}(X)$.

10.3.1.2 Residual Orthogonality and Fitted Subspaces

The defining geometric optimality condition is residual orthogonality: the residual is orthogonal to every column of X, hence to $\mathcal{C}(X)$. This yields the orthogonal decomposition $y = \widehat{y} + r$ with $\widehat{y} \in \mathcal{C}(X)$ and $r \in \mathcal{C}(X)^\perp$, and the Pythagorean decomposition of squared norms. In ML terms, it states that the residual is uncorrelated with each feature direction, a key diagnostic for model adequacy.

Definition 10.3.2 (Fitted Values and Residual) Let $\widehat{w}$ be an OLS solution. Define

$$\widehat{y} := X\widehat{w}, \qquad r := y - \widehat{y}.$$

Proposition 10.3.2 (Residual Orthogonality) *For any OLS solution $\widehat{w}$ with residual $r = y - X\widehat{w}$,*

$$X^\top r = 0, \qquad \textit{equivalently} \qquad r \in \mathcal{N}(X^\top) = \mathcal{C}(X)^\perp.$$

Consequently,

$$\|y\|_2^2 = \|\widehat{y}\|_2^2 + \|r\|_2^2.$$

Remark 10.3.2 (Degrees of Freedom) If P is the hat matrix, then $\mathrm{tr}(P) = \mathrm{rank}(X)$. This quantity acts as an effective number of parameters (degrees of freedom) in many risk estimates and model selection criteria.

10.3.2 Statistical Guarantees (Roadmap)

10.3.2.1 Gauss–Markov Intuition and Variance of Estimators

Statistical guarantees require a data-generating model. In the classical linear model $y = Xw^\star + \varepsilon$, OLS is unbiased and has minimum variance among linear unbiased estimators when errors are homoscedastic and uncorrelated. The variance of $\widehat{w}$ is governed by $(X^\top X)^{-1}$: poor conditioning inflates coefficient uncertainty. This connects geometry to statistics: the Gram matrix shape controls estimator variance, and hence generalization in parametric regimes.

Definition 10.3.3 (Classical Linear Model) Assume

$$y = Xw^\star + \varepsilon,$$

where X is treated as fixed (conditioning on X), $\mathbb{E}[\varepsilon \mid X] = 0$, and

$$\mathrm{Cov}(\varepsilon \mid X) = \sigma^2 I_n$$

(homoscedastic, uncorrelated noise).

Proposition 10.3.3 (Unbiasedness and Covariance of OLS (under CLM)) *If X has full column rank and the classical linear model holds, then*

$$\mathbb{E}[\widehat{w} \mid X] = w^\star, \qquad \mathrm{Cov}(\widehat{w} \mid X) = \sigma^2 (X^\top X)^{-1}.$$

Remark 10.3.3 (Gauss–Markov Theorem (Intuition Statement)) Under the classical linear model assumptions, OLS is the *BLUE* (Best Linear Unbiased Estimator): among all estimators of the form $\widetilde{w} = Ay$ that are unbiased for all $w^\star$, OLS has minimum covariance (in PSD order). We will use this as intuition; a full proof can be placed in an appendix if desired.

10.3.2.2 Confidence Intervals for Coefficients: High-Level Steps

To form coefficient intervals, we need a sampling distribution for $\widehat{w}$. Under Gaussian errors, $\widehat{w}$ is exactly normal conditional on X. With unknown σ^2, we estimate it from residuals, and studentized coefficients follow a t distribution (exact under Gaussianity). In large samples, asymptotic normality can also be used. The practical pipeline: estimate σ^2, compute standard errors from diagonal of $(X^\top X)^{-1}$, form t-intervals, and apply multiple-testing correction if many coefficients are examined.

Remark 10.3.4 (Standard Error and t-Interval Template) Let $\widehat{\sigma}^2 = \frac{1}{n-d}\|r\|_2^2$ (when X full rank and $n > d$). Then the standard error for coefficient j is

$$\widehat{\mathrm{se}}(\widehat{w}_j) = \widehat{\sigma}\sqrt{(X^\top X)^{-1}_{jj}}.$$

An approximate (or exact under Gaussian errors) $(1-\alpha)$ CI is

$$\widehat{w}_j \pm t_{n-d,\,1-\alpha/2}\,\widehat{\mathrm{se}}(\widehat{w}_j),$$

where $t_{n-d,\,1-\alpha/2}$ is a t-quantile with $n-d$ degrees of freedom.

10.3.3 Computation and Numerics

10.3.3.1 Solving via QR/SVD vs. Normal Equations

Computing $\widehat{w} = (X^\top X)^{-1} X^\top y$ via normal equations is tempting but numerically risky: $\kappa(X^\top X) = \kappa(X)^2$ amplifies conditioning and roundoff. Stable solvers use QR factorization (orthogonal transformations preserve norms) or SVD (explicitly reveals rank and filters small singular values). In ML pipelines, this distinction matters when features are correlated, scales differ, or d is large.

Remark 10.3.5 (QR Method (Conceptual)) If $X = QR$ with $Q \in \mathbb{R}^{n\times d}$ having orthonormal columns and $R \in \mathbb{R}^{d\times d}$ upper triangular, then OLS reduces to solving

$$R\widehat{w} = Q^\top y,$$

a stable triangular solve.

Remark 10.3.6 (SVD Method (Conceptual)) If $X = U\Sigma V^\top$, then the minimum-norm least-squares solution is

$$\widehat{w} = V\Sigma^\dagger U^\top y,$$

which remains meaningful under rank deficiency and supports truncated-SVD regularization.

10.3.3.2 Rank Deficiency and Identifiability Issues

When $\operatorname{rank}(X) < d$, the parameter vector is not identifiable: infinitely many w yield the same fitted values. However, the fitted vector $\widehat{y}$ is still uniquely defined as the projection of y onto $\mathcal{C}(X)$. To select a unique coefficient vector, one uses additional criteria: minimum norm (pseudoinverse), ridge regularization, or sparsity constraints. In practice, rank deficiency is common after feature expansions or multicollinearity; numerical methods must detect it and avoid unstable inverses.

Proposition 10.3.4 (Uniqueness of Fitted Values; Non-uniqueness of Coefficients) *For any X, the set* $\arg\min_w \|Xw - y\|_2^2$ *is nonempty, and all minimizers yield the same fitted values $\widehat{y}$. If* $\operatorname{rank}(X) < d$*, the minimizer is not unique: if $\widehat{w}$ is a minimizer and $v \in \mathcal{N}(X)$, then $\widehat{w} + v$ is also a minimizer.*

Remark 10.3.7 (Minimum-Norm Solution) Among all least-squares minimizers, the Moore–Penrose solution $w^\dagger = X^\dagger y$ has minimal Euclidean norm and lies in $\mathcal{C}(X^\top)$.

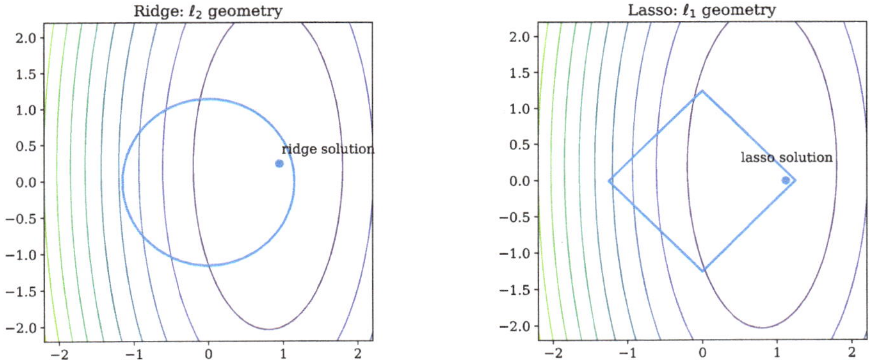

Fig. 10.3 Regularization geometry. Left: ridge corresponds to intersecting loss contours with an ℓ_2 ball, yielding smooth shrinkage. Right: lasso corresponds to intersecting loss contours with an ℓ_1 ball, whose corners promote sparse solutions

10.4 Regularization and Inductive Bias

Regularization formalizes the idea that, among many predictors that fit the training data well, we prefer those with certain *geometric* or *structural* properties (small norm, sparsity, group structure). This is *inductive bias*: a choice about what solutions are plausible before seeing data. Regularization also stabilizes computation by improving conditioning and reducing variance under noise. This section develops ridge (Tikhonov) regularization, lasso and elastic net, and practical selection of hyperparameters and paths (cf. Fig. 10.3).

10.4.1 Ridge Regression

10.4.1.1 Tikhonov View, Shrinkage, and Conditioning

Ridge regression adds an ℓ_2 penalty to OLS (Hoerl and Kennard 2000). Algebraically it replaces inversion of $X^\top X$ by inversion of $X^\top X + \lambda I$, shifting eigenvalues away from 0 and improving conditioning. Geometrically, it computes the best fit subject to an ℓ_2-ball constraint, shrinking coefficients toward 0. Spectrally, ridge is a *filter*: directions corresponding to small singular values of X are damped most strongly, which is exactly where OLS is unstable.

Definition 10.4.1 (Ridge (Tikhonov) Regression) Given $X \in \mathbb{R}^{n \times d}$, $y \in \mathbb{R}^n$, and $\lambda > 0$, ridge solves

$$\widehat{w}_\lambda \in \arg\min_{w \in \mathbb{R}^d} \|Xw - y\|_2^2 + \lambda \|w\|_2^2.$$

(Intercepts are typically unpenalized; handle by centering or augmenting with a separate bias parameter.)

Proposition 10.4.1 (Closed Form) *The ridge objective is strictly convex and has the unique minimizer*

$$\widehat{w}_\lambda = (X^\top X + \lambda I_d)^{-1} X^\top y.$$

Remark 10.4.1 (Constrained (Tikhonov) Viewpoint) Under convexity, the penalized problem is equivalent (for a suitable correspondence between λ and r) to

$$\min_w \|Xw - y\|_2^2 \quad \text{s.t.} \quad \|w\|_2 \le r,$$

so ridge is "least squares inside an ℓ_2 ball." This makes the geometric shrinkage effect explicit.

Proposition 10.4.2 (Spectral Shrinkage/Filter Form) *Let $X = U\Sigma V^\top$ be an SVD with singular values $\sigma_1 \ge \cdots \ge \sigma_r > 0$ and $r = \mathrm{rank}(X)$. Then*

$$\widehat{w}_\lambda = \sum_{i=1}^{r} \frac{\sigma_i}{\sigma_i^2 + \lambda} \left(u_i^\top y\right) v_i.$$

Thus components along v_i are shrunk by factor $\sigma_i/(\sigma_i^2+\lambda)$, with strongest shrinkage when σ_i is small.

Remark 10.4.2 (Conditioning Improvement) If $X^\top X$ has eigenvalues $\{\gamma_i\}$ (nonnegative), then $X^\top X + \lambda I$ has eigenvalues $\{\gamma_i + \lambda\}$. Hence the smallest eigenvalue is at least λ, preventing near-singularity. This is the computational reason ridge stabilizes inversion.

10.4.1.2 Bias–Variance Trade-Off and Effective Degrees of Freedom

Ridge deliberately introduces bias (shrinkage) to reduce variance (noise amplification). The trade-off is visible in closed form: $(X^\top X + \lambda I)^{-1}$ suppresses high-variance directions of OLS. To quantify model complexity under ridge, one uses *effective degrees of freedom*: the trace of the smoothing (hat) matrix. This generalizes the OLS fact $\mathrm{tr}(P) = d$ (or $\mathrm{rank}(X)$), and it plugs into model selection criteria and uncertainty heuristics.

Definition 10.4.2 (Ridge Hat Matrix and Effective Degrees of Freedom) Define the ridge fitted values

$$\widehat{y}_\lambda := X\widehat{w}_\lambda = S_\lambda y, \qquad S_\lambda := X(X^\top X + \lambda I)^{-1} X^\top \in \mathbb{R}^{n\times n}.$$

The *effective degrees of freedom* of ridge is

$$\mathrm{df}(\lambda) := \mathrm{tr}(S_\lambda).$$

Proposition 10.4.3 (Range of Effective Degrees of Freedom) *Let* $r = \mathrm{rank}(X)$. *Then* $\mathrm{df}(\lambda)$ *satisfies*

$$0 < \mathrm{df}(\lambda) < r, \qquad \mathrm{df}(\lambda) \downarrow 0 \textit{ as } \lambda \to \infty, \qquad \mathrm{df}(\lambda) \uparrow r \textit{ as } \lambda \downarrow 0.$$

Moreover, using the SVD of X,

$$\mathrm{df}(\lambda) = \sum_{i=1}^{r} \frac{\sigma_i^2}{\sigma_i^2 + \lambda}.$$

Remark 10.4.3 (Bias–Variance under the Classical Linear Model (Roadmap)) Assume $y = Xw^\star + \varepsilon$ with $\mathbb{E}[\varepsilon] = 0$ and $\mathrm{Cov}(\varepsilon) = \sigma^2 I$. Then

$$\mathbb{E}[\widehat{w}_\lambda] = (X^\top X + \lambda I)^{-1} X^\top X\, w^\star \;\Rightarrow\; \mathrm{Bias}(\widehat{w}_\lambda) = \Big((X^\top X + \lambda I)^{-1} X^\top X - I\Big) w^\star,$$

and

$$\mathrm{Cov}(\widehat{w}_\lambda) = \sigma^2 (X^\top X + \lambda I)^{-1} X^\top X (X^\top X + \lambda I)^{-1}.$$

Thus λ increases bias but decreases covariance; the same eigen-directions control both.

10.4.2 Lasso and Elastic Net

10.4.2.1 Sparsity, Feature Selection, and Geometry of ℓ_1

Lasso replaces ℓ_2 shrinkage by an ℓ_1 penalty, which has corners aligned with coordinate axes (Tibshirani 1996). Geometrically, these corners make optimal solutions land on low-dimensional faces, producing exact zeros (sparsity). Computationally, lasso is convex but nonsmooth; its optimality conditions involve subgradients. In an orthonormal-design regime, the solution has an explicit soft-thresholding form, revealing how sparsity is created.

Definition 10.4.3 (Lasso) Given $X \in \mathbb{R}^{n\times d}$, $y \in \mathbb{R}^n$, and $\lambda > 0$, the lasso solves

$$\widehat{w}_\lambda^{(1)} \in \arg\min_{w\in\mathbb{R}^d} \frac{1}{2}\|Xw - y\|_2^2 + \lambda \|w\|_1.$$

Definition 10.4.4 (Subgradient of ℓ_1) For $w \in \mathbb{R}^d$, the subgradient of $\|w\|_1$ is the set

$$\partial \|w\|_1 = \{s \in \mathbb{R}^d : s_i = \operatorname{sign}(w_i) \text{ if } w_i \neq 0, \text{ and } s_i \in [-1, 1] \text{ if } w_i = 0\}.$$

Proposition 10.4.4 (KKT Optimality for Lasso) *A vector $\widehat{w}$ is optimal for lasso iff there exists $s \in \partial \|\widehat{w}\|_1$ such that*

$$X^\top (X\widehat{w} - y) + \lambda s = 0.$$

Equivalently, for each coordinate j,

$$x_j^\top (y - X\widehat{w}) = \lambda s_j, \quad s_j \in \begin{cases} \{\operatorname{sign}(\widehat{w}_j)\}, & \widehat{w}_j \neq 0, \\ [-1, 1], & \widehat{w}_j = 0, \end{cases}$$

where x_j is the j-th column of X.

Proposition 10.4.5 (Soft-Thresholding in the Orthonormal Design Case) *Assume $X^\top X = I_d$ (orthonormal columns) and define $z = X^\top y$. Then the lasso solution is*

$$\widehat{w}_j = \operatorname{sign}(z_j) \max\{|z_j| - \lambda, 0\}, \qquad j \in [d].$$

Remark 10.4.4 (Geometric Meaning of Sparsity) The condition $|z_j| \leq \lambda \Rightarrow \widehat{w}_j = 0$ shows that coordinates whose correlation with the response is below the threshold are eliminated. The threshold arises from the ℓ_1 corner geometry and the subgradient interval $[-1, 1]$ at 0.

10.4.2.2 Scaling, Correlated Features, and Stability Issues

Because ℓ_1 penalizes coordinates directly, scaling of features changes the meaning of λ per feature: without standardization, lasso can unfairly penalize small-scale features and favor large-scale ones. Correlated features create instability: lasso may arbitrarily select one among a correlated group and set others to zero, leading to high variance in feature selection. Elastic net mitigates this by combining ℓ_1 and ℓ_2 penalties, encouraging group-wise selection and improving numerical stability.

Definition 10.4.5 (Elastic Net) Given $\lambda_1, \lambda_2 \geq 0$, elastic net solves

$$\widehat{w}^{(\text{EN})} \in \arg\min_{w \in \mathbb{R}^d} \frac{1}{2}\|Xw - y\|_2^2 + \lambda_1 \|w\|_1 + \frac{\lambda_2}{2}\|w\|_2^2.$$

Setting $\lambda_2 = 0$ recovers lasso; setting $\lambda_1 = 0$ recovers ridge (up to scaling conventions).

Remark 10.4.5 (Standardization Is Not Optional for ℓ_1 Methods) A common practice is to center y and standardize each feature column of X to unit variance (and often unit ℓ_2 norm) before fitting lasso/elastic net, then transform the coefficients back. This makes the penalty comparable across coordinates and stabilizes selection.

Remark 10.4.6 (Correlated Features and the Grouping Effect) With highly correlated columns, ridge tends to distribute weight across the group (shrink together), while lasso tends to pick one representative and drop others. Elastic net inherits sparsity from ℓ_1 but also encourages correlated features to enter together via the ℓ_2 term.

10.4.3 Model Selection and Tuning

10.4.3.1 Cross-Validation, Information Criteria

Regularization introduces hyperparameters (λ, or (λ_1, λ_2)) that must be chosen to generalize. Cross-validation estimates out-of-sample risk of the *tuning procedure* and is the default in ML. Information criteria (AIC/BIC) are likelihood-based approximations that trade off fit and complexity; in linear-Gaussian models, they can be expressed using residual variance and degrees of freedom (exact for OLS, effective for ridge). This preview provides the core formulas and interpretation.

Remark 10.4.7 (K-fold CV for Tuning (Template)) Choose λ by

$$\widehat{\lambda} \in \arg\min_{\lambda} \frac{1}{K} \sum_{k=1}^{K} \widehat{R}_{\text{val}}^{(k)}(\lambda),$$

where $\widehat{R}_{\text{val}}^{(k)}$ is the validation loss on fold k for the model trained (and preprocessed) without fold k.

Remark 10.4.8 (AIC/BIC (Gaussian Likelihood, Preview)) For a Gaussian regression model with maximized log-likelihood $\ell(\widehat{\theta})$ and parameter dimension p,

$$\text{AIC} = -2\ell(\widehat{\theta}) + 2p, \qquad \text{BIC} = -2\ell(\widehat{\theta}) + p \log n.$$

In OLS, $p = \text{rank}(X)$ (plus intercept if included); for ridge one often uses $p \approx \text{df}(\lambda) = \text{tr}(S_\lambda)$ as an effective complexity proxy.

10.4.3.2 Regularization Paths and Warm Starts (Conceptual)

As λ varies, the solution traces a path $\lambda \mapsto \widehat{w}_\lambda$. For ridge, the path is smooth and can be computed efficiently using linear algebra updates; for lasso, the path is piecewise linear in λ under standard conditions (leading to algorithms like LARS). Computationally, one exploits *warm starts*: solve for a decreasing sequence of λ values, initializing each solve at the previous solution, dramatically reducing cost. This is standard in large-scale ML pipelines.

Remark 10.4.9 (Warm-Start Principle) Let $\lambda_1 > \lambda_2 > \cdots > \lambda_M$. Use $\widehat{w}_{\lambda_{k-1}}$ as the initial point for solving at λ_k. Because solutions vary continuously (and sparsity patterns change gradually), optimization converges faster than from scratch.

Remark 10.4.10 (Ridge Path Identity (Useful for Intuition)) Ridge solutions satisfy

$$\widehat{w}_\lambda = (X^\top X + \lambda I)^{-1} X^\top y,$$

so decreasing λ increases flexibility and approaches OLS (or the minimum-norm least-squares solution) as $\lambda \downarrow 0$, while increasing λ shrinks $\widehat{w}_\lambda$ toward 0.

10.5 Generalized Linear Models (GLMs)

Generalized linear models (GLMs) extend linear regression to non-Gaussian outcomes by combining (McCullagh and Nelder 1989): (i) an exponential-family likelihood for $Y \mid X = x$, (ii) a linear predictor $\eta(x) = w^\top x + b$, and (iii) a link function connecting the mean response to the linear predictor. This yields a unified framework for logistic regression (binary), softmax regression (multiclass), Poisson regression (counts), and more (cf. Fig. 10.4). GLMs are foundational in ML because they provide calibrated probabilistic predictions, convex objectives in many cases, and efficient second-order solvers (IRLS) derived from likelihood curvature.

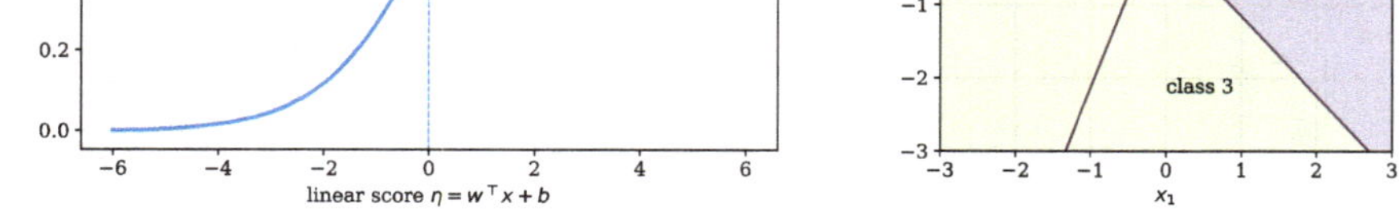

Fig. 10.4 GLM and classification geometry. Left: a linear score is mapped through a sigmoid to produce calibrated probabilities in logistic regression. Right: multiclass linear scores partition feature space into polyhedral decision regions

10.5.1 *Exponential Family and Link Functions*

10.5.1.1 Natural Parameter, Sufficient Statistics (Conceptual)

Exponential families provide likelihoods of the form $\log p(y \mid \theta) = \langle \eta(\theta), T(y) \rangle - A(\theta) + \log h(y)$. They come with sufficient statistics, convex log-partition functions, and clean derivatives (moments from A). GLMs exploit this structure by letting the natural parameter depend on covariates through a linear predictor. Conceptually, this is "linear modeling in the natural parameter space," which yields both interpretability and tractable inference.

Definition 10.5.1 (Exponential Family (One-Dimensional Response)) A conditional distribution of Y given covariates is an exponential family if its density/pmf can be written as

$$p(y \mid \theta) = h(y) \exp\Big(\eta(\theta)\, T(y) - A(\theta)\Big),$$

where $\eta(\theta)$ is the natural parameter, $T(y)$ is the sufficient statistic, and $A(\theta)$ is the log-partition function.

Remark 10.5.1 (Moments from the Log-Partition (Conceptual)) Under standard regularity, derivatives of A w.r.t. the natural parameter give moments:

$$\mathbb{E}[T(Y) \mid \eta] = A'(\eta), \qquad \mathrm{Var}(T(Y) \mid \eta) = A''(\eta).$$

This is the curvature-moment duality behind IRLS and Fisher information.

10.5.1.2 Canonical Links and Interpretation

A GLM models the conditional mean $\mu(x) = \mathbb{E}[Y \mid X = x]$ via a link function g: $g(\mu(x)) = \eta(x) = w^\top x + b$. The canonical link corresponds to using the natural parameter as the linear predictor, which simplifies likelihood and yields convexity in many cases. Interpretability comes from how x changes the linear predictor and hence the mean response; for canonical links, coefficients often have direct log-odds or log-rate interpretations.

Definition 10.5.2 (GLM: Linear Predictor and Link) A GLM specifies:

$$\eta(x) = w^\top x + b \quad \text{(linear predictor)}, \qquad g(\mu(x)) = \eta(x),$$
$$\mu(x) = \mathbb{E}[Y \mid X = x].$$

Here g is the link function and g^{-1} maps linear predictors to mean responses.

Remark 10.5.2 (Canonical Link (Conceptual)) In many exponential families, the canonical link sets $\eta(x)$ equal to the natural parameter. This yields particularly simple gradients/Hessians and often global convexity of the negative log-likelihood.

10.5.2 Logistic and Softmax Regression

10.5.2.1 Likelihood-Based Derivation and Gradients

Logistic regression is the canonical GLM for binary classification: it models $p(y = 1 \mid x)$ using a sigmoid of a linear score. Softmax regression (multinomial logistic) generalizes this to multiclass. Both are trained by maximizing conditional likelihood (equivalently minimizing cross-entropy). The gradients have a clean form: they are feature-weighted residuals between predicted probabilities and labels, which is why these models fit naturally into gradient-based optimization.

Definition 10.5.3 (Logistic Regression (Binary)) Let $y \in \{0, 1\}$ and define the score $\eta = w^\top x + b$. Logistic regression models

$$p_\theta(y = 1 \mid x) = \sigma(\eta) = \frac{1}{1 + e^{-\eta}}, \qquad p_\theta(y \mid x) = \sigma(\eta)^y (1 - \sigma(\eta))^{1-y}.$$

Given data $\{(x_i, y_i)\}_{i=1}^n$, the negative log-likelihood (cross-entropy) is

$$f(w, b) = -\sum_{i=1}^{n} \big(y_i \log \sigma(\eta_i) + (1 - y_i) \log(1 - \sigma(\eta_i))\big), \qquad \eta_i = w^\top x_i + b.$$

Proposition 10.5.1 (Gradient of Logistic NLL) *Let $p_i = \sigma(\eta_i)$. Then*

$$\nabla_w f(w, b) = \sum_{i=1}^{n} (p_i - y_i)\, x_i, \qquad \frac{\partial f}{\partial b} = \sum_{i=1}^{n} (p_i - y_i).$$

Definition 10.5.4 (Softmax Regression (Multiclass)) Let $y \in \{1, \ldots, K\}$ and scores $s(x) = Wx + b \in \mathbb{R}^K$ with $W \in \mathbb{R}^{K \times d}$, $b \in \mathbb{R}^K$. Define probabilities

$$p_\theta(Y = k \mid x) = \frac{e^{s_k(x)}}{\sum_{j=1}^{K} e^{s_j(x)}}.$$

With one-hot labels e_{y_i}, the NLL is

$$f(W, b) = -\sum_{i=1}^{n} \log p_\theta(Y = y_i \mid x_i) = -\sum_{i=1}^{n} \Big(s_{y_i}(x_i) - \log \sum_{j=1}^{K} e^{s_j(x_i)}\Big).$$

Remark 10.5.3 (Gradient Form (Conceptual)) For softmax, gradients are also "probability minus label" times features:

$$\nabla_W f = \sum_{i=1}^{n} (p_i - e_{y_i}) x_i^\top,$$

where $p_i \in \mathbb{R}^K$ is the predicted probability vector at x_i.

10.5.2.2 Decision Boundaries and Calibration

A GLM yields both a hard decision rule and a probabilistic forecast. For logistic regression, the 0.5 threshold corresponds to the hyperplane $w^\top x + b = 0$. For softmax regression, decision boundaries between classes are also hyperplanes: differences of linear scores. Because GLMs produce probabilities from a likelihood model, they are often reasonably calibrated under correct specification, but miscalibration can occur under misspecification or regularization. Calibration should be evaluated and, if needed, corrected.

Proposition 10.5.2 (Decision Boundary in Logistic Regression) *In logistic regression, predicting class* $\widehat{y} = \mathbf{1}\{p_\theta(Y = 1 \mid x) \geq 1/2\}$ *is equivalent to*

$$w^\top x + b \geq 0.$$

Thus, logistic regression has a linear decision boundary.

Proposition 10.5.3 (Pairwise Boundaries in Softmax Regression) *In softmax regression, the boundary between classes a and b is*

$$\{x : s_a(x) = s_b(x)\} = \{x : (W_{a:} - W_{b:})x + (b_a - b_b) = 0\},$$

a hyperplane. Thus multiclass softmax partitions space into polyhedral regions.

Remark 10.5.4 (Calibration Note) Logistic/softmax likelihood training optimizes log-loss, a proper scoring rule, which encourages calibrated probabilities. However, overparameterization, regularization, and dataset shift can degrade calibration; reliability diagrams and ECE are standard diagnostics.

10.5.3 IRLS and Optimization View

10.5.3.1 Newton/IRLS Updates as Weighted Least Squares

The logistic and many GLM objectives are convex but do not have closed-form minimizers. Newton's method uses curvature to accelerate convergence. For GLMs, the Hessian has the form $X^\top WX$ with a data-dependent weight matrix W, so each Newton step solves a weighted least squares problem. This is IRLS. Geometrically, IRLS repeatedly refines a local quadratic approximation and projects onto the column space under a changing metric.

Remark 10.5.5 (Logistic Hessian Structure (Binary)) For logistic regression with design matrix $X \in \mathbb{R}^{n\times d}$ (no intercept shown), let $p_i = \sigma(x_i^\top w)$ and define $W = \operatorname{diag}(p_i(1-p_i))$. Then the Hessian of the NLL is

$$\nabla_w^2 f(w) = X^\top WX,$$

which is PSD, implying convexity.

Remark 10.5.6 (IRLS Step (Conceptual)) A Newton step solves

$$(X^\top WX)\,\Delta w = X^\top(y-p), \qquad w^{(t+1)} = w^{(t)} + \Delta w,$$

where p and W are evaluated at $w^{(t)}$. This is equivalent to a weighted least squares problem on a working response.

10.5.3.2 Regularized GLMs and Practical Convergence Issues

Regularization is essential in high dimensions and under collinearity. For GLMs, adding ℓ_2 regularization preserves smoothness and strong convexity (often improving conditioning and convergence), while ℓ_1 regularization yields sparse logistic regression but introduces nonsmoothness (requiring coordinate descent or proximal methods). Practically, convergence depends on feature scaling, separability (which can make MLE diverge in logistic regression), and numerical stability. Robust workflows standardize features, add regularization, and use line search/damping in Newton methods.

Definition 10.5.5 (Regularized Logistic Regression) A common regularized objective is

$$\min_{w,b}\ f(w,b) + \frac{\lambda}{2}\|w\|_2^2 \quad \text{(ridge/logistic)}, \qquad \text{or}$$

$$\min_{w,b}\ f(w,b) + \lambda\|w\|_1 \quad \text{(lasso/logistic)}.$$

Remark 10.5.7 (Complete Separation (Pathology)) If classes are perfectly separable, the unregularized logistic likelihood has no finite maximizer: $\|w\|_2 \to \infty$ along a separating direction. Regularization (or Bayesian priors) restores a well-defined optimum and finite coefficients.

Remark 10.5.8 (Practical Convergence Safeguards) Standardize features; use damping/line-search for Newton/IRLS; monitor gradient norms and objective decrease; prefer stable solvers (QR/Cholesky with safeguards) for linear systems inside IRLS; and add $\lambda > 0$ to ensure strong convexity.

10.6 Classification Geometry and Performance

Linear classifiers are geometric objects: they partition feature space into decision regions separated by hyperplanes. Performance is therefore governed by geometry (margins, separability) and by probabilistic quality (calibration), and it is summarized by metric curves (ROC/PR) that depend on thresholds and prevalence. This section connects these components into a coherent evaluation viewpoint.

10.6.1 Decision Boundaries and Margins

10.6.1.1 Separability, Margins, and Robustness

Separability is the existence of a hyperplane that classifies all training points correctly. But many separating hyperplanes exist; margin quantifies how far points are from the boundary and directly controls robustness: a perturbation must be large enough to cross the hyperplane. In linear models, the dual-norm relationship $\sup_{\|\delta\|\le\varepsilon} |w^\top \delta| = \varepsilon\|w\|_*$ gives a precise robustness radius. This is why max-margin methods (SVM) and norm control are central: they improve robustness and often generalization.

Definition 10.6.1 (Linear Classifier, Margin, and Separability) Let $y_i \in \{-1, +1\}$ and classifier score $s(x) = w^\top x + b$. The dataset $\{(x_i, y_i)\}_{i=1}^n$ is *linearly separable* if there exists (w, b) such that

$$y_i(w^\top x_i + b) > 0 \quad \forall i.$$

For a fixed (w, b) with $w \neq 0$, the geometric margin on example i is

$$\gamma_i^{\text{geo}} := \frac{y_i(w^\top x_i + b)}{\|w\|_2}, \qquad \gamma^{\text{geo}} := \min_{i\in[n]} \gamma_i^{\text{geo}}.$$

Proposition 10.6.1 (Robustness Radius via Dual Norms) *Fix (w, b) and a norm $\|\cdot\|$ with dual $\|\cdot\|_*$. If $y(w^\top x + b) \ge \gamma > 0$, then for any perturbation δ with*

$$\|\delta\| \le \frac{\gamma}{\|w\|_*},$$

the prediction remains unchanged: $y(w^\top(x + \delta) + b) \ge 0$.

Remark 10.6.1 (Geometric Meaning) For $\|\cdot\| = \|\cdot\|_2$, the robustness radius equals the signed distance to the hyperplane. For $\|\cdot\| = \|\cdot\|_\infty$, it is $\gamma/\|w\|_1$ (coordinatewise-bounded perturbations).

10.6.1.2 Cost-Sensitive Classification and Class Imbalance

In many ML tasks, false positives and false negatives have different costs, and class prevalence can be highly imbalanced. The Bayes-optimal decision threshold is no longer $1/2$; it shifts according to costs and priors. Geometrically, changing the threshold translates the decision boundary (changes b) while keeping the normal vector w fixed. Thus, one can decouple representation/score learning from operating-point selection: train a probabilistic model, then choose a threshold to optimize a cost-sensitive metric on validation data.

Definition 10.6.2 (Cost-Sensitive 0–1 Loss) For binary $Y \in \{0, 1\}$ and decision $\widehat{y} \in \{0, 1\}$, define costs

$$L(\widehat{y}, y) = \begin{cases} 0, & \widehat{y} = y, \\ c_{\mathrm{FP}}, & (\widehat{y}, y) = (1, 0), \\ c_{\mathrm{FN}}, & (\widehat{y}, y) = (0, 1), \end{cases} \qquad c_{\mathrm{FP}}, c_{\mathrm{FN}} > 0.$$

Proposition 10.6.2 (Bayes Threshold under Costs) *Let $\eta(x) := \mathbb{P}(Y = 1 \mid X = x)$. Under the cost-sensitive loss of Definition 10.6.2, a Bayes rule predicts $\widehat{y} = 1$ iff*

$$\eta(x) \geq \frac{c_{\mathrm{FP}}}{c_{\mathrm{FP}} + c_{\mathrm{FN}}}.$$

Remark 10.6.2 (Imbalance and Prevalence) Even with symmetric costs, different base rates across deployment contexts can shift the optimal threshold. This is why calibration and threshold selection should be tied to the deployment distribution, not only the training distribution.

10.6.2 Probabilistic Outputs and Calibration

10.6.2.1 Log-Loss vs. 0–1 Loss and Proper Scoring

0–1 loss evaluates only hard decisions, ignoring probability quality. Log-loss evaluates probabilistic predictions and is a proper scoring rule: it incentivizes honest conditional probabilities. These objectives can yield different models: minimizing log-loss focuses on calibrated probabilities and likelihood fit, while minimizing 0–1 loss focuses on classification boundaries. In practice, logistic regression minimizes log-loss, then a threshold yields 0–1 decisions. Understanding this separation clarifies why calibration can be improved without changing accuracy.

Definition 10.6.3 (Log-Loss and 0–1 Loss) For $y \in \{0, 1\}$ and predicted probability $q \in (0, 1)$,

$$\ell_{\log}(q, y) = -\big(y \log q + (1 - y) \log(1 - q)\big), \qquad \ell_{0\text{–}1}(\widehat{y}, y) = \mathbf{1}\{\widehat{y} \neq y\}.$$

Remark 10.6.3 (Proper Scoring (High-Level)) Log-loss is a proper scoring rule: the true conditional probability minimizes expected log-loss. Thus, likelihood-based training targets probability calibration (under correct specification) rather than only boundary accuracy.

10.6.2.2 Platt Scaling/Isotonic Regression

Even if a model ranks examples well, its raw scores may not be calibrated probabilities. Calibration maps scores to probabilities using a held-out set. Platt scaling fits a logistic map $q = \sigma(as + b)$ (parametric, smooth); isotonic regression fits a monotone piecewise-constant map (nonparametric, flexible). Both are post-estimation problems: fit g minimizing log-loss on validation, then deploy $g \circ s$. Calibration must be performed on data representative of deployment.

Remark 10.6.4 (Platt Scaling (Template)) Given scores s_i and labels y_i, fit parameters (a, b) by minimizing

$$\sum_i -\big(y_i \log \sigma(as_i + b) + (1 - y_i) \log(1 - \sigma(as_i + b))\big),$$

and output calibrated probabilities $\widehat{q} = \sigma(as + b)$.

Remark 10.6.5 (Isotonic Regression (Template)) Fit a nondecreasing function $g : [\min s, \max s] \to [0, 1]$ minimizing $\sum_i (g(s_i) - y_i)^2$ (or log-loss variants) subject to monotonicity. This yields a flexible calibration curve.

10.6.3 Metrics and Thresholding

10.6.3.1 ROC/PR Curves and Operating Points

A probabilistic classifier induces a family of hard classifiers by thresholding: $\widehat{y}_t = \mathbf{1}\{q(x) \geq t\}$. ROC and PR curves summarize performance across all thresholds. ROC focuses on TPR vs. FPR and is relatively insensitive to prevalence; PR focuses on precision vs. recall and is sensitive to prevalence, making it more informative under class imbalance. Selecting an operating point is a decision problem: choose t to optimize a cost function or constraint (e.g., FPR $\leq \alpha$).

Definition 10.6.4 (Confusion Rates at Threshold t) Let $q(x)$ be a score/probability and $\widehat{y}_t = \mathbf{1}\{q(x) \geq t\}$. Define (population) rates:

$$\mathrm{TPR}(t) = \mathbb{P}(\widehat{y}_t = 1 \mid Y = 1), \quad \mathrm{FPR}(t) = \mathbb{P}(\widehat{y}_t = 1 \mid Y = 0),$$

$$\mathrm{Precision}(t) = \mathbb{P}(Y = 1 \mid \widehat{y}_t = 1), \quad \mathrm{Recall}(t) = \mathrm{TPR}(t).$$

Remark 10.6.6 (ROC and PR Curves) The ROC curve plots $\mathrm{TPR}(t)$ vs. $\mathrm{FPR}(t)$ as t varies. The PR curve plots $\mathrm{Precision}(t)$ vs. $\mathrm{Recall}(t)$ as t varies. AUC-ROC and AUC-PR integrate these curves into scalar summaries, but threshold selection is still needed for deployment.

10.6.3.2 Precision/Recall Tradeoffs and Prevalence Effects

Precision depends on prevalence: even a classifier with fixed TPR/FPR will have lower precision when positives are rare. Thus PR curves change with the class prior, while ROC curves do not (in their conditional-rate definition). In ML, this matters when deployment prevalence differs from training (dataset shift) or when evaluating on case-control datasets. Thresholds must therefore be selected with respect to the deployment base rate and costs.

Proposition 10.6.3 (Precision in Terms of TPR/FPR and Prevalence) *Let $\pi = \mathbb{P}(Y = 1)$ be prevalence. Then for any threshold t,*

$$\mathrm{Precision}(t) = \frac{\pi\,\mathrm{TPR}(t)}{\pi\,\mathrm{TPR}(t) + (1 - \pi)\,\mathrm{FPR}(t)}.$$

Remark 10.6.7 (Consequence under Imbalance) If π is small, the denominator is dominated by $(1 - \pi)\mathrm{FPR}(t)$ unless $\mathrm{FPR}(t)$ is extremely small. Thus, high precision in rare-event detection requires extremely low false positive rates or very strong separation.

10.7 Diagnostics, Influence, and Model Criticism

Fitting a linear model is only half the job; the other half is verifying that the model is reliable and interpretable. Diagnostics identify overfitting/underfitting, distributional misspecification, and numerical instability. Influence and leverage quantify how much individual points affect the fit (and thus how fragile conclusions are). Multicollinearity undermines identifiability and inflates uncertainty in coefficients. This section provides the geometric and statistical tools used to critique linear models in ML pipelines.

10.7.1 Generalization and Diagnostics

10.7.1.1 Learning Curves and Validation Curves

Learning curves plot performance as a function of training set size; validation curves plot performance as a function of a hyperparameter (e.g., λ). Both are estimators of generalization behavior: they diagnose whether more data or different regularization would likely help. In linear models, these curves often have interpretable shapes: large train–test gaps indicate variance/overfitting; high errors on both indicate bias/underfitting; U-shaped validation curves reflect bias–variance tradeoffs in λ.

Definition 10.7.1 (Learning Curve) Fix a training procedure (including preprocessing and hyperparameters). For training sizes $n_1 < n_2 < \cdots$, let $\widehat{R}_{\text{train}}(n_k)$ and $\widehat{R}_{\text{val}}(n_k)$ denote training and validation risk estimates when training on n_k samples. The plots $n_k \mapsto \widehat{R}_{\text{train}}(n_k)$ and $n_k \mapsto \widehat{R}_{\text{val}}(n_k)$ form the learning curve.

Definition 10.7.2 (Validation Curve) For hyperparameter values $\lambda \in \Lambda$, let $\widehat{R}_{\text{val}}(\lambda)$ be the validation estimate of risk. The plot $\lambda \mapsto \widehat{R}_{\text{val}}(\lambda)$ is the validation curve.

Remark 10.7.1 (Practical Use) Learning curves suggest whether you are data-limited (validation improves with n) or bias-limited (both curves plateau high). Validation curves guide regularization strength and reveal instability when performance changes sharply with λ.

10.7.1.2 Overfitting, Underfitting, and Model Capacity

Overfitting is variance-dominated behavior: the model adapts to sample noise. Underfitting is bias-dominated behavior: the model class cannot represent the signal. Capacity in linear models is governed by effective degrees of freedom: rank in OLS and $\operatorname{tr}(S_\lambda)$ in ridge. Feature expansions increase capacity; regularization reduces it. This provides a quantitative bridge between geometry (subspace dimension) and generalization (variance of the estimator).

Remark 10.7.2 (Capacity Diagnostics via Train–Validation Gap) A large gap (low train loss, high validation loss) suggests high variance/overfitting. A small gap but both losses high suggests bias/underfitting. A good regime has both low and close losses (subject to noise floor).

Remark 10.7.3 (Effective Degrees of Freedom (Recall)) For ridge, $\operatorname{df}(\lambda) = \operatorname{tr}(S_\lambda) = \sum_i \frac{\sigma_i^2}{\sigma_i^2+\lambda}$. As λ decreases, df increases and variance typically increases.

10.7.2 *Influence and Leverage*

10.7.2.1 Hat Matrix Intuition and Leverage Points

Leverage measures how "far" a point's features are in the design space and how strongly it can affect the fitted values. The hat matrix P maps y to fitted values: $\widehat{y} = Py$. Its diagonal entries $h_{ii} = P_{ii}$ are leverage scores; large leverage means point i can heavily influence its own fitted value. In ML, leverage diagnostics help detect distribution shift, rare subpopulations, and points that dominate the fit.

Definition 10.7.3 (Hat Matrix and Leverage) Assume $X \in \mathbb{R}^{n\times d}$ has full column rank. The hat matrix is

$$P := X(X^\top X)^{-1}X^\top, \qquad \widehat{y} = Py.$$

The *leverage* of point i is $h_{ii} := P_{ii}$.

Proposition 10.7.1 (Basic Leverage Facts) *For the hat matrix P:*

$$0 \le h_{ii} \le 1, \qquad \sum_{i=1}^{n} h_{ii} = \mathrm{tr}(P) = d,$$

and h_{ii} depends only on X (not on y).

Remark 10.7.4 (Geometric Interpretation) h_{ii} is the squared length of the projection of the i-th standard basis vector e_i onto $\mathcal{C}(X)$ in $\mathbb{R}^n$. Equivalently, it quantifies how much the i-th observation influences the fitted values through the geometry of X.

10.7.2.2 Outliers, Cook's Distance (Conceptual)

An outlier is a point with unusually large residual; a leverage point has unusual features. The most problematic are points with both high leverage and large residual: they can drastically change the fitted coefficients. Cook's distance summarizes influence by approximating how much the fitted vector changes when a point is deleted. While we present it conceptually, its logic is simple: combine residual size with leverage (and scale by noise).

Remark 10.7.5 (Studentized Residual (Conceptual)) A common diagnostic is the studentized residual

$$t_i = \frac{r_i}{\widehat{\sigma}\sqrt{1-h_{ii}}},$$

where r_i is the residual and $\widehat{\sigma}^2 = \|r\|_2^2/(n-d)$. Large $|t_i|$ indicates an outlier relative to model noise and leverage.

Remark 10.7.6 (Cook's Distance (Conceptual Form)) Cook's distance for point i is often written as

$$D_i = \frac{r_i^2}{d\,\widehat{\sigma}^2} \cdot \frac{h_{ii}}{(1-h_{ii})^2}.$$

Large D_i indicates that removing point i would substantially change the fitted coefficients/predictions.

10.7.3 Multicollinearity and Identifiability

10.7.3.1 Variance Inflation and Unstable Coefficients

Multicollinearity means features are nearly linearly dependent. Geometrically, $X^\top X$ has small eigenvalues; statistically, $\mathrm{Var}(\widehat{w}) = \sigma^2(X^\top X)^{-1}$ inflates along those directions. Thus coefficients become unstable: small perturbations in data yield large changes in $\widehat{w}$, even if fitted values are relatively stable. In ML, this undermines interpretability and can cause numerical issues, especially with feature expansions and interactions.

Remark 10.7.7 (Variance Inflation (Directional Form)) For any direction $v \in \mathbb{R}^d$,

$$\mathrm{Var}(v^\top \widehat{w} \mid X) = \sigma^2\, v^\top (X^\top X)^{-1} v.$$

If v aligns with a small-eigenvalue direction of $X^\top X$, this variance is large.

Remark 10.7.8 (Practical Symptom) Coefficient signs/magnitudes change dramatically across resamples or minor preprocessing changes, while predictive performance remains similar. This indicates non-identifiability in coefficient space.

10.7.3.2 Remedies: Regularization, Feature Selection, PCA

Remedies aim to remove or stabilize near-null directions: ridge adds λI and improves conditioning; lasso selects a sparse subset of features; PCA regression projects data onto stable principal components before regression; feature selection removes redundant variables. All these are geometric operations on the column space and its conditioning. The right choice depends on goal: interpretability (sparse features), prediction (ridge), or dimension reduction (PCA).

Remark 10.7.9 (Remedy Map)

- **Ridge:** $(X^\top X+\lambda I)^{-1}$ controls variance blow-up in small-eigenvalue directions.
- **Lasso:** reduces dimensionality by setting many coefficients to zero (but can be unstable under correlated features).
- **PCA regression:** regress on top-k principal components of X to remove low-variance/noisy directions.
- **Feature selection:** remove near-duplicate or highly correlated features, or use domain-driven grouping.

10.8 Extensions and Bridges

Linear models are not merely endpoints; they are bridges. By expanding features, one turns nonlinear relationships into linear structure in a larger space. By penalizing roughness, one controls smoothness and avoids overfitting, yielding models that are interpretable yet flexible. Finally, linear models appear as first-order approximations of arbitrary nonlinear predictors via Taylor/Jacobian linearization, which explains both local interpretability and the widespread use of linear probes in representation learning.

10.8.1 Basis Expansion and Generalized Additive Models

10.8.1.1 Splines and Smoothness Penalties (Conceptual)

Splines provide a systematic basis expansion for univariate or low-dimensional functions using piecewise polynomials joined smoothly at knots. A spline model is linear in its basis coefficients, so it remains a linear regression problem after feature mapping. To prevent overfitting, one penalizes roughness (e.g., integrated squared second derivative), which is a quadratic form in coefficients. Thus splines are a prototypical example of "linear model + structured regularization" and serve as a bridge from parametric linear models to nonparametric function estimation.

Remark 10.8.1 (Spline Basis Regression (Template)) For scalar input x, choose spline basis functions $b_1(x), \ldots, b_p(x)$ and model

$$f(x) = \sum_{j=1}^{p} \theta_j b_j(x) = \theta^\top b(x).$$

Fitting θ by least squares is linear regression on the design matrix $\Phi_{ij} = b_j(x_i)$. A smoothness penalty has the form $\theta^\top \Omega \theta$ for a PSD matrix Ω determined by the basis and the chosen roughness functional.

Remark 10.8.2 (Smoothness Penalty Intuition) A common penalty is $\int (f''(x))^2\, dx$, which discourages high curvature. This acts like ridge regularization in a derivative-weighted geometry, selecting smooth functions among those that fit the data.

10.8.1.2 Additivity as an Inductive Bias

Additivity asserts that the response is a sum of one-dimensional (or low-dimensional) effects:

$$f(x) = \sum_{j=1}^{d} f_j(x_j).$$

This reduces complexity relative to a full multivariate function and improves interpretability: each feature contributes via its own shape function. In ML, generalized additive models (GAMs) are strong baselines that capture nonlinearity without high-dimensional interactions. Geometrically, additivity restricts the hypothesis class to a direct sum of function subspaces, making estimation and regularization modular.

Definition 10.8.1 (Generalized Additive Model (GAM) Form) A generalized additive model for regression is

$$\widehat{y} = \beta_0 + \sum_{j=1}^{d} f_j(x_j),$$

where each f_j belongs to a chosen function class (e.g., splines with smoothness penalties).

Remark 10.8.3 (Bias–Variance Tradeoff Under Additivity) Additivity reduces variance by restricting interactions, but can introduce bias if true interactions are important. Thus it is an inductive bias: beneficial when interactions are weak or data are limited, limiting when interactions drive the signal.

10.8.2 Linear Models as First-Order Approximations

10.8.2.1 Local Linearization of Nonlinear Predictors

Every differentiable nonlinear predictor behaves like an affine function in a small neighborhood. This is the precise content of Taylor's theorem: around a point x_0, the function is approximated by its gradient (Jacobian). This provides local interpretability (saliency, sensitivity), explains adversarial perturbations (linearized loss gradients), and motivates local surrogate models such as LIME. In optimization, this is also why gradient-based methods are meaningful: they are driven by local linear approximations.

Proposition 10.8.1 (First-Order Taylor Approximation) *Let $f : \mathbb{R}^d \to \mathbb{R}$ be differentiable at x_0. Then for small $\Delta \in \mathbb{R}^d$,*

$$f(x_0 + \Delta) = f(x_0) + \nabla f(x_0)^\top \Delta + o(\|\Delta\|_2).$$

For vector-valued $f : \mathbb{R}^d \to \mathbb{R}^k$, replace $\nabla f(x_0)$ by the Jacobian $J_f(x_0)$:

$$f(x_0 + \Delta) = f(x_0) + J_f(x_0)\Delta + o(\|\Delta\|_2).$$

Remark 10.8.4 (Sensitivity and Gradients) The gradient $\nabla f(x_0)$ gives the direction of steepest local increase under ℓ_2 geometry. Thus local robustness and adversarial sensitivity are controlled by gradient norms and dual norms.

10.8.2.2 Linear Probes and Representation Evaluation

A linear probe evaluates a learned representation $\phi(x)$ by training only a linear model on top of it. Because linear training is convex and fast, probe performance isolates representational quality: if a linear model performs well, the representation has made the task nearly linearly separable. This diagnostic is widely used in self-supervised learning and transfer learning, where the goal is to learn representations that support many downstream tasks with simple heads.

Definition 10.8.2 (Linear Probe) Given a fixed embedding map $\phi : \mathcal{X} \to \mathbb{R}^m$, a linear probe fits parameters (W, b) to minimize a supervised loss on $(\phi(x_i), y_i)$:

$$\min_{W,b} \sum_{i=1}^{n} \ell\big(W\phi(x_i) + b,\ y_i\big),$$

with ℓ typically squared loss (regression) or cross-entropy (classification).

Remark 10.8.5 (Interpretation) High linear-probe accuracy suggests that the representation organizes the data into geometrically simple structures (e.g., separable by hyperplanes). Low accuracy suggests that either the task is not linearly representable in ϕ-space or that more complex heads (nonlinearities) are needed.

10.9 Exercises

Unless stated otherwise, use the column-vector convention, natural logarithms, and treat all matrices/vectors as real.

Exercise 10.9.1 (Regression Setup: Design Matrix, Hypothesis Class, and Risks) Consider the dataset with $n = 6, d = 2$:

$$x_1 = \begin{bmatrix} 0 \\ 0 \end{bmatrix}, \ y_1 = 0; \quad x_2 = \begin{bmatrix} 1 \\ 0 \end{bmatrix}, \ y_2 = 1; \quad x_3 = \begin{bmatrix} 2 \\ 0 \end{bmatrix}, \ y_3 = 2;$$

$$x_4 = \begin{bmatrix} 0 \\ 1 \end{bmatrix}, \ y_4 = 1; \quad x_5 = \begin{bmatrix} 1 \\ 1 \end{bmatrix}, \ y_5 = 2; \quad x_6 = \begin{bmatrix} 2 \\ 1 \end{bmatrix}, \ y_6 = 3.$$

(i) Form $X \in \mathbb{R}^{6\times 2}$ and $y \in \mathbb{R}^6$ and write the affine model $w^\top x + b$ as a linear model with augmented design $\widetilde{X}$.

(ii) For the parameter $\widetilde{w} = \begin{bmatrix} w \\ b \end{bmatrix} = \begin{bmatrix} 1 \\ 1 \\ 0 \end{bmatrix}$, compute $\widehat{y} = \widetilde{X}\widetilde{w}$ and the empirical squared risk $\widehat{R}_n = \frac{1}{6}\sum_i \frac{1}{2}(\widehat{y}_i - y_i)^2$.

(iii) Assume a population model $Y = w_\star^\top X + b_\star + \varepsilon$ with $\mathbb{E}[\varepsilon \mid X] = 0$ and $\mathrm{Var}(\varepsilon \mid X) = 1$; express the population risk under squared loss in terms of $\mathbb{E}[(w^\top X + b - w_\star^\top X - b_\star)^2]$ plus a noise term.

Exercise 10.9.2 (OLS by Normal Equations and Projection Geometry) Use the dataset in Exercise 10.9.1 with augmented design $\widetilde{X} \in \mathbb{R}^{6\times 3}$.

(i) Compute $\widehat{\widetilde{w}}_{\mathrm{OLS}} = (\widetilde{X}^\top\widetilde{X})^{-1}\widetilde{X}^\top y$ exactly.

(ii) Compute fitted values $\widehat{y} = \widetilde{X}\widehat{\widetilde{w}}$ and residual $r = y - \widehat{y}$.

(iii) Verify residual orthogonality: $\widetilde{X}^\top r = 0$.

(iv) Compute the hat matrix $P = \widetilde{X}(\widetilde{X}^\top\widetilde{X})^{-1}\widetilde{X}^\top$ and verify $P^2 = P$, $P^\top = P$, and $\widehat{y} = Py$.

Exercise 10.9.3 (OLS Statistical Variance under the Classical Linear Model) Let $X \in \mathbb{R}^{5\times 2}$ and $y \in \mathbb{R}^5$ be

$$X = \begin{bmatrix} 1 & 0 \\ 1 & 1 \\ 1 & 2 \\ 1 & 3 \\ 1 & 4 \end{bmatrix}, \qquad y = \begin{bmatrix} 0 \\ 1 \\ 1 \\ 2 \\ 2 \end{bmatrix}.$$

Assume the classical linear model $y = X\beta + \varepsilon$ with $\mathbb{E}[\varepsilon \mid X] = 0$ and $\mathrm{Cov}(\varepsilon \mid X) = \sigma^2 I_5$.

(i) Compute $\widehat{\beta}_{\mathrm{OLS}} = (X^\top X)^{-1}X^\top y$.

(ii) Compute $(X^\top X)^{-1}$ explicitly and hence express $\mathrm{Cov}(\widehat{\beta}_{\mathrm{OLS}} \mid X) = \sigma^2(X^\top X)^{-1}$.

(iii) Estimate σ^2 by $\widehat{\sigma}^2 = \|r\|_2^2/(5-2)$ with $r = y - X\widehat{\beta}$ and compute numerical standard errors for $\widehat{\beta}_0$ and $\widehat{\beta}_1$.

(iv) Give a 95% t-interval for β_1 using $t_{3,0.975}$ (leave $t_{3,0.975}$ symbolically if desired).

Exercise 10.9.4 (Numerics: Normal Equations vs. QR) Let

$$X = \begin{bmatrix} 1 & 1 \\ 1 & 1.001 \\ 1 & 0.999 \\ 1 & 1.002 \end{bmatrix}, \qquad y = \begin{bmatrix} 1 \\ 1 \\ 1 \\ 1 \end{bmatrix}.$$

(i) Compute $X^\top X$ and its condition number $\kappa_2(X^\top X)$ (you may compute eigenvalues explicitly).
(ii) Compute the OLS solution by normal equations $w_{\mathrm{NE}} = (X^\top X)^{-1} X^\top y$ (carry at least 6 decimals).
(iii) Compute a QR factorization $X = QR$ using Gram–Schmidt (conceptual is fine) and solve $Rw_{\mathrm{QR}} = Q^\top y$.
(iv) Compare w_{NE} and w_{QR} numerically and explain (one paragraph) why normal equations are sensitive here.

Exercise 10.9.5 (Rank Deficiency and Minimum-Norm Solution) Let

$$X = \begin{bmatrix} 1 & 1 & 2 \\ 1 & 2 & 4 \\ 1 & 3 & 6 \\ 1 & 4 & 8 \end{bmatrix} \in \mathbb{R}^{4\times 3}, \qquad y = \begin{bmatrix} 1 \\ 2 \\ 3 \\ 4 \end{bmatrix}.$$

(i) Compute $\operatorname{rank}(X)$ and find a nonzero $v \in \mathcal{N}(X)$.
(ii) Show that if $\widehat{w}$ is a least-squares minimizer then $\widehat{w} + tv$ is also a minimizer for any $t \in \mathbb{R}$.
(iii) Compute the minimum-norm least-squares solution $w^\dagger = X^\dagger y$ by computing an SVD of X *or* by solving the constrained problem $\min \|w\|_2$ subject to $w \in \arg\min_u \|Xu - y\|_2^2$ (show all steps).

Exercise 10.9.6 (Ridge Regression: Closed Form, Shrinkage, and df(λ)) Let

$$X = \begin{bmatrix} 1 & 0 \\ 1 & 1 \\ 1 & 2 \\ 1 & 3 \\ 1 & 4 \end{bmatrix}, \qquad y = \begin{bmatrix} 0 \\ 1 \\ 1 \\ 2 \\ 2 \end{bmatrix}.$$

For $\lambda \in \{0.1, 1, 10\}$:

(i) Compute $\widehat{w}_\lambda = (X^\top X + \lambda I_2)^{-1} X^\top y$.

(ii) Compute the ridge hat matrix $S_\lambda = X(X^\top X + \lambda I_2)^{-1}X^\top$ and its effective degrees of freedom $\mathrm{df}(\lambda) = \mathrm{tr}(S_\lambda)$.
(iii) Verify numerically that $\mathrm{df}(\lambda)$ decreases as λ increases.
(iv) Compute $\|\widehat{w}_\lambda\|_2$ and $\|X\widehat{w}_\lambda - y\|_2$ and comment (one paragraph) on bias vs. variance intuition.

Exercise 10.9.7 (Lasso in Orthonormal Design via Soft-Thresholding) Assume $X \in \mathbb{R}^{4\times 4}$ satisfies $X^\top X = I_4$ and $z = X^\top y$ is given by

$$z = \begin{bmatrix} 2.0 \\ -0.4 \\ 1.2 \\ -3.5 \end{bmatrix}.$$

For $\lambda \in \{0.5, 1.0, 2.5\}$:

(i) Compute the lasso solution using soft-thresholding $\widehat{w}_j = \mathrm{sign}(z_j)\max\{|z_j| - \lambda, 0\}$.
(ii) For each λ, compute the number of nonzeros and $\|\widehat{w}\|_1$.
(iii) Interpret (one paragraph) how λ controls sparsity and why this corresponds to ℓ_1 geometry.

Exercise 10.9.8 (Elastic Net vs. Lasso under Correlated Features) Let

$$X = \begin{bmatrix} 1 & 1 \\ 1 & 1.01 \\ 1 & 0.99 \\ 1 & 1.02 \\ 1 & 0.98 \end{bmatrix}, \qquad y = \begin{bmatrix} 1 \\ 1 \\ 1 \\ 1 \\ 1 \end{bmatrix}.$$

(i) Compute $X^\top y$ and $X^\top X$ and argue why the two columns are highly correlated.
(ii) Consider elastic net with $(\lambda_1, \lambda_2) = (0.1, 1)$ and write the optimality/KKT system for the minimizer (you may leave the ℓ_1 subgradient conditions explicitly as intervals).
(iii) Contrast (one paragraph) the expected qualitative behavior of solutions for lasso ($\lambda_1 = 0.1, \lambda_2 = 0$) vs. elastic net $(0.1, 1)$ under such correlation.

Exercise 10.9.9 (Logistic Regression: Likelihood, Gradient, and One Newton Step) Binary classification data (x_i, y_i) for $i \in [6]$:

$$x = (-2, -1, 0, 1, 2, 3), \qquad y = (0, 0, 0, 1, 1, 1).$$

Model: $p_\theta(Y = 1 \mid x) = \sigma(\theta x)$ with $\sigma(t) = 1/(1 + e^{-t})$ (no intercept).

(i) Write the negative log-likelihood $f(\theta)$.
(ii) Compute $f'(\theta)$ and $f''(\theta)$ explicitly in terms of $\sigma(\theta x_i)$.

(iii) Starting from $\theta^{(0)} = 0$, compute one Newton step $\theta^{(1)} = \theta^{(0)} - f'(\theta^{(0)})/f''(\theta^{(0)})$ numerically.
(iv) Using $\theta^{(1)}$, compute predicted probabilities for all x_i and the empirical log-loss.

Exercise 10.9.10 (Softmax Regression: One Gradient Computation) Consider $K = 3$ classes, $d = 2$ features, and parameter matrix $W \in \mathbb{R}^{3\times 2}$ with no bias:

$$W = \begin{bmatrix} 1 & 0 \\ 0 & 1 \\ -1 & -1 \end{bmatrix}.$$

Data (x_i, y_i) for $i \in [4]$:

$$x_1 = \begin{bmatrix} 1 \\ 0 \end{bmatrix}, \ y_1 = 1; \quad x_2 = \begin{bmatrix} 0 \\ 1 \end{bmatrix}, \ y_2 = 2; \quad x_3 = \begin{bmatrix} 1 \\ 1 \end{bmatrix}, \ y_3 = 3;$$

$$x_4 = \begin{bmatrix} 2 \\ 1 \end{bmatrix}, \ y_4 = 1.$$

(i) For each i, compute scores $s_i = Wx_i \in \mathbb{R}^3$ and probabilities $p_i = \text{softmax}(s_i)$.
(ii) Compute the cross-entropy loss $L(W) = -\sum_{i=1}^{4} \log p_i(y_i)$.
(iii) Compute the gradient $\nabla_W L = \sum_{i=1}^{4} (p_i - e_{y_i}) x_i^\top$ explicitly.

Exercise 10.9.11 (Classification Metrics and Threshold Selection) For $n = 12$ examples, true labels $y_i \in \{0, 1\}$ and predicted probabilities q_i are

$$y = (1, 0, 1, 0, 1, 0, 0, 1, 0, 1, 0, 1),$$

$$q = (0.95, 0.70, 0.60, 0.55, 0.52, 0.50, 0.45, 0.40, 0.35, 0.30, 0.20, 0.10).$$

(i) For thresholds $t \in \{0.5, 0.4, 0.3\}$, compute confusion matrices and report (TPR, FPR), precision, and recall.
(ii) Plot (by listing points) the ROC points for these thresholds and identify which threshold yields FPR ≤ 0.25 with maximal TPR.
(iii) Compute the empirical prevalence $\pi = \frac{1}{12}\sum_i y_i$ and, for each threshold, verify the identity Precision $= \frac{\pi\,\text{TPR}}{\pi\,\text{TPR} + (1-\pi)\,\text{FPR}}$ using your computed rates.

Exercise 10.9.12 (Calibration and Proper Scoring: Log-Loss vs. 0–1) Using the same (y, q) as Exercise 10.9.11:

(i) Compute average log-loss $\frac{1}{n}\sum_i -\big(y_i \log q_i + (1-y_i)\log(1-q_i)\big)$ and average Brier score $\frac{1}{n}\sum_i (q_i - y_i)^2$.
(ii) Compute 0–1 error at threshold $t = 0.5$.

(iii) Construct a reliability table with bins $[0, 0.25)$, $[0.25, 0.5)$, $[0.5, 0.75)$, $[0.75, 1]$ and compute $(\widehat{q}_b, \widehat{p}_b)$ for each bin.

(iv) Give one paragraph interpreting how a model can have similar 0–1 error but different log-loss due to calibration.

Exercise 10.9.13 (Influence and Leverage: Hat Matrix Computation) Let

$$X = \begin{bmatrix} 1 & 0 \\ 1 & 1 \\ 1 & 2 \\ 1 & 3 \\ 1 & 4 \end{bmatrix}, \qquad y = \begin{bmatrix} 0 \\ 1 \\ 1 \\ 2 \\ 10 \end{bmatrix}.$$

(i) Compute the OLS hat matrix $P = X(X^\top X)^{-1}X^\top$ and the leverage scores $h_{ii} = P_{ii}$.

(ii) Compute the OLS fit $\widehat{y} = Py$ and residuals $r = y - \widehat{y}$.

(iii) Identify which point has the largest leverage and which has the largest residual magnitude.

(iv) Compute Cook's distance proxy

$$D_i = \frac{r_i^2}{2\widehat{\sigma}^2} \cdot \frac{h_{ii}}{(1 - h_{ii})^2}, \qquad \widehat{\sigma}^2 = \frac{\|r\|_2^2}{n-2},$$

and determine the most influential point.

Exercise 10.9.14 (Feature Expansion: Polynomial Regression and Multicollinearity) Consider scalar inputs $t \in \mathbb{R}$ and polynomial features $\phi_2(t) = (1, t, t^2)^\top$. Data: $t = (0, 1, 2, 3, 4)$ and $y = (1, 1.9, 3.8, 6.1, 8.2)$.

(i) Construct $\Phi \in \mathbb{R}^{5\times 3}$ with rows $\phi_2(t_i)^\top$ and compute the OLS coefficients $\widehat{\theta} = (\Phi^\top \Phi)^{-1}\Phi^\top y$.

(ii) Compute $\kappa_2(\Phi^\top \Phi)$ (via eigenvalues) and comment (one paragraph) on multicollinearity induced by t and t^2.

(iii) Compute ridge coefficients for $\lambda = 1$ and compare to OLS.

Exercise 10.9.15 (Linearization: Local Surrogate of a Nonlinear Regressor) Let $f : \mathbb{R}^2 \to \mathbb{R}$ be

$$f(x) = \sin(x_1) + x_1 x_2 + \frac{1}{2}x_2^2.$$

(i) Compute the gradient $\nabla f(x)$ and the first-order Taylor approximation

$$f(x) \approx f(x_0) + \nabla f(x_0)^\top (x - x_0)$$

at $x_0 = (1, -1)^\top$.

(ii) Use this linearization to predict $f(x)$ at $x = (1.1, -0.9)^\top$ and compare to the exact value (compute numerically to 4 decimals).
(iii) Interpret (one paragraph) how this connects to linear models as local approximations and to gradient-based sensitivity.

Chapter 11
Kernels, Dimensionality Reduction, and High-Dimensional Tricks

A large part of modern ML can be viewed as *geometry engineering*: choose (or learn) a representation in which similarity aligns with semantics, and then apply algorithms whose behavior is governed by distances, inner products, and low-dimensional structure. Kernel methods formalize this by replacing explicit feature maps with similarity functions, while dimensionality reduction and randomized tricks exploit high-dimensional phenomena (concentration, low-rank structure) to make learning scalable. This chapter builds the conceptual bridge from Euclidean geometry to similarity-based learning and provides the computational tools needed to operate in high dimensions (Schölkopf and Smola 2002).

11.1 Why Similarity and Geometry Drive Modern ML

11.1.1 From Explicit Features to Similarity Functions

11.1.1.1 Distance-Based Reasoning and Nearest Neighbors

Many ML procedures can be written using only pairwise distances: nearest neighbors, clustering, manifold learning, and retrieval. This is powerful because it decouples learning from explicit parametric forms: the algorithm depends on the geometry of the data cloud. However, it also makes the method sensitive to the choice of metric and to feature scaling/normalization. In high dimensions, naive distance-based intuition can fail due to concentration, making representation learning and metric design essential.

Definition 11.1.1 (Metric-Induced Nearest Neighbor Rule) Let $(\mathcal{X}, d)$ be a metric space, and let labeled samples $(x_i, y_i)_{i=1}^n$ with $x_i \in \mathcal{X}$. The 1-nearest

P. Singh, B. Raman, *Foundations of Machine Learning and AI*, Studies in Big Data 199, https://doi.org/10.1007/978-3-032-30336-3_11

neighbor rule is

$$\widehat{y}(x) = y_{i^\star}, \qquad i^\star \in \arg\min_{i\in[n]} d(x, x_i).$$

More generally, k-NN predicts by majority vote (classification) or averaging (regression) over the k closest points.

Remark 11.1.1 (Geometry Is the Model) In nearest-neighbor methods, the hypothesis class is implicitly "all labelings consistent with local neighborhoods" induced by d. Thus, choosing d (and preprocessing that defines d) is the modeling step.

Example 11.1.1 (Cosine vs. Euclidean on Normalized Embeddings) If embeddings are normalized to unit norm, $\|u\|_2 = \|v\|_2 = 1$, then

$$\|u - v\|_2^2 = 2(1 - \langle u, v\rangle),$$

so ranking by Euclidean distance is equivalent to ranking by cosine similarity. This is why modern retrieval systems often normalize embeddings and use dot-product search.

11.1.1.2 Kernel Idea: Implicit Feature Spaces

Kernel methods replace explicit coordinates by a similarity function $k(x, z)$ that behaves like an inner product in some (possibly infinite-dimensional) feature space. This enables linear algorithms (ridge regression, SVM, PCA) to become nonlinear in input space while remaining convex and computationally structured. The key requirement is *positive semidefiniteness*: the kernel matrix must be a Gram matrix. Once this holds, all computations can be done using the kernel matrix without ever computing the feature map.

Definition 11.1.2 (Kernel and Feature Map) A function $k : \mathcal{X} \times \mathcal{X} \to \mathbb{R}$ is a (Mercer) *kernel* if there exists an inner-product space $(\mathcal{H}, \langle\cdot,\cdot\rangle_{\mathcal{H}})$ and a map $\phi : \mathcal{X} \to \mathcal{H}$ such that

$$k(x, z) = \langle \phi(x), \phi(z)\rangle_{\mathcal{H}} \qquad \forall x, z \in \mathcal{X}.$$

Definition 11.1.3 (Kernel (Gram) Matrix) Given data $x_1, \ldots, x_n \in \mathcal{X}$, define the kernel matrix $K \in \mathbb{R}^{n\times n}$ by

$$K_{ij} = k(x_i, x_j).$$

Proposition 11.1.1 (PSD Condition (Kernel Feasibility)) *A symmetric function k yields a valid kernel matrix for all finite samples iff for every n and every choice of* *points $x_1, \ldots, x_n$, the matrix K is positive semidefinite:*

$$c^\top Kc \geq 0 \quad \forall c \in \mathbb{R}^n.$$

Equivalently, K is a Gram matrix of some vectors in an inner-product space.

Remark 11.1.2 (Kernel Trick (Operational)) If an algorithm uses data only through inner products $\langle x_i, x_j \rangle$, then replacing them by $k(x_i, x_j)$ yields a nonlinear method. The computation typically reduces to linear algebra on K and does not require explicit $\phi(x)$.

11.1.2 High-Dimensional Geometry in Practice

11.1.2.1 Distance Concentration and Metric Pitfalls

High-dimensional spaces exhibit concentration: norms and pairwise distances can become nearly constant across points when data are roughly isotropic. This reduces distance contrast and can degrade nearest-neighbor behavior. The pitfall is not that distance-based methods never work, but that they require *structure* (clusters, low-dimensional manifolds, sparsity) or a learned metric that restores contrast. Diagnostically, if nearest and farthest distances are similar, you should suspect metric mismatch or representation issues.

Remark 11.1.3 (Distance Concentration Heuristic) If for most pairs (i, j) we have $\|x_i - x_j\|_2 \approx \mu$ with small relative spread, then NN decisions can be unstable: the difference between the nearest and farthest neighbor distances is too small compared to noise and finite-sample variability. This motivates dimensionality reduction and metric learning.

Remark 11.1.4 (Feature Scaling Is Metric Design) If features have different units/scales, Euclidean distance is dominated by large-scale coordinates. Standardization/whitening changes the geometry and can dramatically change nearest-neighbor structure.

11.1.2.2 Why Representations and Normalization Matter

Representation learning aims to map raw inputs into a space where geometry matches the task: same-class points are close, different-class points are separated by margins, and dot products correlate with semantic similarity. Normalization (centering, whitening, ℓ_2-normalization) removes nuisance degrees of freedom (mean/scale) and stabilizes distances and angles. In kernel methods, normalization is implicit in kernel choice; in embedding methods, it is explicit and often essential for stable retrieval and calibration.

Remark 11.1.5 (Normalization as Geometric Control)

- Centering removes translation effects and aligns PCA/kernel PCA with variance structure.
- Whitening turns covariance ellipsoids into spheres, improving conditioning and distance interpretability.
- ℓ_2-normalization projects to the sphere, making similarity primarily angular (cosine-based).

Each is a deliberate modification of the metric.

Remark 11.1.6 (Practical Workflow) When a method depends on distances or similarities:

1. standardize/normalize features,
2. check distance contrast (nearest/farthest ratio, histogram of pairwise distances),
3. if contrast is poor, consider dimensionality reduction (PCA, random projections) or learned representations/metrics,
4. only then trust nearest-neighbor or kernel-based geometry.

11.2 Kernel Methods and RKHS

Kernel methods turn linear algorithms into nonlinear ones by replacing explicit feature vectors with a similarity function (kernel) that acts like an inner product in a (possibly infinite-dimensional) feature space. The mathematical backbone is the theory of Reproducing Kernel Hilbert Spaces (RKHS): a Hilbert space of functions where evaluation is continuous and inner products are represented by kernels (Aronszajn 1950). This section develops the kernel trick, PSD Gram geometry, the RKHS viewpoint (reproducing property, representer theorem), core algorithms (kernel ridge and kernel logistic), and practical kernel design choices.

11.2.1 Kernel Trick

11.2.1.1 Feature Maps and Implicit Inner Products

Many linear algorithms depend on data only through inner products $\langle x_i, x_j \rangle$ (Gram matrices). If we can compute an alternative inner product $k(x_i, x_j)$ corresponding to an implicit feature map $\phi(x)$, we can run the same algorithm in feature space without explicitly computing ϕ. This is the kernel trick: replace inner products by kernel evaluations. The result is nonlinear decision boundaries in input space with convex optimization structure retained.

Definition 11.2.1 (Feature Map and Kernel) Let $\phi : \mathcal{X} \to \mathcal{H}$ map inputs into an inner-product space $(\mathcal{H}, \langle\cdot,\cdot\rangle_{\mathcal{H}})$. Define

$$k(x, z) := \langle \phi(x), \phi(z) \rangle_{\mathcal{H}}.$$

Then k is a kernel and the matrix $K_{ij} = k(x_i, x_j)$ is the Gram matrix in feature space.

Remark 11.2.1 (Kernel Trick (Operational)) If an algorithm can be written using only dot products $\langle \phi(x_i), \phi(x_j) \rangle_{\mathcal{H}}$, we can implement it using only $k(x_i, x_j)$, without computing ϕ.

11.2.1.2 PSD Kernels and Gram Matrices Revisited

Not every similarity function is a valid inner product. Validity requires positive semidefiniteness (PSD): for any finite sample, the kernel matrix must be PSD. This is exactly the Gram matrix feasibility condition discussed earlier. PSD ensures geometric consistency: distances derived from k behave like squared Euclidean distances in feature space, and optimization problems remain convex where expected.

Definition 11.2.2 (PSD Kernel) A symmetric function $k : \mathcal{X} \times \mathcal{X} \to \mathbb{R}$ is *positive semidefinite* if for all n and all $x_1, \ldots, x_n \in \mathcal{X}$, the kernel matrix $K \in \mathbb{R}^{n\times n}$ with $K_{ij} = k(x_i, x_j)$ satisfies

$$c^\top K c \geq 0 \quad \forall c \in \mathbb{R}^n.$$

Remark 11.2.2 (Gram Interpretation) PSD means K is a Gram matrix of some vectors $\phi(x_i)$ in a Hilbert space: $K_{ij} = \langle \phi(x_i), \phi(x_j) \rangle$. This implies symmetry and nonnegative eigenvalues.

11.2.2 Reproducing Kernel Hilbert Spaces (RKHS)

11.2.2.1 Reproducing Property and Representer Theorem (Conceptual)

RKHS theory elevates kernels from "similarities" to *function spaces*. An RKHS is a Hilbert space of functions $f : \mathcal{X} \to \mathbb{R}$ where evaluation at a point is a continuous linear functional. This continuity implies the reproducing property: evaluating f at x equals an inner product with a kernel section $k(x, \cdot)$. The representer theorem then states that many regularized empirical risk minimizers in an RKHS lie in the span of kernel sections at the training points. This reduces infinite-dimensional optimization to finite-dimensional linear algebra.

Definition 11.2.3 (RKHS and Reproducing Property) A Hilbert space $\mathcal{H}$ of real-valued functions on $\mathcal{X}$ is a *Reproducing Kernel Hilbert Space* if there exists a function $k : \mathcal{X} \times \mathcal{X} \to \mathbb{R}$ such that:

1. For each $x \in \mathcal{X}$, $k(x, \cdot) \in \mathcal{H}$.
2. (*Reproducing property*) For all $f \in \mathcal{H}$ and all $x \in \mathcal{X}$,

$$f(x) = \langle f,\ k(x, \cdot)\rangle_{\mathcal{H}}.$$

The function k is called the reproducing kernel of $\mathcal{H}$.

Remark 11.2.3 (Kernel Sections as Feature Vectors) In an RKHS, the object $k(x, \cdot)$ plays the role of the feature vector for x. The reproducing property is the analogue of $f(x) = \langle w, \phi(x)\rangle$ in finite-dimensional linear models.

Theorem 11.2.1 (Representer Theorem (Conceptual Statement)) *Consider the regularized empirical risk problem*

$$\min_{f \in \mathcal{H}} \sum_{i=1}^{n} \ell\big(f(x_i), y_i\big) + \lambda \|f\|_{\mathcal{H}}^2, \qquad \lambda > 0,$$

where $\mathcal{H}$ is an RKHS with kernel k and ℓ depends on f only through values at the sample points. Then there exists an optimizer of the form

$$f^\star(\cdot) = \sum_{i=1}^{n} \alpha_i\, k(x_i, \cdot)$$

for some coefficients $\alpha \in \mathbb{R}^n$.

11.2.2.2 RKHS Norm as Smoothness/Complexity Control

The RKHS norm plays the role of a complexity measure: small $\|f\|_{\mathcal{H}}$ corresponds to functions that are "smooth" with respect to the kernel geometry. This is the functional analogue of ℓ_2 regularization in linear regression. Different kernels induce different notions of smoothness: RBF kernels prefer very smooth functions, polynomial kernels permit higher-frequency variation, and string kernels encode smoothness in discrete combinatorial spaces. Thus, choosing a kernel is choosing both a similarity measure and a regularization geometry.

Remark 11.2.4 (Norm Control and Generalization) Regularization $\lambda \|f\|_{\mathcal{H}}^2$ restricts the effective hypothesis class to a ball in $\mathcal{H}$, which controls capacity and enables generalization bounds via Rademacher complexity in RKHS.

11.2.3 Representer Theorem and Algorithms

11.2.3.1 Kernel Ridge Regression Derivation

Kernel ridge regression (KRR) is ridge regression in an RKHS. The representer theorem reduces the optimization over functions to optimization over coefficients $\alpha \in \mathbb{R}^n$. The solution is obtained by solving a linear system involving the kernel matrix K. This mirrors primal ridge $(X^\top X + \lambda I)^{-1} X^\top y$, but now the geometry is encoded by K. Computationally, KRR costs $O(n^3)$ naive, motivating approximations (Nyström, random features) for large n.

Definition 11.2.4 (Kernel Ridge Regression (KRR)) Given data $(x_i, y_i)_{i=1}^n$ and kernel k, KRR solves

$$\min_{f \in \mathcal{H}} \sum_{i=1}^{n} (f(x_i) - y_i)^2 + \lambda \|f\|_{\mathcal{H}}^2.$$

Proposition 11.2.1 (KRR Solution in Dual Coefficients) *Let $K \in \mathbb{R}^{n\times n}$ with $K_{ij} = k(x_i, x_j)$ and let $y \in \mathbb{R}^n$. An optimal solution has the form $f(\cdot) = \sum_{i=1}^n \alpha_i k(x_i, \cdot)$ where $\alpha \in \mathbb{R}^n$ solves*

$$(K + \lambda I_n)\alpha = y.$$

Then predictions at training points are $\widehat{y} = K\alpha$ and for a new point x,

$$f(x) = k_x^\top \alpha, \qquad (k_x)_i = k(x_i, x).$$

Remark 11.2.5 (Connection to Ridge in Feature Space) If $k(x, z) = \langle \phi(x), \phi(z) \rangle$ for a finite-dimensional ϕ, KRR equals ridge regression on the feature matrix Φ. The dual system uses $K = \Phi\Phi^\top$.

11.2.3.2 Kernel Logistic Classification: High-Level Pipeline

Kernel logistic regression combines the kernel representation with logistic likelihood. The representer theorem again yields $f(\cdot) = \sum_i \alpha_i k(x_i, \cdot)$, but coefficients have no closed form: one optimizes a convex objective in α using Newton/IRLS-like methods in the dual. This yields nonlinear decision boundaries while retaining convexity (for binary logistic). Practical implementation requires numerical stability: add λI regularization and use line search/trust regions.

Remark 11.2.6 (Kernel Logistic Objective (Conceptual)) For binary labels $y_i \in \{0, 1\}$, define $f_i = (K\alpha)_i$ and probabilities $p_i = \sigma(f_i)$. A regularized objective is

$$\min_{\alpha \in \mathbb{R}^n} \sum_{i=1}^{n} \Big(- y_i \log p_i - (1 - y_i) \log(1 - p_i) \Big) + \frac{\lambda}{2} \alpha^\top K \alpha,$$

where $\alpha^\top K \alpha = \|f\|_{\mathcal{H}}^2$ in RKHS terms.

Remark 11.2.7 (Optimization Pipeline) Compute K, initialize α, iterate Newton/IRLS steps using gradient and Hessian involving K and diagonal weights $p_i(1 - p_i)$, apply regularization and damping, stop by tolerance, and then predict with $f(x) = k_x^\top \alpha$.

11.2.4 Kernel Design

11.2.4.1 Linear, Polynomial, RBF, and String Kernels (Examples)

Kernel choice encodes similarity and inductive bias. Linear kernels correspond to ordinary linear models. Polynomial kernels capture interactions up to a given degree. RBF kernels encode locality and yield very smooth functions, often strong defaults. String kernels compare discrete sequences via shared substrings, enabling kernel methods in NLP without explicit embeddings. The key is PSD: each kernel must yield PSD Gram matrices.

Definition 11.2.5 (Common Kernels) For $x, z \in \mathbb{R}^d$:

$$k_{\text{lin}}(x, z) = x^\top z, \qquad k_{\text{poly}}(x, z) = (x^\top z + c)^p, \quad c \geq 0, \; p \in \mathbb{N},$$

$$k_{\text{rbf}}(x, z) = \exp\left(- \frac{\|x - z\|_2^2}{2\sigma^2} \right), \quad \sigma > 0.$$

For strings s, t over an alphabet, a (conceptual) n-gram kernel can be defined by

$$k_{\text{str}}(s, t) = \sum_{u \in \mathcal{U}_n} \text{count}_u(s) \, \text{count}_u(t),$$

where $\mathcal{U}_n$ is the set of length-n substrings and $\text{count}_u(\cdot)$ counts occurrences.

Remark 11.2.8 (PSD Sanity Check) Many kernels are PSD by construction (inner products in explicit feature maps). For example, polynomial kernels correspond to monomial feature expansions; string kernels correspond to bag-of-n-grams features. RBF is PSD as a radial basis kernel; we will accept this as a standard fact and use it operationally.

11.2.4.2 Choosing Bandwidths and Scaling Features

Kernel hyperparameters (e.g., RBF bandwidth σ) determine the geometry: small σ makes the kernel highly local (risking overfitting), large σ makes it nearly constant (underfitting). Feature scaling matters enormously because RBF depends on Euclidean distance: unscaled features distort distances and thus similarities. A good workflow standardizes/whitens features first, then tunes σ via validation/CV, often using heuristics such as the median distance rule as a starting point. In high dimensions, distance concentration can make bandwidth selection delicate, reinforcing the need for normalization.

Remark 11.2.9 (RBF Extremes)

- $\sigma \to 0$: $k_{\mathrm{rbf}}(x_i, x_j) \to \mathbf{1}\{i = j\}$ (kernel matrix approaches identity) $\Rightarrow$ can interpolate/noise-fit.
- $\sigma \to \infty$: $k_{\mathrm{rbf}}(x, z) \to 1$ (kernel matrix nearly rank-1) $\Rightarrow$ underfits.

Thus σ controls effective complexity similarly to λ.

Remark 11.2.10 (Scaling Workflow) Standardize features (zero mean, unit variance) or whiten (unit covariance) before using distance-based kernels. Then tune σ (and λ) by CV; report the preprocessing and tuning protocol because it is part of the model.

11.3 Support Vector Machines (Preview via Geometry)

Support Vector Machines (SVMs) are the archetypal *geometric* classifiers: they choose a separating hyperplane with the largest margin, yielding robustness and strong generalization behavior in many settings. Their dual formulation depends only on inner products, so they kernelize naturally via Gram matrices. This section gives a geometry-first preview: margin maximization, slack variables for nonseparable data, and the kernelized dual/decision function highlighting support vectors.

11.3.1 Maximum-Margin Principle

11.3.1.1 Hard-Margin and Soft-Margin Intuition

If data are linearly separable, there are infinitely many separating hyperplanes. The maximum-margin principle selects the one that keeps all points farthest from the boundary in Euclidean distance. This makes classification robust: small perturbations of inputs are less likely to flip the label. When data are not perfectly separable (common in ML), soft-margin SVM introduces slack variables to allow some violations, balancing margin size against training errors. This yields a convex optimization problem with a clear geometric trade-off.

Definition 11.3.1 (Hard-Margin SVM (Primal)) Let $(x_i, y_i)_{i=1}^n$ with $x_i \in \mathbb{R}^d$ and $y_i \in \{-1, +1\}$. If the data are separable, the hard-margin SVM solves

$$\min_{w,b} \frac{1}{2}\|w\|_2^2 \quad \text{s.t.} \quad y_i(w^\top x_i + b) \geq 1 \;\; \forall i \in [n].$$

Remark 11.3.1 (Geometric Meaning of the Objective) Under the constraints $y_i(w^\top x_i + b) \geq 1$, the geometric margin is $1/\|w\|_2$. Thus minimizing $\|w\|_2^2$ is equivalent to maximizing the margin.

Definition 11.3.2 (Soft-Margin SVM (Primal)) For potentially nonseparable data, the soft-margin SVM introduces slack variables $\xi_i \geq 0$ and solves

$$\min_{w,b,\xi} \frac{1}{2}\|w\|_2^2 + C\sum_{i=1}^n \xi_i \quad \text{s.t.} \quad y_i(w^\top x_i + b) \geq 1 - \xi_i, \;\; \xi_i \geq 0, \;\; \forall i.$$

Here $C > 0$ controls the margin–violation trade-off.

Remark 11.3.2 (Loss Viewpoint (Hinge Loss)) Eliminating ξ yields the equivalent unconstrained form

$$\min_{w,b} \frac{1}{2}\|w\|_2^2 + C\sum_{i=1}^n \max\{0, \; 1 - y_i(w^\top x_i + b)\},$$

where the term $\max\{0, 1 - \cdot\}$ is the hinge loss. This links SVM to regularized empirical risk minimization.

11.3.1.2 Margins, Slack Variables, and Regularization

Soft margin makes explicit the regularization structure: $\|w\|_2^2$ controls complexity (and robustness), while slacks measure violations. Parameter C is effectively an inverse regularization strength: large C emphasizes fitting constraints, small C emphasizes large margin (strong regularization). This parallels ridge/logistic regularization tuning. Geometrically, the solution depends mostly on points near the boundary (or violating it), which become support vectors in the dual.

Remark 11.3.3 (Interpretation of C)

- $C \to \infty$: prioritize minimizing violations; approaches hard-margin when separable.
- $C \to 0$: prioritize maximizing margin; allows many violations (underfitting risk).

Thus C is tuned by validation/CV as a regularization hyperparameter.

Remark 11.3.4 (Robustness Link) For a correctly classified point with functional margin $y(w^\top x + b)$, the Euclidean robustness radius is roughly $\frac{y(w^\top x+b)}{\|w\|_2}$; maximizing margin improves this across the dataset.

11.3.2 Kernelized SVM

11.3.2.1 Dual Formulation and Gram Matrices

The computational and conceptual power of SVM comes from its dual. The dual depends only on dot products $x_i^\top x_j$, so it kernelizes by replacing them with $k(x_i, x_j)$. The dual variables α_i indicate which points are active constraints; only a subset has $\alpha_i > 0$. These points—the support vectors—fully determine the classifier. This is a sparse representation in the sample space.

Definition 11.3.3 (SVM Dual (Kernel-Ready Form)) Let K be the Gram matrix with $K_{ij} = x_i^\top x_j$ (or $K_{ij} = k(x_i, x_j)$ for a kernel). The (soft-margin) SVM dual is

$$\max_{\alpha \in \mathbb{R}^n} \sum_{i=1}^{n} \alpha_i - \frac{1}{2} \sum_{i=1}^{n} \sum_{j=1}^{n} \alpha_i \alpha_j y_i y_j K_{ij} \quad \text{s.t.} \quad 0 \le \alpha_i \le C, \ \sum_{i=1}^{n} \alpha_i y_i = 0.$$

Remark 11.3.5 (Kernelization) Replacing $K_{ij} = x_i^\top x_j$ by $K_{ij} = k(x_i, x_j)$ yields a nonlinear classifier in input space while keeping the same dual form. PSD of K ensures convexity of the dual.

11.3.2.2 Decision Function and Support Vectors

Once the dual coefficients are found, prediction at a new point is a weighted sum of kernel similarities to training points. Only support vectors (with $\alpha_i > 0$) contribute, giving computational sparsity at inference time. Geometrically, support vectors are the points that lie on or within the margin (including violations); they define the separating surface. In ML practice, the number of support vectors controls inference cost and can increase with noisy data or weak regularization.

Definition 11.3.4 (Kernel SVM Decision Function) Given a solution α and bias b, the decision function is

$$f(x) = \sum_{i=1}^{n} \alpha_i y_i \, k(x_i, x) + b, \qquad \widehat{y}(x) = \operatorname{sign}(f(x)).$$

The *support vectors* are indices i with $\alpha_i > 0$.

Remark 11.3.6 (Support Vectors and KKT Intuition) At optimum, points with $\alpha_i = 0$ are strictly outside the margin; points with $0 < \alpha_i < C$ lie on the margin; points with $\alpha_i = C$ are margin violators or inside-margin points (soft-margin case). Thus the classifier is determined by the geometry near the boundary.

Remark 11.3.7 (Practical Tuning) Kernel SVM has two key hyperparameters: C (regularization) and kernel parameters (e.g., RBF bandwidth σ). Both must be tuned jointly by CV; feature scaling is essential for distance-based kernels.

11.4 Nonlinear Dimensionality Reduction

Dimensionality reduction is a geometric compression problem: represent high-dimensional data in a lower-dimensional space while preserving structure relevant to downstream tasks. Linear methods (PCA) preserve global variance in Euclidean geometry. Nonlinear methods aim to capture curved manifolds or neighborhood relations: they replace Euclidean structure with kernel similarities or graph-based geometry, or they learn representations via reconstruction objectives (autoencoders). This section covers spectral nonlinear methods (kernel PCA, Laplacian eigenmaps), manifold visualization tools (t-SNE/UMAP), and an autoencoder viewpoint that bridges linear algebra to deep representation learning (cf. Fig. 11.1).

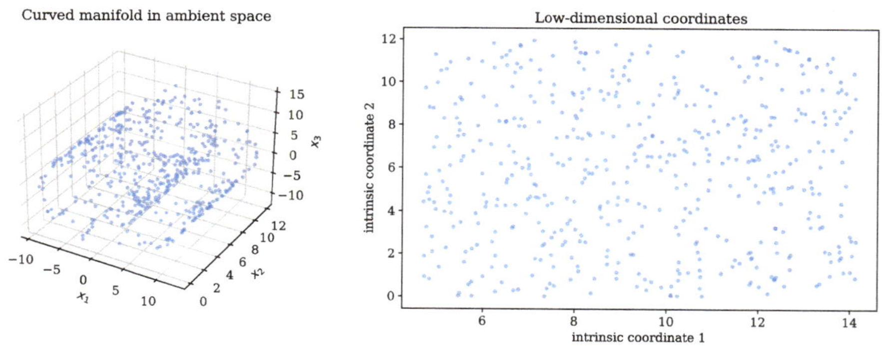

Fig. 11.1 Nonlinear dimensionality reduction viewpoints. Left: a nonlinear manifold in ambient space that cannot be captured by a single linear subspace. Right: spectral or neighborhood-based methods seek low-dimensional coordinates that preserve local structure rather than raw Euclidean geometry

11.4.1 Spectral Methods

11.4.1.1 Kernel PCA: Relation to PCA and Covariance

Kernel PCA performs PCA not in input space but in an implicit feature space $\phi(x)$ induced by a kernel (Schölkopf et al. 1998). It replaces the covariance matrix $X_c^\top X_c$ with a centered Gram matrix in feature space. The eigenvectors of this kernel matrix yield principal directions in RKHS, enabling nonlinear components in input space. Computationally, it is an eigen-decomposition of an $n \times n$ matrix, which is feasible for moderate n and motivates approximations for large n.

Definition 11.4.1 (Centered Kernel Matrix) Given data $x_1, \ldots, x_n$ and kernel matrix K with $K_{ij} = k(x_i, x_j)$, define the centering matrix

$$H := I_n - \frac{1}{n}\mathbf{1}\mathbf{1}^\top.$$

The centered kernel matrix is

$$K_c := HKH.$$

Remark 11.4.1 (Why Centering Is Necessary) PCA is defined on centered data. In feature space, centering corresponds to subtracting the mean feature vector $\bar{\phi} = \frac{1}{n}\sum_i \phi(x_i)$, which is implemented exactly by $K_c = HKH$ without explicit ϕ.

Definition 11.4.2 (Kernel PCA (Spectral Problem)) Kernel PCA solves the eigenproblem

$$K_c v = \lambda v,$$

and uses the top m eigenvectors $v_1, \ldots, v_m$ (with eigenvalues $\lambda_1 \geq \cdots \geq \lambda_m$) to form an embedding. A common coordinate for point x_i along component ℓ is

$$z_{i\ell} := \frac{1}{\sqrt{\lambda_\ell}}(v_\ell)_i,$$

(up to convention and normalization).

Remark 11.4.2 (Connection to Ordinary PCA) If $k(x, z) = x^\top z$ and data are centered, then $K = XX^\top$ and kernel PCA reduces to PCA expressed through the Gram matrix. Thus kernel PCA is PCA in an implicit feature space.

11.4.1.2 Laplacian Eigenmaps: Graph-Geometry Intuition

Laplacian eigenmaps build a graph from data (edges connect neighbors with weights indicating similarity) and seek an embedding that preserves local neighborhoods: nearby points in the graph should stay nearby in the embedding. The graph Laplacian encodes discrete geometry: it penalizes variations across edges. Spectral embedding then finds low-frequency eigenvectors (smooth functions on the graph), analogous to Fourier modes on manifolds. This method is a cornerstone of manifold learning and motivates many graph neural network intuitions.

Definition 11.4.3 (Weighted Graph and Laplacian) Let $W \in \mathbb{R}^{n\times n}$ be a symmetric weight matrix with $W_{ij} \geq 0$ and $W_{ii} = 0$. Define the degree matrix $D = \operatorname{diag}(d_1, \ldots, d_n)$ with $d_i = \sum_{j=1}^{n} W_{ij}$. The (unnormalized) graph Laplacian is

$$L := D - W.$$

Remark 11.4.3 (Quadratic Form as Smoothness Penalty) For $f \in \mathbb{R}^n$ (a scalar value on nodes), one has

$$f^\top L f = \frac{1}{2}\sum_{i,j} W_{ij}(f_i - f_j)^2,$$

so minimizing $f^\top L f$ forces f to vary slowly across strongly weighted edges.

Definition 11.4.4 (Laplacian Eigenmaps (Spectral Embedding)) To obtain an m-dimensional embedding, compute the eigenvectors corresponding to the smallest nonzero eigenvalues of L (or of a normalized Laplacian variant). Stack these eigenvectors as columns to form coordinates in $\mathbb{R}^m$.

Remark 11.4.4 (Normalization Variants) Common choices include the symmetric normalized Laplacian $L_{\text{sym}} = I - D^{-1/2} W D^{-1/2}$ and random-walk Laplacian $L_{\text{rw}} = I - D^{-1}W$. These adjust for degree variability and often improve embeddings on nonuniform graphs.

11.4.2 Manifold Visualization Methods

11.4.2.1 t-SNE/UMAP Objectives: Guiding Principles

t-SNE and UMAP are primarily visualization tools: they prioritize preserving *local* neighborhood structure when embedding into 2D/3D. They define probabilistic neighborhood graphs in high dimension and low dimension and then optimize a divergence between them. The guiding principle is: local neighbors should stay neighbors; global distances are secondary. This is why they can reveal clusters well, but cannot be trusted for global geometry (e.g., distances between clusters).

Remark 11.4.5 (t-SNE Guiding Principle (Conceptual)) t-SNE defines high-dimensional neighbor probabilities P_{ij} (Gaussian affinities with perplexity controlling scale) and low-dimensional probabilities Q_{ij} (Student-t affinities), and minimizes $\mathrm{KL}(P \| Q)$. This penalizes losing true neighbors heavily, emphasizing local structure.

Remark 11.4.6 (UMAP Guiding Principle (Conceptual)) UMAP constructs a weighted kNN graph approximating a fuzzy simplicial set in high dimension and optimizes a cross-entropy-like objective to match a low-dimensional fuzzy graph. Like t-SNE, it emphasizes local neighborhoods but often preserves more global structure depending on hyperparameters.

11.4.2.2 Cautions: Distortions, Hyperparameters, and Misuse

Nonlinear visualization embeddings are not faithful metric embeddings of the original data. They can distort densities, cluster sizes, and inter-cluster distances; results are sensitive to hyperparameters (perplexity, learning rate, number of neighbors, min-dist) and to randomness/initialization. Thus, one must not over-interpret 2D plots as ground truth geometry. In ML practice, these methods are best used for exploratory analysis, with robustness checks across hyperparameters and seeds.

Remark 11.4.7 (Common Misuse Patterns)

- interpreting distances between clusters as meaningful (often not),
- concluding separability from a single embedding without checking stability across settings,
- comparing embeddings across runs without aligning/controlling randomness,
- using t-SNE/UMAP as features for downstream learning without careful validation (can work but is not guaranteed).

Remark 11.4.8 (Robust Usage Checklist) Run multiple seeds, vary key hyperparameters, and verify whether qualitative conclusions (cluster presence, neighbor relations) are stable. Use quantitative metrics (kNN preservation, trustworthiness/continuity) when possible.

11.4.3 Autoencoder Viewpoint

11.4.3.1 Linear Autoencoders and PCA Equivalence (Conceptual)

An autoencoder is an encoder–decoder pair trained to reconstruct inputs through a low-dimensional bottleneck. With linear encoder/decoder and squared loss, the optimal solution spans the same subspace as PCA: it learns the principal subspace. This provides a deep bridge: PCA is the simplest autoencoder, and nonlinear autoencoders generalize PCA by replacing linear maps with neural networks.

Remark 11.4.9 (Linear Autoencoder Template) Let encoder $E(x) = W_e x$ with $W_e \in \mathbb{R}^{m\times d}$ and decoder $D(z) = W_d z$ with $W_d \in \mathbb{R}^{d\times m}$. Minimize $\sum_i \|x_i - W_d W_e x_i\|_2^2$ with $m < d$. The optimal reconstruction subspace matches the top-m principal subspace of the data (under centering), hence PCA equivalence.

11.4.3.2 Nonlinear Embeddings and Reconstruction Objectives

Nonlinear autoencoders replace W_e, W_d by neural networks and can learn curved manifolds. The bottleneck coordinates become a learned embedding optimized for reconstruction, not necessarily for classification. Regularization (weight decay, sparsity penalties, contractive terms, denoising objectives) shapes the learned geometry and prevents trivial identity mappings. This preview sets up later chapters on deep learning and representation learning.

Definition 11.4.5 (Autoencoder Objective (Conceptual)) Given data $x_i \in \mathbb{R}^d$, an encoder E_θ and decoder D_ϕ, the reconstruction objective is

$$\min_{\theta,\phi} \sum_{i=1}^{n} \|x_i - D_\phi(E_\theta(x_i))\|_2^2$$

(or a likelihood-based reconstruction loss for non-Gaussian data). The latent code $z_i = E_\theta(x_i)$ is the learned embedding.

Remark 11.4.10 (Why Reconstruction Alone may be Insufficient) Reconstruction can preserve nuisance variation not relevant to a downstream task. Task-aware representation learning modifies the objective (contrastive losses, supervised heads) to align geometry with semantics.

11.5 Matrix Sketching and Random Projections

Random projections and matrix sketching exploit high-dimensional concentration to reduce computational cost while approximately preserving geometric structure (distances, inner products, subspaces) (Halko et al. 2011). The Johnson–Lindenstrauss (JL) phenomenon provides the prototypical guarantee: a small random embedding dimension preserves pairwise distances for a finite set. Sketching primitives (Gaussian, CountSketch, hashed projections) extend this to linear-algebraic tasks such as least squares. Random features then use sketching ideas to approximate kernels by explicit low-dimensional embeddings. This section presents these ideas at a geometric/computational level (cf. Fig. 11.2).

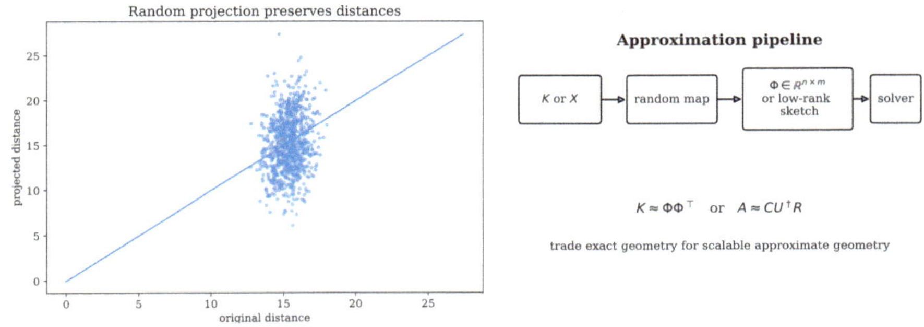

Fig. 11.2 Randomized high-dimensional tricks. Left: a Johnson–Lindenstrauss projection approximately preserves pairwise distances after compression. Right: random features and low-rank sketches replace expensive kernel or matrix computations by smaller approximate linear-algebra problems

11.5.1 Johnson–Lindenstrauss

11.5.1.1 Distance Preservation: Statement and Implications

JL is the fundamental claim that "dimension is not destiny" for finite datasets: N points in high-dimensional space can be embedded into $m = O(\varepsilon^{-2} \log N)$ dimensions so that all pairwise distances are preserved up to $(1 \pm \varepsilon)$. This implies that algorithms depending only on distances (kNN, clustering, linear separators with margin) can often be run in compressed space. In ML, JL motivates random projections as a baseline dimensionality reduction method with provable guarantees independent of ambient d.

Theorem 11.5.1 (Johnson–Lindenstrauss (JL) Lemma) *Fix* $\varepsilon \in (0, 1)$ *and points* $x_1, \ldots, x_N \in \mathbb{R}^d$. *There exists a linear map* $A : \mathbb{R}^d \to \mathbb{R}^m$ *with*

$$m = O\Big(\varepsilon^{-2} \log N\Big)$$

such that for all $i, j \in [N]$,

$$(1 - \varepsilon)\|x_i - x_j\|_2^2 \le \|Ax_i - Ax_j\|_2^2 \le (1 + \varepsilon)\|x_i - x_j\|_2^2.$$

Moreover, a random matrix A *with i.i.d. entries* $\mathcal{N}(0, 1/m)$ *satisfies this with high probability.*

Remark 11.5.1 (Algorithmic Implication) If your algorithm depends on pairwise Euclidean distances up to relative error ε, you can first project to $m = O(\varepsilon^{-2} \log N)$ and run the algorithm in $\mathbb{R}^m$. This can reduce both computation and storage when d is huge.

11.5.1.2 Random Features and Compressed Representations

JL can be viewed as constructing random features: $z = Ax$ is a randomized embedding whose dot products and distances approximate those in the original space (after appropriate normalization). This is a general theme: represent a complex object by a lower-dimensional randomized sketch that preserves the quantities you care about. In ML, this underlies random projections for text (hashed features), compressing high-dimensional sparse vectors, and building scalable approximate nearest neighbor indices.

Remark 11.5.2 (Dot Product Preservation (Related Intuition)) If A approximately preserves norms of x and y and of $x - y$, then it also approximately preserves inner products via

$$\langle x, y\rangle = \frac{1}{2}\big(\|x\|_2^2 + \|y\|_2^2 - \|x - y\|_2^2\big).$$

Thus JL-type embeddings preserve both distances and dot products (up to controlled error) on a finite set.

11.5.2 Sketching Primitives

11.5.2.1 CountSketch/Hashed Projections: Intuition

Gaussian projections are dense and expensive for very large sparse data. CountSketch and hashing-based projections are sparse, fast, and streaming-friendly: they map a high-dimensional sparse vector to a lower-dimensional sketch by randomly hashing coordinates into buckets with random signs. This preserves inner products in expectation and preserves norms with high probability on structured sets. These sketches power scalable linear models, approximate second-moment estimation, and randomized numerical linear algebra.

Definition 11.5.1 (CountSketch (Conceptual)) Choose hash functions $h : [d] \to [m]$ and sign function $s : [d] \to \{-1, +1\}$. Define the sketch $S : \mathbb{R}^d \to \mathbb{R}^m$ by

$$(Sx)_j := \sum_{i:\, h(i)=j} s(i)\, x_i, \qquad j \in [m].$$

Equivalently, S is an $m \times d$ sparse matrix with exactly one nonzero per column: $S_{h(i),i} = s(i)$.

Remark 11.5.3 (Why It Is Fast) If x is sparse with $\mathrm{nnz}(x)$ nonzeros, then computing Sx takes $O(\mathrm{nnz}(x))$ time: each nonzero coordinate updates one bucket. This is crucial for high-dimensional sparse ML (text, recommender systems).

11.5.2.2 Sketched Least Squares: Speed–Accuracy Tradeoffs

Least squares is a core subroutine. Sketching accelerates it by compressing rows: solve a smaller problem $\min_w \|S(Xw - y)\|_2^2$ where $S \in \mathbb{R}^{m\times n}$ is a sketch matrix with $m \ll n$. If S approximately preserves norms on the column space of X (a subspace embedding), then the sketched solution approximates the true OLS solution. This yields a tunable speed–accuracy tradeoff: larger sketch dimension m yields better accuracy but higher cost.

Definition 11.5.2 (Sketched Least Squares) Given $X \in \mathbb{R}^{n\times d}$ and $y \in \mathbb{R}^n$, choose a sketch matrix $S \in \mathbb{R}^{m\times n}$ with $m \ll n$. The sketched problem is

$$\widetilde{w} \in \arg\min_{w\in\mathbb{R}^d} \|S(Xw - y)\|_2^2 = \arg\min_{w} \|(SX)w - Sy\|_2^2.$$

Remark 11.5.4 (Subspace Embedding Condition (Conceptual)) A sufficient condition is that for all $v \in \mathbb{R}^d$,

$$(1-\varepsilon)\|Xv\|_2^2 \le \|SXv\|_2^2 \le (1+\varepsilon)\|Xv\|_2^2.$$

Then the sketched objective approximates the original objective uniformly over w, implying $\widetilde{w}$ yields near-optimal residual norm.

Remark 11.5.5 (When Sketching Helps Most) Sketching is most helpful when n is huge and d is moderate, or when X is sparse and one can compute SX cheaply. For very large d, one may sketch columns or use random feature methods instead.

11.5.3 Random Features for Kernels

11.5.3.1 Random Fourier Features (Conceptual)

Kernel methods can be too expensive because they scale with n via the kernel matrix. Random Fourier features (RFF) approximate certain shift-invariant kernels (notably RBF) by an explicit randomized feature map $\phi : \mathbb{R}^d \to \mathbb{R}^m$ such that $\phi(x)^\top\phi(z) \approx k(x, z)$ (Rahimi and Recht 2007). Then one can train a linear model in $\mathbb{R}^m$ instead of a kernel model, reducing computational cost from $O(n^2)$ to $O(nm)$. This is a sketching idea: approximate the Gram matrix by a low-rank factor $\Phi\Phi^\top$.

Remark 11.5.6 (RFF Template for Shift-Invariant Kernels) For a shift-invariant kernel $k(x, z) = k(x - z)$ with Fourier transform measure $p(\omega)$ (Bochner's theorem), sample $\omega_1, \ldots, \omega_m \sim p$ and $b_1, \ldots, b_m \sim \text{Uniform}(0, 2\pi)$, and define

$$\phi(x) = \sqrt{\frac{2}{m}}\big(\cos(\omega_1^\top x + b_1), \ldots, \cos(\omega_m^\top x + b_m)\big)^\top.$$

Then $\mathbb{E}[\phi(x)^\top \phi(z)] = k(x, z)$ and concentration yields approximation with high probability for large m.

11.5.3.2 Approximation Accuracy vs. Feature Dimension

RFF accuracy improves with feature dimension m by concentration: the Monte Carlo estimate of the kernel inner product converges as $O(1/\sqrt{m})$ (typically up to log factors when uniform over many pairs). This creates a second tuning knob: larger m improves approximation but increases computation and memory. In practice, one tunes m jointly with regularization λ and kernel bandwidth σ (for RBF), and monitors downstream task performance rather than only kernel approximation error.

Remark 11.5.7 (Rule-of-Thumb Scaling) For a fixed pair (x, z), $\phi(x)^\top \phi(z)$ concentrates around $k(x, z)$ at rate $\sim O(1/\sqrt{m})$. For N points and uniform control over all $O(N^2)$ pairs, one typically needs m to scale like $\log N$ times the desired accuracy factor.

Remark 11.5.8 (Practical Tuning) Start from a moderate m (e.g., a few thousand), standardize features, tune kernel bandwidth (e.g., via median distance heuristic), and then increase m until validation performance saturates. This treats m as a computational budget parameter rather than a purely theoretical requirement.

11.6 Low-Rank Models and Factorizations

Low-rank structure is one of the most powerful inductive biases in modern ML. It asserts that high-dimensional observations are governed by a small number of latent factors, enabling denoising, compression, scalable computation, and improved generalization. This section introduces low-rank bias via nuclear norm, explains matrix completion as the canonical application (recommender systems), and previews Nyström/CUR approximations as practical low-rank surrogates for large Gram/data matrices.

11.6.1 Low-Rank Inductive Bias

11.6.1.1 Nuclear Norm as Convex Surrogate

Rank minimization is natural but combinatorial and nonconvex. The nuclear norm $\|M\|_*$ (sum of singular values) is the convex envelope of rank on the unit spectral-norm ball and serves as the standard convex surrogate for low-rank structure, analogous to ℓ_1 as a surrogate for sparsity. It yields convex optimization problems with global optima and admits efficient proximal/thresholding methods via singular-value shrinkage. In kernel and representation learning, nuclear norms encourage low-dimensional latent structure while remaining tractable.

Definition 11.6.1 (Rank and Nuclear Norm) Let $M \in \mathbb{R}^{m\times n}$ have singular values $\sigma_1 \geq \cdots \geq \sigma_r > 0$ with $r = \operatorname{rank}(M)$. The *nuclear norm* is

$$\|M\|_* := \sum_{i=1}^{r} \sigma_i .$$

The spectral norm is $\|M\|_2 = \sigma_1$ and the Frobenius norm is $\|M\|_F = (\sum_i \sigma_i^2)^{1/2}$.

Remark 11.6.1 (Convex Surrogate Viewpoint) Minimizing rank is hard. Minimizing $\|M\|_*$ promotes low-rank solutions in a convex way, just as ℓ_1 promotes sparse solutions.

Remark 11.6.2 (Proximal Operator) The proximal operator of $\lambda\|\cdot\|_*$ is singular-value soft-thresholding: if $M = U\Sigma V^\top$, then

$$\operatorname{prox}_{\lambda\|\cdot\|_*}(M) = U\,(\Sigma - \lambda I)_+\,V^\top,$$

where $(\cdot)_+$ thresholds singular values at 0. This is the exact analogue of coordinate soft-thresholding for ℓ_1.

11.6.1.2 Bias–Variance and Denoising Interpretation

Low-rank regularization is a denoising prior: it assumes signal lies in a low-dimensional subspace and noise spreads across many directions. Truncating small singular values removes directions dominated by noise. This introduces bias (discarded components) but reduces variance (noise amplification), mirroring ridge/lasso tradeoffs. In practice, low-rank shrinkage improves generalization when the true signal is approximately low rank, and it provides compression by storing only latent factors.

Remark 11.6.3 (Spectral Denoising Intuition) Given a noisy matrix $Y = M^\star + E$, SVD of Y splits energy across singular directions. If $M^\star$ is low rank, its energy concentrates in top singular values, while noise spreads. Shrinking/truncating small singular values estimates $M^\star$ by filtering noise.

11.6.2 Matrix Completion

11.6.2.1 Recommender Systems: Observation Model

Matrix completion models partially observed user–item ratings as samples from an underlying low-rank matrix. The observation pattern is sparse and highly nonuniform (some users rate many items, others few). The core statistical question is identifiability: can the missing entries be inferred from the observed ones under a low-rank prior? The core computational question is scalability: factorized parameterizations $M = UV^\top$ allow efficient learning from sparse observations.

Definition 11.6.2 (Matrix Completion Observation Model) Let $M^\star \in \mathbb{R}^{m\times n}$ be an unknown matrix (e.g., user–item preferences). We observe a subset of entries $\Omega \subseteq [m] \times [n]$ and values

$$Y_{ij} = M^\star_{ij} + \varepsilon_{ij}, \qquad (i, j) \in \Omega,$$

where ε_{ij} is noise (often assumed mean-zero). The goal is to estimate $M^\star$ (or its unobserved entries) from $\{(i, j, Y_{ij})\}_{(i,j)\in\Omega}$.

Remark 11.6.4 (Factorization Estimator (Nonconvex But Scalable)) A common estimator parameterizes $M = UV^\top$ with $U \in \mathbb{R}^{m\times r}$, $V \in \mathbb{R}^{n\times r}$ and solves

$$\min_{U,V} \sum_{(i,j)\in\Omega} (U_i^\top V_j - Y_{ij})^2 + \lambda\big(\|U\|_F^2 + \|V\|_F^2\big),$$

where U_i is row i of U and V_j is row j of V. This is the workhorse behind recommender systems.

11.6.2.2 Identifiability Intuition and Practical Pitfalls

Low rank alone is not sufficient: the matrix must be "spread out" (incoherent) so that information is not concentrated in a few entries, and the sampling pattern must cover the matrix adequately. Nonuniform sampling, cold-start users/items, and selection bias (observations not missing at random) are major practical pitfalls. Identifiability failures appear as many equally good completions of observed entries with very different missing entries. Thus evaluation must be carefully designed and assumptions must be stated explicitly.

Remark 11.6.5 (Incoherence Intuition) If singular vectors align with coordinate axes (high coherence), then a few entries carry most information. Missing those entries makes completion impossible. Incoherence asserts that singular vectors are not too spiky, distributing information across entries.

Remark 11.6.6 (Sampling Pitfalls)

- **Non-MCAR missingness:** users rate items they like/dislike, so observed entries are biased.
- **Cold start:** users/items with few observations cannot be reliably inferred from low-rank structure alone.

- **Popularity bias:** dense columns/rows dominate training, hurting generalization to sparse ones.

These issues are statistical, not merely algorithmic.

11.6.3 Nyström and CUR Approximations

11.6.3.1 Low-Rank Approximation of Gram Matrices

Kernel methods scale poorly with n because they operate on an $n \times n$ Gram matrix K. Nyström approximates K by sampling a subset of columns (landmarks) and forming a low-rank surrogate that enables fast approximate kernel learning. CUR approximations similarly approximate a matrix using sampled columns and rows, preserving interpretability (actual columns/rows) and sparsity patterns. These approximations are sketches specialized to low-rank structure.

Definition 11.6.3 (Nyström Approximation (Conceptual Form)) Let $K \in \mathbb{R}^{n\times n}$ be a PSD Gram matrix. Choose a landmark set $S \subseteq [n]$ with $|S| = c$. Let $C = K_{:,S} \in \mathbb{R}^{n\times c}$ and $W = K_{S,S} \in \mathbb{R}^{c\times c}$. A Nyström approximation is

$$K \approx \widetilde{K} := CW^{\dagger}C^{\top},$$

which has rank at most c.

Definition 11.6.4 (CUR Approximation (Conceptual)) For a general matrix $A \in \mathbb{R}^{m\times n}$, choose a subset of columns C and rows R and their intersection U. A CUR approximation has the form

$$A \approx C\,U^{\dagger}\,R,$$

where C consists of selected columns of A, R consists of selected rows of A, and U is the corresponding submatrix.

Remark 11.6.7 (Why Nyström Works) If K is approximately low rank and the landmark columns capture the dominant eigenspace, then $CW^{\dagger}C^{\top}$ approximates K well in spectral/Frobenius norms. This reduces kernel learning to linear algebra on $c \times c$ matrices rather than $n \times n$.

11.6.3.2 When Approximations Preserve Learning Performance

Approximating a matrix is useful only if downstream learning is preserved. For kernel ridge regression, replacing K by $\widetilde{K}$ changes predictions; stability depends on spectral approximation quality, regularization strength, and noise level. Generally, stronger regularization makes learning less sensitive to approximation error. In practice, one selects the sketch size c by validation: increase c until performance saturates.

Remark 11.6.8 (Regularization Improves Approximation Robustness) In kernel ridge, the solution involves $(K + \lambda I)^{-1}$. If λ is not too small, then small perturbations in K lead to controlled perturbations in $(K + \lambda I)^{-1}$, so predictions are stable under Nyström/CUR approximations.

Remark 11.6.9 (Practical Selection Rule) Treat the sketch size c (or CUR ranks) as a compute budget hyperparameter: start small, validate, and increase until the validation metric stops improving. Always report the approximation method and c alongside results, as they are part of the model.

11.7 Nearest Neighbor Geometry and Approximate Search

Nearest-neighbor methods are the purest form of geometry-driven learning: they predict using only distances (or similarities) between the query and stored examples. Their strengths are simplicity, nonparametric flexibility, and strong performance when the metric aligns with semantics. Their weaknesses are computational cost and sensitivity to high-dimensional geometry (distance concentration). This section previews kNN prediction as a metric-dependent procedure, the curse of dimensionality, and approximate nearest neighbor (ANN) techniques such as locality-sensitive hashing and indexing/quantization pipelines.

11.7.1 Distance-Based Prediction

11.7.1.1 *k*NN Regression/Classification and Metric Dependence

kNN replaces model fitting by a geometric rule: aggregate labels of the k closest samples. Thus, the metric is the model. Different norms or learned metrics can change predictions dramatically, especially in high-dimensional feature spaces. Weighted kNN further makes geometry explicit by decaying influence with distance. This aligns with kernel smoothing: kNN is a local averaging estimator whose bias/variance depends on neighborhood size.

Definition 11.7.1 (*k*NN Set) Let $(\mathcal{X}, d)$ be a metric space and data $(x_i, y_i)_{i=1}^n$. For a query $x \in \mathcal{X}$, let $N_k(x) \subseteq [n]$ be the indices of the k nearest points:

$$N_k(x) \in \arg \min_{S \subseteq [n],\, |S|=k} \sum_{i \in S} d(x, x_i),$$

with ties resolved by a fixed rule.

Definition 11.7.2 (*k*NN Regression and Classification)

- (**Regression**) For $y_i \in \mathbb{R}$,

$$\widehat{y}(x) = \frac{1}{k} \sum_{i \in N_k(x)} y_i, \quad \text{or weighted} \quad \widehat{y}(x) = \frac{\sum_{i \in N_k(x)} w_i(x)\, y_i}{\sum_{i \in N_k(x)} w_i(x)},$$

  with weights $w_i(x)$ decreasing in $d(x, x_i)$.
- (**Classification**) For $y_i \in \{1, \ldots, K\}$,

$$\widehat{y}(x) = \arg\max_{c \in [K]} \sum_{i \in N_k(x)} \mathbf{1}\{y_i = c\},$$

  or weighted analogues.

Remark 11.7.1 (Metric Dependence Is Inductive Bias) If d changes (e.g., standardization, whitening, cosine distance on normalized embeddings), the neighborhoods $N_k(x)$ change. Thus, kNN performance depends more on representation/metric design than on algorithmic complexity.

11.7.1.2 Curse of Dimensionality and Why Features Matter

In high dimensions, neighborhoods behave poorly: balls concentrate near their boundaries, distances to nearest and farthest neighbors become similar, and the number of points needed to cover the space grows exponentially with dimension. This is the curse of dimensionality for nonparametric methods. However, in ML practice, data often lie near low-dimensional manifolds or have structured representations, so effective dimension can be much smaller. Feature learning and normalization are therefore not optional for NN methods; they are the mechanism that reduces effective dimension and restores contrast.

Remark 11.7.2 (A Geometric Symptom: Distance Contrast Collapse) If for a typical query x,

$$\frac{\max_i d(x, x_i) - \min_i d(x, x_i)}{\min_i d(x, x_i)}$$

is small, then the nearest neighbor is not much closer than a random point, making NN predictions unstable. This often happens in high-dimensional isotropic settings.

Remark 11.7.3 (Why Representation Learning Helps) A learned embedding $\phi(x)$ aims to make semantically similar points close and dissimilar points far under a simple metric, which is exactly what NN methods need. Thus, embedding learning and NN retrieval form a coherent pipeline:

$$x \mapsto \phi(x) \in \mathbb{R}^m \quad \text{then NN search under } \ell_2 \text{ or cosine.}$$

11.7.2 Approximate Nearest Neighbors

11.7.2.1 Locality-Sensitive Hashing (LSH) Intuition

Exact NN search can be too slow for large n and moderate-to-high dimensions. ANN methods trade exactness for speed: return a neighbor that is close enough with high probability. LSH is the prototypical randomized ANN technique: it constructs hash functions such that nearby points collide (same hash) with higher probability than far points. Querying then checks only candidates in the same buckets, reducing search cost. The key is to design hashes aligned with the chosen metric (e.g., random hyperplanes for cosine similarity).

Definition 11.7.3 (Locality-Sensitive Hash Family (Conceptual)) A family $\mathcal{H}$ of hash functions $h : \mathcal{X} \to \mathcal{U}$ is *locality-sensitive* for a distance d if there exist radii $r < cr$ and probabilities $p_1 > p_2$ such that for h drawn uniformly from $\mathcal{H}$:

$$d(x, y) \le r \Rightarrow \mathbb{P}(h(x) = h(y)) \ge p_1, \quad d(x, y) \ge cr \Rightarrow \mathbb{P}(h(x) = h(y)) \le p_2.$$

Here $c > 1$ is the approximation factor.

Remark 11.7.4 (Random Hyperplane LSH for Cosine Similarity (Intuition)) For unit vectors, choose a random vector a and define $h_a(x) = \text{sign}(a^\top x)$. Then $\mathbb{P}(h_a(x) = h_a(y))$ increases with cosine similarity $\langle x, y \rangle$: nearby directions collide more often than orthogonal ones. Concatenating multiple such bits yields a locality-sensitive code.

11.7.2.2 Indexing, Quantization, and Retrieval Pipelines (Conceptual)

Industrial retrieval systems combine multiple approximations: coarse indexing to prune candidates, vector quantization or product quantization (PQ) to compress embeddings, and re-ranking to refine results with a more accurate score. These are geometric engineering pipelines: they exploit cluster structure, low-dimensional subspaces, and distance decompositions. In ML, the pipeline is part of the model: training objectives (contrastive losses) are often chosen to produce embeddings that are ANN-friendly.

Remark 11.7.5 (A Canonical ANN Pipeline (Conceptual))

1. **Embedding:** map items and queries to vectors $\phi(\cdot)$, often normalized.
2. **Coarse index:** cluster embeddings (e.g., k-means) and assign each item to a centroid (inverted file system).
3. **Quantization:** compress residual vectors using PQ or scalar quantization to store many vectors efficiently.
4. **Search:** for a query, scan a small number of nearest centroids, compute approximate distances to candidates using compressed codes.

5. **Re-rank:** compute exact distances for top candidates and optionally apply a stronger model for final ranking.

Remark 11.7.6 (Metric and Normalization Matter) Most ANN methods assume a specific metric (often ℓ_2 or inner product). If the embedding space is not normalized/standardized appropriately, ANN quality degrades sharply. Thus, feature normalization is both a geometric and a systems requirement.

11.8 High-Dimensional Phenomena (What Breaks, What Helps)

High-dimensional geometry is simultaneously an ally and an adversary in ML. It helps by enabling stable random projections and predictable norms, but it breaks naive intuitions about neighborhoods and distances: most points become nearly equidistant and random directions become nearly orthogonal. These effects produce practical pathologies such as *hubness* in nearest-neighbor search. The remedy is geometry control: normalization, regularization, and metric learning reshape the space so that distances and angles reflect semantics rather than artifacts of scale and dimensionality.

11.8.1 Concentration of Measure (Informal)

11.8.1.1 Near-Orthogonality and Cosine Similarity Behavior

When vectors are drawn from roughly isotropic distributions, their directions behave like random points on a high-dimensional sphere. Then inner products concentrate near 0 and angles concentrate near $\pi/2$. Operationally: unrelated embeddings have cosine similarity close to 0, and meaningful similarity signals must exceed a small noise baseline of order $1/\sqrt{d}$. This concentration underlies the effectiveness of cosine similarity and normalized dot products in embedding systems, but it also means that small deviations in cosine similarity can be meaningful and that calibration of similarity scores is important.

Remark 11.8.1 (Baseline Scale) For independent unit vectors $u, v \in \mathbb{S}^{d-1}$ chosen uniformly at random,

$$\langle u, v\rangle = \cos(u, v) \approx 0, \qquad \mathbb{E}[\langle u, v\rangle^2] = \frac{1}{d}.$$

Thus typical cosine similarity magnitude is on the order of $1/\sqrt{d}$.

Remark 11.8.2 (Practical Implication for Embedding Thresholds) In high d, a cosine similarity of (say) 0.2 can represent a very strong alignment relative to the random baseline, even though it may look "small" numerically. Thresholds should be chosen empirically, ideally after normalization and calibration.

11.8.1.2 Distance Concentration and Hubness

Distance concentration means that for many distributions in large d, pairwise distances cluster tightly around a mean value. This reduces contrast between nearest and farthest neighbors, making nearest-neighbor decisions unstable. Hubness is a further phenomenon: some points become "hubs" that appear in many nearest-neighbor lists, often due to geometry plus finite-sample effects and anisotropy. In retrieval, hubness causes popularity bias and degraded diversity. These issues are not merely algorithmic; they are geometric and statistical and must be addressed by representation design.

Remark 11.8.3 (Distance Concentration Heuristic) If for a typical query x,

$$\|x - x_{(1)}\|_2 \approx \|x - x_{(k)}\|_2 \approx \|x - x_{(n)}\|_2,$$

then nearest-neighbor ranking is dominated by noise and minor preprocessing changes. This is common when raw features are high-dimensional and poorly scaled.

Remark 11.8.4 (Hubness (Conceptual Definition)) Given a dataset $\{x_i\}_{i=1}^n$ and a fixed k, define the hubness score of point x_j as

$$H_k(j) := \big|\{i : j \in N_k(x_i)\}\big|,$$

the number of times x_j appears in other points' kNN lists. Large $H_k(j)$ indicates a hub. Hubness is amplified by anisotropy and by concentration of distances.

Remark 11.8.5 (Practical Symptoms)

- retrieval returns the same items for many distinct queries,
- nearest neighbors are unstable across random seeds or minor preprocessing changes,
- kNN graphs show a few nodes with extremely high in-degree.

11.8.2 Regularization and Normalization as Geometry Control

11.8.2.1 Feature Normalization and Angular Metrics

Normalization reshapes geometry by removing nuisance variation (mean and scale) and by changing the metric. Standardization/whitening addresses coordinate scale and correlation; ℓ_2 normalization projects to a sphere, making similarity primarily angular (cosine). These operations counteract distance concentration by reducing radial variability and by emphasizing directional information, which is often more semantically meaningful for embeddings. In kernel and NN methods, such preprocessing is not optional: it defines the effective geometry of the algorithm.

Remark 11.8.6 (Three Common Normalizations)

- **Standardization:** $x \mapsto D_\sigma^{-1}(x - \mu)$ (unit variance per feature).
- **Whitening:** $x \mapsto \Sigma^{-1/2}(x - \mu)$ (unit covariance).
- **ℓ_2 normalization:** $x \mapsto x/\|x\|_2$ (unit sphere).

Each changes the metric: Euclidean distance after whitening equals Mahalanobis distance before whitening.

Remark 11.8.7 (Angular Distances) On the unit sphere, Euclidean distance and cosine similarity are equivalent orderings:

$$\|u - v\|_2^2 = 2(1 - \langle u, v\rangle).$$

Thus many retrieval systems reduce to nearest-neighbor search under an angular metric.

11.8.2.2 Metric Learning Objectives

When a fixed metric is inadequate, metric learning aims to learn a representation or a distance so that geometry aligns with labels or semantics. In linear metric learning, one learns a PSD matrix $M \succeq 0$ defining Mahalanobis distance $\|x - z\|_M$. In deep metric learning, one learns an embedding ϕ using contrastive/triplet losses so that same-class points cluster and different-class points separate by a margin. These objectives explicitly fight hubness and concentration by enforcing separation and structuring the embedding distribution.

Definition 11.8.1 (Mahalanobis Metric Learning (Linear)) A Mahalanobis distance is

$$d_M(x, z)^2 = (x - z)^\top M(x - z), \qquad M \succeq 0.$$

Learning M (or its factor L with $M = L^\top L$) from labeled pairs/triplets is linear metric learning.

Definition 11.8.2 (Triplet Loss (Metric Learning, Preview)) Given anchor a, positive p, negative n, and margin $\alpha > 0$, a standard objective is

$$\mathcal{L}_{\text{tri}} = \max\{0,\ \|\phi(a) - \phi(p)\|_2^2 - \|\phi(a) - \phi(n)\|_2^2 + \alpha\}.$$

Minimizing this enforces that positives are closer than negatives by margin α in embedding space.

Remark 11.8.8 (Geometric Effect) Metric learning objectives sculpt the geometry: they reduce intra-class distances and increase inter-class distances, increasing distance contrast, and mitigating hubness. They also tend to produce embeddings that are ANN-friendly for retrieval pipelines.

11.9 Exercises

Unless stated otherwise, all logarithms are natural and all norms are Euclidean.

Exercise 11.9.1 (Kernel Matrix Validity and PSD Check) Consider the three points in $\mathbb{R}^2$:

$$x_1 = \begin{bmatrix} 0 \\ 0 \end{bmatrix}, \quad x_2 = \begin{bmatrix} 1 \\ 0 \end{bmatrix}, \quad x_3 = \begin{bmatrix} 0 \\ 2 \end{bmatrix}.$$

(i) Form the **linear** Gram matrix K^{lin} with $K_{ij}^{\text{lin}} = x_i^\top x_j$ and compute its eigenvalues. (ii) Form the **polynomial** Gram matrix K^{poly} with $k(x, z) = (x^\top z+1)^2$ and compute K^{poly} explicitly and verify it is PSD by eigenvalues or principal minors. (iii) Define an ad-hoc "similarity" matrix

$$S = \begin{bmatrix} 1 & 2 & -1 \\ 2 & 1 & 2 \\ -1 & 2 & 1 \end{bmatrix}.$$

Decide whether S can be a Gram matrix of real vectors by checking PSD; if it is not PSD, compute its most negative eigenvalue and explain (one paragraph) why this invalidates it as an inner-product table.

Exercise 11.9.2 (Kernel Distance and Centering Trick) Let $K \in \mathbb{R}^{4\times4}$ be a PSD kernel matrix for points $x_1, \ldots, x_4$:

$$K = \begin{bmatrix} 4 & 1 & 0 & 2 \\ 1 & 3 & 1 & 0 \\ 0 & 1 & 2 & 1 \\ 2 & 0 & 1 & 5 \end{bmatrix}.$$

(i) Compute the pairwise squared distances in feature space $d_k(x_i, x_j)^2 = K_{ii} + K_{jj} - 2K_{ij}$ for all $1 \le i < j \le 4$. (ii) Form the centering matrix $H = I_4 - \frac{1}{4}\mathbf{1}\mathbf{1}^\top$ and compute the centered Gram matrix $K_c = HKH$. (iii) Using K_c, compute centered squared distances again via $d_{k,c}(x_i, x_j)^2 = (K_c)_{ii} + (K_c)_{jj} - 2(K_c)_{ij}$ and comment (one paragraph) on why distances are translation-invariant but inner products are not (hence centering matters).

Exercise 11.9.3 (Kernel Ridge Regression by Solving $(K+\lambda I)\alpha = y$) Given the kernel matrix

$$K = \begin{bmatrix} 2 & 1 & 0 \\ 1 & 2 & 1 \\ 0 & 1 & 2 \end{bmatrix}, \qquad y = \begin{bmatrix} 1 \\ 0 \\ 1 \end{bmatrix}, \qquad \lambda = 1,$$

(i) solve $(K + \lambda I_3)\alpha = y$ to obtain α exactly, (ii) compute fitted values $\widehat{y} = K\alpha$, (iii) compute the RKHS norm surrogate $\alpha^\top K\alpha$ and the objective value

$$\|K\alpha - y\|_2^2 + \lambda\, \alpha^\top K\alpha,$$

and (iv) for a new point x with kernel vector $k_x = \big(k(x_1, x), k(x_2, x), k(x_3, x)\big)^\top = \begin{bmatrix} 1 \\ 1 \\ 0 \end{bmatrix}$, compute the prediction $f(x) = k_x^\top \alpha$.

Exercise 11.9.4 (Representer Theorem in a Finite-Dimensional RKHS Model) Let $\phi(x_i) \in \mathbb{R}^2$ be explicit features of $n = 4$ points given by

$$\phi(x_1) = \begin{bmatrix} 1 \\ 0 \end{bmatrix}, \ \phi(x_2) = \begin{bmatrix} 1 \\ 1 \end{bmatrix}, \ \phi(x_3) = \begin{bmatrix} 0 \\ 1 \end{bmatrix}, \ \phi(x_4) = \begin{bmatrix} 2 \\ 1 \end{bmatrix}, \quad y = \begin{bmatrix} 1 \\ 0 \\ 1 \\ 2 \end{bmatrix}.$$

Consider ridge in feature space:

$$\min_{w \in \mathbb{R}^2} \|\Phi w - y\|_2^2 + \lambda \|w\|_2^2, \qquad \lambda = 1,$$

where $\Phi \in \mathbb{R}^{4\times 2}$ stacks $\phi(x_i)^\top$ as rows. (i) Solve for $w^\star$ explicitly. (ii) Form the Gram matrix $K = \Phi\Phi^\top$ and solve the dual system $(K + \lambda I)\alpha = y$. (iii) Verify that the predictions satisfy $\widehat{y} = \Phi w^\star = K\alpha$ and that $w^\star = \Phi^\top \alpha$. (This is the representer theorem in a finite-dimensional feature space.)

Exercise 11.9.5 (Kernel Logistic Regression: One Newton Step in the Dual) Let $y \in \{0, 1\}^3$ be $y = (1, 0, 1)^\top$ and let the kernel matrix be

$$K = \begin{bmatrix} 1 & 1/2 & 0 \\ 1/2 & 1 & 1/2 \\ 0 & 1/2 & 1 \end{bmatrix}.$$

Consider the kernel logistic model with scores $f = K\alpha$ and probabilities $p = \sigma(f)$ elementwise, and objective

$$J(\alpha) = \sum_{i=1}^{3} \Big(-y_i \log p_i - (1 - y_i) \log(1 - p_i) \Big) + \frac{\lambda}{2} \alpha^\top K \alpha, \qquad \lambda = 1.$$

Starting from $\alpha^{(0)} = 0$, compute: (i) $f^{(0)}$, $p^{(0)}$, (ii) the gradient $g = K(p - y) + \lambda K\alpha$ at $\alpha^{(0)}$, (iii) the Hessian $H = KWK + \lambda K$ where $W = \mathrm{diag}(p_i(1 - p_i))$ at $\alpha^{(0)}$, and (iv) one Newton step $\alpha^{(1)} = \alpha^{(0)} - H^{-1} g$. (You must carry exact fractions for K and at least 4 decimal places for $\sigma(0) = 1/2$ computations.)

Exercise 11.9.6 (RBF Kernel Bandwidth Sensitivity) Let points in $\mathbb{R}$ be $x_1 = 0$, $x_2 = 1$, $x_3 = 3$ and use the RBF kernel

$$k_\sigma(x, z) = \exp\left(-\frac{(x - z)^2}{2\sigma^2} \right).$$

(i) Compute the kernel matrices $K(\sigma)$ for $\sigma \in \{0.5, 1, 2\}$. (ii) For each σ, compute the eigenvalues of $K(\sigma)$ numerically (to 3 decimals). (iii) Explain (one paragraph) how the limiting behaviors $\sigma \to 0$ and $\sigma \to \infty$ correspond to $K \approx I$ and $K \approx \mathbf{1}\mathbf{1}^\top$, and relate this to overfitting/underfitting.

Exercise 11.9.7 (Kernel PCA: Centering and Top Component) Let a kernel matrix be

$$K = \begin{bmatrix} 1 & 1 & 0 & 0 \\ 1 & 1 & 0 & 0 \\ 0 & 0 & 1 & 1 \\ 0 & 0 & 1 & 1 \end{bmatrix}.$$

(i) Compute $K_c = HKH$ with $H = I_4 - \frac{1}{4}\mathbf{1}\mathbf{1}^\top$. (ii) Compute the eigenvalues/eigenvectors of K_c and identify the top nonzero eigenvalue λ_1 and a corresponding unit eigenvector v_1. (iii) Using the convention $z_{i1} = (v_1)_i / \sqrt{\lambda_1}$, compute the 1D kernel PCA embedding coordinates $(z_{11}, z_{21}, z_{31}, z_{41})$. (iv) Interpret geometrically what structure this embedding reveals about the four points.

Exercise 11.9.8 (Laplacian Eigenmaps on a Small Graph) Consider the weighted graph on $n = 5$ nodes with adjacency (weight) matrix

$$W = \begin{bmatrix} 0 & 1 & 0 & 0 & 0 \\ 1 & 0 & 1 & 0 & 0 \\ 0 & 1 & 0 & 1 & 0 \\ 0 & 0 & 1 & 0 & 1 \\ 0 & 0 & 0 & 1 & 0 \end{bmatrix}.$$

(i) Compute the degree matrix D and the unnormalized Laplacian $L = D - W$. (ii) Compute the two smallest eigenvalues of L and corresponding eigenvectors. (iii) Use the second-smallest eigenvector (Fiedler vector) to produce a 1D embedding of the nodes and explain (one paragraph) how it orders the chain structure.

Exercise 11.9.9 (Random Projection: Explicit JL Distortion Computation) Let $A \in \mathbb{R}^{3\times 5}$ be

$$A = \frac{1}{\sqrt{3}} \begin{bmatrix} 1 & 0 & -1 & 2 & 0 \\ 0 & 2 & 1 & 0 & -1 \\ 1 & 1 & 0 & -1 & 2 \end{bmatrix}.$$

Let $x_1, x_2, x_3 \in \mathbb{R}^5$ be

$$x_1 = \begin{bmatrix} 1 \\ 0 \\ 2 \\ -1 \\ 1 \end{bmatrix}, \quad x_2 = \begin{bmatrix} 0 \\ 1 \\ -1 \\ 2 \\ 0 \end{bmatrix}, \quad x_3 = \begin{bmatrix} 2 \\ -1 \\ 0 \\ 1 \\ -2 \end{bmatrix}.$$

(i) Compute all pairwise distances $\|x_i - x_j\|_2$. (ii) Compute $y_i = Ax_i$ and all pairwise distances $\|y_i - y_j\|_2$. (iii) For each pair (i, j) compute the distortion ratio $\|y_i - y_j\|_2/\|x_i - x_j\|_2$ and state the maximum relative distortion.

Exercise 11.9.10 (CountSketch/Hashing by Explicit Bucket Computation) Let $d = 8, m = 3$, and define hash $h : [8] \to [3]$ and sign $s : [8] \to \{-1, +1\}$ by

$$h = (1, 2, 3, 1, 2, 3, 1, 2), \qquad s = (+1, -1, +1, -1, +1, -1, +1, -1),$$

meaning $h(1) = 1, h(2) = 2, \ldots$ and similarly for s. For the sparse vector $x \in \mathbb{R}^8$ with

$$x = (2, 0, -1, 3, 0, 0, 4, -2)^\top,$$

compute the sketch $z = Sx \in \mathbb{R}^3$ where $(Sx)_j = \sum_{i:h(i)=j} s(i)x_i$. Then compute $\|x\|_2^2$ and $\|z\|_2^2$ and comment (one paragraph) on whether this single sketch preserved the norm and why such preservation is only probabilistic.

Exercise 11.9.11 (Sketched Least Squares on a Tiny Instance) Let

$$X = \begin{bmatrix} 1 & 0 \\ 1 & 1 \\ 1 & 2 \\ 1 & 3 \\ 1 & 4 \\ 1 & 5 \end{bmatrix} \in \mathbb{R}^{6\times 2}, \qquad y = \begin{bmatrix} 0 \\ 1 \\ 1 \\ 2 \\ 2 \\ 3 \end{bmatrix} \in \mathbb{R}^6.$$

(i) Compute the exact OLS solution $\widehat{w} = (X^\top X)^{-1}X^\top y$. (ii) Let the sketch matrix $S \in \mathbb{R}^{3\times 6}$ select rows $\{1, 3, 6\}$ (a deterministic "sketch"):

$$S = \begin{bmatrix} 1\,0\,0\,0\,0\,0 \\ 0\,0\,1\,0\,0\,0 \\ 0\,0\,0\,0\,0\,1 \end{bmatrix}.$$

Compute $\widetilde{w} = \arg\min_w \|(SX)w - Sy\|_2^2$ and compare $\widetilde{w}$ to $\widehat{w}$. (iii) Compute the residual norms $\|X\widehat{w} - y\|_2$ and $\|X\widetilde{w} - y\|_2$ and interpret the trade-off.

Exercise 11.9.12 (Random Fourier Features: Explicit Feature Map and Kernel Approximation) Let $d = 1$ and target kernel $k(x, z) = \exp(-\frac{(x-z)^2}{2\sigma^2})$ with $\sigma = 1$. Use $m = 4$ random features with frequencies and phases fixed as

$$\omega = (0,\ 1,\ -1,\ 2), \qquad b = \left(0,\ \frac{\pi}{3},\ \frac{\pi}{2},\ \pi\right),$$

and define

$$\phi(x) = \sqrt{\frac{2}{m}}\big(\cos(\omega_1 x + b_1), \ldots, \cos(\omega_m x + b_m)\big)^\top.$$

For $x = 0$ and $z = 1$: (i) compute $\phi(x)$, $\phi(z)$, and $\phi(x)^\top\phi(z)$, (ii) compute the true kernel value $k(x, z) = e^{-1/2}$, (iii) compute the absolute approximation error $|\phi(x)^\top\phi(z) - k(x, z)|$.

Exercise 11.9.13 (Matrix Completion: Low-Rank Factorization with Sparse Observations) Let an unknown 3×3 matrix $M^\star$ be rank-1 of the form $M^\star = uv^\top$. You observe entries

$$M^\star_{11} = 2, \quad M^\star_{12} = 4, \quad M^\star_{21} = 3, \quad M^\star_{22} = 6, \quad M^\star_{33} = 5.$$

(i) Determine whether these observations are consistent with a rank-1 matrix; if yes, find one pair (u, v) producing a consistent completion. (ii) Use your (u, v) to fill in all missing entries of $M^\star$. (iii) Explain (one paragraph) whether the completion is unique and what additional observation(s) would remove scaling ambiguities.

Exercise 11.9.14 (Nyström Approximation of a Gram Matrix) Let

$$K = \begin{bmatrix} 4\,2\,0\,0 \\ 2\,3\,1\,0 \\ 0\,1\,2\,1 \\ 0\,0\,1\,2 \end{bmatrix} \in \mathbb{R}^{4\times 4}$$

(assume $K \succeq 0$). Choose landmark set $S = \{1, 3\}$ (so $c = 2$). (i) Form $C = K_{:,S} \in \mathbb{R}^{4\times 2}$ and $W = K_{S,S} \in \mathbb{R}^{2\times 2}$. (ii) Compute $\widetilde{K} = CW^{-1}C^\top$ explicitly (use exact arithmetic). (iii) Compute $\|K - \widetilde{K}\|_F$ and comment (one paragraph) on which entries are well-approximated and why landmark choice matters.

Exercise 11.9.15 (Hubness Score Computation on a Toy Embedding Set) Let $n = 6$ points in $\mathbb{R}^2$ be

$$x_1 = (0, 0),\ x_2 = (1, 0),\ x_3 = (0, 1),\ x_4 = (1, 1),\ x_5 = (10, 10),\ x_6 = (10, 11).$$

Using Euclidean distance and $k = 1$, compute the nearest neighbor of each point (break ties by choosing the smallest index), then compute the hubness scores $H_1(j) = |\{i : j \in N_1(x_i)\}|$ for all $j \in [6]$. Identify any hubs and explain (one paragraph) how cluster structure creates hubness even in low dimension.

Chapter 12
Optimization for Statistical Learning

Machine learning is often presented as a modeling problem, but in practice, it is equally an optimization problem: we choose parameters by minimizing an objective constructed from data, regularization, and constraints. The geometry of the objective controls what algorithms work, how fast they converge, and whether the solution generalizes. This chapter develops optimization principles and algorithms that are repeatedly used across ML, emphasizing how statistical considerations (noise, generalization, model complexity) interact with computational constraints.

12.1 Why Optimization Is Central to ML

12.1.1 Learning as Optimization of an Objective

12.1.1.1 Empirical Risk Minimization and Regularization

Most ML training procedures are instances of empirical risk minimization (ERM): choose a hypothesis that minimizes average loss on the sample. Regularization modifies ERM to encode inductive bias, stabilize computation, and improve generalization. In probabilistic models, ERM with log-loss is maximum likelihood; with priors, it becomes MAP. Thus, optimization is the computational language that unifies learning paradigms (Bottou et al. 2018) (cf. Fig. 12.1).

Definition 12.1.1 (ERM) Let $\mathcal{H}$ be a hypothesis class and $\ell(h(x), y)$ a loss. Given data $\{(x_i, y_i)\}_{i=1}^n$, ERM chooses

$$\widehat{h} \in \arg\min_{h \in \mathcal{H}} \widehat{R}_n(h), \qquad \widehat{R}_n(h) := \frac{1}{n} \sum_{i=1}^{n} \ell(h(x_i), y_i).$$

P. Singh, B. Raman, *Foundations of Machine Learning and AI*, Studies in Big Data 199, https://doi.org/10.1007/978-3-032-30336-3_12

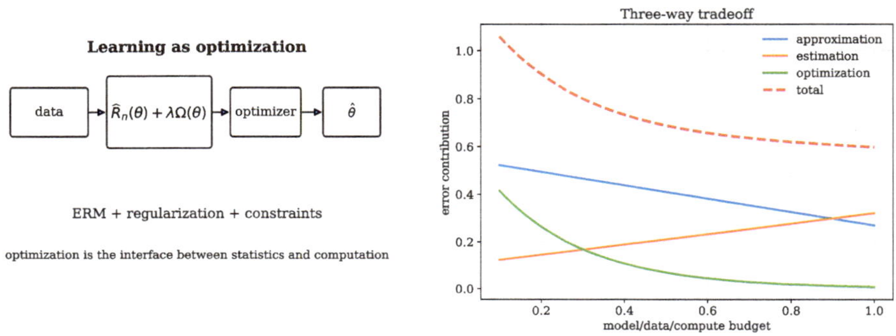

Fig. 12.1 Optimization as the computational language of learning: ERM plus regularization defines the objective, while approximation, estimation, and optimization errors trade off under a fixed compute budget

Definition 12.1.2 (Regularized ERM) Given a regularizer $\Omega(h)$ and $\lambda \geq 0$, regularized ERM chooses

$$\widehat{h}_\lambda \in \arg\min_{h \in \mathcal{H}} \widehat{R}_n(h) + \lambda\, \Omega(h).$$

In parametric models $h = h_\theta$, this becomes minimizing over θ: $\min_\theta \widehat{R}_n(\theta) + \lambda\Omega(\theta)$.

Remark 12.1.1 (Constraint Form) Many penalties have an equivalent constrained form:

$$\min_h \widehat{R}_n(h) \quad \text{s.t.} \quad \Omega(h) \leq r,$$

for an appropriate correspondence between r and λ under convexity. This connects optimization directly to geometry of feasible sets.

12.1.1.2 From Objectives to Algorithms: What Can Go Wrong

Writing down an objective does not guarantee that optimization will succeed. Problems arise from nonconvexity (local minima/saddles), ill-conditioning (slow convergence), nonsmoothness (subgradients, plateaus), stochasticity (noisy gradients), and numerical issues (overflow/underflow). Additionally, optimizing too aggressively can harm generalization (overfitting), while optimizing too little can underfit. Thus, algorithm design must be informed by objective geometry and by the statistical precision of the task.

Remark 12.1.2 (Common Failure Modes)

- **Ill-conditioning:** gradients zig-zag, requiring tiny step sizes.

- **Nonsmoothness:** coordinate-wise kinks (ℓ_1) require proximal/coordinate methods.
- **Nonconvexity:** multiple minima, saddle points, sensitivity to initialization.
- **Stochastic noise:** SGD variance dominates, requiring schedules/averaging.
- **Numerics:** overflow in exponentials (softmax), catastrophic cancellation, poor scaling.

Remark 12.1.3 (Algorithmic Remedies) Preconditioning/normalization, regularization, line search, momentum, adaptive steps, proximal operators, and second-order curvature approximations are all geometric tools to address these failure modes.

12.1.2 Three-Way Tradeoff: Data, Model, Compute

12.1.2.1 Approximation vs. Estimation vs. Optimization Error (Revisited)

Test performance is governed by three interacting errors: approximation (model class too small), estimation (finite data), and optimization (not minimizing the training objective exactly). A larger model reduces approximation error but increases estimation and optimization difficulty. A larger dataset reduces estimation error but increases computation. Thus, ML design is about balancing these errors given computational and memory constraints.

Remark 12.1.4 (Three-Error Decomposition (Conceptual)) Let $R(\theta)$ be population risk, $\widehat{R}_n(\theta)$ empirical risk, and $\widehat{\theta}_T$ parameters after T optimization steps. A useful conceptual decomposition is

$$R(\widehat{\theta}_T) - \inf_\theta R(\theta) \approx \underbrace{\inf_\theta R(\theta) - R^\star}_{\text{approximation}} + \underbrace{R(\widehat{\theta}_n) - \inf_\theta R(\theta)}_{\text{estimation}} + \underbrace{R(\widehat{\theta}_T) - R(\widehat{\theta}_n)}_{\text{optimization}},$$

where $\widehat{\theta}_n$ is an (ideal) empirical minimizer and $R^\star$ is Bayes risk.

Remark 12.1.5 (Early Stopping as Balancing) Early stopping often reduces compute and can improve generalization by preventing optimization error from dropping far below the estimation noise floor. This is a compute–statistics tradeoff encoded as a stopping rule.

12.1.2.2 Time–Accuracy–Memory Regimes in Practice

Real ML systems run under constraints: limited time (training/inference latency), limited memory (GPU/TPU RAM), and limited data throughput (I/O). Optimization algorithms must be chosen with these regimes in mind. Batch methods use more computation per step but fewer steps; stochastic methods use cheap steps but many iterations. Second-order methods use curvature but require memory/time for solving linear systems. Sketching and low-precision arithmetic trade accuracy for speed. Understanding these regimes clarifies why different algorithms dominate in different ML subfields.

Remark 12.1.6 (Regime Examples)

- **Large n, moderate d:** SGD/mini-batch methods dominate; full-batch Hessians are too expensive.
- **Moderate n, moderate d, convex:** second-order/quasi-Newton can be highly effective.
- **Huge d with sparsity:** coordinate descent and sparse-aware methods excel (lasso, linear models).
- **Streaming/online:** one-pass methods and incremental updates are required; memory is the primary constraint.

Remark 12.1.7 (Memory vs. Curvature) Storing curvature information can accelerate convergence but costs memory: full Hessian is $O(d^2)$, quasi-Newton is $O(dm)$, diagonal/adaptive methods are $O(d)$. Thus curvature exploitation is limited by memory budgets.

12.2 Optimization Problem Classes in ML

ML objectives come in recurring structural classes: unconstrained or constrained, convex or nonconvex, and finite-sum (dataset) or expectation (population/streaming). Recognizing the class is not cosmetic: it determines which algorithms are appropriate, what guarantees are possible, and which failure modes to expect. This section formalizes these classes and connects them to typical ML pipelines.

12.2.1 Unconstrained vs. Constrained Formulations

12.2.1.1 Penalty Methods vs. Constraints; Lagrangian Viewpoint

Many ML problems can be written either with explicit constraints (e.g., norm bounds, probability simplex constraints) or as unconstrained problems with penalties (regularization). The two forms are closely related via Lagrange multipliers. The penalty form is often easier to optimize (especially with gradients), while the constraint form is geometrically transparent. Understanding their equivalence clarifies how λ controls solution geometry and why MAP corresponds to regularization (Boyd and Vandenberghe 2004).

Definition 12.2.1 (Constrained Optimization) A constrained optimization problem has the form

$$\min_{w \in \mathbb{R}^d} f(w) \quad \text{s.t.} \quad w \in \mathcal{C},$$

where $\mathcal{C} \subseteq \mathbb{R}^d$ is the feasible set.

Definition 12.2.2 (Penalized (Regularized) Optimization) A penalized problem has the form

$$\min_{w \in \mathbb{R}^d} f(w) + \lambda\, \Omega(w), \qquad \lambda \geq 0,$$

where Ω encodes preferences (e.g., norms) or constraints (via indicator functions).

Definition 12.2.3 (Lagrangian (Inequality Constraints)) For constraints $g_j(w) \leq 0\ (j = 1, \ldots, m)$, the Lagrangian is

$$\mathcal{L}(w, \lambda) = f(w) + \sum_{j=1}^{m} \lambda_j g_j(w), \qquad \lambda_j \geq 0.$$

Lagrange multipliers λ_j measure the sensitivity of the optimum to tightening constraints.

Remark 12.2.1 (Penalty vs. Constraint Equivalence (Conceptual)) Under convexity and mild regularity, problems $\min_w f(w) + \lambda\Omega(w)$ and $\min_w f(w)$ s.t. $\Omega(w) \leq r$ trace the same solution set as λ and r vary. Thus λ is a dual variable controlling an implicit constraint radius.

12.2.1.2 Feasible Sets: Balls, Simplices, Boxes, and Polytopes

Constraints in ML are typically geometric: norm balls (regularization), simplices (probabilities/attention weights), boxes (parameter bounds, clipping), and polytopes (linear inequalities in constrained learning, fairness constraints). These sets are convex, enabling Euclidean projections and proximal methods. The geometry of the feasible set affects both the solution structure (sparsity/corners) and algorithm design (projection complexity).

Definition 12.2.4 (Canonical Feasible Sets) Common constraint sets $\mathcal{C} \subseteq \mathbb{R}^d$ include:

- ℓ_2 ball: $B_2(r) = \{w : \|w\|_2 \leq r\}$,
- ℓ_1 ball: $B_1(r) = \{w : \|w\|_1 \leq r\}$,
- box: $[\ell, u] = \{w : \ell_i \leq w_i \leq u_i\ \forall i\}$,
- simplex: $\Delta^{d-1} = \{p \in \mathbb{R}^d : p_i \geq 0, \sum_i p_i = 1\}$,
- polytope: $\{w : Aw \leq b\}$ for $A \in \mathbb{R}^{m \times d}$.

Remark 12.2.2 (Geometry Controls Structure) ℓ_1 and simplex constraints have corners and often produce sparse/extreme solutions; ℓ_2 balls are smooth and produce shrinkage; box constraints yield clipping-type projections.

12.2.2 Convex vs. Nonconvex Objectives

12.2.2.1 Global Minima, Local Minima, and Saddle Points

Convexity is the dividing line between landscapes with guaranteed global optimality and those with potentially many traps. In convex problems, any local minimum is global, and first-order methods have predictable convergence. In nonconvex problems, local minima and saddle points can exist; yet in modern ML (especially overparameterized deep learning), certain nonconvex landscapes can still be tractable in practice. Understanding the distinction clarifies what guarantees you can claim and what diagnostics you need.

Definition 12.2.5 (Convex Function) A function $f : \mathbb{R}^d \to \mathbb{R}$ is *convex* if for all x, y and $\theta \in [0, 1]$,

$$f(\theta x + (1 - \theta)y) \leq \theta f(x) + (1 - \theta)f(y).$$

If the inequality is strict for $x \neq y$ and $\theta \in (0, 1)$, f is strictly convex.

Definition 12.2.6 (Stationary Point, Local/Global Minimum, Saddle) Assume f is differentiable. A point $x^\star$ is *stationary* if $\nabla f(x^\star) = 0$. It is a *local minimum* if $f(x^\star) \leq f(x)$ for all x in a neighborhood of $x^\star$, a *global minimum* if $f(x^\star) \leq f(x)$ for all x, and a *saddle point* if it is stationary but not a local minimum (there exist descent directions).

Remark 12.2.3 (Second-Order Intuition) If f is twice differentiable, then at a local minimum the Hessian is PSD, while at a strict saddle it has a negative eigenvalue. This is why curvature-aware methods can escape saddles more reliably.

12.2.2.2 Why Nonconvex Is Common in Deep Learning

Deep learning objectives are nonconvex due to composition of nonlinearities and overparameterization. Even simple two-layer networks produce nonconvex losses. However, nonconvexity alone does not imply impossibility: overparameterization can create benign landscapes where many minima are near-global and SGD finds good solutions. Symmetries (permutation of hidden units, scaling in homogeneous networks) create flat directions. Thus the right mental model is: deep learning is nonconvex, but structure (initialization, normalization, implicit regularization) often makes it workable.

Remark 12.2.4 (Sources of Nonconvexity)

- composition of nonlinear maps ($f \circ g$) makes objectives nonconvex in parameters,
- bilinearities (e.g., matrix factorizations $UV^\top$) are nonconvex,
- symmetries yield multiple equivalent parameterizations (flat directions, manifolds of optima).

12.2.3 *Finite-Sum and Expectation Objectives*

12.2.3.1 Finite-Sum Form: $\frac{1}{n}\sum_{i=1}^{n} \ell_i(w)$

Most supervised learning objectives are finite sums over samples: empirical risk is the average of per-example losses. This structure enables incremental methods: compute gradients on mini-batches, use variance reduction, exploit sparsity, and distribute computation. The finite-sum form also clarifies the cost model: full gradient costs $O(n)$ evaluations per step, while SGD costs $O(1)$ or $O(b)$ per step.

Definition 12.2.7 (Finite-Sum Objective) An empirical risk objective typically has the form

$$\min_{w \in \mathbb{R}^d} f(w), \qquad f(w) = \frac{1}{n}\sum_{i=1}^{n} \ell_i(w),$$

where $\ell_i(w) = \ell(w; z_i)$ is the loss on sample z_i.

Remark 12.2.5 (Gradient Structure) If each ℓ_i is differentiable, then

$$\nabla f(w) = \frac{1}{n}\sum_{i=1}^{n} \nabla \ell_i(w).$$

This makes unbiased stochastic gradients possible by sampling i uniformly.

12.2.3.2 Population Risk: $\mathbb{E}[\ell(w; Z)]$ and Stochastic Gradients

The true objective of learning is population risk: expectation of loss under the unknown data distribution. Finite-sum objectives approximate this expectation using samples. In streaming or large-scale settings, we may never form a fixed dataset; instead we optimize the expectation directly using stochastic gradients computed from fresh samples. The central guarantee needed is unbiasedness: $\mathbb{E}[\nabla \ell(w; Z)] = \nabla \mathbb{E}[\ell(w; Z)]$ under regularity. This connects optimization to probability: convergence rates depend on variance and tail behavior of stochastic gradients.

Definition 12.2.8 (Population (Expected) Objective) Let $Z \sim \mathcal{D}$ and loss $\ell(w; Z)$. The population objective is

$$F(w) := \mathbb{E}[\ell(w; Z)].$$

Empirical risk is $f_n(w) = \frac{1}{n}\sum_{i=1}^{n} \ell(w; Z_i)$ for i.i.d. samples Z_i.

Definition 12.2.9 (Stochastic Gradient) A random vector $g(w; Z)$ is an *unbiased stochastic gradient* of F at w if

$$\mathbb{E}[g(w; Z)] = \nabla F(w).$$

The canonical choice is $g(w; Z) = \nabla_w \ell(w; Z)$ when differentiation and expectation can be interchanged.

Remark 12.2.6 (Variance Matters) Even if unbiased, stochastic gradients can have large variance or heavy tails, slowing convergence and destabilizing training. This motivates mini-batching, variance reduction, gradient clipping, and robust objectives.

12.3 Differential Viewpoint: Gradients, Hessians, and Curvature

Optimization is geometry in motion. Gradients encode local sensitivity and the direction of steepest increase with respect to an underlying inner product; Hessians encode curvature and local anisotropy. Together they yield the local quadratic model that underlies Newton, preconditioning, and condition-number reasoning. This section develops gradients and directional derivatives, Hessians and Taylor models, and the core matrix-calculus rules that make ML optimization derivations systematic (trace tricks, Jacobians, and backprop as Jacobian-vector products).

12.3.1 Gradients and Directional Derivatives

12.3.1.1 Geometric Meaning of the Gradient

The gradient is not just a vector of partial derivatives; it is the Riesz representation of the differential: it is the unique vector whose inner product with any direction equals the directional derivative. Thus, the gradient depends on the chosen inner product (geometry). In Euclidean geometry, it points in the direction of steepest ascent and its norm is the maximal directional derivative over unit directions. In ML, gradients determine update directions (GD/SGD), sensitivity to perturbations (adversarial directions), and influence of parameters on loss.

Definition 12.3.1 (Directional Derivative) Let $f : \mathbb{R}^d \to \mathbb{R}$ and $w \in \mathbb{R}^d$. The directional derivative of f at w in direction $u \in \mathbb{R}^d$ is

$$Df(w)[u] := \lim_{t \downarrow 0} \frac{f(w + tu) - f(w)}{t},$$

when the limit exists. If the two-sided limit exists (as $t \to 0$), it defines the derivative along u.

Definition 12.3.2 (Gradient as Riesz Representation) Assume f is differentiable at w. The *gradient* $\nabla f(w) \in \mathbb{R}^d$ is the unique vector satisfying

$$Df(w)[u] = \langle \nabla f(w), u \rangle \qquad \forall u \in \mathbb{R}^d,$$

where $\langle \cdot, \cdot \rangle$ is the Euclidean inner product.

Proposition 12.3.1 (Steepest Ascent Characterization) *If f is differentiable at w, then*

$$\max_{\|u\|_2 \le 1} Df(w)[u] = \|\nabla f(w)\|_2,$$

and a maximizer is $u^\star = \nabla f(w)/\|\nabla f(w)\|_2$ when $\nabla f(w) \neq 0$.

Remark 12.3.1 (Why Gradients Depend on Geometry) If we change the inner product (e.g., to a Mahalanobis geometry), the vector representing the same differential changes. This is the motivation behind natural gradients and preconditioning: choose a geometry in which "steepest" is meaningful and well-conditioned.

12.3.1.2 Directional Derivatives and Steepest Descent

Gradient descent is steepest descent in Euclidean geometry: it moves opposite the direction of steepest ascent. More generally, the steepest descent direction depends on the chosen norm: the dual norm controls the maximum directional derivative. This connects optimization to robustness and regularization: norms used in penalties correspond to dual norms governing sensitivity. In ML, this viewpoint explains why different optimizers behave differently under rescaling and why preconditioning alters step geometry.

Proposition 12.3.2 (Steepest Descent under a General Norm (Duality)) *Let $\|\cdot\|$ be any norm with dual $\|\cdot\|_*$. If f is differentiable at w, then*

$$\max_{\|u\| \le 1} Df(w)[u] = \|\nabla f(w)\|_*.$$

Thus the steepest descent *direction under $\|\cdot\|$ is any $u^\star$ achieving $Df(w)[u^\star] = -\|\nabla f(w)\|_*$ with $\|u^\star\| \le 1$.*

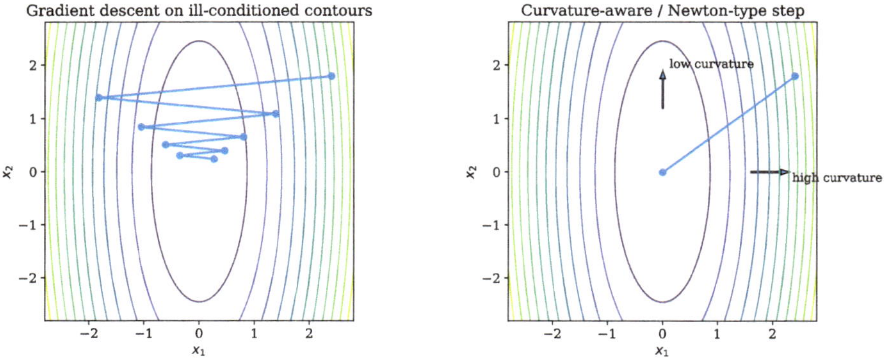

Fig. 12.2 Optimization geometry. Left: on an ill-conditioned quadratic, gradient descent zig-zags across elongated level sets. Right: curvature-aware steps align with the local Hessian geometry and move more directly toward the minimizer

Remark 12.3.2 (Euclidean Specialization) For $\|\cdot\| = \|\cdot\|_2$, we have $\|\cdot\|_* = \|\cdot\|_2$, recovering Proposition 12.3.1.

12.3.2 Hessians and Local Quadratic Models

12.3.2.1 Curvature, Conditioning, and Anisotropy

The Hessian captures how gradients change: it is the local curvature tensor of the objective. In convex problems, it is PSD and encodes anisotropy of level sets; in nonconvex problems, it can have negative eigenvalues (saddles). Conditioning is the spectral spread of the Hessian: large condition number means long, narrow valleys where gradient descent zig-zags. This is why second-order methods and preconditioning matter: they attempt to "sphere-ify" the curvature (cf. Fig. 12.2).

Definition 12.3.3 (Hessian) Let $f : \mathbb{R}^d \to \mathbb{R}$ be twice differentiable at w. The Hessian is the matrix of second derivatives

$$\nabla^2 f(w) := \left[\frac{\partial^2 f}{\partial w_i \, \partial w_j}(w) \right]_{i,j=1}^{d} \in \mathbb{R}^{d \times d}.$$

Remark 12.3.3 (Quadratic Form and Curvature) For a direction u, the second directional derivative is

$$\frac{d^2}{dt^2} f(w + tu)\Big|_{t=0} = u^\top \nabla^2 f(w)\, u,$$

so eigenvectors of $\nabla^2 f(w)$ are principal curvature directions.

Definition 12.3.4 (Condition Number (Local)) If f is twice differentiable and $\nabla^2 f(w) \succ 0$, define the local condition number

$$\kappa(w) := \frac{\lambda_{\max}(\nabla^2 f(w))}{\lambda_{\min}(\nabla^2 f(w))}.$$

Large κ indicates strong anisotropy and slow convergence for first-order methods without preconditioning.

12.3.2.2 Taylor Expansions and Local Models

Taylor expansion is the formal justification for local optimization models. A first-order model yields gradient descent; a second-order model yields Newton's method. The second-order Taylor approximation is a quadratic whose minimizer is the Newton step. This explains both fast local convergence and the danger: if the Hessian is ill-conditioned or indefinite, the quadratic model can be inaccurate or lead to unstable steps. Damping, trust regions, and line search are safeguards.

Theorem 12.3.1 (Second-Order Taylor Expansion) *If f is twice continuously differentiable near w, then for small Δ,*

$$f(w+\Delta) = f(w) + \langle \nabla f(w), \Delta \rangle + \frac{1}{2}\Delta^\top \nabla^2 f(w)\Delta + o(\|\Delta\|_2^2).$$

Remark 12.3.4 (Newton Step as Quadratic Minimizer) Minimizing the quadratic model

$$m(\Delta) = f(w) + \langle \nabla f(w), \Delta \rangle + \tfrac{1}{2}\Delta^\top H \Delta, \quad H = \nabla^2 f(w),$$

yields $\Delta^\star = -H^{-1}\nabla f(w)$ when H is invertible. This is the Newton direction.

12.3.3 Matrix Calculus Essentials for ML

12.3.3.1 Gradients of Quadratic Forms, Trace Objectives, and Norms

ML objectives often involve matrix parameters and quadratic terms: least squares, ridge, covariance penalties, Frobenius norms. Trace identities provide a clean route: rewrite scalar objectives as traces, take differentials, and identify gradients by matching coefficients. This avoids index-heavy calculus and scales well to complex models. Here we record the most-used derivatives in real-valued matrix calculus.

Proposition 12.3.3 (Quadratic and Least-Squares Gradients) *Let* $x \in \mathbb{R}^d$, $A \in \mathbb{R}^{d\times d}$, $b \in \mathbb{R}^d$, $C \in \mathbb{R}^{n\times d}$, *and* $y \in \mathbb{R}^n$. *Then*

$$\nabla_x(b^\top x) = b, \qquad \nabla_x\Big(\frac{1}{2}\|x\|_2^2\Big) = x,$$

$$\nabla_x\Big(\frac{1}{2}x^\top Ax\Big) = \frac{1}{2}(A + A^\top)x \quad (\textit{and} = Ax \textit{ if } A = A^\top),$$

$$\nabla_x\Big(\frac{1}{2}\|Cx - y\|_2^2\Big) = C^\top(Cx - y).$$

Proposition 12.3.4 (Matrix-Parameter Least Squares) *Let* $W \in \mathbb{R}^{k\times d}$, $X \in \mathbb{R}^{n\times d}$, $Y \in \mathbb{R}^{n\times k}$ *and consider*

$$f(W) = \frac{1}{2}\|XW - Y\|_F^2 = \frac{1}{2}\mathrm{tr}\big((XW - Y)^\top(XW - Y)\big).$$

Then

$$\nabla_W f(W) = X^\top(XW - Y).$$

Remark 12.3.5 (Trace Trick Reminder) Two identities used repeatedly are

$$\|A\|_F^2 = \mathrm{tr}(A^\top A), \qquad \mathrm{tr}(ABC) = \mathrm{tr}(BCA) = \mathrm{tr}(CAB),$$

whenever dimensions match. They enable moving parameters next to differentials in gradient derivations.

12.3.3.2 Chain Rule, Jacobians, and Backprop Viewpoint

Backpropagation is the chain rule on a computation graph. For vector-valued maps, derivatives are Jacobians, and gradients propagate via transposed Jacobians (reverse mode). In large-scale ML, we never form full Jacobians; instead we compute Jacobian-vector products (JVPs) and vector-Jacobian products (VJPs). This is the computational meaning of the chain rule in high dimensions and explains why shape discipline is essential.

Definition 12.3.5 (Jacobian (Shape)) For $f : \mathbb{R}^d \to \mathbb{R}^k$ differentiable at w, the Jacobian is $J_f(w) \in \mathbb{R}^{k\times d}$. If $g : \mathbb{R}^k \to \mathbb{R}^p$, then

$$J_{g\circ f}(w) = J_g(f(w))\, J_f(w).$$

Proposition 12.3.5 (Reverse-Mode (VJP) Identity) *Let* $L : \mathbb{R}^k \to \mathbb{R}$ *be differentiable and define* $y = f(w)$. *Then*

$$\nabla_w L(f(w)) = J_f(w)^\top \nabla_y L(y).$$

Thus gradients propagate backward by multiplying by transposed Jacobians.

Remark 12.3.6 (Backprop as Repeated VJPs) In a layered model $f = f_L \circ \cdots \circ f_1$, reverse-mode differentiation computes ∇ by a sequence of VJPs, never forming full Jacobians explicitly. This is both memory- and compute-efficient.

12.4 Convex Analysis Primer

Convex analysis provides the cleanest optimization theory used in ML. It explains when global minima are guaranteed, how fast first-order methods converge, and how to handle nonsmooth regularizers (lasso, hinge loss, total variation) via subgradients and proximal operators. This section collects the core definitions and lemmas that will be used repeatedly: convex sets/functions, strong convexity, smoothness (L-Lipschitz gradients), condition numbers, and subdifferentials.

12.4.1 Convex Sets and Convex Functions

12.4.1.1 Epigraphs, Jensen's Inequality, and Basic Examples

Convexity is geometry: a set is convex if it contains line segments; a function is convex if its epigraph is convex. This epigraph viewpoint is powerful because it turns function convexity into set convexity and enables separation arguments. Jensen's inequality is the probabilistic expression of convexity and is constantly used in ML (risk bounds, variational methods). Examples to keep in mind: norms are convex; squared norms are strongly convex; log-sum-exp is convex (softmax). These examples underpin regularization, logistic regression, and many convex surrogates.

Definition 12.4.1 (Convex Set) A set $C \subseteq \mathbb{R}^d$ is *convex* if for all $x, y \in C$ and all $\theta \in [0, 1]$,

$$\theta x + (1 - \theta)y \in C.$$

Definition 12.4.2 (Epigraph and Convex Function) The epigraph of a function $f : \mathbb{R}^d \to \mathbb{R} \cup \{+\infty\}$ is

$$\text{epi}(f) := \{(x, t) \in \mathbb{R}^d \times \mathbb{R} : \ f(x) \leq t\}.$$

The function f is *convex* iff $\text{epi}(f)$ is a convex set.

Proposition 12.4.1 (First-Order Convexity Inequality) *If f is convex and differentiable, then for all $x, y \in \mathbb{R}^d$,*

$$f(y) \geq f(x) + \langle \nabla f(x), \; y - x \rangle.$$

Equivalently, every tangent hyperplane underestimates a convex function.

Theorem 12.4.1 (Jensen's Inequality (Recall)) *If φ is convex and X integrable, then $\varphi(\mathbb{E}[X]) \leq \mathbb{E}[\varphi(X)]$.*

Remark 12.4.1 (Basic Convex Examples in ML)

- Norms $x \mapsto \|x\|$ are convex.
- Quadratics $x \mapsto \frac{1}{2}\|Ax - b\|_2^2$ are convex; if A has full column rank they are strongly convex.
- $\log \sum_{k=1}^{K} e^{s_k}$ is convex in s (log-sum-exp), hence softmax cross-entropy is convex in linear scores.

12.4.1.2 Strong Convexity and Its Consequences

Strong convexity is convexity with curvature: the function lies above its tangents by a quadratic margin. This yields uniqueness of the minimizer and faster convergence rates for gradient methods (linear convergence under smoothness). In ML, adding ℓ_2 regularization often makes objectives strongly convex (ridge, regularized logistic), improving stability and conditioning. Strong convexity is also a quantitative form of identifiability: the objective penalizes deviations from the optimum in all directions.

Definition 12.4.3 (Strong Convexity) A function $f : \mathbb{R}^d \to \mathbb{R}$ is *μ-strongly convex* (with $\mu > 0$) if for all x, y and $\theta \in [0, 1]$,

$$f(\theta x + (1 - \theta)y) \leq \theta f(x) + (1 - \theta) f(y) - \frac{\mu}{2}\theta(1 - \theta)\|x - y\|_2^2.$$

If f is differentiable, this is equivalent to

$$f(y) \geq f(x) + \langle \nabla f(x), y - x \rangle + \frac{\mu}{2}\|y - x\|_2^2.$$

Proposition 12.4.2 (Uniqueness of Minimizer) *If f is μ-strongly convex on $\mathbb{R}^d$, then it has at most one global minimizer.*

Remark 12.4.2 (Quadratic Lower Bound Around Optimum) If f is μ-strongly convex and $x^\star$ is its minimizer, then

$$f(x) - f(x^\star) \geq \frac{\mu}{2}\|x - x^\star\|_2^2.$$

Thus function suboptimality controls distance to the optimum.

12.4.2 Smoothness and Lipschitz Conditions

12.4.2.1 L-Smooth Functions and Descent Lemmas

Smoothness controls how rapidly gradients change. An L-smooth function has Lipschitz-continuous gradient; its Taylor approximation has a quadratic error term bounded by L. This yields the descent lemma, which is the basic inequality behind gradient descent step-size rules: with step size $\eta \leq 1/L$, each step decreases the objective by a quantifiable amount. In ML, smoothness constants reflect curvature and scaling; poor scaling increases L and slows training.

Definition 12.4.4 (L-Smoothness) A differentiable function f is *L-smooth* if its gradient is L-Lipschitz:

$$\|\nabla f(x) - \nabla f(y)\|_2 \leq L\|x - y\|_2 \qquad \forall x, y.$$

Proposition 12.4.3 (Descent Lemma) *If f is L-smooth, then for all x, y,*

$$f(y) \leq f(x) + \langle \nabla f(x), y - x\rangle + \frac{L}{2}\|y - x\|_2^2.$$

Remark 12.4.3 (Gradient Descent Decrease (Template)) For gradient descent $x^+ = x - \eta\nabla f(x)$, if $\eta \leq 1/L$, then

$$f(x^+) \leq f(x) - \frac{\eta}{2}\|\nabla f(x)\|_2^2.$$

This is the core monotonicity guarantee used in convex optimization analysis.

12.4.2.2 Condition Numbers and Convergence Speed

Condition number quantifies anisotropy: the ratio between largest and smallest curvature. For μ-strongly convex and L-smooth functions, $\kappa = L/\mu$ controls convergence speed of gradient descent: large κ implies slow convergence (long narrow valleys), small κ implies fast convergence. In ML, conditioning is influenced by feature scaling, regularization, and preprocessing (whitening). This is the optimization analogue of statistical variance inflation under collinearity.

Definition 12.4.5 (Condition Number) If f is μ-strongly convex and L-smooth, define the condition number

$$\kappa := \frac{L}{\mu}.$$

For a quadratic $f(x) = \frac{1}{2}x^\top Hx - b^\top x$ with $H \succ 0$, one has $L = \lambda_{\max}(H)$ and $\mu = \lambda_{\min}(H)$, so $\kappa = \kappa(H)$.

Remark 12.4.4 (Convergence Speed Intuition) Gradient descent converges linearly for L-smooth, μ-strongly convex f, with a rate depending on κ. Thus improving conditioning (reducing κ) via preconditioning/normalization can be as important as tuning step size.

12.4.3 Subgradients and Nonsmooth Convexity

12.4.3.1 Subdifferentials and Optimality Conditions

Nonsmooth convex objectives arise constantly in ML: ℓ_1 regularization, hinge loss, total variation. Gradients do not exist everywhere, but convexity still provides a substitute: subgradients. The subdifferential generalizes the tangent-plane underestimator property: $g \in \partial f(x)$ if it defines a global supporting hyperplane. Optimality is characterized by $0 \in \partial f(x^\star)$, replacing $\nabla f(x^\star) = 0$. This provides a rigorous foundation for proximal methods and coordinate descent.

Definition 12.4.6 (Subgradient and Subdifferential) Let $f : \mathbb{R}^d \to \mathbb{R} \cup \{+\infty\}$ be convex. A vector $g \in \mathbb{R}^d$ is a *subgradient* of f at x if for all y,

$$f(y) \geq f(x) + \langle g, y - x \rangle.$$

The set of all subgradients at x is the *subdifferential* $\partial f(x)$.

Proposition 12.4.4 (First-Order Optimality for Convex Functions) *If f is convex, then $x^\star$ is a global minimizer of f iff*

$$0 \in \partial f(x^\star).$$

Remark 12.4.5 (Differentiable Case) If f is differentiable at x, then $\partial f(x) = \{\nabla f(x)\}$, so the condition reduces to $\nabla f(x^\star) = 0$.

12.4.3.2 Nonsmooth Penalties: ℓ_1, Hinge, and TV

The most common nonsmooth convex functions in ML create corners that induce structure: ℓ_1 corners induce sparsity, hinge corners induce margin-based classification, and total variation corners induce piecewise-constant structure. Their subgradients have simple forms involving sign patterns, active constraints, or discrete differences. This preview links these penalties to geometric constraints and foreshadows proximal operators as the computational tool to handle them efficiently.

Remark 12.4.6 (ℓ_1 Subdifferential) For $f(x) = \|x\|_1$, the subdifferential is

$$(\partial\|x\|_1)_i = \begin{cases} \{\text{sign}(x_i)\}, & x_i \neq 0, \\ [-1, 1], & x_i = 0. \end{cases}$$

The interval at 0 is the algebraic source of soft-thresholding and sparsity.

Remark 12.4.7 (Hinge Loss Subgradient (Binary)) For $f(t) = \max\{0, 1 - t\}$, a subgradient is

$$\partial f(t) = \begin{cases} \{-1\}, & t < 1, \\ [-1, 0], & t = 1, \\ \{0\}, & t > 1. \end{cases}$$

Thus only margin-violating points contribute nonzero subgradients.

Remark 12.4.8 (Total Variation (TV) Preview) For a 1D signal $x \in \mathbb{R}^d$, the TV seminorm is $\|x\|_{\text{TV}} = \sum_{i=1}^{d-1} |x_{i+1} - x_i|$. It is convex and nonsmooth; its subgradients depend on sign patterns of discrete differences. TV regularization encourages piecewise-constant solutions and is used in denoising and certain structured ML problems.

12.5 Optimality Conditions and Duality

Optimization in ML is not just running an algorithm; it is reasoning about what an optimum *means*. Optimality conditions tell us when a point is optimal (or stationary), how constraints shape solutions, and how dual variables encode sensitivity and geometry. Duality provides alternative formulations that can be computationally more efficient, reveal structure (sparsity, margins), and connect regularization to norms and conjugates. This section presents first-order optimality, KKT conditions, and a preview of Fenchel duality in ML.

12.5.1 First-Order Optimality

12.5.1.1 Stationarity, Gradient Zero, and Subgradient Containment

For smooth unconstrained problems, optimality is expressed as $\nabla f(x^\star) = 0$. For nonsmooth convex problems, gradients may not exist and the correct condition becomes $0 \in \partial f(x^\star)$. For constrained problems, stationarity must incorporate feasibility and normal-cone geometry (KKT). In ML, these conditions appear as normal equations (least squares), KKT for lasso, and margin constraints in SVM.

Definition 12.5.1 (Stationary Point) Let $f : \mathbb{R}^d \to \mathbb{R}$ be differentiable. A point $x^\star$ is *stationary* if $\nabla f(x^\star) = 0$.

Definition 12.5.2 (First-Order Optimality for Convex (Subgradient)) If f is convex (possibly nonsmooth), then $x^\star$ is optimal iff

$$0 \in \partial f(x^\star),$$

where $\partial f(x^\star)$ is the subdifferential.

Remark 12.5.1 (Interpretation of $0 \in \partial f(x^\star)$) There exists a supporting hyperplane at $x^\star$ with slope 0, meaning $f(y) \geq f(x^\star)$ for all y. This is the convex analogue of "no descent direction."

12.5.1.2 Convex Optimality and Global Minima

Convexity removes local traps: any point satisfying first-order optimality is globally optimal. This is the foundational reason convex ML models (ridge, logistic regression, SVM) admit strong guarantees. It also enables certificates: verifying optimality can be done by checking a first-order condition. In nonconvex problems, stationarity is necessary but not sufficient; however, convexity gives sufficiency.

Theorem 12.5.1 (First-Order Condition Is Sufficient for Convexity) *Let f be convex and differentiable. Then $x^\star$ is a global minimizer of f iff*

$$\nabla f(x^\star) = 0.$$

More generally, if f is convex (not necessarily differentiable), $x^\star$ is a global minimizer iff $0 \in \partial f(x^\star)$.

Remark 12.5.2 (Convexity as a Guarantee) This theorem is why convex models are reliable: finding a stationary point (by gradient methods) means you have found the global optimum.

12.5.2 Lagrangians and KKT Conditions

12.5.2.1 Equality/Inequality Constraints and Complementary Slackness

Constraints encode geometry: feasible sets are often convex polytopes, balls, or manifolds. The Lagrangian packages the constraints into a single scalar function with multipliers. KKT conditions generalize gradient-zero conditions to constrained problems by adding multiplier-weighted constraint gradients. Complementary slackness captures which constraints are active: a constraint either binds (active) with positive multiplier, or is slack with zero multiplier. In ML, KKT underlies SVM support vectors, lasso sparsity patterns, and simplex projections.

Definition 12.5.3 (Constrained Problem and Lagrangian) Consider

$$\min_{x \in \mathbb{R}^d} f(x) \quad \text{s.t.} \quad g_i(x) \le 0 \ (i = 1, \dots, m), \qquad h_j(x) = 0 \ (j = 1, \dots, p),$$

with differentiable f, g_i, h_j. The Lagrangian is

$$\mathcal{L}(x, \lambda, \nu) = f(x) + \sum_{i=1}^{m} \lambda_i g_i(x) + \sum_{j=1}^{p} \nu_j h_j(x), \qquad \lambda_i \ge 0.$$

Theorem 12.5.2 (KKT Conditions (Convex Case, Template)) *Assume f and g_i are convex and h_j are affine, and a constraint qualification holds (e.g., Slater's condition). A point $x^\star$ is optimal iff there exist multipliers $\lambda^\star \ge 0$ and $\nu^\star$ such that:*

1. (Primal feasibility) $g_i(x^\star) \le 0$ *and* $h_j(x^\star) = 0$.
2. (Dual feasibility) $\lambda_i^\star \ge 0$.
3. (Stationarity)

$$\nabla f(x^\star) + \sum_{i=1}^{m} \lambda_i^\star \nabla g_i(x^\star) + \sum_{j=1}^{p} \nu_j^\star \nabla h_j(x^\star) = 0.$$

4. (Complementary slackness) $\lambda_i^\star g_i(x^\star) = 0$ *for all i.*

Remark 12.5.3 (Complementary Slackness as "Active Set" Logic) If $g_i(x^\star) < 0$ (constraint inactive), then $\lambda_i^\star = 0$. If $\lambda_i^\star > 0$, then $g_i(x^\star) = 0$ (constraint active). Thus multipliers identify which constraints shape the optimum.

12.5.2.2 Geometric Interpretation of Multipliers

Multipliers are geometric objects: they combine normals of active constraints to balance the gradient of the objective. At the optimum, the negative gradient lies in the cone generated by active constraint normals (plus the subspace from equality constraints). This is why multipliers quantify sensitivity: they measure how the optimal value changes if constraints are tightened. In ML, this interpretation explains support vectors: constraints with nonzero multipliers are the points that define the max-margin boundary.

Remark 12.5.4 (Normals and Cones) For active inequality constraints $g_i(x^\star) = 0$, the vectors $\nabla g_i(x^\star)$ are normals to constraint boundaries. The stationarity condition says $-\nabla f(x^\star)$ is a conic combination of these normals (plus equality normals), i.e., the objective cannot decrease without leaving the feasible region.

Remark 12.5.5 (Sensitivity (Conceptual)) Under regularity, the multiplier $\lambda_i^\star$ often equals the derivative of the optimal value with respect to the constraint bound: tightening the constraint slightly increases the optimal value by approximately $\lambda_i^\star$ times the tightening amount.

12.5.3 Duality in ML

12.5.3.1 Fenchel Conjugates and Dual Norms (Bridge to Geometry)

Fenchel conjugates provide a systematic way to derive dual problems for convex objectives. They unify many ML dualities: hinge loss $\leftrightarrow$ SVM dual, ℓ_1 norm $\leftrightarrow$ ℓ_∞ constraints, log-sum-exp $\leftrightarrow$ entropy terms. Dual norms appear naturally as conjugates of indicator functions of norm balls. Thus, duality is geometry: it swaps constraints and penalties and reveals which directions are controlled.

Definition 12.5.4 (Fenchel Conjugate) For a function $f : \mathbb{R}^d \to \mathbb{R} \cup \{+\infty\}$, its Fenchel conjugate is

$$f^*(y) := \sup_{x \in \mathbb{R}^d} \{\langle y, x\rangle - f(x)\}.$$

Proposition 12.5.1 (Fenchel–Young Inequality) *For all $x, y \in \mathbb{R}^d$,*

$$f(x) + f^*(y) \geq \langle x, y\rangle,$$

with equality iff $y \in \partial f(x)$ (under convexity).

Remark 12.5.6 (Dual Norms as Support Functions) For a norm $\|\cdot\|$, the dual norm satisfies

$$\|y\|_* = \sup_{\|x\|\leq 1} \langle x, y\rangle,$$

which is exactly a support function and a special case of a Fenchel conjugate (indicator of the unit ball).

12.5.3.2 Primal–Dual Views of Regularization

Many regularized ML problems have informative duals. The primal emphasizes parameter geometry (norm penalties); the dual emphasizes constraints on correlations with data and often lives in sample space. This is computationally crucial: when $n \ll d$ or $d \ll n$, one form may be dramatically cheaper. Conceptually, primal–dual pairs reveal which constraints are active (sparsity, support vectors) and provide optimality certificates via duality gaps.

Remark 12.5.7 (Ridge Primal vs. Dual (Conceptual)) Ridge regression has a primal solution involving $(X^\top X + \lambda I)^{-1}$ and a dual involving $(XX^\top + \lambda I)^{-1}$. Choosing between them depends on whether d or n is smaller. Kernel ridge is exactly the dual form with $XX^\top$ replaced by a kernel matrix K.

Remark 12.5.8 (Duality Gap) For convex problems with strong duality, the difference between primal objective value and dual objective value provides a computable certificate of suboptimality, used to monitor convergence in solvers.

12.6 First-Order Methods (Deterministic)

First-order methods use only gradients (and sometimes simple proximal/projection operators). They scale to large dimensions, are easy to implement, and form the backbone of ML optimization. Their behavior is governed by smoothness (L-Lipschitz gradients), convexity, and conditioning. This section develops deterministic gradient descent, accelerated variants (momentum/Nesterov), and coordinate/block-coordinate methods for high-dimensional and structured problems.

12.6.1 Gradient Descent (GD)

12.6.1.1 Step Sizes: Fixed, Diminishing, and Adaptive Schedules

Step size (learning rate) is the main control knob of gradient descent. If η is too large, iterates can diverge; if too small, convergence is slow. For L-smooth objectives, $\eta \leq 1/L$ guarantees descent. Diminishing step sizes handle noisy or merely convex settings and are essential in stochastic contexts. Adaptive schedules (time-varying η_t) are pragmatic heuristics to cope with unknown curvature. Even in deterministic settings, backtracking line search is an adaptive rule that finds a safe step size using only function values and gradients.

Definition 12.6.1 (Gradient Descent) Given a differentiable objective $f : \mathbb{R}^d \to \mathbb{R}$, gradient descent iterates

$$x_{t+1} = x_t - \eta_t \nabla f(x_t), \qquad t = 0, 1, 2, \ldots,$$

with step size $\eta_t > 0$.

Remark 12.6.1 (Fixed Step Size) A common choice is constant $\eta_t = \eta$. For L-smooth convex f, choosing $\eta \in (0, 1/L]$ yields monotone decrease of $f(x_t)$. For strongly convex f, a constant step size yields linear convergence.

Remark 12.6.2 (Diminishing Step Size (Template)) In settings without strong convexity or with noise (SGD), one uses η_t decreasing, e.g., $\eta_t = \eta_0/\sqrt{t+1}$ or $\eta_t = \eta_0/(t+1)$, to ensure convergence while averaging out noise. Deterministic convex problems often prefer constant or line-search steps for faster convergence.

Remark 12.6.3 (Backtracking Line Search (Conceptual)) Given x_t and a candidate η, shrink $\eta \leftarrow \beta\eta$ until the Armijo condition holds:

$$f(x_t - \eta\nabla f(x_t)) \leq f(x_t) - c\,\eta\|\nabla f(x_t)\|_2^2,$$

for constants $c \in (0, 1)$ and $\beta \in (0, 1)$. This adapts to unknown L and ensures descent.

12.6.1.2 Convergence Under Convexity and Smoothness (Roadmap)

Convergence rates depend on problem class. For convex L-smooth functions, GD achieves $O(1/t)$ suboptimality in function value. For μ-strongly convex and L-smooth functions, GD achieves linear (geometric) convergence with rate governed by the condition number $\kappa = L/\mu$. These statements are roadmap-level here; proofs use the descent lemma and convexity inequalities. In ML, this explains why regularization (which can induce strong convexity) accelerates optimization.

Theorem 12.6.1 (GD Rate for Convex L-Smooth (Roadmap Statement)) *Let f be convex and L-smooth, and let $x^\star$ be a minimizer. With constant step size $\eta = 1/L$, gradient descent satisfies*

$$f(x_t) - f(x^\star) \leq \frac{L\|x_0 - x^\star\|_2^2}{2t} \qquad \textit{for } t \geq 1.$$

Theorem 12.6.2 (GD Rate for μ-Strongly Convex and L-Smooth (Roadmap Statement)) *Let f be μ-strongly convex and L-smooth. With step size $\eta \in (0, 2/L)$, gradient descent converges linearly:*

$$\|x_t - x^\star\|_2 \leq \rho^t\|x_0 - x^\star\|_2, \qquad f(x_t) - f(x^\star) \leq \rho^{2t}\big(f(x_0) - f(x^\star)\big),$$

where $\rho \in (0, 1)$ depends on η, μ, L (optimally $\rho \approx 1 - \mu/L$ for well-chosen η).

Remark 12.6.4 (Condition Number Controls Speed) When $\kappa = L/\mu$ is large, $\rho \approx 1 - 1/\kappa$ is close to 1, so convergence is slow. This motivates preconditioning and acceleration.

12.6.2 Accelerated Methods

12.6.2.1 Momentum and Heavy-Ball Intuition

Momentum adds an inertial term: move in a direction that combines the current gradient with past updates. Geometrically, momentum smooths oscillations in directions of high curvature and amplifies progress along low-curvature directions. The heavy-ball method is the canonical deterministic momentum scheme for smooth optimization. In ML, momentum is ubiquitous because it improves practical convergence in ill-conditioned problems without requiring second-order information.

Definition 12.6.2 (Heavy-Ball (Momentum) Method) Given step size $\eta > 0$ and momentum parameter $\beta \in [0, 1)$, define

$$v_{t+1} = \beta v_t + \nabla f(x_t), \qquad x_{t+1} = x_t - \eta v_{t+1},$$

with $v_0 = 0$. Equivalently,

$$x_{t+1} = x_t - \eta \nabla f(x_t) + \beta(x_t - x_{t-1})$$

(with suitable initialization).

Remark 12.6.5 (Quadratic Intuition) For a quadratic $f(x) = \frac{1}{2}x^\top Hx$, momentum can be analyzed eigen-direction by eigen-direction. The method can significantly reduce zig-zagging when H is ill-conditioned by coupling steps across iterations.

12.6.2.2 Nesterov Acceleration (Conceptual)

Nesterov acceleration modifies momentum by evaluating the gradient at a look-ahead point, achieving provably faster rates in convex optimization: $O(1/t^2)$ for smooth convex problems, and improved linear rates under strong convexity. Conceptually, it is a carefully tuned momentum scheme that corrects overshooting. In ML practice, "Nesterov momentum" is a standard variant of SGD/momentum and often improves convergence in deep networks as well.

Definition 12.6.3 (Nesterov Accelerated Gradient (NAG), Template) One common Nesterov scheme is

$$y_t = x_t + \beta_t(x_t - x_{t-1}), \qquad x_{t+1} = y_t - \eta \nabla f(y_t),$$

with suitable schedules for β_t (and fixed η under known L) to achieve accelerated rates in convex settings.

Remark 12.6.6 (Rate Statement (Conceptual)) For convex L-smooth objectives, Nesterov acceleration achieves $f(x_t) - f(x^\star) = O(L\|x_0 - x^\star\|_2^2/t^2)$, improving GD's $O(1/t)$. We treat this as a roadmap result; full proof is standard but lengthy.

12.6.3 Coordinate and Block-Coordinate Methods

12.6.3.1 Coordinate Descent for Sparse/High-Dimensional Problems

When d is huge and the objective has separable structure (especially with ℓ_1 penalties), updating one coordinate at a time can be far cheaper than full gradient steps. Coordinate descent exploits sparsity and cheap partial derivatives; it is the workhorse for lasso/elastic net in large-scale linear models. Its success relies on the fact that many coordinates can be updated exactly or with closed-form proximal steps (soft-thresholding). Randomized coordinate selection often yields strong convergence guarantees in convex settings.

Definition 12.6.4 (Coordinate Descent (Template)) Let $f : \mathbb{R}^d \to \mathbb{R}$. Coordinate descent updates one coordinate at a time: choose an index $j_t \in [d]$ and set

$$x_{t+1} \in \arg\min_{u\in\mathbb{R}} f\big(x_t + (u - x_{t,j_t})e_{j_t}\big),$$

keeping all other coordinates fixed.

Remark 12.6.7 (Lasso Coordinate Update (Conceptual)) For lasso, coordinate descent updates reduce to soft-thresholding of a partial residual correlation. This yields extremely fast solvers for sparse/high-dimensional linear models.

12.6.3.2 Block Updates and Structured Parameterizations

Many ML parameters are naturally grouped: layers in networks, blocks in multi-task models, feature groups in group lasso. Block-coordinate methods update a block of variables at a time, exploiting structure while maintaining scalability. They connect to alternating minimization in matrix factorization and to block proximal methods in structured regularization. The geometry is modular: update along a subspace corresponding to the block, which can reduce complexity and align with problem structure.

Definition 12.6.5 (Block-Coordinate Descent) Partition variables into blocks $x = (x^{(1)}, \dots, x^{(m)})$. A block-coordinate method selects a block index b_t and updates

$$x_{t+1}^{(b_t)} \in \arg\min_{u} f\big(x_t^{(1)}, \dots, x_t^{(b_t-1)}, u, x_t^{(b_t+1)}, \dots, x_t^{(m)}\big),$$

while keeping other blocks fixed.

Remark 12.6.8 (Examples in ML)

- Group lasso: block updates align with feature groups.
- Matrix factorization: alternating updates in U and V are block-coordinate steps.
- Deep networks: layer-wise or module-wise optimization heuristics are block-structured updates (conceptually).

12.7 Stochastic Optimization

Large-scale ML objectives are typically finite sums or expectations over data. Computing full gradients can be too expensive, so we use stochastic gradients computed from a single sample or mini-batch. Stochasticity introduces noise: it can slow convergence but can also act as implicit regularization and help escape saddles. This section develops SGD and mini-batching, learning-rate schedules, variance reduction (SVRG/SAGA), and adaptive methods (AdaGrad/RMSProp/Adam), emphasizing the geometry behind their updates and the practical caveats (cf. Fig. 12.3).

12.7.1 SGD and Mini-Batching

12.7.1.1 Unbiased Gradients, Noise, and Stability

SGD replaces $\nabla f(w) = \frac{1}{n}\sum_{i=1}^{n} \nabla \ell_i(w)$ by an unbiased estimator $\nabla \ell_{i_t}(w)$. Unbiasedness alone is not enough: the variance (and tail behavior) of gradient noise controls stability and convergence. Mini-batching reduces variance by averaging gradients across a batch, trading compute for stability. In ML practice, stability

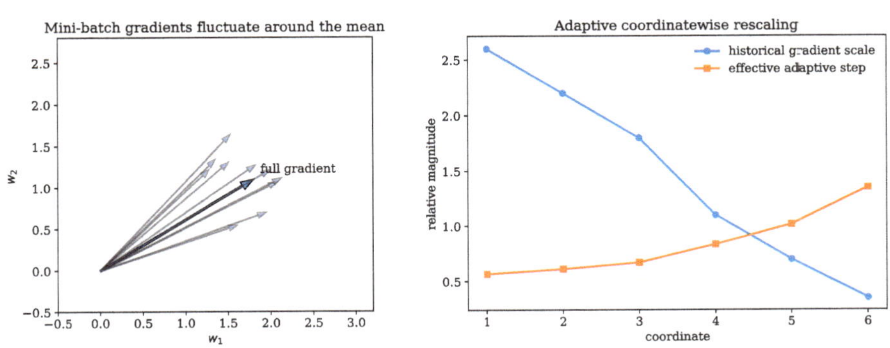

Fig. 12.3 Stochastic optimization geometry. Left: mini-batch gradients fluctuate around the full gradient, with batch size controlling noise. Right: adaptive methods implement coordinatewise rescaling, shrinking steps in frequently large-gradient directions and enlarging them in rare directions

issues manifest as divergence, exploding updates, or sensitivity to initialization and scaling; these are often addressed by normalization, clipping, and careful learning-rate design.

Definition 12.7.1 (Finite-Sum Objective and Stochastic Gradient) Let

$$f(w) = \frac{1}{n} \sum_{i=1}^{n} \ell_i(w),$$

with differentiable ℓ_i. At iteration t, sample an index i_t uniformly from $[n]$ and define

$$g_t := \nabla \ell_{i_t}(w_t).$$

Then $\mathbb{E}[g_t \mid w_t] = \nabla f(w_t)$ (unbiasedness).

Definition 12.7.2 (SGD Update) Stochastic gradient descent updates

$$w_{t+1} = w_t - \eta_t\, g_t,$$

with step size $\eta_t > 0$.

Definition 12.7.3 (Mini-Batch SGD) Let $B_t \subseteq [n]$ be a mini-batch of size b sampled (approximately) uniformly. Define

$$g_t := \frac{1}{b} \sum_{i \in B_t} \nabla \ell_i(w_t),$$

and update $w_{t+1} = w_t - \eta_t g_t$. Then $\mathbb{E}[g_t \mid w_t] = \nabla f(w_t)$ and $\mathrm{Var}(g_t)$ typically decreases with b.

Remark 12.7.1 (Noise Decomposition) Write $g_t = \nabla f(w_t) + \xi_t$, where $\mathbb{E}[\xi_t \mid w_t] = 0$. Convergence depends on controlling the conditional second moment $\mathbb{E}[\|\xi_t\|_2^2 \mid w_t]$ (and tails).

Remark 12.7.2 (Stability Tools (Practical)) Common stabilizers include feature standardization, normalization layers, gradient clipping, and adding ℓ_2 regularization (which induces strong convexity in convex settings). These are geometric controls on step magnitudes and curvature.

12.7.1.2 Learning-Rate Schedules, Warmup, and Decay

The learning rate is the primary knob in SGD: too large yields instability; too small yields slow progress. Schedules reflect a statistical-computational tradeoff: early phases benefit from large steps to explore/learn quickly; later phases require smaller steps to average out noise and refine around a minimizer. Warmup prevents early instability when gradients are poorly scaled; decay reduces noise-induced oscillations. In convex theory, diminishing step sizes are required for convergence to the exact optimum under persistent noise; in deep learning, piecewise-constant or cosine schedules are effective heuristics.

Remark 12.7.3 (Canonical Schedules)

- **Diminishing:** $\eta_t = \eta_0/\sqrt{t+1}$ or $\eta_t = \eta_0/(t+1)$.
- **Step decay:** multiply η by a factor (e.g., 0.1) at planned epochs.
- **Cosine decay (heuristic):** η_t follows a cosine from $\eta_{\max}$ to $\eta_{\min}$ over the training horizon.
- **Warmup:** start with small η and increase linearly for first T_w steps.

Remark 12.7.4 (Why Decay Helps) As w_t approaches a minimizer, gradient magnitudes shrink but noise may not. Reducing η_t lowers the stationary variance of the iterate distribution and improves final accuracy.

12.7.2 Variance Reduction Methods

12.7.2.1 SVRG/SAGA Intuition: Controlling Gradient Noise

Variance reduction exploits finite-sum structure to reduce stochastic gradient noise without using full gradients every step. SVRG periodically computes a full gradient at a reference point and uses it as a control variate; SAGA maintains a table of past gradients. These methods can achieve fast linear convergence for strongly convex finite-sum problems, bridging the gap between SGD and deterministic GD. They are especially valuable when high accuracy is needed and data fit in memory, such as logistic regression and convex ERM.

Remark 12.7.5 (Control Variate Principle) If g is a noisy estimator of $\nabla f(w)$ and c is a correlated quantity with known expectation, then $g - c + \mathbb{E}[c]$ can have the same mean but smaller variance. SVRG and SAGA implement this idea for gradients.

Remark 12.7.6 (SVRG Update (Conceptual)) Pick a reference point $\tilde{w}$ and compute $\nabla f(\tilde{w})$ once. Then for a random index i, use

$$g(w) = \nabla \ell_i(w) - \nabla \ell_i(\tilde{w}) + \nabla f(\tilde{w}),$$

which is unbiased and typically has smaller variance near $\tilde{w}$.

Remark 12.7.7 (SAGA Update (Conceptual)) Maintain stored gradients $\{\nabla \ell_i(\phi_i)\}$ at last-seen points ϕ_i. Update a random index i and use a gradient estimator that combines the fresh gradient with the average of stored ones. This yields a variance-reduced unbiased estimator without full-gradient recomputation each epoch.

12.7.2.2 When Variance Reduction Helps in Practice

Variance reduction methods shine in convex, strongly convex finite-sum problems where high precision is desired: they converge faster than SGD to very low error. In deep learning, benefits are less consistent because objectives are nonconvex, data are huge, and full-gradient computations are expensive. Nevertheless, in the last-phase fine-tuning of convex heads (linear probes, logistic regression), variance reduction can be highly effective. The practical decision is cost model: VR requires extra memory (SAGA) or periodic full passes (SVRG).

Remark 12.7.8 (Practical Rule-of-Thumb) Use variance reduction when: (i) the problem is convex/strongly convex (or locally behaves so), (ii) n is not astronomically large, (iii) you need high accuracy (beyond what SGD plateaus at), (iv) full passes or gradient tables are affordable. Otherwise, SGD/Adam-style methods are usually preferred.

12.7.3 Adaptive Gradient Methods

12.7.3.1 AdaGrad, RMSProp, Adam: Intuition and Geometry

Adaptive methods rescale gradients coordinatewise using past gradient magnitudes (Duchi et al. 2011). Geometrically, they implement a time-varying diagonal preconditioner: steps are larger in rarely updated directions and smaller in frequently updated ones. AdaGrad accumulates squared gradients and yields diminishing effective learning rates, benefiting sparse features. RMSProp uses an exponential moving average to prevent rates from shrinking too fast. Adam combines RMSProp-style scaling with momentum on the first moment, yielding strong practical performance in deep learning.

Definition 12.7.4 (AdaGrad (Diagonal Preconditioning)) Given gradients $g_t \in \mathbb{R}^d$, AdaGrad maintains

$$G_t = \sum_{s=0}^{t} g_s \odot g_s,$$

and updates coordinatewise

$$w_{t+1} = w_t - \eta \frac{g_t}{\sqrt{G_t} + \epsilon},$$

where division and square root are elementwise and $\epsilon > 0$ is a stabilizer.

Definition 12.7.5 (RMSProp (EMA Scaling)) Maintain an exponential moving average of squared gradients:

$$v_t = \beta v_{t-1} + (1-\beta)(g_t \odot g_t),$$

and update

$$w_{t+1} = w_t - \eta \frac{g_t}{\sqrt{v_t} + \epsilon}.$$

Definition 12.7.6 (Adam (Momentum + RMSProp)) Maintain first and second moment EMAs:

$$m_t = \beta_1 m_{t-1} + (1-\beta_1) g_t, \qquad v_t = \beta_2 v_{t-1} + (1-\beta_2)(g_t \odot g_t),$$

apply bias correction $\widehat{m}_t = m_t/(1-\beta_1^t)$, $\widehat{v}_t = v_t/(1-\beta_2^t)$, and update

$$w_{t+1} = w_t - \eta \frac{\widehat{m}_t}{\sqrt{\widehat{v}_t} + \epsilon}.$$

Remark 12.7.9 (Geometric Viewpoint) Adaptive methods use a diagonal metric D_t so that the step is roughly $w_{t+1} = w_t - \eta D_t^{-1/2} g_t$. Thus they are preconditioned SGD with a data-dependent diagonal approximation to curvature.

12.7.3.2 Generalization Behavior and Caveats (Conceptual)

Adaptive methods often optimize training loss quickly but can generalize differently than SGD with momentum. One reason is that they change the implicit regularization: coordinatewise scaling biases the solution toward different norms. They can also be sensitive to hyperparameters ($\beta_1, \beta_2, \epsilon$) and to nonstationary gradient noise. Empirically, switching from Adam to SGD in late training sometimes improves test accuracy ("Adam-to-SGD" schedules). The correct takeaway is not that adaptive methods are bad, but that their geometry differs and must be tuned and evaluated with care.

Remark 12.7.10 (Caveats (Practical))

- Adaptive preconditioning changes the effective notion of step size; standard learning-rate heuristics may not transfer.
- For some problems, Adam-like methods can converge to different solutions than SGD, affecting generalization.
- Weight decay must be implemented correctly (decoupled weight decay in AdamW is standard) to match intended regularization.

Remark 12.7.11 (Robust Practice) Report optimizer details (algorithm, schedule, weight decay), run multiple seeds, and validate both training speed and test performance. Treat the optimizer as part of the model selection procedure.

12.8 Second-Order and Quasi-Second-Order Methods

Second-order methods exploit curvature to accelerate optimization and improve stability in ill-conditioned problems. They approximate the objective locally by a quadratic and choose steps that account for anisotropy of level sets. In ML, full Newton steps can be too expensive for large d, so quasi-Newton methods (BFGS/L-BFGS) provide practical curvature approximations. Gauss–Newton and Fisher-based approximations leverage special structure (least squares, likelihood models) to obtain PSD curvature surrogates that are both computationally feasible and statistically meaningful.

12.8.1 Newton and Damped Newton

12.8.1.1 Quadratic Models and Line Search

Newton's method arises from minimizing the second-order Taylor model of the objective. When the local quadratic approximation is accurate, Newton converges rapidly (often quadratically) near the optimum. However, far from the optimum, the quadratic model can be poor, so steps can overshoot or even increase the objective. Damped Newton (or line-search Newton) stabilizes by scaling the Newton direction with a step size chosen to guarantee sufficient decrease.

Definition 12.8.1 (Newton Direction) Let $f : \mathbb{R}^d \to \mathbb{R}$ be twice differentiable. The Newton direction at x is

$$p_N(x) := -\big(\nabla^2 f(x)\big)^{-1} \nabla f(x),$$

when $\nabla^2 f(x)$ is invertible.

Definition 12.8.2 (Damped Newton/Line-Search Newton) A damped Newton update is

$$x_{t+1} = x_t + \eta_t p_t, \qquad p_t = p_N(x_t),$$

where $\eta_t \in (0, 1]$ is chosen by line search (e.g., Armijo or Wolfe conditions) to ensure sufficient decrease of f.

Remark 12.8.1 (Quadratic Model Viewpoint) The second-order Taylor model at x_t is

$$m_t(\Delta) = f(x_t) + \langle \nabla f(x_t), \Delta \rangle + \frac{1}{2}\Delta^\top \nabla^2 f(x_t)\Delta.$$

If $\nabla^2 f(x_t) \succ 0$, the minimizer is $\Delta^\star = p_N(x_t)$. Line search chooses a safe fraction of this step when the quadratic model is only locally accurate.

12.8.1.2 Hessian Definiteness and Trustworthiness

Newton's method depends critically on the Hessian being positive definite (for minimization). If the Hessian is indefinite (common in nonconvex problems), the Newton direction may not be a descent direction. Damped steps alone may not fix this. Trust-region methods and modified Hessians (adding λI, using PSD approximations) restore trustworthiness. In ML, this is why Gauss–Newton and Fisher approximations are popular: they produce PSD curvature matrices aligned with least squares/likelihood geometry.

Remark 12.8.2 (Descent Direction Condition) If $\nabla^2 f(x) \succ 0$, then $p_N(x)$ is a descent direction:

$$\langle \nabla f(x), p_N(x) \rangle = -\nabla f(x)^\top (\nabla^2 f(x))^{-1} \nabla f(x) < 0 \quad \text{when } \nabla f(x) \neq 0.$$

If $\nabla^2 f(x)$ is indefinite, this inner product can be positive, so Newton can increase f.

Remark 12.8.3 (Trust-Region Idea (Conceptual)) Instead of taking the full Newton step, solve

$$\min_{\|\Delta\|_2 \le \rho} \; m_t(\Delta),$$

where ρ is a trust radius adjusted based on how well the model predicts actual decrease. This makes Newton robust even when curvature is unreliable.

12.8.2 *Quasi-Newton Methods*

12.8.2.1 BFGS/L-BFGS Intuition and Memory Tradeoffs

Quasi-Newton methods approximate curvature from gradients without forming Hessians. BFGS builds an inverse-Hessian approximation H_t satisfying the secant condition $H_t y_t = s_t$ using rank-two updates, and preserves positive definiteness under mild conditions. L-BFGS stores only a small window of past (s_t, y_t) pairs, making it suitable for large-scale problems. In convex ML (logistic regression, kernel methods, ridge-type objectives), L-BFGS can be extremely effective.

Definition 12.8.3 (Secant Condition) Let $s_t = x_{t+1} - x_t$ and $y_t = \nabla f(x_{t+1}) - \nabla f(x_t)$. A quasi-Newton inverse-Hessian approximation H_{t+1} satisfies the secant condition if

$$H_{t+1} y_t = s_t.$$

Remark 12.8.4 (BFGS Update (Conceptual)) BFGS updates H_t by a rank-two formula designed to: (i) satisfy the secant condition, and (ii) preserve symmetry and positive definiteness when $y_t^\top s_t > 0$ (often enforced by line search). We treat the explicit formula as standard and focus on its meaning: curvature is learned from gradient differences.

Remark 12.8.5 (L-BFGS Memory Tradeoff) L-BFGS stores only the last m pairs (s_t, y_t) and applies them via a two-loop recursion to compute $H_t \nabla f(x_t)$. Memory is $O(md)$ rather than $O(d^2)$, making it practical when d is large but gradients are available.

12.8.2.2 When Quasi-Newton Works Well for ML

Quasi-Newton excels when the objective is smooth and moderately sized, and when one can compute full gradients reasonably often. This is common in convex ERM (logistic regression, CRFs, kernel logistic) and in fine-tuning a small head. In deep learning with massive data and noisy gradients, L-BFGS is less common because curvature estimates from mini-batches can be unstable, but it can still work in deterministic or large-batch regimes. The practical choice depends on compute: quasi-Newton reduces iterations but increases per-iteration cost.

Remark 12.8.6 (Rule-of-Thumb) Prefer (L-)BFGS when: (i) f is smooth and (approximately) convex, (ii) full gradients are available or large batches are feasible, (iii) high accuracy is required, (iv) d is large but not so large that even storing m vectors is infeasible. Otherwise, first-order methods dominate.

12.8.3 Gauss–Newton and Fisher Approximations

12.8.3.1 Least Squares Structure and Curvature Approximations

Many ML objectives are sums of squared residuals:

$$f(w) = \frac{1}{2} \sum_{i=1}^{n} r_i(w)^2.$$

The Hessian includes two terms: a PSD term $J^\top J$ (Jacobian Gram) and a residual-weighted second-derivative term. Gauss–Newton drops the second term, yielding a PSD approximation that is accurate near solutions where residuals are small. This yields stable Newton-like steps for nonlinear least squares and motivates IRLS-type updates in GLMs.

Definition 12.8.4 (Gauss–Newton Approximation) Let $r(w) \in \mathbb{R}^n$ be the residual vector with Jacobian $J(w) \in \mathbb{R}^{n\times d}$. For $f(w) = \frac{1}{2}\|r(w)\|_2^2$, the gradient is $\nabla f(w) = J(w)^\top r(w)$ and the Hessian is

$$\nabla^2 f(w) = J(w)^\top J(w) + \sum_{i=1}^{n} r_i(w)\,\nabla^2 r_i(w).$$

The Gauss–Newton approximation uses

$$H_{\mathrm{GN}}(w) := J(w)^\top J(w) \succeq 0.$$

Remark 12.8.7 (Why This Is Trustworthy) $J^\top J$ is PSD by construction, so it avoids indefinite-curvature issues. When residuals are small, the dropped term is small, making Gauss–Newton accurate near good fits.

12.8.3.2 Natural Gradient Intuition via Fisher Geometry

For likelihood objectives, the Fisher information provides a PSD curvature surrogate tied to statistical geometry. Natural gradient scales the gradient by $\mathcal{I}(\theta)^{-1}$, giving reparameterization-invariant updates. In many models, Fisher resembles a Gauss–Newton matrix (e.g., for least squares and for certain exponential families), which explains why Fisher-based methods behave like stable second-order methods while remaining computationally tractable via approximations.

Remark 12.8.8 (Fisher as Expected Curvature) For negative log-likelihood $f(\theta) = -\sum_i \log p_\theta(Z_i)$, the observed Hessian can be noisy/indefinite. The Fisher information

$$\mathcal{I}(\theta) = \mathbb{E}\big[\nabla_\theta \log p_\theta(Z) \nabla_\theta \log p_\theta(Z)^\top\big]$$

is PSD and often approximates the expected Hessian, providing a statistically meaningful metric for steps.

Remark 12.8.9 (Natural Gradient Update) A natural-gradient step is

$$\theta_{t+1} = \theta_t - \eta\,\mathcal{I}(\theta_t)^{-1}\nabla f(\theta_t),$$

which can be interpreted as steepest descent in the Fisher-Riemannian geometry of the model family.

12.9 Line Search, Trust Regions, and Practical Step Selection

Even with a good descent direction, optimization can fail if step sizes are chosen poorly. Line search and trust-region methods are practical mechanisms to select steps that are *provably safe* (sufficient decrease) and *robust* when curvature is uncertain (Nocedal and Wright 2006). In convex problems they provide strong global convergence guarantees; in nonconvex ML problems they act as stabilizers that prevent catastrophic steps and help navigate ill-conditioning.

12.9.1 Line Search Principles

12.9.1.1 Armijo and Wolfe Conditions (Conceptual)

Line search chooses a step length $\alpha > 0$ along a direction p by checking how well the objective decreases. The Armijo condition enforces *sufficient decrease*: the new value must improve by at least a fraction of the linear prediction. Wolfe conditions add a *curvature* requirement to avoid steps that are too short (which would waste progress) or too long (which might overshoot). These are conceptually simple but powerful: they adapt to unknown smoothness constants and work well in deterministic smooth optimization.

Definition 12.9.1 (Line Search Along a Direction) Given current point x and a search direction p with $\langle \nabla f(x), p \rangle < 0$ (descent direction), line search seeks $\alpha > 0$ such that $x^+ = x + \alpha p$ yields a desirable decrease in f.

Definition 12.9.2 (Armijo (Sufficient Decrease) Condition) Fix constants $c_1 \in (0, 1)$. A step length $\alpha > 0$ satisfies the Armijo condition if

$$f(x + \alpha p) \;\le\; f(x) + c_1 \, \alpha \, \langle \nabla f(x), p \rangle.$$

Since $\langle \nabla f(x), p \rangle < 0$, the RHS is smaller than $f(x)$, so Armijo enforces decrease consistent with the local linear model.

Definition 12.9.3 (Wolfe Conditions) Fix $0 < c_1 < c_2 < 1$. A step length $\alpha > 0$ satisfies the (weak) Wolfe conditions if:

1. (Armijo) $f(x + \alpha p) \le f(x) + c_1 \alpha \langle \nabla f(x), p \rangle$,
2. (Curvature) $\langle \nabla f(x + \alpha p), p \rangle \ge c_2 \, \langle \nabla f(x), p \rangle$.

The curvature condition prevents steps that are too small (which would not reduce the directional derivative sufficiently).

Remark 12.9.1 (Strong Wolfe (Common in Quasi-Newton)) A widely used variant is the strong Wolfe condition:

$$|\langle \nabla f(x + \alpha p), p \rangle| \le c_2 \, |\langle \nabla f(x), p \rangle|,$$

which is particularly useful for ensuring stability of BFGS updates.

12.9.1.2 Backtracking and Robustness

Backtracking is the simplest line search: start with a candidate step size and shrink it geometrically until Armijo holds. It is robust because it requires only function values and gradients at the current point and rejects overly aggressive steps. In deterministic convex optimization, backtracking adapts automatically to unknown L. In ML, pure backtracking is less common in large-batch SGD (function evaluations are expensive/noisy), but it is very useful for deterministic solvers (logistic regression, kernel methods) and for second-order methods where a bad step can be disastrous.

Definition 12.9.4 (Backtracking Line Search) Choose parameters $\beta \in (0, 1)$ and $c_1 \in (0, 1)$ and an initial step $\alpha_0 > 0$. Set $\alpha = \alpha_0$ and repeatedly update $\alpha \leftarrow \beta\alpha$ until

$$f(x + \alpha p) \le f(x) + c_1 \alpha \langle \nabla f(x), p \rangle.$$

Return α and update $x^+ = x + \alpha p$.

Remark 12.9.2 (Why Backtracking Terminates (Smooth Case)) If f is L-smooth and $p = -\nabla f(x)$, then any $\alpha \le 1/L$ satisfies a sufficient decrease bound. Backtracking will eventually reach such an α by geometric shrinking.

Remark 12.9.3 (Robustness in Practice) Backtracking guards against:

- underestimated curvature (step too large),
- poor local quadratic approximation (Newton steps far from optimum),
- numerical blow-ups (e.g., overflow in exp/softmax),
- nonconvex “bad” regions where naive steps increase loss.

12.9.2 Trust-Region Methods

12.9.2.1 Quadratic Models with Radius Control

Trust-region methods invert the line-search logic. Instead of selecting a step length along a direction, they define a local model (typically quadratic) and restrict the step to lie within a ball of radius Δ. The radius represents how far we “trust” the model. If the model predicts well, expand the radius; if it predicts poorly, shrink it. This approach is highly robust for Newton-like methods and is well-suited to nonconvex optimization because it can avoid taking large steps in directions where curvature is unreliable.

Definition 12.9.5 (Trust-Region Subproblem) Given current point x, gradient $g = \nabla f(x)$, and a symmetric matrix B (e.g., Hessian or approximation), define the quadratic model

$$m(\Delta x) = f(x) + \langle g, \Delta x \rangle + \frac{1}{2} \Delta x^\top B \, \Delta x.$$

A trust-region step solves

$$\min_{\Delta x \in \mathbb{R}^d} \; m(\Delta x) \quad \text{s.t.} \quad \|\Delta x\|_2 \le \Delta,$$

where $\Delta > 0$ is the trust-region radius.

Remark 12.9.4 (Cauchy Point (Cheap Descent Step)) Even without solving the subproblem exactly, one can obtain a guaranteed decrease by taking the Cauchy step: move along $-\nabla f(x)$ and clip to the trust region. This yields robust progress with minimal computation.

12.9.2.2 Connection to Damping and Curvature

Damped Newton can be seen as a line-search stabilization of Newton, while trust regions stabilize by limiting step magnitude directly. There is also an algebraic link: solving the trust-region subproblem often yields a step resembling a *damped* Newton step, equivalent to minimizing the quadratic model plus a quadratic penalty $\frac{\lambda}{2}\|\Delta x\|_2^2$ for some $\lambda \ge 0$. This is the Levenberg–Marquardt idea: add λI to the Hessian to ensure positive definiteness and control step size. In ML, this perspective explains why adding ridge-like damping can stabilize second-order updates.

Remark 12.9.5 (Levenberg–Marquardt/Damped Curvature Idea) A common stabilized Newton-like direction solves

$$(B + \lambda I) \, \Delta x = -g,$$

with $\lambda \ge 0$ chosen so that $\|\Delta x\|_2 \le \Delta$ and the predicted decrease matches actual decrease. This is a bridge between trust regions (radius control) and damping (curvature regularization).

Remark 12.9.6 (Why Curvature Matters) If B is indefinite (nonconvex), the unconstrained quadratic model is unbounded below. The trust-region constraint restores well-posedness and yields a step that can exploit negative curvature safely. This is one reason trust-region methods are robust in nonconvex landscapes.

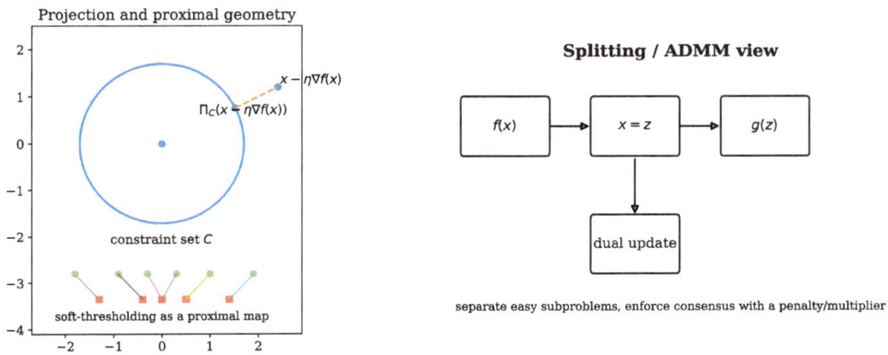

Fig. 12.4 Geometry of constrained and composite optimization. Left: projected gradient maps an unconstrained step back to the nearest feasible point. Right: variable splitting and proximal updates separate a smooth term from a nonsmooth or constrained term while enforcing agreement through consensus

12.10 Constraints, Projections, and Proximal Methods

Many ML objectives combine a smooth data-fit term with constraints or nonsmooth penalties: lasso (ℓ_1), SVM hinge loss, simplex constraints for probabilities, box constraints, and total variation. Euclidean projection and proximal operators provide the correct geometric primitives for such problems. Projected gradient descent handles hard constraints; proximal-gradient methods handle composite objectives of the form "smooth + nonsmooth" (Parikh and Boyd 2014). This section bridges geometry (non-expansive projections) to algorithms (PGD, ISTA/FISTA) and highlights common ML applications (cf. Fig. 12.4).

12.10.1 Projections onto Convex Sets

12.10.1.1 Projection Theorem and Non-expansiveness (Bridge to Geometry)

Projection onto a closed convex set is the closest-point operator in Euclidean geometry. Convexity guarantees existence and uniqueness, and projection is non-expansive (1-Lipschitz): it cannot increase distances. These properties are the stability backbone of constrained optimization methods. They also provide Pythagorean-type inequalities that generalize orthogonal projection onto subspaces.

Definition 12.10.1 (Euclidean Projection) Let $C \subseteq \mathbb{R}^d$ be nonempty. The Euclidean projection of x onto C is

$$\Pi_C(x) \in \arg\min_{z \in C} \|z - x\|_2.$$

Theorem 12.10.1 (Projection Theorem for Closed Convex Sets) *If C is nonempty, closed, and convex, then $\Pi_C(x)$ exists and is unique for every x. Moreover, with $p = \Pi_C(x)$, for all $z \in C$,*

$$\langle x - p,\ z - p\rangle \le 0 \quad \textit{and} \quad \|x - z\|_2^2 \ge \|x - p\|_2^2 + \|p - z\|_2^2.$$

Theorem 12.10.2 (Non-expansiveness) *If C is nonempty, closed, and convex, then for all $x, y \in \mathbb{R}^d$,*

$$\|\Pi_C(x) - \Pi_C(y)\|_2 \le \|x - y\|_2.$$

Remark 12.10.1 (Stability Message) Non-expansiveness means projection is numerically safe: it does not amplify perturbations. This is crucial when iterates are noisy (stochastic gradients) or when constraints are enforced repeatedly.

12.10.1.2 Projected Gradient Descent

Projected gradient descent (PGD) handles constrained problems $\min_{x \in C} f(x)$ by taking an unconstrained gradient step followed by projection back onto C. The method is simple, widely applicable, and inherits stability from projection. Under convexity and smoothness, PGD enjoys convergence guarantees similar to GD. In ML, PGD appears in constrained optimization (simplex/box constraints) and also as the standard algorithm for generating adversarial examples (projected ascent on loss under norm constraints), highlighting the dual role of projections in learning and robustness.

Definition 12.10.2 (Projected Gradient Descent) For minimizing f over a convex set C, PGD iterates

$$x_{t+1} = \Pi_C\big(x_t - \eta_t \nabla f(x_t)\big),$$

with step sizes $\eta_t > 0$.

Remark 12.10.2 (Common Projection Sets)

- Box constraints: projection is coordinatewise clipping.
- ℓ_2 ball: projection rescales if $\|x\|_2 > r$.
- Simplex: projection is a sorting/thresholding operation (KKT-based).

Projection cost is often the main per-iteration cost in PGD.

12.10.2 *Proximal Operators*

12.10.2.1 Prox as Generalized Projection; Moreau Envelope (Conceptual)

Proximal operators generalize projection by replacing hard constraints with soft penalties. They solve a regularized "denoise" subproblem: stay close to the current point while reducing penalty. This captures shrinkage and thresholding phenomena in sparse learning. The Moreau envelope smooths a nonsmooth function by infimal convolution with a quadratic, yielding a differentiable surrogate with controlled smoothing. These are the geometric primitives behind ISTA/FISTA and many modern optimization algorithms.

Definition 12.10.3 (Proximal Operator) Let $g : \mathbb{R}^d \to (-\infty, \infty]$ be proper (typically convex) and $\lambda > 0$. The proximal operator is

$$\operatorname{prox}_{\lambda g}(x) := \arg\min_{z \in \mathbb{R}^d} \left\{ \frac{1}{2}\|z - x\|_2^2 + \lambda g(z) \right\},$$

when the minimizer exists and is unique.

Remark 12.10.3 (Projection Is a Prox) If $g = \iota_C$ is the indicator of a closed convex set C (0 on C, $+\infty$ outside), then $\operatorname{prox}_{\lambda \iota_C}(x) = \Pi_C(x)$ for any $\lambda > 0$. Thus constraints are penalties in disguise.

Definition 12.10.4 (Moreau Envelope (Conceptual)) For convex g and parameter $\lambda > 0$, the Moreau envelope is

$$e_\lambda g(x) := \min_z \left\{ g(z) + \frac{1}{2\lambda}\|z - x\|_2^2 \right\}.$$

It is a smoothed version of g; its minimizer is exactly $\operatorname{prox}_{\lambda g}(x)$.

12.10.2.2 Soft-Thresholding and Shrinkage Mappings

The proximal operator of ℓ_1 is soft-thresholding: a coordinatewise shrinkage that sets small values to zero. This is the computational heart of lasso and sparse recovery. The proximal operator of ℓ_2^2 is uniform shrinkage (ridge-like). These explicit mappings are the reason proximal methods are practical: each iteration reduces to a gradient step plus a cheap denoising step.

Proposition 12.10.1 (Soft-Thresholding) *For* $g(z) = \|z\|_1$ *and* $\lambda > 0$,

$$\operatorname{prox}_{\lambda \|\cdot\|_1}(x) = \big(\operatorname{sign}(x_i)\max\{|x_i| - \lambda, 0\}\big)_{i=1}^d.$$

Proposition 12.10.2 (Ridge Shrinkage as Prox) *For* $g(z) = \frac{1}{2}\|z\|_2^2$ *and* $\lambda > 0$,

$$\mathrm{prox}_{\lambda g}(x) = \frac{1}{1+\lambda} x.$$

Remark 12.10.4 (Group Shrinkage) For group lasso $g(z) = \sum_g \|z_G\|_2$, the prox performs block soft-thresholding: each group vector is shrunk toward zero and can be set exactly to zero if its norm is below threshold.

12.10.3 Composite Optimization

12.10.3.1 Proximal Gradient and ISTA/FISTA (Conceptual)

Many ML objectives have the composite form $F(x) = f(x) + g(x)$ where f is smooth (data-fit) and g is convex but possibly nonsmooth (regularizer/constraint). Proximal gradient methods take a gradient step on f and a proximal step on g. ISTA is the basic method with $O(1/t)$ rate in convex settings; FISTA accelerates to $O(1/t^2)$ by Nesterov-style momentum. These methods are the optimization backbone for lasso, elastic net (with prox variants), and many structured regularizers.

Definition 12.10.5 (Composite Objective) A composite optimization problem has the form

$$\min_{x \in \mathbb{R}^d} F(x) = f(x) + g(x),$$

where f is differentiable with L-Lipschitz gradient and g is convex, proper, and possibly nonsmooth.

Definition 12.10.6 (Proximal Gradient (ISTA) Update) With step size $\eta \in (0, 1/L]$, ISTA iterates

$$x_{t+1} = \mathrm{prox}_{\eta g}\big(x_t - \eta \nabla f(x_t)\big).$$

Remark 12.10.5 (FISTA (Conceptual)) FISTA adds Nesterov-style extrapolation: compute an extrapolated point y_t from (x_t, x_{t-1}), then apply the ISTA prox-gradient step at y_t. This yields an accelerated $O(1/t^2)$ function-value rate in convex problems.

12.10.3.2 Applications: Lasso, Group Lasso, Elastic Net

Composite methods map directly onto popular ML estimators: lasso is squared loss $+$ ℓ_1 penalty, group lasso is squared loss $+$ group norm penalty, elastic net adds ℓ_2^2. The smooth term f provides gradients; the nonsmooth term g provides a prox with closed form (coordinate or block thresholding). Thus, each iteration is computationally simple and scales to high-dimensional problems.

Remark 12.10.6 (Lasso via ISTA) For lasso,

$$f(w) = \frac{1}{2}\|Xw - y\|_2^2, \qquad g(w) = \lambda\|w\|_1,$$

ISTA update becomes

$$w_{t+1} = \text{prox}_{\eta\lambda\|\cdot\|_1}\big(w_t - \eta X^\top(Xw_t - y)\big),$$

i.e., gradient step on least squares plus soft-thresholding.

Remark 12.10.7 (Elastic Net as Composite (One View)) Elastic net can be treated as $f(w) = \frac{1}{2}\|Xw - y\|_2^2 + \frac{\lambda_2}{2}\|w\|_2^2$ (smooth) and $g(w) = \lambda_1\|w\|_1$ (nonsmooth), so prox-gradient uses ridge-modified gradients and the same soft-thresholding prox.

Remark 12.10.8 (Group Lasso via Block Prox) Group lasso uses $g(w) = \lambda\sum_g \|w_{G_g}\|_2$, whose prox is block soft-thresholding. Thus ISTA/FISTA naturally produce block-sparse solutions.

12.11 ADMM and Splitting Methods

Many ML objectives are sums of terms that are individually easy but difficult together: a smooth loss plus a nonsmooth regularizer, or a data-fit term plus complicated constraints. Splitting methods exploit this structure by introducing auxiliary variables so that each subproblem becomes simple. ADMM (Alternating Direction Method of Multipliers) is a particularly effective splitting method: it combines the decomposability of dual ascent with the numerical stability of augmented Lagrangians. This section explains why splitting helps, presents the consensus form for distributed settings, and derives the ADMM updates with practical tuning intuition.

12.11.1 Why Splitting Helps

12.11.1.1 Decomposing Objectives and Constraints

A typical composite objective $\min_x f(x) + g(x)$ may be hard because f and g pull in different algorithmic directions: f is smooth and suited to gradients, g is nonsmooth and suited to proximal steps. Splitting introduces a copy variable z and enforces $x = z$, turning the problem into one with separable terms. The constraint is then handled by Lagrange multipliers and a quadratic penalty, yielding subproblems that often have closed form (e.g., least squares for f and soft-thresholding for g). This is the central idea behind ADMM.

Definition 12.11.1 (Variable Splitting) A composite problem

$$\min_{x \in \mathbb{R}^d} f(x) + g(x)$$

can be rewritten by introducing z as

$$\min_{x,z} f(x) + g(z) \quad \text{s.t.} \quad x - z = 0.$$

This separates the roles of f and g into different variables.

Remark 12.11.1 (Algorithmic Benefit) After splitting, one can update x using methods tailored to f (often a smooth quadratic/least squares subproblem), and update z using methods tailored to g (often a proximal operator), while enforcing $x \approx z$ through penalties and multipliers.

12.11.1.2 Consensus Form and Distributed Settings

In distributed ML, data are partitioned across machines. Each machine has a local objective $f_i(x)$. Consensus optimization introduces local copies x_i and enforces that they agree with a global variable z. ADMM handles this naturally: each worker solves a local subproblem in parallel, then a global averaging step enforces consensus. This is one reason ADMM is popular in large-scale convex ML and in federated-like settings.

Definition 12.11.2 (Consensus Optimization) Let $f_i : \mathbb{R}^d \to \mathbb{R}$ for $i = 1, \ldots, m$. The consensus form is

$$\min_{x_1,\ldots,x_m,z} \sum_{i=1}^{m} f_i(x_i) \quad \text{s.t.} \quad x_i - z = 0 \ \forall i.$$

Here z is the consensus (global) parameter and each x_i is a local copy.

Remark 12.11.2 (Parallelism) Given z and multipliers, the x_i-updates decouple across i and can be computed in parallel, then aggregated to update z.

12.11.2 ADMM Mechanics

12.11.2.1 Augmented Lagrangian and Alternating Updates

Plain Lagrangian methods can be unstable because dual ascent may oscillate. The augmented Lagrangian adds a quadratic penalty to enforce constraints more strongly, improving conditioning. ADMM alternates minimizing the augmented Lagrangian over x and z (block coordinate descent in primal variables), followed by a dual ascent step. The result is a robust algorithm with convergence guarantees for convex problems and strong empirical performance.

Definition 12.11.3 (Augmented Lagrangian (Scaled Form)) For

$$\min_{x,z} f(x) + g(z) \quad \text{s.t.} \quad x - z = 0,$$

the augmented Lagrangian is

$$\mathcal{L}_\rho(x, z, u) = f(x) + g(z) + \frac{\rho}{2}\|x - z + u\|_2^2 - \frac{\rho}{2}\|u\|_2^2,$$

where u is the scaled dual variable and $\rho > 0$ is the penalty parameter.

Definition 12.11.4 (ADMM Updates) Given (x^t, z^t, u^t), ADMM iterates:

$$x^{t+1} \in \arg\min_x \left(f(x) + \frac{\rho}{2}\|x - z^t + u^t\|_2^2\right),$$

$$z^{t+1} \in \arg\min_z \left(g(z) + \frac{\rho}{2}\|x^{t+1} - z + u^t\|_2^2\right),$$

$$u^{t+1} = u^t + x^{t+1} - z^{t+1}.$$

Remark 12.11.3 (Prox Interpretation) If the z-subproblem is exactly a proximal operator, then

$$z^{t+1} = \text{prox}_{g/\rho}(x^{t+1} + u^t),$$

and similarly, many x-updates reduce to ridge/least-squares solves when f is quadratic. Thus ADMM often alternates "least squares" and "prox" steps.

Example 12.11.1 (Lasso via ADMM (Conceptual Specialization)) For lasso,

$$f(x) = \frac{1}{2}\|Xx - y\|_2^2, \qquad g(z) = \lambda\|z\|_1.$$

Then the x-update solves a ridge-like system

$$(X^\top X + \rho I)x^{t+1} = X^\top y + \rho(z^t - u^t),$$

and the z-update is soft-thresholding:

$$z^{t+1} = \text{prox}_{(\lambda/\rho)\|\cdot\|_1}(x^{t+1} + u^t).$$

12.11.2.2 Convergence Intuition and Parameter Tuning

For convex problems, ADMM converges under mild conditions. Intuitively, the quadratic penalty forces primal feasibility ($x \approx z$), while dual updates enforce constraint satisfaction and balance the two objectives. The parameter ρ controls a tradeoff: too small and constraints are weak (slow primal convergence); too large and subproblems become ill-conditioned (slow numerical solves). Practical tuning monitors primal and dual residuals and adapts ρ to balance them. In ML, ADMM is particularly effective when subproblems have closed form or can be solved efficiently and when splitting yields parallelism.

Definition 12.11.5 (Primal and Dual Residuals (Diagnostics)) In the constraint $x = z$ setting, define

$$r^{t+1} := x^{t+1} - z^{t+1} \quad \text{(primal residual)}, \qquad s^{t+1} := \rho(z^{t+1} - z^t) \quad \text{(dual residual)}.$$

Convergence is often monitored by checking $\|r^{t+1}\|_2$ and $\|s^{t+1}\|_2$ against tolerances.

Remark 12.11.4 (Heuristic Tuning of ρ) A common heuristic is to adjust ρ so that primal and dual residual norms are of similar magnitude: increase ρ if $\|r\|$ is much larger than $\|s\|$, decrease ρ if $\|s\|$ is much larger than $\|r\|$. This balances constraint enforcement and dual progress.

Remark 12.11.5 (When ADMM Is a Good Idea) ADMM is attractive when: (i) the problem splits into easy subproblems, (ii) constraints are simple but coupling is hard, (iii) distributed/parallel computation is needed, (iv) you want robust convergence on convex objectives. For large nonconvex deep learning, SGD-family methods are more common, but ADMM-style ideas still appear in constrained training and federated optimization.

12.12 Nonconvex Optimization in Deep Learning

Deep learning objectives are nonconvex due to layered composition, overparameterization, and symmetries. Yet gradient-based methods often succeed in practice. The right way to understand this is not "nonconvex is hopeless," but "nonconvex has structure." Training dynamics are governed by the behavior of critical points (especially saddles), initialization and scaling (which set curvature and signal propagation), and implicit regularization (which biases which minima are found). This section provides a rigorous, geometry-first preview of these ideas.

12.12.1 Critical Points and Saddle Behavior

12.12.1.1 Saddles vs. Minima; Negative Curvature Directions

In high dimensions, saddle points are far more common than poor local minima. A saddle is a stationary point with at least one direction of negative curvature (Hessian has a negative eigenvalue), meaning one can decrease the objective by moving along that direction. This motivates curvature-aware methods and explains why many "stuck" behaviors are actually slow escape from flat or saddle regions. In deep networks, symmetries produce flat directions (zero curvature), while composition produces both sharp and flat curvature along different directions.

Definition 12.12.1 (Critical Point and Strict Saddle) For differentiable $f : \mathbb{R}^d \to \mathbb{R}$, a point $x^\star$ is *critical* if $\nabla f(x^\star) = 0$. If f is twice differentiable, $x^\star$ is a *strict saddle* if $\nabla f(x^\star) = 0$ and

$$\lambda_{\min}(\nabla^2 f(x^\star)) < 0.$$

Remark 12.12.1 (Negative Curvature Direction) If v is a unit eigenvector of $\nabla^2 f(x^\star)$ with eigenvalue $\lambda < 0$, then locally

$$f(x^\star + tv) \approx f(x^\star) + \frac{1}{2}\lambda t^2,$$

so small steps along $\pm v$ decrease f (to second order). This is the geometric signature of a saddle.

Remark 12.12.2 (Why Saddles Are Prevalent) In large d, random matrices have mixed-sign spectra, and layered models generate high-dimensional parameter spaces with many flat and curved directions. Thus saddle-like regions are abundant, and optimization must cope with them.

12.12.1.2 Escaping Saddles: Noise and Perturbations (Conceptual)

Deterministic gradient descent can linger near saddles because gradients are small and curvature is flat in many directions. Stochasticity helps: SGD noise perturbs iterates, injecting energy that can push them out of saddle basins. Explicit perturbations (random restarts, adding noise when gradients are small) can also help. The conceptual takeaway is that noise is not merely a nuisance; it can be algorithmically beneficial in nonconvex landscapes.

Remark 12.12.3 (Noise as Implicit Saddle Escape) Write SGD as $x_{t+1} = x_t - \eta(\nabla f(x_t) + \xi_t)$ with zero-mean noise ξ_t. Near a saddle, $\nabla f(x_t) \approx 0$ but the noise term can push x_t into a negative-curvature direction, after which gradients grow and descent accelerates.

Remark 12.12.4 (Perturbed Gradient Methods (Conceptual)) A standard conceptual strategy is: if $\|\nabla f(x_t)\|$ is small for many steps, add a random perturbation Δ and continue gradient steps. This increases the probability of leaving strict-saddle neighborhoods.

12.12.2 Initialization and Scaling

12.12.2.1 Why Initialization Affects Training Dynamics

Initialization sets the starting region of the nonconvex landscape and determines early signal propagation. Bad scaling leads to vanishing/exploding activations and gradients, effectively destroying optimization. Variance-preserving initializations aim to keep activations and backpropagated gradients at stable magnitudes across layers, which is a geometric conditioning control mechanism. Initialization also interacts with symmetry: if initialized symmetrically, units can remain redundant; random initialization breaks symmetry.

Remark 12.12.5 (Scale Propagation Intuition) In deep compositions, gradients are products of Jacobians. If typical Jacobian singular values are < 1, gradients vanish exponentially with depth; if > 1, they explode. Initialization schemes aim to keep singular values near 1 on average, stabilizing training.

Remark 12.12.6 (Symmetry Breaking) Neurons with identical initial weights receive identical gradients and remain identical. Random initialization breaks this symmetry so units can specialize.

12.12.2.2 Normalization Layers as Conditioning Control

Normalization layers (BatchNorm, LayerNorm, RMSNorm, etc.) modify the geometry of the optimization problem: they rescale activations to control variance and reduce internal covariate shift-like effects. From an optimization perspective, they improve conditioning by stabilizing gradient magnitudes and reducing anisotropy. They also introduce scale invariances (reparameterization effects), which can change how regularization and learning rates behave. Thus, normalization is both a statistical and a geometric tool.

Remark 12.12.7 (Normalization as Adaptive Reparameterization) A normalization layer can be viewed as applying a data-dependent affine transformation to intermediate representations. This changes the effective metric seen by gradient descent, often making the landscape smoother and easier to traverse.

Remark 12.12.8 (Practical Consequence) With normalization, larger learning rates can be stable, and training becomes less sensitive to initialization scale, though not immune. This is one reason normalization is ubiquitous in deep learning.

12.12.3 Implicit Regularization and Generalization

12.12.3.1 Early Stopping as Regularization

Optimization itself can act as regularization: among many minimizers, algorithms tend to prefer certain ones. Early stopping is the simplest instance: stop before fully fitting noise. In linear models, early stopping in gradient descent corresponds to a form of spectral filtering similar to ridge (it suppresses small-eigenvalue directions early). In deep learning, early stopping controls overfitting and interacts with learning-rate schedules, often improving test performance.

Remark 12.12.9 (Early Stopping as a Complexity Control) If training loss keeps decreasing but validation loss stops improving, continuing optimization primarily fits noise. Stopping at the validation optimum is therefore a regularization choice (a hyperparameter).

12.12.3.2 Optimization Bias and Margin/Flatness Intuitions (Conceptual)

In nonconvex problems, there are many minima with similar training loss but different generalization. Empirically, solutions found by SGD often generalize better than those found by certain adaptive methods or full-batch deterministic solvers, suggesting an *implicit bias* of the algorithm. Two influential geometric intuitions are: (1) margin: for classification, larger margins tend to generalize better and be more robust; (2) flatness: minima in flatter regions (low curvature) are thought to be more stable to perturbations and thus generalize better, though "flatness" must be defined carefully because it depends on parameterization.

Remark 12.12.10 (Margin Intuition (Classification)) For linear classifiers, maximizing margin improves robustness and can improve generalization. In deep networks, training with cross-entropy often continues to increase margins even after achieving zero training error, suggesting a link between optimization dynamics and margin growth.

Remark 12.12.11 (Flatness Intuition and Caveat) Flat minima are often associated with small Hessian eigenvalues and stability to parameter perturbations. However, naive flatness measures are not invariant to reparameterization (rescaling layers can change Hessian). Thus, "flatness" is a useful intuition but must be used with care.

12.13 Optimization Meets Statistics

Optimization does not happen in a vacuum: we optimize an objective built from finite data. Thus, there is a natural *statistical precision* limit beyond which further optimization yields diminishing returns. Conversely, inadequate optimization can dominate error if we stop too early. This section explains how to choose optimization accuracy relative to statistical uncertainty, and previews why stability of the learning algorithm (sensitivity to data perturbations) connects optimization dynamics to generalization, including the roles of noise and batch size.

12.13.1 Choosing Optimization Accuracy

12.13.1.1 Statistical Precision vs. Optimization Tolerance

A training objective (empirical risk) is a random approximation to the population risk. Even if we found its exact minimizer, the resulting predictor has estimation error due to finite data. Therefore, it is often wasteful to drive optimization error far below the statistical error floor. The right target is to optimize until the remaining optimization gap is comparable to expected generalization gap. This provides a principled stopping rule and explains why approximate solvers and early stopping can be optimal in compute-constrained regimes.

Definition 12.13.1 (Optimization Error vs. Estimation Error (Conceptual)) Let $F(w) = \mathbb{E}[\ell(w; Z)]$ be population objective and $f_n(w) = \frac{1}{n}\sum_{i=1}^{n} \ell(w; Z_i)$ empirical objective. Let $w_n^\star \in \arg\min_w f_n(w)$ and let w_T be the iterate after T optimization steps. Define

$$\text{Optimization gap:} \quad \text{OptGap}_T := f_n(w_T) - f_n(w_n^\star),$$

$$\text{Estimation gap (for a solution):} \quad \text{EstGap}(w) := F(w) - \inf_u F(u).$$

The total excess risk $F(w_T) - \inf F$ can be decomposed conceptually into estimation and optimization contributions.

Remark 12.13.1 (Practical Target Rule-of-Thumb) Stop optimization when improvements in validation loss (or an estimate of generalization gap) are on the same order as the noise in validation estimates, or when the training objective decrease per epoch becomes negligible relative to expected statistical fluctuations.

Remark 12.13.2 (Strongly Convex Case Intuition) In strongly convex problems, the empirical minimizer is unique and optimization error decreases geometrically. If estimation error scales like $O(1/n)$ or $O(1/\sqrt{n})$, then targeting optimization error much smaller than that scale is often unnecessary.

12.13.1.2 When "Over-optimizing" Hurts or Does Nothing

Over-optimizing means fitting the empirical objective beyond what the data can support. In flexible models, this can overfit noise: training loss keeps decreasing while test loss worsens. Even in well-specified convex problems, further optimization beyond the statistical floor may not improve test performance because the population objective is already within the unavoidable estimation error. Thus, additional compute can be wasted or harmful. This motivates early stopping, regularization, and validation-based stopping rules.

Remark 12.13.3 (Two Regimes)

- **Harmless over-optimization:** the model is well-regularized and generalization is stable; extra optimization yields negligible test change.
- **Harmful over-optimization:** the model is high-capacity and under-regularized; extra optimization fits noise and increases test error.

The difference is controlled by capacity, noise level, and regularization/early stopping.

Remark 12.13.4 (Validation Curve Symptom) If validation loss is U-shaped over training time (decreases then increases), the optimal stopping time is a statistical model-selection parameter. Repeatedly choosing it based on the same validation set is itself a selection problem (nested validation is the clean evaluation remedy).

12.13.2 Stability and Generalization

12.13.2.1 Algorithmic Stability Intuition

Algorithmic stability asks: how much does the learned predictor change if we change one training example? If changing one point changes the output only slightly, the algorithm generalizes well because it is not overfitting idiosyncrasies. Stability provides a path to generalization bounds that does not require uniform convergence over large hypothesis classes. Optimization interacts with stability: strongly convex regularized objectives produce stable solutions; certain optimization procedures (like SGD with noise) can also increase stability in practice.

Remark 12.13.5 (Stability as Sensitivity Control) Let $\mathcal{D}$ and $\mathcal{D}'$ differ by one example, and let $\widehat{h}(\mathcal{D})$ and $\widehat{h}(\mathcal{D}')$ be trained predictors. Stability is the property that predictions or losses change little:

$$|\ell(\widehat{h}(\mathcal{D}); z) - \ell(\widehat{h}(\mathcal{D}'); z)| \text{ is small on average.}$$

This implies that empirical risk is a reliable estimate of population risk.

Remark 12.13.6 (Regularization Improves Stability) In convex ERM, adding $\lambda \|w\|_2^2$ makes the objective strongly convex. Strong convexity forces solutions to move continuously with data perturbations, yielding stability and thus generalization.

12.13.2.2 Noise, Batch Size, and Generalization Behavior

SGD injects noise through mini-batch sampling. This noise interacts with the loss landscape and can bias the algorithm toward certain minima (implicit regularization). Batch size controls noise magnitude: small batches have high noise (more exploration, sometimes better generalization), large batches have low noise (more deterministic, can converge faster but sometimes to sharper minima with worse generalization). These are not universal laws, but robust empirical patterns that connect optimization stochasticity to generalization.

Remark 12.13.7 (Batch Size as a Noise Knob) Mini-batch gradient noise variance typically decreases as batch size increases. Thus batch size trades compute efficiency (parallelism) against noise-induced exploration and implicit regularization.

Remark 12.13.8 (Practical Implication) Changing optimizer, batch size, and learning-rate schedule can change test performance even when training loss is similar, because these choices change both the optimization trajectory and the implicit regularization. Hence, optimizer configuration is part of model selection.

12.14 Large-Scale and Distributed Optimization

Modern ML is often constrained not by the existence of an optimization algorithm, but by systems reality: data are huge, models are large, and training must be parallelized across devices. Distributed optimization introduces new bottlenecks (communication, synchronization, stragglers) and new failure modes (stale gradients, drift), and it motivates approximation strategies such as gradient compression and curvature approximations that are feasible at scale. This section previews the core ideas: data-parallel SGD, communication-aware training, and scalable second-order methods via Hessian-vector products and K-FAC.

12.14.1 Data-Parallel Training

12.14.1.1 Synchronous vs. Asynchronous SGD (Conceptual)

In data-parallel training, multiple workers compute gradients on different mini-batches and aggregate them to update shared parameters. Synchronous SGD aggregates gradients from all workers at each step, producing a clean equivalent of larger batch training but suffering from stragglers. Asynchronous SGD allows workers to update without waiting, improving throughput but introducing staleness: gradients may be computed on outdated parameters. The key tradeoff is statistical vs. systems efficiency: sync is stable and easier to analyze; async is faster but can degrade convergence unless carefully controlled.

Definition 12.14.1 (Data-Parallel SGD (Conceptual)) Let there be M workers. At iteration t, worker m computes a stochastic gradient $g_t^{(m)}$ on its mini-batch. An aggregated gradient is

$$g_t := \frac{1}{M} \sum_{m=1}^{M} g_t^{(m)}.$$

The parameter update is $w_{t+1} = w_t - \eta_t g_t$.

Remark 12.14.1 (Synchronous SGD) In synchronous SGD, all M gradients are computed on the current w_t and aggregated (e.g., via all-reduce) before updating. This is equivalent to a mini-batch of size $M \cdot b$ when each worker uses batch size b, assuming identical learning rate handling.

Remark 12.14.2 (Asynchronous SGD and Staleness) In asynchronous SGD, worker gradients may be computed using parameters $w_{t-\tau}$ for some delay τ. Stale gradients can point in suboptimal directions if the objective changes rapidly, especially with large learning rates. Practical async methods require bounded staleness, smaller learning rates, or momentum corrections.

12.14.1.2 Communication Costs and Gradient Compression

At scale, communication often dominates compute: transmitting a full gradient of size d each step is expensive. Gradient compression reduces bandwidth by quantization (lower precision), sparsification (send only top-k coordinates), or sketching (CountSketch-style). However, compression introduces error; to maintain convergence one often uses error feedback (accumulate the dropped residual). In ML systems, compression is a compute–communication–accuracy tradeoff, similar in spirit to matrix sketching.

Remark 12.14.3 (Communication Model) If each step requires sending d floating-point numbers and the network bandwidth is limited, then scaling to many workers yields diminishing returns unless communication is reduced. Thus, throughput depends on both compute and communication latency.

Remark 12.14.4 (Compression Primitives)

- **Quantization:** represent gradients with fewer bits (e.g., 8-bit, 1-bit sign).
- **Sparsification:** transmit only a subset of coordinates (e.g., top-k magnitude).
- **Sketching:** transmit a hashed sketch of the gradient (CountSketch) that preserves key information.
- **Error feedback:** keep a residual of dropped information and add it to the next gradient before compressing again.

12.14.2 Second-Order at Scale

12.14.2.1 Hessian-Vector Products and Implicit Methods

Full Hessians are infeasible in large models, but Hessian-vector products (HVPs) can often be computed at a cost comparable to a gradient using automatic differentiation (Pearlmutter trick). This enables *implicit* second-order methods: conjugate gradient for Newton steps, truncated Newton, and trust-region subproblem solvers that use only HVPs. These methods can exploit curvature without forming matrices, bridging second-order benefits with large-scale feasibility.

Definition 12.14.2 (Hessian-Vector Product) Let $f : \mathbb{R}^d \to \mathbb{R}$ be twice differentiable. For a vector $v \in \mathbb{R}^d$, the Hessian-vector product is

$$\mathrm{HVP}(w; v) := \nabla^2 f(w)\, v.$$

In modern autodiff systems, $\mathrm{HVP}(w; v)$ can often be computed in $O(d)$ time (up to constants), without forming $\nabla^2 f(w)$.

Remark 12.14.5 (Implicit Newton Step) Newton direction p solves $\nabla^2 f(w)\, p = -\nabla f(w)$. Instead of inverting the Hessian, one can approximately solve this linear system using conjugate gradient, where each iteration uses only HVPs. This is the core idea of truncated/implicit Newton methods.

12.14.2.2 K-FAC/Curvature Approximations (Conceptual)

Natural gradient uses Fisher information as a metric, but computing Fisher/inverse is expensive in deep networks. K-FAC (Kronecker-Factored Approximate Curvature) approximates the Fisher (or Gauss–Newton) matrix with a block-diagonal structure, where each block is approximated by a Kronecker product of smaller matrices derived from layer activations and gradients. This yields efficient approximate natural-gradient steps for deep nets and is a prime example of exploiting structure (Kronecker products) to make second-order geometry feasible.

Remark 12.14.6 (K-FAC Intuition (Conceptual)) For a layer with weights W, Fisher blocks often involve expectations of outer products of activations and backpropagated errors. K-FAC approximates these blocks as

$$F_W \approx A \otimes G,$$

where A depends on activation second moments and G depends on gradient/error second moments. Then $(A \otimes G)^{-1} = A^{-1} \otimes G^{-1}$ (Kronecker rule), making inversion tractable.

Remark 12.14.7 (When Curvature Approximations Help) Curvature methods (K-FAC, Gauss–Newton) can reduce sensitivity to learning-rate tuning and improve convergence in ill-conditioned networks, especially in large-batch regimes. They incur overhead per step, so they are most effective when fewer steps are needed and compute budget allows.

12.15 Implementation and Debugging Toolkit

Optimization failures in ML are often not conceptual but operational: a sign error in gradients, a silent shape mismatch, overflow in exponentials, or an unstable learning rate. A practical toolkit is therefore as important as theory. This section provides a rigorous, implementation-oriented view of automatic differentiation (reverse-mode AD), gradient checking, numerical stability techniques, and diagnostics that isolate common failure modes (loss not decreasing, exploding gradients, overfitting, and selection bias in early stopping).

12.15.1 Automatic Differentiation

12.15.1.1 Reverse-Mode AD as Vector-Jacobian Products

Reverse-mode AD (backpropagation) is efficient when the output is scalar (a loss) and the parameter dimension is large. It computes gradients by repeatedly applying vector-Jacobian products (VJPs) along the computation graph. The key principle is the chain rule in linear-algebra form: for $y = f(x)$ and scalar loss $L(y)$, $\nabla_x L = J_f(x)^\top \nabla_y L$. This avoids forming Jacobians explicitly and makes gradients scalable to millions or billions of parameters.

Definition 12.15.1 (Vector-Jacobian Product (VJP)) Let $f : \mathbb{R}^d \to \mathbb{R}^k$ be differentiable at x with Jacobian $J_f(x) \in \mathbb{R}^{k \times d}$. For a vector $v \in \mathbb{R}^k$, the vector-

Jacobian product is

$$\mathrm{VJP}_f(x; v) := J_f(x)^\top v \in \mathbb{R}^d.$$

Reverse-mode AD computes gradients by chaining VJPs.

Proposition 12.15.1 (Backprop Identity) *Let $L : \mathbb{R}^k \to \mathbb{R}$ be differentiable and define $y = f(x)$. Then*

$$\nabla_x (L \circ f)(x) = J_f(x)^\top \nabla_y L(y).$$

Remark 12.15.1 (Why Reverse-Mode Is Efficient) Computing $\nabla_x(L \circ f)$ for scalar L costs (up to constants) about the cost of evaluating f once, independent of d in the sense that it does not require k separate passes. This is why backprop is the default in deep learning.

12.15.1.2 Gradient Checking and Finite Differences

Automatic differentiation can still be wrong due to implementation bugs (incorrect broadcasting, missing detach, wrong reduction). Gradient checking validates an implementation by comparing AD gradients to finite-difference approximations of directional derivatives. This is most reliable on small problems in double precision, with deterministic computation and carefully chosen step sizes. The goal is not to replace AD but to verify correctness during development.

Definition 12.15.2 (Finite-Difference Directional Derivative) Let $f : \mathbb{R}^d \to \mathbb{R}$ and choose a direction u with $\|u\|_2 = 1$. The central finite-difference approximation at x with step $h > 0$ is

$$D_h f(x)[u] := \frac{f(x + hu) - f(x - hu)}{2h}.$$

If f is smooth, then $D_h f(x)[u] \approx \langle \nabla f(x), u \rangle$ for small h.

Remark 12.15.2 (Practical Gradient-Check Protocol) Use float64, disable dropout/augmentation, fix random seeds, and check random directions u rather than every coordinate. Compare relative error:

$$\frac{|D_h f(x)[u] - \langle \nabla f(x), u \rangle|}{\max\{1, |D_h f(x)[u]|, |\langle \nabla f(x), u \rangle|\}}.$$

If errors are large, reduce problem size, inspect shapes, and verify each component operation.

Remark 12.15.3 (Step Size Choice) Too large h yields truncation error; too small h yields catastrophic cancellation. A typical stable range in double precision is $h \in [10^{-6}, 10^{-4}]$, but it depends on scale of x and f.

12.15.2 *Numerical Stability in Optimization*

12.15.2.1 Exploding/Vanishing Gradients and Rescaling

Exploding/vanishing gradients arise when gradients are products of many Jacobians (deep nets, RNNs). If typical singular values are >1, gradients explode; if <1, they vanish. Rescaling strategies target this directly: variance-preserving initialization, normalization layers, residual connections, and careful activation choices. In classical convex problems, analogous issues appear as ill-conditioning and poor scaling: gradients become extremely anisotropic and require tiny learning rates. The unifying concept is conditioning.

Remark 12.15.4 (A Diagnostic Scalar) Track $\|\nabla f(w_t)\|_2$ across iterations:

- rapidly growing norms suggest exploding gradients or too-large learning rate,
- near-zero norms early suggest vanishing gradients, saturation, or dead activations,
- oscillatory norms suggest step size too large relative to curvature.

Remark 12.15.5 (Rescaling Remedies)

- standardize/whiten inputs; normalize intermediate activations (BatchNorm/LayerNorm),
- use residual connections; choose non-saturating activations where appropriate,
- reduce learning rate or use warmup; add weight decay to control parameter growth,
- use gradient clipping to cap rare explosions.

12.15.2.2 Clipping, Normalization, and Safe Computation

Stability failures often come from arithmetic: overflow in exp, division by tiny numbers, subtracting nearly equal values. Safe computation rewrites expressions (log-sum-exp trick), adds stabilizers (ϵ), and uses clipping or normalization. Clipping gradients (global norm or per-parameter) prevents rare spikes from destroying the iterate. Normalization controls scale drift and improves conditioning. These are engineering manifestations of geometric control on step sizes and curvature.

Remark 12.15.6 (Log-Sum-Exp Trick) To compute $\log \sum_i e^{s_i}$ stably, use

$$\log \sum_{i=1}^{K} e^{s_i} = m + \log \sum_{i=1}^{K} e^{s_i - m}, \qquad m := \max_i s_i,$$

which prevents overflow when some s_i are large.

Remark 12.15.7 (Gradient Clipping (Global Norm)) Given gradient g, clip to threshold $\tau > 0$ by

$$g_{\text{clip}} = \begin{cases} g, & \|g\|_2 \leq \tau, \\ \tau \dfrac{g}{\|g\|_2}, & \|g\|_2 > \tau. \end{cases}$$

This caps step magnitude in Euclidean geometry and can prevent catastrophic updates.

Remark 12.15.8 (ϵ-Stabilizers) In adaptive methods, denominators like $\sqrt{v_t}$ can be tiny; add $\epsilon > 0$ to avoid division blow-up: $g/(\sqrt{v_t} + \epsilon)$. Similarly, when normalizing by variance, add ϵ in the square root.

12.15.3 Practical Diagnostics

12.15.3.1 Loss Curves, Gradient Norms, and Step-Size Issues

Optimization debugging is largely about reading curves: loss vs. iteration, gradient norms vs. iteration, parameter norms, and learning rates. Each failure mode has a signature: divergence indicates too-large steps or numerical overflow; plateaus can indicate vanishing gradients, poor conditioning, or overly small learning rate; noisy loss indicates too-small batch or too-large step; sudden NaNs indicate instability. The goal is to isolate whether the problem is (i) objective/implementation, (ii) step size/schedule, (iii) conditioning/scaling, or (iv) data issues.

Remark 12.15.9 (Canonical Curve Signatures)

- **Loss increases immediately:** learning rate too high, sign error, or wrong objective.
- **Loss decreases then diverges:** step size too aggressive as curvature changes; add decay or line search.
- **Loss flat from start:** vanishing gradients, saturated activations, dead units, or data/label mismatch.
- **Loss highly noisy:** gradient variance high; increase batch size, reduce learning rate, or add momentum/averaging.

Remark 12.15.10 (Minimal Instrumentation) Log: training loss, validation loss, learning rate, $\|\nabla f\|_2$, parameter norm $\|w\|_2$, and (if applicable) gradient clipping fraction. These few signals usually locate the issue quickly.

12.15.3.2 Overfitting Signals and Early Stopping Protocols

Overfitting is diagnosed by divergence of training and validation curves: training continues to improve while validation degrades. Early stopping selects a stopping time based on validation, acting as regularization; but it is also hyperparameter selection and can itself overfit validation. A robust protocol uses patience, smoothing, and a clean outer test set or nested validation for final reporting. In ML, early stopping interacts with learning-rate schedules, so one should treat it as part of the optimization configuration.

Remark 12.15.11 (Early Stopping Protocol (Practical Template)) Monitor validation loss every E steps, keep the best checkpoint, and stop if validation does not improve for P checks (patience). Report the chosen checkpoint rule and ensure the test set is used only once at the end.

Remark 12.15.12 (Selection Bias Warning) If you tune learning rate, architecture, and stopping time on the same validation set repeatedly, you effectively perform multiple comparisons. For reliable reporting, use nested CV or a final untouched test set.

12.16 Exercises

Unless stated otherwise, all norms are Euclidean and all logarithms are natural.

Exercise 12.16.1 (Lipschitz Gradient Constant and Safe Step Size) Let $f(x) = \frac{1}{2}x^\top Hx - b^\top x$ on $\mathbb{R}^2$ with

$$H = \begin{bmatrix} 6 & 2 \\ 2 & 4 \end{bmatrix}, \qquad b = \begin{bmatrix} 2 \\ 1 \end{bmatrix}.$$

(i) Compute $\nabla f(x)$ and $\nabla^2 f(x)$.
(ii) Compute the eigenvalues of H and hence determine the smallest L such that f is L-smooth and the largest μ such that f is μ-strongly convex.
(iii) Compute the condition number $\kappa = L/\mu$ and interpret its optimization meaning.
(iv) For gradient descent with fixed step size η, determine the largest η that guarantees descent for all x (use the L-smooth bound).

Exercise 12.16.2 (Gradient Descent Iterates on a Quadratic) Consider the quadratic from Exercise 12.16.1. Let $x_0 = \begin{bmatrix} 0 \\ 0 \end{bmatrix}$ and choose $\eta = 1/L$ using your computed L.

(i) Compute x_1, x_2, x_3 explicitly.
(ii) Compute $f(x_t)$ for $t = 0, 1, 2, 3$ and verify monotone decrease.

(iii) Compute the exact minimizer $x^\star = H^{-1}b$ and report $\|x_3 - x^\star\|_2$.

Exercise 12.16.3 (Armijo Backtracking on a Concrete Function) Let $f(x) = \frac{1}{4}(x_1^4+x_2^4)+\frac{1}{2}(x_1^2+2x_2^2)$ on $\mathbb{R}^2$ and start at $x_0 = (1, -1)$. Take the steepest descent direction $p_0 = -\nabla f(x_0)$. Use Armijo backtracking with $c_1 = 10^{-4}$, $\beta = \frac{1}{2}$, and initial step $\alpha_0 = 1$: repeatedly halve α until Armijo holds. Compute $\nabla f(x_0)$, evaluate $f(x_0+\alpha p_0)$ for $\alpha \in \{1, \frac{1}{2}, \frac{1}{4}, \dots\}$ until acceptance, and report the accepted α and the new point $x_1 = x_0 + \alpha p_0$.

Exercise 12.16.4 (Trust-Region Step for a Quadratic Model) Let $f(x) = \frac{1}{2}x^\top Hx - b^\top x$ in $\mathbb{R}^2$ with

$$H = \begin{bmatrix} 10 & 0 \\ 0 & 1 \end{bmatrix}, \qquad b = \begin{bmatrix} 1 \\ 1 \end{bmatrix}, \qquad x_0 = \begin{bmatrix} 0 \\ 0 \end{bmatrix}.$$

Consider the trust-region subproblem at x_0 with radius $\Delta = 0.3$ using the quadratic model (which is exact here).

(i) Compute the unconstrained Newton step $p_N = -H^{-1}\nabla f(x_0)$ and check whether it satisfies $\|p_N\|_2 \le \Delta$.
(ii) If not feasible, compute the constrained minimizer of $m(\Delta x) = f(x_0) + g^\top \Delta x + \frac{1}{2}\Delta x^\top H \Delta x$ subject to $\|\Delta x\|_2 \le \Delta$, by solving $(H + \lambda I)\Delta x = -g$ for $\lambda \ge 0$ such that $\|\Delta x\|_2 = \Delta$. Report λ and Δx numerically to 4 decimals.

Exercise 12.16.5 (Projected Gradient Descent on a Simplex) Let $f(p) = \frac{1}{2}\|Ap - b\|_2^2$ with

$$A = \begin{bmatrix} 1 & 2 & 0 \\ 0 & 1 & 1 \end{bmatrix}, \qquad b = \begin{bmatrix} 1 \\ 1 \end{bmatrix},$$

and constrain p to the simplex $\Delta^2 = \{p \in \mathbb{R}^3 : p_i \ge 0, \sum_i p_i = 1\}$. Start at $p_0 = (1/3, 1/3, 1/3)$ and take step size $\eta = 0.5$. Perform two projected-gradient iterations

$$p_{t+1} = \Pi_{\Delta^2}\big(p_t - \eta \nabla f(p_t)\big),$$

computing $\nabla f(p_t)$ explicitly and projecting onto Δ^2 via KKT (sorting-thresholding). Report p_1 and p_2.

Exercise 12.16.6 (Proximal Gradient (ISTA) for Lasso: One Full Iteration) Let $f(w) = \frac{1}{2}\|Xw - y\|_2^2$ and $g(w) = \lambda\|w\|_1$ with

$$X = \begin{bmatrix} 1 & 0 \\ 1 & 1 \\ 1 & 2 \end{bmatrix}, \qquad y = \begin{bmatrix} 1 \\ 2 \\ 2 \end{bmatrix}, \qquad \lambda = 0.4.$$

Start at $w_0 = (0, 0)$.

(i) Compute $\nabla f(w_0) = X^\top (X w_0 - y)$.
(ii) Use step size $\eta = 0.25$ and compute the ISTA update

$$w_1 = \text{prox}_{\eta g}\big(w_0 - \eta \nabla f(w_0)\big)$$

using soft-thresholding.
(iii) Compute the objective values $F(w_0) = f(w_0) + g(w_0)$ and $F(w_1)$ and verify decrease.

Exercise 12.16.7 (ADMM for Lasso: Two Iterations by Hand) Solve the lasso problem from Exercise 12.16.6 via ADMM splitting $\min_x \frac{1}{2}\|Xx - y\|_2^2 + \lambda\|z\|_1$ subject to $x = z$. Use $\rho = 1$, initialize $x^0 = z^0 = u^0 = 0$ (scaled dual u), and perform two ADMM iterations:

$$x^{t+1} = \arg\min_x \frac{1}{2}\|Xx - y\|_2^2 + \frac{\rho}{2}\|x - z^t + u^t\|_2^2,$$

$$z^{t+1} = \text{prox}_{\lambda/\rho\|\cdot\|_1}(x^{t+1} + u^t), \qquad u^{t+1} = u^t + x^{t+1} - z^{t+1}.$$

Compute x^1, z^1, u^1 and x^2, z^2, u^2 explicitly (solve the 2×2 linear system for x each time). Also compute primal residual $\|r^{t+1}\|_2 = \|x^{t+1} - z^{t+1}\|_2$ and dual residual $\|s^{t+1}\|_2 = \rho\|z^{t+1} - z^t\|_2$ for $t = 0, 1$.

Exercise 12.16.8 (Stochastic Gradient Unbiasedness and Variance under Mini-Batching) Consider a finite-sum objective $f(w) = \frac{1}{n}\sum_{i=1}^n \ell_i(w)$ with gradients $g_i = \nabla \ell_i(w)$ at a fixed w. Assume $\frac{1}{n}\sum_{i=1}^n g_i = \nabla f(w) = \mu$ and define empirical second moment

$$S := \frac{1}{n}\sum_{i=1}^n \|g_i - \mu\|_2^2.$$

(i) If i is sampled uniformly from $[n]$, show $\mathbb{E}[g_i] = \mu$ and compute $\mathbb{E}\|g_i - \mu\|_2^2 = S$.
(ii) If a mini-batch B of size b is sampled *with replacement* and $g_B = \frac{1}{b}\sum_{i\in B} g_i$, show $\mathbb{E}[g_B] = \mu$ and compute $\mathbb{E}\|g_B - \mu\|_2^2 = S/b$.
(iii) Evaluate S numerically for the following gradients in $\mathbb{R}^2$ (at a fixed w):

$$g_1 = \begin{bmatrix} 2 \\ 0 \end{bmatrix}, \quad g_2 = \begin{bmatrix} 0 \\ 2 \end{bmatrix}, \quad g_3 = \begin{bmatrix} 2 \\ 2 \end{bmatrix}, \quad g_4 = \begin{bmatrix} 0 \\ 0 \end{bmatrix},$$

and then compute the variance for $b = 2$.

Exercise 12.16.9 (AdaGrad vs. SGD: Coordinatewise Geometry) Let gradients in $\mathbb{R}^3$ be

$$g_0 = (3, 0, 1), \quad g_1 = (3, 0, 1), \quad g_2 = (0, 4, 1)$$

(treated as column vectors). Start from $w_0 = (0, 0, 0)$, choose $\eta = 1$ and $\epsilon = 0$ for simplicity.

(i) Run three steps of vanilla SGD: $w_{t+1} = w_t - \eta g_t$ and report w_3.
(ii) Run three steps of AdaGrad with accumulator $G_t = \sum_{s=0}^{t} g_s \odot g_s$ and update $w_{t+1} = w_t - \eta\, g_t / \sqrt{G_t}$ (elementwise). Report w_3.
(iii) Explain in one paragraph why AdaGrad takes smaller steps in frequently updated coordinates and larger steps in rare coordinates, using your computed iterates.

Exercise 12.16.10 (Hessian-Vector Product by Finite Differences) Let $f(x) = \log(1 + e^{a^\top x}) + \frac{1}{2}\|x\|_2^2$ with $a = (1, 2)^\top$ and $x = (0.2, -0.1)^\top$.

(i) Compute $\nabla f(x)$ analytically.
(ii) For $v = (1, -1)^\top$, approximate the Hessian-vector product $Hv = \nabla^2 f(x)v$ using the finite-difference formula

$$\nabla^2 f(x)v \approx \frac{\nabla f(x + hv) - \nabla f(x - hv)}{2h},$$

with $h = 10^{-4}$.
(iii) Compute Hv analytically by deriving $\nabla^2 f(x)$ and compare the two results numerically.

Exercise 12.16.11 (Newton Step and Damping on Logistic Loss) Consider logistic regression with parameter $w \in \mathbb{R}^2$ and data

$$X = \begin{bmatrix} 1 & 0 \\ 1 & 1 \\ 1 & 2 \\ 1 & 3 \end{bmatrix}, \qquad y = \begin{bmatrix} 0 \\ 0 \\ 1 \\ 1 \end{bmatrix}.$$

Model $p_i = \sigma(x_i^\top w)$ and NLL

$$f(w) = -\sum_{i=1}^{4} \big(y_i \log p_i + (1 - y_i)\log(1 - p_i)\big).$$

Start at $w_0 = (0, 0)$.

(i) Compute gradient $g_0 = \nabla f(w_0)$ and Hessian $H_0 = \nabla^2 f(w_0)$ using $H = X^\top W X$ with $W_{ii} = p_i(1 - p_i)$.

(ii) Compute the Newton direction $p_N = -H_0^{-1} g_0$.
(iii) Apply a damped step $w_1 = w_0 + \eta p_N$ with $\eta = 0.5$ and compute $f(w_0)$ and $f(w_1)$ numerically (to 4 decimals) to verify decrease.

Exercise 12.16.12 (Coordinate Descent for Lasso: One Coordinate Update) Consider the lasso objective $F(w) = \frac{1}{2}\|Xw - y\|_2^2 + \lambda\|w\|_1$ with

$$X = \begin{bmatrix} 1 & 0 & 0 \\ 1 & 1 & 0 \\ 1 & 0 & 1 \\ 1 & 1 & 1 \end{bmatrix}, \qquad y = \begin{bmatrix} 1 \\ 2 \\ 2 \\ 3 \end{bmatrix}, \qquad \lambda = 0.5.$$

Initialize $w^{(0)} = (0, 0, 0)$. Perform a single coordinate descent update for coordinate $j = 2$ (holding w_1, w_3 fixed at 0): derive the one-dimensional objective in w_2, compute the optimal update using soft-thresholding of the partial residual correlation, and report the updated vector $w^{(1)}$.

Exercise 12.16.13 (Momentum vs. GD on a Badly Conditioned Quadratic) Let $f(x) = \frac{1}{2}x^\top Hx$ with

$$H = \begin{bmatrix} 100 & 0 \\ 0 & 1 \end{bmatrix}, \qquad x_0 = (1, 1), \quad x_{-1} = x_0.$$

(i) Run 5 steps of gradient descent with $\eta = 0.01$ and report x_5.
(ii) Run 5 steps of heavy-ball momentum with the same $\eta = 0.01$ and $\beta = 0.9$ (use $x_{-1} = x_0$), and report x_5.
(iii) Compare the progress in each coordinate and explain (one paragraph) how momentum reduces zig-zagging in high-curvature directions.

Exercise 12.16.14 (Gradient Checking by Random Directions) Let $f(w) = \frac{1}{2}\|Aw - b\|_2^2 + \frac{1}{4}\|w\|_2^4$ with

$$A = \begin{bmatrix} 1 & 2 \\ 0 & -1 \\ 3 & 1 \end{bmatrix}, \qquad b = \begin{bmatrix} 1 \\ 0 \\ -1 \end{bmatrix}, \qquad w = (0.3, -0.2).$$

(i) Compute $\nabla f(w)$ analytically.
(ii) Let $u = \frac{1}{\sqrt{5}}(1, 2)$ and compute the directional derivative approximation

$$D_h f(w)[u] = \frac{f(w + hu) - f(w - hu)}{2h}$$

for $h = 10^{-5}$.
(iii) Compute $\langle \nabla f(w), u \rangle$ and report the relative error between the finite-difference and gradient inner product.

Appendix A
Linear Algebra Reference

A.1 Matrix Identities Used Throughout the Book

A.1.1 Trace, Transpose, and Frobenius Identities

A.1.1.1 $\mathrm{tr}(AB) = \mathrm{tr}(BA)$ and Consequences

Trace identities are the main "algebraic compressor" of the book: they let us rewrite long coordinate sums as compact expressions that are invariant under cyclic reordering. This is indispensable in (i) spectral arguments (trace as sum of eigenvalues), (ii) matrix calculus (move factors around to isolate differentials), and (iii) probabilistic identities (trace of covariance). The guiding rule is *cyclicity, not commutativity*: you may rotate factors inside $\mathrm{tr}(\cdot)$, but you may not arbitrarily swap them unless dimensions permit and the cyclic form is respected.

Definition A.1.1 (Trace) For a square matrix $A \in \mathbb{R}^{d\times d}$,

$$\mathrm{tr}(A) := \sum_{i=1}^{d} A_{ii}.$$

Proposition A.1.1 (Basic Trace Properties) *For $A, B \in \mathbb{R}^{d\times d}$ and $\alpha \in \mathbb{R}$,*

$$\mathrm{tr}(A + B) = \mathrm{tr}(A) + \mathrm{tr}(B), \qquad \mathrm{tr}(\alpha A) = \alpha\, \mathrm{tr}(A), \qquad \mathrm{tr}(A^\top) = \mathrm{tr}(A).$$

Theorem A.1.1 (Cyclicity of Trace) *Let $A \in \mathbb{R}^{m\times n}$ and $B \in \mathbb{R}^{n\times m}$. Then*

$$\mathrm{tr}(AB) = \mathrm{tr}(BA).$$

P. Singh, B. Raman, *Foundations of Machine Learning and AI*, Studies in Big Data 199, https://doi.org/10.1007/978-3-032-30336-3

More generally, if $A_1, \ldots, A_k$ are matrices such that the product $A_1 A_2 \cdots A_k$ is square (hence traceable), then

$$\operatorname{tr}(A_1 A_2 \cdots A_k) = \operatorname{tr}(A_2 \cdots A_k A_1) = \cdots = \operatorname{tr}(A_k A_1 \cdots A_{k-1}).$$

Proof For the two-factor case, write $AB \in \mathbb{R}^{m \times m}$ and compute

$$\operatorname{tr}(AB) = \sum_{i=1}^{m} (AB)_{ii} = \sum_{i=1}^{m} \sum_{j=1}^{n} A_{ij} B_{ji} = \sum_{j=1}^{n} \sum_{i=1}^{m} B_{ji} A_{ij} = \sum_{j=1}^{n} (BA)_{jj} = \operatorname{tr}(BA).$$

The k-factor statement follows by repeated application to $(A_1)(A_2 \cdots A_k)$. □

Cyclicity in Longer Products

Theorem A.1.1 implies that in any traceable product one may *rotate* factors without changing the trace. A frequently used special case is

$$\operatorname{tr}(ABC) = \operatorname{tr}(BCA) = \operatorname{tr}(CAB),$$

provided the products are defined (e.g., $A \in \mathbb{R}^{m \times n}$, $B \in \mathbb{R}^{n \times p}$, $C \in \mathbb{R}^{p \times m}$). This permits "moving" a matrix next to the term you want to differentiate or simplify.

Remark A.1.1 (Caution: Cyclicity Is Not Commutativity) Even though $\operatorname{tr}(AB) = \operatorname{tr}(BA)$, it is generally false that $AB = BA$. Also, you cannot reorder factors arbitrarily: $\operatorname{tr}(ABCD)$ equals $\operatorname{tr}(BCDA)$ but not necessarily $\operatorname{tr}(ACBD)$.

Trace Invariance under Similarity

Similarity transforms correspond to changing coordinates of a linear operator. Trace is invariant under similarity, hence is a coordinate-free quantity: it equals the sum of eigenvalues (with multiplicity).

Proposition A.1.2 (Trace Invariance under Similarity) *Let $A \in \mathbb{R}^{d \times d}$ and let $S \in \mathbb{R}^{d \times d}$ be invertible. Then*

$$\operatorname{tr}(S^{-1} A S) = \operatorname{tr}(A).$$

In particular, if A and B are similar, then $\operatorname{tr}(A) = \operatorname{tr}(B)$.

Proof By cyclicity,

$$\operatorname{tr}(S^{-1} A S) = \operatorname{tr}(A S S^{-1}) = \operatorname{tr}(A).$$

□

Remark A.1.2 (Spectral Meaning) If A is diagonalizable (in particular if A is symmetric), then $\mathrm{tr}(A) = \sum_{i=1}^{d} \lambda_i(A)$. This is used throughout the book in PCA (sum of variances) and in effective-dimension quantities.

Quadratic Forms as Traces: $x^\top Ax = \mathrm{tr}(Axx^\top)$

Quadratic forms can be rewritten as traces, enabling clean matrix calculus and geometric identities.

Proposition A.1.3 (Quadratic Form Trace Identity) *For $A \in \mathbb{R}^{d\times d}$ and $x \in \mathbb{R}^d$,*

$$x^\top Ax = \mathrm{tr}(x^\top Ax) = \mathrm{tr}(Axx^\top).$$

More generally, for $X \in \mathbb{R}^{d\times n}$,

$$\mathrm{tr}(X^\top AX) = \sum_{j=1}^{n} x_j^\top Ax_j,$$

where x_j are the columns of X.

Proof The scalar $x^\top Ax$ equals its trace. Then, by cyclicity,

$$\mathrm{tr}(x^\top Ax) = \mathrm{tr}(Axx^\top).$$

The matrix form follows by writing $X = [x_1 \ \cdots \ x_n]$ and expanding $X^\top AX$ entrywise. □

Remark A.1.3 (A Ubiquitous Pattern) Whenever you see $\sum_i x_i^\top Ax_i$, try rewriting it as $\mathrm{tr}(X^\top AX)$. This conversion is the starting point for many derivations (e.g., covariance identities and ridge objectives).

A.1.1.2 $\langle A, B\rangle_F = \mathrm{tr}(A^\top B)$

The Frobenius inner product turns matrices into a Euclidean space: it makes $\mathbb{R}^{m\times n}$ a Hilbert space. This allows us to speak about angles, orthogonality, and projections of matrices exactly as we do for vectors. The identity $\langle A, B\rangle_F = \mathrm{tr}(A^\top B)$ is the bridge between entrywise sums and trace algebra, and it is the standard tool for deriving gradients and proving orthogonal invariance properties.

Definition A.1.2 (Frobenius Inner Product and Frobenius Norm) For $A, B \in \mathbb{R}^{m\times n}$, define

$$\langle A, B\rangle_F := \sum_{i=1}^{m}\sum_{j=1}^{n} A_{ij}B_{ij}.$$

The induced norm is

$$\|A\|_F := \sqrt{\langle A, A\rangle_F}.$$

Proposition A.1.4 (Trace Form of the Frobenius Inner Product) *For $A, B \in \mathbb{R}^{m\times n}$,*

$$\langle A, B\rangle_F = \operatorname{tr}(A^\top B).$$

Proof Compute

$$\operatorname{tr}(A^\top B) = \sum_{j=1}^{n}(A^\top B)_{jj} = \sum_{j=1}^{n}\sum_{i=1}^{m}(A^\top)_{ji}B_{ij} = \sum_{j=1}^{n}\sum_{i=1}^{m}A_{ij}B_{ij} = \langle A, B\rangle_F.$$

□

Frobenius Norm and Energy: $\|A\|_F^2 = \operatorname{tr}(A^\top A)$

Setting $B = A$ in Proposition A.1.4 yields

$$\|A\|_F^2 = \langle A, A\rangle_F = \operatorname{tr}(A^\top A).$$

If A has singular values $\sigma_1, \dots, \sigma_r$, then $\|A\|_F^2 = \sum_{i=1}^r \sigma_i^2$ (spectral energy identity).

Orthogonal Invariance: $\|QA\|_F = \|A\|_F$ and $\|AQ\|_F = \|A\|_F$

Orthogonal transformations preserve Frobenius norm.

Proposition A.1.5 (Orthogonal Invariance of Frobenius Norm) *Let $A \in \mathbb{R}^{m\times n}$. If $Q \in \mathbb{R}^{m\times m}$ and $P \in \mathbb{R}^{n\times n}$ are orthogonal, then*

$$\|QA\|_F = \|A\|_F, \qquad \|AP\|_F = \|A\|_F, \qquad \|QAP\|_F = \|A\|_F.$$

Proof Using trace form,

$$\|QA\|_F^2 = \operatorname{tr}\big((QA)^\top(QA)\big) = \operatorname{tr}(A^\top Q^\top QA) = \operatorname{tr}(A^\top A) = \|A\|_F^2.$$

Similarly,

$$\|AP\|_F^2 = \operatorname{tr}\big((AP)^\top(AP)\big) = \operatorname{tr}(P^\top A^\top AP) = \operatorname{tr}(A^\top APP^\top) = \operatorname{tr}(A^\top A).$$

□

Projection Identities via Frobenius Inner Products

The Frobenius inner product lets us treat matrix approximation as projection in a Euclidean space. A central application is: the best approximation in Frobenius norm over a subspace is obtained by orthogonal projection.

Proposition A.1.6 (Orthogonal Projection in Frobenius Geometry) *Let $\mathcal{S} \subseteq \mathbb{R}^{m\times n}$ be a linear subspace and let $\Pi_{\mathcal{S}}$ denote orthogonal projection with respect to $\langle\cdot,\cdot\rangle_F$. Then for any $A \in \mathbb{R}^{m\times n}$,*

$$\Pi_{\mathcal{S}}(A) \in \arg\min_{B\in\mathcal{S}} \|A - B\|_F,$$

and the residual $R := A - \Pi_{\mathcal{S}}(A)$ satisfies

$$\langle R, B\rangle_F = 0 \qquad \forall B \in \mathcal{S}.$$

Consequently,

$$\|A\|_F^2 = \|\Pi_{\mathcal{S}}(A)\|_F^2 + \|A - \Pi_{\mathcal{S}}(A)\|_F^2.$$

Remark A.1.4 (Low-Rank Approximation as a Projection) The set of rank-k matrices is not a subspace, but the Eckart–Young theorem shows that truncated SVD behaves like an optimal projection onto the *nonlinear* manifold of rank-k matrices under $\|\cdot\|_F$. The Pythagorean-style decomposition above is still a guiding intuition.

A.1.2 Inverse/Update Identities

Inverse-update identities are indispensable in ML because many problems require solving or differentiating systems that change by *low-rank* or *diagonal* modifications: adding ridge λI, adding a few new samples, updating a Gaussian posterior, or adjusting a kernel matrix. Forming an inverse explicitly is both numerically and computationally suboptimal; instead, we use identities that reduce a large inverse to a smaller inverse (often $k \times k$), and we compute log-determinants via stable factorizations. The guiding principle is: *rewrite first, factor/solve second.*

A.1.2.1 Woodbury Identity and Fast Rank-*k* Updates

Woodbury is the master identity for inverting a matrix plus a low-rank update. It explains the primal/dual equivalence in ridge regression and Gaussian processes: depending on whether $n \ll d$ or $d \ll n$, one can invert a smaller matrix. It also provides an exact and efficient update rule when a covariance/precision matrix is modified by a low-rank term. Numerically, Woodbury is safe when used to *avoid* explicit inverses: one should implement it with solves (Cholesky/QR) rather than by forming A^{-1}.

Theorem A.1.2 (Woodbury Matrix Identity) *Let $A \in \mathbb{R}^{d \times d}$ be invertible, $U \in \mathbb{R}^{d \times k}$, $V \in \mathbb{R}^{k \times d}$, and $C \in \mathbb{R}^{k \times k}$ be invertible. Then*

$$(A + UCV)^{-1} = A^{-1} - A^{-1}U\big(C^{-1} + VA^{-1}U\big)^{-1}VA^{-1}.$$

Proof Let $M := A + UCV$ and define

$$B := A^{-1} - A^{-1}U\big(C^{-1} + VA^{-1}U\big)^{-1}VA^{-1}.$$

We verify $MB = I$. First expand:

$$MB = (A + UCV)A^{-1} - (A + UCV)A^{-1}U\big(C^{-1} + VA^{-1}U\big)^{-1}VA^{-1}.$$

Since $(A + UCV)A^{-1} = I + UCVA^{-1}$ and $(A + UCV)A^{-1}U = U + UCVA^{-1}U = U\big(I + CVA^{-1}U\big)$, we get

$$MB = I + UCVA^{-1} - U\big(I + CVA^{-1}U\big)\big(C^{-1} + VA^{-1}U\big)^{-1}VA^{-1}.$$

Use the identity $I + CVA^{-1}U = C\big(C^{-1} + VA^{-1}U\big)$ to obtain

$$\big(I + CVA^{-1}U\big)\big(C^{-1} + VA^{-1}U\big)^{-1} = C.$$

Hence the second term becomes $UCVA^{-1}$, canceling $UCVA^{-1}$, leaving $MB = I$. □

Sherman–Morrison Rank-One Special Case

Taking $k = 1$ with $U = u \in \mathbb{R}^d$, $V = v^\top \in \mathbb{R}^{1 \times d}$, and $C = [1]$ yields:

Corollary A.1.1 (Sherman–Morrison) *Let $A \in \mathbb{R}^{d \times d}$ be invertible and $u, v \in \mathbb{R}^d$. If $1 + v^\top A^{-1}u \neq 0$, then*

$$(A + uv^\top)^{-1} = A^{-1} - \frac{A^{-1}uv^\top A^{-1}}{1 + v^\top A^{-1}u}.$$

Remark A.1.5 (When the Update Becomes Singular) If $1 + v^\top A^{-1}u = 0$, then $A + uv^\top$ is singular. This provides a fast singularity test for rank-one updates.

Ridge Dual/Primal Swap: $(X^\top X + \lambda I)^{-1} X^\top = X^\top (XX^\top + \lambda I)^{-1}$

This identity lets you compute ridge predictions either in the *primal* $(d \times d)$ or *dual* $(n \times n)$ space. It is an immediate consequence of Woodbury (or can be verified by multiplication).

Proposition A.1.7 (Primal–Dual Ridge Identity) *Let* $X \in \mathbb{R}^{n \times d}$ *and* $\lambda > 0$. *Then*

$$(X^\top X + \lambda I_d)^{-1} X^\top = X^\top (XX^\top + \lambda I_n)^{-1}.$$

Proof Consider the Woodbury identity with $A = \lambda I_d$, $U = X^\top$, $C = I_n$, $V = X$:

$$(\lambda I_d + X^\top X)^{-1} = \lambda^{-1} I_d - \lambda^{-1} X^\top \left(I_n + X\lambda^{-1} X^\top\right)^{-1} X\lambda^{-1}.$$

Right-multiply by $X^\top$ and simplify:

$$(\lambda I_d + X^\top X)^{-1} X^\top = \lambda^{-1} X^\top - \lambda^{-1} X^\top \left(I_n + \lambda^{-1} XX^\top\right)^{-1} \lambda^{-1} XX^\top.$$

Factor $\lambda^{-1} X^\top$ on the left:

$$(\lambda I_d + X^\top X)^{-1} X^\top = \lambda^{-1} X^\top \left[I_n - \left(I_n + \lambda^{-1} XX^\top\right)^{-1} \lambda^{-1} XX^\top\right].$$

Use $I - (I + M)^{-1} M = (I + M)^{-1}$ (check by multiplying by $I + M$) with $M = \lambda^{-1} XX^\top$ to get

$$(\lambda I_d + X^\top X)^{-1} X^\top = \lambda^{-1} X^\top \left(I_n + \lambda^{-1} XX^\top\right)^{-1} = X^\top (XX^\top + \lambda I_n)^{-1}.$$

□

Remark A.1.6 (Computational Rule of Thumb) If $d \ll n$, solve in the primal via $d \times d$ systems. If $n \ll d$, solve in the dual via $n \times n$ systems. The identity above guarantees both produce the same predictions in exact arithmetic.

Stable Update Patterns and Common Pitfalls
Stable Patterns (Recommended)

- **Avoid explicit inverses.** Implement Woodbury via linear solves: compute $A^{-1}U$ by solving $AZ = U$ (e.g., with LU/Cholesky), and compute VA^{-1} similarly if needed.
- **Factor once, reuse many times.** If A is fixed across updates, cache its factorization and reuse it for repeated solves.
- **Exploit SPD structure.** If $A \succ 0$ and $C \succ 0$, use Cholesky for both A and the small matrix $C^{-1} + VA^{-1}U$.

Pitfalls (Avoid)

- Forming A^{-1} explicitly and then multiplying by U (slow and unstable).
- Using Woodbury when $C^{-1} + VA^{-1}U$ is ill-conditioned without regularization (can magnify noise).
- Forgetting symmetry/PSD structure: for SPD updates prefer symmetric forms (e.g., $A + UU^\top$) and symmetric solvers.

A.1.2.2 Matrix Determinant Lemma and Log-Determinants

Determinants and log-determinants appear throughout ML: Gaussian log-likelihoods, Bayesian evidence, kernel ridge/GP marginal likelihoods, and information measures. Direct determinant computation is numerically fragile; the correct practice is to compute $\log\det(\cdot)$ via stable factorizations (Cholesky for SPD). The determinant lemma complements Woodbury: it reduces a large determinant under a low-rank update to a smaller determinant, often turning an $d \times d$ log-det into a $k \times k$ log-det.

Theorem A.1.3 (Matrix Determinant Lemma) *Let $A \in \mathbb{R}^{d\times d}$ be invertible, $U \in \mathbb{R}^{d\times k}$, $V \in \mathbb{R}^{k\times d}$, and $C \in \mathbb{R}^{k\times k}$ be invertible. Then*

$$\det(A + UCV) = \det(C^{-1} + VA^{-1}U)\ \det(C)\ \det(A).$$

In particular, if $C = I_k$,

$$\det(A + UV) = \det(I_k + VA^{-1}U)\ \det(A).$$

Proof Use the identity $\det(I + PQ) = \det(I + QP)$ (valid for compatible rectangular P, Q), with $P = A^{-1}UC$ and $Q = V$. Then

$$\begin{aligned}\det(A + UCV) &= \det(A)\det(I + A^{-1}UCV) = \det(A)\det(I + PQ)\\ &= \det(A)\det(I + QP).\end{aligned}$$

Now

$$I + QP = I + VA^{-1}UC = C\big(C^{-1} + VA^{-1}U\big),$$

so $\det(I + QP) = \det(C)\det(C^{-1} + VA^{-1}U)$, yielding the claim. □

Log-Det Decomposition for Gaussian Objectives

Taking logs in Theorem A.1.3 gives the decomposition

$$\log\det(A + UCV) = \log\det(A) + \log\det(C) + \log\det\big(C^{-1} + VA^{-1}U\big),$$

which is the standard route to reducing Gaussian log-determinants with low-rank structure. A common special case is $A = \lambda I$:

$$\log\det(\lambda I + X^\top X) = d\log\lambda + \log\det\left(I + \frac{1}{\lambda}XX^\top\right),$$

reducing a $d \times d$ log-det to an $n \times n$ log-det (or vice versa) depending on which is smaller.

Determinants under Triangular Factors: $\det(L) = \prod_i L_{ii}$

Proposition A.1.8 (Determinant of a Triangular Matrix) *If $T \in \mathbb{R}^{d\times d}$ is triangular (upper or lower), then*

$$\det(T) = \prod_{i=1}^{d} T_{ii}.$$

Remark A.1.7 (Immediate Corollaries) If $A = LU$ (LU factorization with L unit lower triangular), then $\det(A) = \det(L)\det(U) = \prod_i U_{ii}$. If $A = LL^\top$ (Cholesky, SPD), then $\det(A) = \det(L)^2 = \big(\prod_i L_{ii}\big)^2$.

Numerically Stable Log-Det Computation via Cholesky

For SPD $A \succ 0$, Cholesky $A = LL^\top$ yields

$$\log\det(A) = 2\sum_{i=1}^{d}\log L_{ii}.$$

This is numerically stable because it avoids forming the determinant directly (which can under/overflow) and uses only diagonal entries of the Cholesky factor.

Remark A.1.8 (Practical Stability Checklist)

- Use Cholesky for $A \succ 0$; if Cholesky fails, add jitter $A \leftarrow A + \lambda I$.
- Never compute $\det(A)$ and then take log in floating point; compute $\log\det(A)$ via factors.

- If using determinant lemma, still compute small log-dets via stable factorizations (e.g., Cholesky of the $k \times k$ matrix).

A.1.3 Kronecker, Vec, and Reshaping

Vectorization and Kronecker products provide a disciplined way to turn "matrix unknown" problems into standard linear algebra. They are essential whenever we (i) solve for a matrix parameter (matrix least squares), (ii) model matrix-valued random variables (separable covariances, matrix normal distributions), or (iii) reason about reshaping and parameter sharing in neural layers. Block-matrix identities and Schur complements complement this toolkit by enabling modular elimination and revealing conditional structure (especially in Gaussian models and constrained optimization).

A.1.3.1 $\mathrm{vec}(AXB) = (B^\top \otimes A)\mathrm{vec}(X)$

The identity $\mathrm{vec}(AXB) = (B^\top \otimes A)\mathrm{vec}(X)$ is the single most important reshaping rule in the book. It converts a bilinear transformation in X into a linear transformation in $\mathrm{vec}(X)$ with a Kronecker-structured matrix. This allows one to reuse standard tools (least squares, conditioning, projections) for matrix-valued unknowns and to expose structure for efficient computation.

Definition A.1.3 (vec(·) (Column-Stacking)) For $X \in \mathbb{R}^{m\times n}$,

$$\mathrm{vec}(X) := \begin{bmatrix} X_{:1} \\ X_{:2} \\ \vdots \\ X_{:n} \end{bmatrix} \in \mathbb{R}^{mn}.$$

Thus X_{ij} appears at index $i + (j-1)m$ in $\mathrm{vec}(X)$.

Definition A.1.4 (Kronecker Product) For $A \in \mathbb{R}^{p\times q}$ and $B \in \mathbb{R}^{r\times s}$,

$$A \otimes B \in \mathbb{R}^{(pr)\times(qs)}$$

is the block matrix whose (i, j) block equals $a_{ij}B$.

Theorem A.1.4 (Vec–Kronecker Identity) *Let $A \in \mathbb{R}^{p\times m}$, $X \in \mathbb{R}^{m\times n}$, and $B \in \mathbb{R}^{n\times q}$. Then*

$$\mathrm{vec}(AXB) = (B^\top \otimes A)\,\mathrm{vec}(X).$$

Remark A.1.9 (Shape Sanity Check) $AXB \in \mathbb{R}^{p\times q}$, so $\operatorname{vec}(AXB) \in \mathbb{R}^{pq}$. Also $B^\top \otimes A \in \mathbb{R}^{(pq)\times(mn)}$ and $\operatorname{vec}(X) \in \mathbb{R}^{mn}$.

Commutation Matrix and $\operatorname{vec}(X^\top)$

Transposition corresponds to a fixed permutation of entries under vectorization.

Definition A.1.5 (Commutation Matrix) For integers $m, n \geq 1$, the *commutation matrix* $K_{m,n} \in \mathbb{R}^{mn\times mn}$ is the unique permutation matrix such that

$$K_{m,n} \operatorname{vec}(X) = \operatorname{vec}(X^\top) \qquad \forall X \in \mathbb{R}^{m\times n}.$$

Proposition A.1.9 (Basic Properties of the Commutation Matrix) *$K_{m,n}$ is orthogonal (a permutation matrix) and satisfies*

$$K_{m,n}^{-1} = K_{m,n}^\top = K_{n,m}.$$

Moreover, for compatible A, B,

$$K_{m,n}(A \otimes B)K_{p,q} = (B \otimes A)$$

when dimensions match (precise indexing follows from reshaping conventions).

Remark A.1.10 (Practical Use) $K_{m,n}$ is rarely formed explicitly. Its role is conceptual: it lets you reason about transposes under vec without index gymnastics. In implementations, $K_{m,n}\operatorname{vec}(X)$ corresponds to a reshape+transpose+reshape operation.

Kronecker Algebra: Transpose, Inverse, and Mixed-Product Rules

These rules are used repeatedly for simplifying vectorized systems and deriving fast updates.

Proposition A.1.10 (Core Kronecker Algebra) *Let A, B, C, D be matrices of compatible sizes.*

1. ***Transpose:*** $(A \otimes B)^\top = A^\top \otimes B^\top$.
2. ***Mixed-product:*** $(A \otimes B)(C \otimes D) = (AC) \otimes (BD)$.
3. ***Inverse:*** *If A and B are invertible, then* $(A \otimes B)^{-1} = A^{-1} \otimes B^{-1}$.
4. ***Spectral norm:*** $\|A \otimes B\|_2 = \|A\|_2 \, \|B\|_2$.
5. ***Determinant (square case):*** *If $A \in \mathbb{R}^{p\times p}$ and $B \in \mathbb{R}^{q\times q}$, then*

$$\det(A \otimes B) = \det(A)^q \, \det(B)^p.$$

Remark A.1.11 (Why Mixed-Product Matters) The mixed-product rule is the algebraic reason why separable operators can be applied efficiently and why vectorized forms remain structured. It also underlies Kronecker-factored curvature approximations (e.g., K-FAC) in optimization.

Separable Covariance: $\mathrm{Cov}(\mathrm{vec}(Z)) = \Sigma_c \otimes \Sigma_r$

Separable (Kronecker) covariance models are the standard structured model for matrix-valued random variables.

Definition A.1.6 (Separable (Kronecker) Covariance Model) Let $Z \in \mathbb{R}^{m\times n}$ be a random matrix (mean zero for simplicity). A *separable covariance* model assumes

$$\mathrm{Cov}(\mathrm{vec}(Z)) = \Sigma_c \otimes \Sigma_r,$$

where $\Sigma_r \in \mathbb{R}^{m\times m}$ (row covariance) and $\Sigma_c \in \mathbb{R}^{n\times n}$ (column covariance) are PSD.

Remark A.1.12 (Parameter Reduction) A general covariance of $\mathrm{vec}(Z) \in \mathbb{R}^{mn}$ has $(mn)(mn+1)/2$ parameters. The separable model uses only $m(m+1)/2 + n(n+1)/2$ parameters, a dramatic reduction when m, n are large.

Remark A.1.13 (Whitening under Separability) If $\Sigma_r \succ 0$ and $\Sigma_c \succ 0$, then the linear transform

$$\widetilde{Z} := \Sigma_r^{-1/2}\, Z\, \Sigma_c^{-1/2}$$

satisfies $\mathrm{Cov}(\mathrm{vec}(\widetilde{Z})) = I_{mn}$ under the separable model (verify using Theorem A.1.4).

A.1.3.2 Block-Matrix Multiplication and Schur Complements

Block-matrix algebra is the language of modular models: multi-view features, multi-task couplings, KKT systems, and Gaussian conditioning. The Schur complement is the key operator: it describes the effective system after eliminating a block. In ML, it appears in (i) constrained least squares, (ii) Gaussian posterior conditioning, and (iii) efficient updates of ridge/GP systems.

Block Gaussian Elimination Template

Consider the block linear system

$$\begin{bmatrix} A & B \\ C & D \end{bmatrix} \begin{bmatrix} x \\ y \end{bmatrix} = \begin{bmatrix} u \\ v \end{bmatrix},$$

where $A \in \mathbb{R}^{p\times p}$ and $D \in \mathbb{R}^{q\times q}$.

Proposition A.1.11 (Block Elimination (Schur Complement Solve)) *Assume A is invertible and define the Schur complement*

$$S_{/A} := D - CA^{-1}B.$$

Then any solution satisfies

$$S_{/A}y = v - CA^{-1}u, \qquad x = A^{-1}(u - By).$$

Analogously, if D is invertible, using $S_{/D} := A - BD^{-1}C$ *yields*

$$S_{/D}x = u - BD^{-1}v, \qquad y = D^{-1}(v - Cx).$$

Schur Complement Inversion Formula (Reference)

The inverse of a block matrix can be written explicitly in terms of a Schur complement.

Proposition A.1.12 (Block Inverse via Schur Complement) *Let*

$$M = \begin{bmatrix} A & B \\ C & D \end{bmatrix},$$

with A invertible and $S_{/A} = D - CA^{-1}B$ *invertible. Then*

$$M^{-1} = \begin{bmatrix} A^{-1} + A^{-1}BS_{/A}^{-1}CA^{-1} & -A^{-1}BS_{/A}^{-1} \\ -S_{/A}^{-1}CA^{-1} & S_{/A}^{-1} \end{bmatrix}.$$

Similarly, if D and $S_{/D} = A - BD^{-1}C$ *are invertible, an analogous formula holds.*

Remark A.1.14 (Computational Message) You should *not* implement block inverses by explicitly forming these expressions with inverses. Instead, interpret them as sequences of solves with A and $S_{/A}$ (or D and $S_{/D}$), using LU/Cholesky as appropriate.

KKT Systems and Constrained Least Squares (Reference)

Many constrained optimization problems yield block KKT systems. A canonical case is equality-constrained least squares:

$$\min_{x \in \mathbb{R}^n} \frac{1}{2}\|Ax - b\|_2^2 \quad \text{s.t.} \quad Cx = d,$$

with $A \in \mathbb{R}^{m\times n}$ and $C \in \mathbb{R}^{p\times n}$. The first-order optimality conditions lead to the KKT system

$$\begin{bmatrix} A^\top A & C^\top \\ C & 0 \end{bmatrix} \begin{bmatrix} x \\ \lambda \end{bmatrix} = \begin{bmatrix} A^\top b \\ d \end{bmatrix},$$

where λ are Lagrange multipliers. Schur complements provide the standard elimination route (eliminate x or λ) and connect to projected solves and null-space methods.

A.2 Proof Templates for Linear Algebra Arguments

The book repeatedly uses a small number of proof archetypes. Writing them explicitly helps students recognize what "move" to make when encountering a new lemma: either (i) move to an orthonormal eigenbasis to reduce matrix statements to scalar inequalities, or (ii) interpret an optimization problem geometrically as a projection and use orthogonality/Pythagoras. The templates below are not tied to a specific theorem; they are reusable proof skeletons.

A.2.1 Orthonormal-Basis (Spectral) Proofs

A.2.1.1 Diagonalize $A = Q\Lambda Q^\top$ and Reduce to Scalars

Template (When to Use)

Use this template whenever A is symmetric (or more generally normal/unitarily diagonalizable) and the claim involves quadratic forms $x^\top Ax$, PSD/PD conditions, eigenvalue bounds, norms, or conditioning. The goal is to rewrite everything in an orthonormal eigenbasis where A becomes diagonal.

Step-by-Step Skeleton

1. **Diagonalize.** For symmetric $A \in \mathbb{R}^{d\times d}$, write

$$A = Q\Lambda Q^\top, \qquad Q^\top Q = I_d, \quad \Lambda = \mathrm{diag}(\lambda_1, \ldots, \lambda_d).$$

2. **Change variables.** For any $x \in \mathbb{R}^d$, define $z := Q^\top x$ so that $x = Qz$ and $\|z\|_2 = \|x\|_2$ (orthogonal invariance).
3. **Rewrite the expression.** Typical rewrites:

$$x^\top Ax = x^\top Q\Lambda Q^\top x = z^\top \Lambda z = \sum_{i=1}^{d} \lambda_i z_i^2,$$

$$\|Ax\|_2^2 = x^\top A^\top Ax = x^\top A^2 x \quad (\text{if } A = A^\top), \quad \|A\|_2 = \max_{\|x\|_2=1} \|Ax\|_2 = \max_i |\lambda_i|.$$

4. **Reduce to scalar inequalities.** Use $\sum_i z_i^2 = \|z\|_2^2$ and standard scalar bounds such as

$$\min_i \lambda_i \le \sum_i \lambda_i z_i^2 \le \max_i \lambda_i \quad \text{when } \sum_i z_i^2 = 1.$$

5. **Translate back.** Conclude the desired matrix statement by interpreting the scalar result in the original coordinates.

Remark A.2.1 (Typical Outcomes) This template proves, with minimal effort:

- $A \succeq 0 \iff \lambda_i \ge 0$ for all i.
- $\lambda_{\max}(A) = \max_{\|x\|=1} x^\top Ax$ and $\lambda_{\min}(A) = \min_{\|x\|=1} x^\top Ax$.
- For SPD A, $\kappa_2(A) = \lambda_{\max}(A)/\lambda_{\min}(A)$ and ellipsoid geometry of level sets.
- Orthogonal invariance identities such as $\|Qx\|_2 = \|x\|_2$ and $\|QAQ^\top\|_2 = \|A\|_2$.

A.2.1.2 Rayleigh Quotient Maximization under $\|x\|_2 = 1$

Template (When to Use)

Use this template when a problem asks for an extremal eigenvalue/eigenvector, principal component direction, or any constrained maximization/minimization of a symmetric quadratic form over the unit sphere.

Two Standard Routes

Route A: Spectral Decomposition Route (Fast)

1. Diagonalize $A = Q\Lambda Q^\top$ and set $z = Q^\top x$ with $\|z\|_2 = 1$.
2. Rewrite

$$x^\top Ax = z^\top \Lambda z = \sum_{i=1}^{d} \lambda_i z_i^2, \qquad \sum_{i=1}^{d} z_i^2 = 1.$$

3. Conclude

$$\max_{\|x\|_2=1} x^\top Ax = \max_{\|z\|_2=1} \sum_i \lambda_i z_i^2 = \lambda_{\max}(A),$$

achieved at $z = e_{i^\star}$ for $i^\star \in \arg\max_i \lambda_i$, hence $x = q_{i^\star}$ (corresponding eigenvector).

Route B: Lagrange Multipliers Route (Structural)

1. Form $\mathcal{L}(x, \mu) = x^\top Ax - \mu(x^\top x - 1)$.
2. Stationarity gives

$$\nabla_x \mathcal{L} = 2Ax - 2\mu x = 0 \quad \Longrightarrow \quad Ax = \mu x.$$

   Thus critical points are eigenvectors and the objective equals the eigenvalue: $x^\top Ax = \mu$ on $\|x\|_2 = 1$.
3. Identify maxima/minima by selecting the largest/smallest eigenvalue (using spectral ordering or second-order arguments).

Remark A.2.2 (PCA Connection) For covariance $\Sigma \succeq 0$, maximizing $u^\top \Sigma u$ over $\|u\|_2 = 1$ yields the first principal component direction. Additional PCs follow by imposing orthogonality constraints, leading to Courant–Fischer min–max forms.

A.2.2 Projection-Based Proofs

A.2.2.1 Orthogonality Condition: $r \perp \mathcal{U}$ Implies Optimality

Template (When to Use)

Use this template whenever you are minimizing Euclidean distance to a subspace (least squares, orthogonal projection), or more generally proving optimality of an approximation by showing the residual is orthogonal to the feasible subspace.

Skeleton (Subspace Case)

Let $\mathcal{U} \subseteq \mathbb{R}^d$ be a subspace and suppose you propose $u^\star \in \mathcal{U}$ as the best approximation to x. Let $r := x - u^\star$.

1. **Show orthogonality.** Prove $r \perp \mathcal{U}$, i.e., $\langle r, u \rangle = 0$ for all $u \in \mathcal{U}$.
2. **Compare any candidate.** For any $u \in \mathcal{U}$, write

$$x - u = (x - u^\star) + (u^\star - u) = r + (u^\star - u), \quad \text{with } (u^\star - u) \in \mathcal{U}.$$

3. **Use Pythagoras.** Since $r \perp (u^\star - u)$,

$$\|x - u\|_2^2 = \|r\|_2^2 + \|u^\star - u\|_2^2 \geq \|r\|_2^2 = \|x - u^\star\|_2^2.$$

4. **Conclude optimality and uniqueness.** This proves $u^\star$ is a minimizer; uniqueness follows because equality forces $u = u^\star$.

Remark A.2.3 (Normal Equations Are Orthogonality) For least squares $\min_w \|Xw - y\|_2^2$, the residual $r = y - X\widehat{w}$ is orthogonal to $\mathcal{C}(X)$. Equivalently $X^\top r = 0$ (normal equations). This is exactly the template above in $\mathbb{R}^n$ with $\mathcal{U} = \mathcal{C}(X)$.

A.2.2.2 Pythagorean Decomposition and Energy Splitting Arguments

Template (When to Use)

Use this template whenever you need to split "total energy" into orthogonal components: least squares error decomposition, PCA explained variance, projection residual bounds, or orthogonal invariance statements.

Skeleton

1. **Identify an orthogonal decomposition.** Write $x = a + b$ with $\langle a, b\rangle = 0$. Commonly, $a = Px$ and $b = (I - P)x$ for an orthogonal projector P.
2. **Apply Pythagoras.** Conclude

$$\|x\|_2^2 = \|a\|_2^2 + \|b\|_2^2.$$

3. **Sum over samples (if needed).** For data $\{x_i\}$,

$$\sum_i \|x_i\|_2^2 = \sum_i \|a_i\|_2^2 + \sum_i \|b_i\|_2^2,$$

   interpreting the terms as explained vs. residual energy/variance.
4. **Optimize by minimizing residual energy.** Many optimality results reduce to choosing a subspace/projection that minimizes $\sum_i \|b_i\|_2^2$ (or equivalently maximizes $\sum_i \|a_i\|_2^2$).

Remark A.2.4 (PCA in One Line) With centered data X_c, choose a rank-k orthogonal projector $P = Q_k Q_k^\top$ to minimize $\|X_c - X_c P\|_F^2 = \|X_c(I - P)\|_F^2$. Eckart–Young shows the minimizer is given by the top-k right singular vectors/eigenvectors. The proof is a sophisticated application of the same energy-splitting template.

A.2.3 *SVD-Based Proofs*

SVD-based proofs are the default template for *rectangular* matrices and for best-approximation statements. They turn matrix inequalities into statements about singular values by expanding in orthonormal singular vector bases. This approach is both conceptually clean (geometry of ellipsoids) and computationally aligned with practice (low-rank truncation, PCA). Two core moves recur: (i) expand A as a sum of rank-one outer products and (ii) express norms and approximation errors as explicit functions of $\{\sigma_i\}$.

A.2.3.1 Write $A = \sum_i \sigma_i u_i v_i^\top$ and Truncate

Template (When to Use)

Use this when proving or applying: low-rank approximation, PCA reconstruction error, nuclear-norm heuristics, or any statement that compares A to a rank-k matrix. The key is the orthogonal rank-one expansion: each singular component is independent in Frobenius geometry.

Skeleton Let $A \in \mathbb{R}^{m\times n}$ have SVD $A = U\Sigma V^\top$ with rank r and singular values $\sigma_1 \geq \cdots \geq \sigma_r > 0$.

1. **Rank-one expansion.** Write

$$A = \sum_{i=1}^{r} \sigma_i u_i v_i^\top,$$

   with orthonormal $\{u_i\} \subset \mathbb{R}^m$ and $\{v_i\} \subset \mathbb{R}^n$.
2. **Define truncation.** For $k < r$, define the rank-k truncation

$$A_k := \sum_{i=1}^{k} \sigma_i u_i v_i^\top.$$

3. **Residual decomposition.** Then

$$A - A_k = \sum_{i=k+1}^{r} \sigma_i u_i v_i^\top.$$

   Because the rank-one terms are orthogonal under $\langle\cdot,\cdot\rangle_F$, this decomposition behaves like an orthogonal sum.
4. **Conclude optimality (outline).** Any rank-k matrix B cannot capture more than k singular directions. Formalize this by expressing $U^\top BV$ and applying singular-value inequalities, yielding Eckart–Young.

Remark A.2.5 (Orthogonality of Singular Components) For $i \neq j$,

$$\langle u_i v_i^\top,\ u_j v_j^\top\rangle_F = \operatorname{tr}\big((v_i u_i^\top)(u_j v_j^\top)\big) = \operatorname{tr}\big(v_i (u_i^\top u_j) v_j^\top\big) = 0,$$

so the rank-one components form an orthonormal family after normalization. This is why Frobenius energies add over singular components.

A.2.3.2 Relate $\|A\|_2$, $\|A\|_F$ and Errors to Singular Values

Template (When to Use)

Use this for: bounding operator amplification, computing reconstruction error, comparing truncations, and reading "energy" from spectra. Most such results reduce to: spectral norm is the largest singular value; Frobenius norm is the ℓ_2 norm of the singular value vector.

Core Identities If A has singular values $\sigma_1 \geq \cdots \geq \sigma_r > 0$, then

$$\|A\|_2 = \sigma_1, \qquad \|A\|_F^2 = \sum_{i=1}^{r} \sigma_i^2, \qquad \operatorname{rank}(A) = r.$$

For truncation A_k,

$$\|A - A_k\|_2 = \sigma_{k+1}, \qquad \|A - A_k\|_F^2 = \sum_{i=k+1}^{r} \sigma_i^2.$$

Remark A.2.6 (Energy Capture and Scree Intuition) The fraction of energy captured by the top k singular modes is

$$\mathrm{EC}(k) = \frac{\sum_{i=1}^{k} \sigma_i^2}{\sum_{i=1}^{r} \sigma_i^2}.$$

In PCA, σ_i^2 (scaled by $1/n$) equals variance along component i; truncation error equals discarded variance.

Remark A.2.7 (Two Complementary Error Lenses)

- Spectral-norm error $\|A - A_k\|_2$ is a worst-case distortion bound (controls maximum amplification error).
- Frobenius-norm error $\|A - A_k\|_F$ is an average-energy error bound (controls total squared reconstruction error).

Choosing which to use depends on whether you need worst-case guarantees or mean-squared fidelity.

A.2.4 Matrix Inequality Templates

Many ML proofs require bounding complicated matrix expressions without computing eigenvalues exactly. Two standard tools recur: (i) norm submultiplicativity (to upper bound products) and (ii) PSD ordering (to compare quadratic forms). Together they allow fast derivations of stability bounds, Lipschitz constants, and conditioning estimates.

A.2.4.1 Use $\|A^\top A\|_2 = \|A\|_2^2$ and Submultiplicativity

Template (When to Use)

Use this for: bounding gradients/Lipschitz constants in least squares, bounding perturbations, showing conditioning of normal equations, and controlling Jacobian products in deep networks.

Proposition A.2.1 (Submultiplicativity and Basic Spectral-Norm Rules) *For compatible matrices A, B,*

$$\|AB\|_2 \leq \|A\|_2 \, \|B\|_2, \qquad \|A^\top\|_2 = \|A\|_2, \qquad \|A^\top A\|_2 = \|A\|_2^2.$$

Remark A.2.8 (Canonical Applications)

- Least squares smoothness: for $f(w) = \frac{1}{2}\|Xw - y\|_2^2$, $\|\nabla f(w) - \nabla f(v)\|_2 = \|X^\top X(w - v)\|_2 \leq \|X^\top X\|_2 \|w - v\|_2 = \|X\|_2^2 \|w - v\|_2$.
- Normal equations conditioning: $\kappa_2(X^\top X) = \|X^\top X\|_2 \|(X^\top X)^{-1}\|_2 = \|X\|_2^2 \|X^\dagger\|_2^2 = \kappa_2(X)^2$ (full column rank case).

A.2.4.2 PSD Ordering: $0 \preceq B \preceq A \Rightarrow x^\top Bx \leq x^\top Ax$

Template (When to Use)

Use PSD ordering to compare energies, covariances, and curvature matrices; to establish monotonicity of functions of PSD matrices; and to upper/lower bound quadratic forms without diagonalizing.

Definition A.2.1 (PSD Order) For symmetric matrices $A, B \in \mathbb{R}^{d \times d}$, write $A \succeq B$ if $A - B \succeq 0$. Equivalently, $A \succeq B$ iff $x^\top (A - B)x \geq 0$ for all x.

Proposition A.2.2 (Quadratic-Form Monotonicity) *If $0 \preceq B \preceq A$, then for all $x \in \mathbb{R}^d$,*

$$x^\top Bx \leq x^\top Ax.$$

Moreover, $0 \preceq A \preceq \alpha I$ *iff* $\|A\|_2 \leq \alpha$ *(for* $A \succeq 0$*).*

Proof $0 \preceq B \preceq A$ means $A - B \succeq 0$, so $x^\top (A - B)x \geq 0$, hence $x^\top Bx \leq x^\top Ax$. If $A \succeq 0$, then $\|A\|_2 = \lambda_{\max}(A)$, and $A \preceq \alpha I$ iff $\lambda_{\max}(A) \leq \alpha$. □

Remark A.2.9 (Canonical Applications)

- Ridge monotonicity: $X^\top X \preceq X^\top X + \lambda I$ implies $(X^\top X + \lambda I)^{-1} \preceq (X^\top X)^\dagger$ (in an appropriate generalized sense), reflecting reduced sensitivity.
- Covariance comparisons: if $\Sigma_1 \preceq \Sigma_2$, then $\mathrm{Var}(u^\top X)$ is larger under Σ_2 for every direction u: $u^\top \Sigma_1 u \leq u^\top \Sigma_2 u$.

A.3 Matrix Calculus Cheat Sheet

This cheat sheet records the derivative identities used repeatedly in the book. The aim is *shape-safe* differentiation: every gradient has the same shape as the variable it differentiates with respect to, and the identities are stated in a consistent convention. We use the Euclidean/Frobenius inner products to identify differentials with gradients: for a scalar function $f(X)$ with $X \in \mathbb{R}^{m \times n}$, the gradient $\nabla_X f$ is the unique matrix satisfying

$$df = \langle \nabla_X f, dX \rangle_F = \mathrm{tr}\big((\nabla_X f)^\top dX\big).$$

Throughout, transpose placement matters; whenever uncertain, verify by a differential/trace derivation.

A.3.1 Gradients of Common Objectives

A.3.1.1 Quadratics, Least Squares, and Ridge

Vector Case: $\nabla_x \frac{1}{2}\|Ax - b\|_2^2 = A^\top (Ax - b)$

Let $A \in \mathbb{R}^{m \times d}$, $b \in \mathbb{R}^m$, and define

$$f(x) = \frac{1}{2}\|Ax - b\|_2^2 = \frac{1}{2}(Ax - b)^\top (Ax - b), \qquad x \in \mathbb{R}^d.$$

Then

$$\nabla_x f(x) = A^\top (Ax - b), \qquad \nabla_x^2 f(x) = A^\top A.$$

A useful special case is the quadratic form: if $M \in \mathbb{R}^{d\times d}$ is symmetric, then

$$\nabla_x\left(\frac{1}{2}x^\top Mx\right) = Mx, \qquad \nabla_x(x^\top Mx) = (M + M^\top)x.$$

Matrix Case: $\nabla_W \frac{1}{2}\|WX - Y\|_F^2 = (WX - Y)X^\top$

Let $W \in \mathbb{R}^{k\times d}$ be the variable, and let $X \in \mathbb{R}^{d\times n}$ and $Y \in \mathbb{R}^{k\times n}$ be fixed. Define

$$F(W) = \frac{1}{2}\|WX - Y\|_F^2 = \frac{1}{2}\operatorname{tr}\big((WX - Y)^\top(WX - Y)\big).$$

Then

$$\nabla_W F(W) = (WX - Y)X^\top, \quad \nabla_W^2 F \text{ (as a linear operator)} : \ \Delta W \mapsto \Delta W\,(XX^\top).$$

Equivalently, vectorizing gives $\operatorname{vec}(\nabla_W F) = \big(XX^\top \otimes I_k\big)\operatorname{vec}(W) - \operatorname{vec}(YX^\top)$.

Ridge/Tikhonov Gradients and Normal Equations

For the ridge objective (vector case)

$$f_\lambda(x) = \frac{1}{2}\|Ax - b\|_2^2 + \frac{\lambda}{2}\|x\|_2^2, \qquad \lambda > 0,$$

the gradient and optimality condition are

$$\nabla f_\lambda(x) = A^\top(Ax - b) + \lambda x, \qquad \nabla f_\lambda(x^\star) = 0 \iff (A^\top A + \lambda I)x^\star = A^\top b.$$

For the matrix ridge objective

$$F_\lambda(W) = \frac{1}{2}\|WX - Y\|_F^2 + \frac{\lambda}{2}\|W\|_F^2,$$

we have

$$\nabla_W F_\lambda(W) = (WX - Y)X^\top + \lambda W, \ \nabla_W F_\lambda(W^\star) = 0 \iff W^\star(XX^\top + \lambda I_d) = YX^\top.$$

(When $XX^\top + \lambda I_d$ is invertible, $W^\star = YX^\top(XX^\top + \lambda I_d)^{-1}$.)

A.3.1.2 Log-det and Trace Objectives

Trace and log-determinant objectives appear in Gaussian likelihoods, entropy-like regularizers, and second-order methods. These identities require careful domain assumptions (typically $X \succ 0$) and are best derived using differentials and the cyclicity of trace. We state the formulas in the real-matrix setting; for symmetric variables one may simplify transposes accordingly.

$\nabla_X \mathrm{tr}(AX) = A^\top$

Let $A \in \mathbb{R}^{m\times n}$ be fixed and $X \in \mathbb{R}^{m\times n}$ be the variable. Then

$$f(X) = \mathrm{tr}(A^\top X) \quad \Rightarrow \quad \nabla_X f = A.$$

Equivalently, in the form stated in the outline:

$$\mathrm{tr}(AX)\ (\text{with } A \in \mathbb{R}^{n\times m},\ X \in \mathbb{R}^{m\times n}) \quad \Rightarrow \quad \nabla_X \mathrm{tr}(AX) = A^\top.$$

$\nabla_X \log\det(X) = X^{-\top}$ for $X \succ 0$

Let $X \in \mathbb{R}^{d\times d}$ be invertible (in applications, typically SPD so that $\log\det(X)$ is real). Define $f(X) = \log\det(X)$. Then

$$\nabla_X f(X) = X^{-\top}.$$

In particular, if X is symmetric, then $X^{-\top} = X^{-1}$.

Trace-Logdet Composite Patterns in Gaussian Likelihoods

A canonical Gaussian negative log-likelihood (up to constants) for covariance $\Sigma \succ 0$ has the form

$$\mathcal{L}(\Sigma) = \frac{1}{2}\Big(\log\det(\Sigma) + \mathrm{tr}(\Sigma^{-1}S)\Big),$$

where $S \succeq 0$ is a sample covariance/second-moment matrix. The differential identities yield the gradient

$$\nabla_\Sigma \mathcal{L}(\Sigma) = \frac{1}{2}\Big(\Sigma^{-\top} - \Sigma^{-\top}S\Sigma^{-\top}\Big),$$

and in the symmetric case this simplifies to

$$\nabla_\Sigma \mathcal{L}(\Sigma) = \frac{1}{2}\Big(\Sigma^{-1} - \Sigma^{-1} S \Sigma^{-1}\Big).$$

A frequently used chain pattern is: if $X = X(\theta)$ depends on parameters, then

$$\nabla_\theta \log\det(X(\theta)) = \operatorname{tr}\!\big(X(\theta)^{-1} \nabla_\theta X(\theta)\big),$$

interpreting $\nabla_\theta X$ componentwise and using trace to collapse dimensions.

A.3.2 *Differentials and Chain Rules*

Differential calculus is the most reliable way to derive matrix gradients without memorizing tables. The workflow is: (i) write f in trace form, (ii) take the differential df using product/transpose rules, (iii) rearrange into $df = \operatorname{tr}(G^\top dX)$, and (iv) read off $\nabla_X f = G$ under the Frobenius pairing. This section records the minimal rule set and the shape checks that prevent common mistakes in ML derivations.

A.3.2.1 Differential Notation for Clean Derivations

Rules: $d(AB) = dA\, B + A\, dB$, $d(X^\top) = (dX)^\top$

The following rules hold for differentials (Fréchet differentials) whenever the products are well-defined:

$$d(\alpha A + \beta B) = \alpha\, dA + \beta\, dB, \quad d(AB) = dA\, B + A\, dB,$$
$$d(A^{-1}) = -A^{-1}(dA)A^{-1},$$

$$d(X^\top) = (dX)^\top, \quad d\operatorname{tr}(A) = \operatorname{tr}(dA),$$
$$d\log\det(A) = \operatorname{tr}(A^{-1} dA) \text{ (for invertible } A).$$

The inverse rule is particularly important in Gaussian log-likelihoods and second-order methods.

Remark A.3.1 (Product Rule Extends to Longer Products) For ABC, one has

$$d(ABC) = dA\, BC + A\, dB\, C + AB\, dC.$$

This is the algebraic backbone of backprop-style differentiation in matrix form.

Convert $df = \mathrm{tr}(G^\top dX)$ to $\nabla_X f = G$

We use the Frobenius pairing $\langle A, B\rangle_F = \mathrm{tr}(A^\top B)$. For a scalar function $f : \mathbb{R}^{m\times n} \to \mathbb{R}$, the gradient $\nabla_X f \in \mathbb{R}^{m\times n}$ is defined by

$$df = \langle \nabla_X f, dX\rangle_F = \mathrm{tr}\big((\nabla_X f)^\top dX\big) \quad \text{for all } dX.$$

Hence, once you have manipulated df into the canonical form

$$df = \mathrm{tr}(G^\top dX),$$

you may immediately read off

$$\nabla_X f = G.$$

Example A.3.1 (One-Line Derivation: $\nabla_X \|X\|_F^2 = 2X$) Let $f(X) = \|X\|_F^2 = \mathrm{tr}(X^\top X)$. Then

$$\begin{aligned} df &= \mathrm{tr}(d(X^\top)X + X^\top dX) = \mathrm{tr}((dX)^\top X) + \mathrm{tr}(X^\top dX) = \mathrm{tr}(X^\top dX) \\ &\quad + \mathrm{tr}(X^\top dX) = \mathrm{tr}((2X)^\top dX), \end{aligned}$$

so $\nabla_X f = 2X$.

Remark A.3.2 (Common Rearrangement Trick) Use cyclicity of trace to move dX to the far right:

$$\mathrm{tr}(A\, dX\, B) = \mathrm{tr}(BA\, dX),$$

so that the coefficient of dX becomes explicit.

A.3.2.2 Jacobians and Hessians: Shapes and Sanity Checks

Jacobian Shapes for $f : \mathbb{R}^d \to \mathbb{R}^k$

Let $f = (f_1, \ldots, f_k) : \mathbb{R}^d \to \mathbb{R}^k$ be differentiable at x. The Jacobian is the $k \times d$ matrix

$$J_f(x) = \left[\frac{\partial f_i}{\partial x_j}(x)\right]_{i\in[k],\, j\in[d]} \in \mathbb{R}^{k\times d}.$$

The differential form is

$$df = J_f(x)\, dx,$$

where $dx \in \mathbb{R}^d$ and $df \in \mathbb{R}^k$. This identity is the most reliable *shape check*: J_f must map dx (dimension d) into df (dimension k).

Remark A.3.3 (Chain Rule (Jacobian Form)) For $g : \mathbb{R}^k \to \mathbb{R}^p$ and $h = g \circ f$, one has

$$J_h(x) = J_g(f(x))\, J_f(x),$$

with shapes $(p \times k)(k \times d) = (p \times d)$.

Hessian Symmetry for Scalar Twice-Differentiable f

If $f : \mathbb{R}^d \to \mathbb{R}$ is twice continuously differentiable, its Hessian is

$$\nabla^2 f(x) = \left[\frac{\partial^2 f}{\partial x_i\, \partial x_j}(x) \right]_{i,j \in [d]} \in \mathbb{R}^{d \times d},$$

and satisfies symmetry $\nabla^2 f(x) = (\nabla^2 f(x))^\top$ (Clairaut/Schwarz theorem). In differential form,

$$d(\nabla f(x)) = \nabla^2 f(x)\, dx.$$

A quick sanity check: the Hessian maps $dx \in \mathbb{R}^d$ to $d(\nabla f) \in \mathbb{R}^d$, hence is $d \times d$.

Backprop as Vector-Jacobian Products (Reverse Mode)

For a scalar loss $\ell : \mathbb{R}^k \to \mathbb{R}$ and intermediate $y = f(x) \in \mathbb{R}^k$, the chain rule yields

$$\nabla_x\, \ell(f(x)) = J_f(x)^\top\, \nabla_y \ell(y).$$

This is a *vector-Jacobian product* (VJP): multiply a vector in $\mathbb{R}^k$ by the transpose Jacobian to obtain a vector in $\mathbb{R}^d$. Reverse-mode automatic differentiation computes such products efficiently without forming $J_f(x)$ explicitly.

Remark A.3.4 (VJP vs. JVP)

- JVP (forward-mode): given $v \in \mathbb{R}^d$, compute $J_f(x)v \in \mathbb{R}^k$.
- VJP (reverse-mode): given $u \in \mathbb{R}^k$, compute $J_f(x)^\top u \in \mathbb{R}^d$.

Backprop is reverse-mode: it propagates cotangents by VJPs.

Remark A.3.5 (Shape Checklist for Multilayer Compositions) If $x_0 \in \mathbb{R}^{d_0}$ and $x_\ell = f_\ell(x_{\ell-1}) \in \mathbb{R}^{d_\ell}$, then $J_{f_\ell}(x_{\ell-1}) \in \mathbb{R}^{d_\ell \times d_{\ell-1}}$. Backprop multiplies by $J_{f_\ell}^\top \in \mathbb{R}^{d_{\ell-1} \times d_\ell}$, mapping gradients from layer ℓ to layer $\ell - 1$. If these shapes do not match, the derivation (or implementation) is wrong.

Appendix B
Geometry Reference: Norms, Duality, and Projections

B.1 Norm Facts and Inequalities

Norm inequalities are the “unit conversion rules” of geometry: they tell you how the same vector looks under different notions of size. In finite dimensions, all norms are equivalent, but the equivalence constants often scale with dimension and therefore matter greatly in ML. Dual norms convert geometric constraints into worst-case linear deviations, which is central in margins, robustness, and regularization analysis. This section records the main inequalities and duality facts used throughout the book.

B.1.1 Hölder, Minkowski, and Norm Conversions

B.1.1.1 ℓ_p vs. ℓ_q Inequalities

Two families of facts recur: (i) *monotonicity* of ℓ_p norms (as p increases, the norm decreases), and (ii) *dimension-dependent conversion bounds* (how much you lose when switching p). These are indispensable for translating robustness radii between ℓ_2 and ℓ_∞, comparing regularizers, and interpreting high-dimensional effects (e.g., the ubiquitous $\sqrt{d}$ factor).

Definition B.1.1 (ℓ_p Norms) For $x \in \mathbb{R}^d$ and $1 \le p < \infty$,

$$\|x\|_p := \Big(\sum_{i=1}^{d} |x_i|^p\Big)^{1/p}, \qquad \|x\|_\infty := \max_{i\in[d]} |x_i|.$$

P. Singh, B. Raman, *Foundations of Machine Learning and AI*, Studies in Big Data 199, https://doi.org/10.1007/978-3-032-30336-3

Theorem B.1.1 (Hölder Inequality) *Let $p, q \in [1, \infty]$ satisfy $\frac{1}{p} + \frac{1}{q} = 1$ (with $1/\infty := 0$). Then for all $x, y \in \mathbb{R}^d$,*

$$|\langle x, y\rangle| \leq \|x\|_p \|y\|_q.$$

Theorem B.1.2 (Minkowski Inequality) *For any $p \in [1, \infty]$ and all $x, y \in \mathbb{R}^d$,*

$$\|x + y\|_p \leq \|x\|_p + \|y\|_p.$$

Monotonicity: $p \leq q \Rightarrow \|x\|_q \leq \|x\|_p$

Proposition B.1.1 (ℓ_p Monotonicity) *If $1 \leq p \leq q \leq \infty$, then for all $x \in \mathbb{R}^d$,*

$$\|x\|_q \leq \|x\|_p.$$

Remark B.1.1 (Interpretation) As p increases, the norm increasingly emphasizes the largest coordinates and discounts the small ones; hence the numerical value decreases. In particular, $\|x\|_\infty \leq \|x\|_2 \leq \|x\|_1$.

Dimension Factors: $\|x\|_p \leq d^{(1/p-1/q)} \|x\|_q$

Proposition B.1.2 (ℓ_p–ℓ_q Conversion) *If $1 \leq p \leq q \leq \infty$, then for all $x \in \mathbb{R}^d$,*

$$\|x\|_p \leq d^{\left(\frac{1}{p}-\frac{1}{q}\right)} \|x\|_q.$$

Equivalently,

$$d^{-\left(\frac{1}{p}-\frac{1}{q}\right)} \|x\|_p \leq \|x\|_q \leq \|x\|_p.$$

Remark B.1.2 (Where the Factor Comes From) The exponent $\frac{1}{p} - \frac{1}{q}$ is exactly the rate at which equal-magnitude coordinates accumulate: if $|x_i| = c$ for all i, then $\|x\|_p = d^{1/p}c$ and $\|x\|_q = d^{1/q}c$, so $\|x\|_p/\|x\|_q = d^{1/p-1/q}$. Thus the bound is tight in general.

Common Special Cases: $\ell_1, \ell_2, \ell_\infty$

From Proposition B.1.2 with $(p, q) = (1, 2)$ and $(p, q) = (2, \infty)$, we obtain the standard bounds:

$$\|x\|_2 \leq \|x\|_1 \leq \sqrt{d}\, \|x\|_2, \quad \|x\|_\infty \leq \|x\|_2 \leq \sqrt{d}\, \|x\|_\infty, \quad \|x\|_\infty \leq \|x\|_1 \leq d\, \|x\|_\infty.$$

These are used constantly for translating robustness radii and regularization strengths.

B.1.1.2 Dual Norms and Support Functions

Dual norms encode worst-case alignment of a vector with a norm-bounded perturbation. They are the correct tool for margin/robustness calculations because linear functionals (dot products) are the local first-order behavior of models. Geometrically, the dual norm is the support function of the primal unit ball, and the identity $\|x\| = \sup_{\|y\|_* \le 1} \langle x, y \rangle$ shows that a norm can be recovered from its dual.

Dual Definition: $\|y\|_* = \sup_{\|x\| \le 1} \langle x, y \rangle$

Definition B.1.2 (Dual Norm) Let $\|\cdot\|$ be any norm on $\mathbb{R}^d$ and let $B = \{x : \|x\| \le 1\}$ be its unit ball. The *dual norm* is

$$\|y\|_* := \sup_{\|x\| \le 1} \langle x, y \rangle = \sup_{x \in B} \langle x, y \rangle.$$

Definition B.1.3 (Support Function) For a nonempty set $K \subseteq \mathbb{R}^d$, its support function is

$$h_K(y) := \sup_{x \in K} \langle x, y \rangle.$$

Thus, for the primal unit ball B, we have $\|y\|_* = h_B(y)$.

Theorem B.1.3 (Dual Representation of a Norm) *Let $\|\cdot\|$ be a norm with dual $\|\cdot\|_*$. Then for all $x \in \mathbb{R}^d$,*

$$\|x\| = \sup_{\|y\|_* \le 1} \langle x, y \rangle.$$

Generalized Cauchy–Schwarz: $\langle x, y \rangle \le \|x\| \|y\|_*$

Theorem B.1.4 (Generalized Cauchy–Schwarz) *For any norm $\|\cdot\|$ and its dual $\|\cdot\|_*$, and for all $x, y \in \mathbb{R}^d$,*

$$|\langle x, y \rangle| \le \|x\| \, \|y\|_*.$$

Proof If $x = 0$ the claim is trivial. Otherwise, let $\tilde{x} = x/\|x\|$ so that $\|\tilde{x}\| \leq 1$. Then by Definition B.1.2,

$$\langle x, y\rangle = \|x\| \langle \tilde{x}, y\rangle \leq \|x\| \sup_{\|u\|\leq 1} \langle u, y\rangle = \|x\| \|y\|_*.$$

Apply the same bound to $-y$ to obtain the absolute value form. □

Remark B.1.3 (Worst-Case Linear Deviation) A direct corollary is

$$\sup_{\|\delta\|\leq\varepsilon} |y^\top \delta| = \varepsilon \|y\|_*,$$

which is the basic robustness/margin computation for linearized models.

Dual Pairs: (ℓ_1, ℓ_∞), (ℓ_2, ℓ_2), Group Norms

Proposition B.1.3 (Standard Dual Pairs) *For $y \in \mathbb{R}^d$,*

$$(\|\cdot\|_1)^* = \|\cdot\|_\infty, \qquad (\|\cdot\|_\infty)^* = \|\cdot\|_1, \qquad (\|\cdot\|_2)^* = \|\cdot\|_2.$$

Proposition B.1.4 (Group ℓ_1/ℓ_2 Dual) *Let $\{G_1, \ldots, G_m\}$ partition $[d]$ and define*

$$\|x\|_{1,2} := \sum_{j=1}^{m} \|x_{G_j}\|_2.$$

Then the dual norm is

$$\|y\|_{*,\,1,2} = \max_{j\in[m]} \|y_{G_j}\|_2.$$

Remark B.1.4 (Geometric Interpretation) Duality swaps "sum" and "max" at the appropriate granularity: ℓ_1 sums coordinates while its dual ℓ_∞ takes the maximum coordinate; group ℓ_1/ℓ_2 sums group magnitudes, while its dual takes the maximum group magnitude. This is why ℓ_1 regularization is naturally paired with ℓ_∞ robustness models, and group sparsity with groupwise perturbations.

B.2 Inner Products, Angles, and Subspaces

Angles and subspaces are the geometric language behind embeddings, similarity search, PCA, least squares, and stability analyses. Cosine similarity captures directional alignment (scale-invariant similarity), and correlation is exactly cosine similarity after centering. Subspace angles (principal angles) quantify how close two learned representations or feature spans are, while projector distances provide robust operator-level metrics used in perturbation theory and algorithmic diagnostics.

B.2.1 Cosine Similarity and Correlation Geometry

B.2.1.1 Cosine Similarity on the Sphere and Link to Euclidean Distance

In high-dimensional representation learning, the norm of an embedding is often less meaningful than its direction. Cosine similarity compares directions and is invariant to positive rescaling. On the unit sphere, cosine similarity and Euclidean distance are equivalent orderings: minimizing Euclidean distance is the same as maximizing cosine similarity. This equivalence is routinely used in retrieval systems that normalize embeddings.

Definition B.2.1 (Cosine Similarity and Angle) For nonzero $x, y \in \mathbb{R}^d$,

$$\cos(x, y) := \frac{\langle x, y\rangle}{\|x\|_2 \, \|y\|_2} \in [-1, 1], \qquad \theta(x, y) := \arccos(\cos(x, y)) \in [0, \pi].$$

Proposition B.2.1 (Scale Invariance) *For any $\alpha, \beta > 0$ and nonzero x, y,*

$$\cos(\alpha x, \beta y) = \cos(x, y).$$

Proposition B.2.2 (Cosine–Distance Identity on the Unit Sphere) *If $\|x\|_2 = \|y\|_2 = 1$, then*

$$\|x - y\|_2^2 = 2 - 2\langle x, y\rangle = 2\big(1 - \cos(x, y)\big).$$

Equivalently,

$$\langle x, y\rangle = 1 - \frac{1}{2}\|x - y\|_2^2.$$

Remark B.2.1 (Nearest-Neighbor Equivalence under Normalization) If all embeddings are normalized to unit norm, then

$$\arg\min_j \|x - y_j\|_2 \;=\; \arg\max_j \langle x, y_j\rangle \;=\; \arg\max_j \cos(x, y_j),$$

so dot-product search equals cosine-similarity search equals Euclidean nearest-neighbor search.

B.2.1.2 Centering Matrix and Correlation as Cosine of Centered Vectors

Correlation removes the mean and compares directional alignment of fluctuations. This is exactly cosine similarity in the centered subspace orthogonal to the all-ones vector. The centering matrix H is an orthogonal projector; it performs the geometric operation "subtract the mean" and makes this relationship precise. This view is useful in statistics (correlation/covariance), in kernel centering, and in understanding preprocessing steps.

Definition B.2.2 (Centering Matrix) Let $\mathbf{1} \in \mathbb{R}^n$ be the all-ones vector and define

$$H := I_n - \frac{1}{n}\mathbf{1}\mathbf{1}^\top.$$

Then $H^\top = H$, $H^2 = H$, and $H\mathbf{1} = 0$. For $u \in \mathbb{R}^n$, $Hu = u - \bar{u}\mathbf{1}$ where $\bar{u} = \frac{1}{n}\mathbf{1}^\top u$.

Proposition B.2.3 (Centering as Orthogonal Projection) *H is the orthogonal projector onto the subspace*

$$\mathbf{1}^\perp := \{u \in \mathbb{R}^n : \mathbf{1}^\top u = 0\}.$$

Equivalently, every $u \in \mathbb{R}^n$ decomposes as $u = Hu + \bar{u}\mathbf{1}$ with $Hu \perp \mathbf{1}$.

Definition B.2.3 (Correlation as Cosine Similarity of Centered Vectors) For $u, v \in \mathbb{R}^n$ with nonzero centered versions $\tilde{u} := Hu$ and $\tilde{v} := Hv$, define

$$\operatorname{corr}(u, v) := \frac{\langle \tilde{u}, \tilde{v} \rangle}{\|\tilde{u}\|_2 \, \|\tilde{v}\|_2} = \cos(\tilde{u}, \tilde{v}).$$

Remark B.2.2 (Invariance to Adding Constants) For any constants $a, b \in \mathbb{R}$, $H(u + a\mathbf{1}) = Hu$ and $H(v + b\mathbf{1}) = Hv$, so $\operatorname{corr}(u + a\mathbf{1}, v + b\mathbf{1}) = \operatorname{corr}(u, v)$. This is the geometric meaning of "correlation ignores offsets."

B.2.2 Subspace Angles and Distances

B.2.2.1 Principal Angles via SVD of $Q_U^\top Q_V$

When comparing learned representations, what matters is often not individual vectors but entire subspaces (e.g., PCA subspaces, feature spans, embedding subspaces across domains). Principal angles provide a complete, basis-invariant measure of alignment. Computationally, they are obtained via an SVD of $Q_U^\top Q_V$ where Q_U, Q_V are orthonormal bases of the subspaces. The singular values are cosines of principal angles.

Definition B.2.4 (Principal Angles) Let $\mathcal{U}, \mathcal{V} \subseteq \mathbb{R}^d$ have dimensions k and ℓ, and set $m = \min\{k, \ell\}$. Choose orthonormal basis matrices $Q_U \in \mathbb{R}^{d\times k}$ and $Q_V \in \mathbb{R}^{d\times \ell}$. Let the singular values of $Q_U^\top Q_V \in \mathbb{R}^{k\times \ell}$ be

$$\sigma_1 \geq \cdots \geq \sigma_m \geq 0.$$

The *principal angles* $\theta_1 \leq \cdots \leq \theta_m$ between $\mathcal{U}$ and $\mathcal{V}$ are defined by

$$\cos\theta_i := \sigma_i, \qquad i = 1, \ldots, m,$$

with $\theta_i \in [0, \pi/2]$.

Remark B.2.3 (Key Interpretations)

- θ_1 measures the best alignment: small θ_1 means the subspaces nearly intersect.
- θ_m measures the worst alignment among the matched directions.
- All $\theta_i = 0$ (and $k = \ell$) iff $\mathcal{U} = \mathcal{V}$.
- All $\theta_i = \pi/2$ iff the subspaces are orthogonal.

B.2.2.2 Projection Metrics: $\|P_U - P_V\|_2$ and $\sin\Theta$ Relations (Reference)

Principal angles quantify alignment in a coordinate-free way; projector differences quantify alignment at the operator level. These metrics are central in perturbation theory (Davis–Kahan) and in numerical diagnostics: if P_U and P_V are close in operator norm, then the subspaces are close, and projections/reconstructions are stable.

Definition B.2.5 (Orthogonal Projectors onto Subspaces) Let $Q_U \in \mathbb{R}^{d\times k}$ have orthonormal columns spanning $\mathcal{U}$. The orthogonal projector onto $\mathcal{U}$ is

$$P_U := Q_U Q_U^\top.$$

Similarly define P_V for $\mathcal{V}$.

Proposition B.2.4 (Projector Distance Equals sin **of Largest Principal Angle)** *Let $\mathcal{U}, \mathcal{V} \subseteq \mathbb{R}^d$ have the same dimension k with principal angles $0 \leq \theta_1 \leq \cdots \leq \theta_k \leq \pi/2$. Then*

$$\|P_U - P_V\|_2 = \sin(\theta_k).$$

More generally, one has the operator-norm bound

$$\|P_U - P_V\|_2 \leq \sin(\theta_{\max}), \quad \textit{where } \theta_{\max} := \max_i \theta_i,$$

and equality holds when $\dim(\mathcal{U}) = \dim(\mathcal{V})$.

Remark B.2.4 (Practical Consequence: Projection Stability) For any $x \in \mathbb{R}^d$,

$$\|P_U x - P_V x\|_2 \le \|P_U - P_V\|_2 \|x\|_2 = \sin(\theta_{\max}) \|x\|_2.$$

Thus, if the largest principal angle is small, projections onto the two subspaces are uniformly close. This is exactly the stability statement used in PCA perturbation analyses.

Remark B.2.5 (Connection to Davis–Kahan) Davis–Kahan bounds $\sin(\theta_{\max})$ by $\|E\|_2/\text{gap}$ for perturbations of symmetric matrices. Combining with Proposition B.2.4 yields bounds on projector differences $\|P_U - P_V\|_2$.

B.3 Projection Results

Projection operators are the geometric backbone of least squares, PCA, and many stability arguments. The central idea is that orthogonal projection is the unique closest-point map in Euclidean geometry. This section records projector formulas, residual orthogonality characterizations, and the key operator-norm metrics used to quantify subspace distances and perturbation stability.

B.3.1 Orthogonal Projections onto Subspaces

B.3.1.1 Projector Formulas and Residual Orthogonality

Every orthogonal projector is determined by its range subspace $\mathcal{U}$. In practice, $\mathcal{U}$ is represented either (i) by a basis matrix A whose columns span $\mathcal{U}$, or (ii) by an orthonormal basis matrix Q (preferred numerically). The defining optimality condition is residual orthogonality: $r = x - Px$ is orthogonal to $\mathcal{U}$. This single condition yields the normal equations in regression and explains the Pythagorean error split.

Definition B.3.1 (Orthogonal Projector onto a Subspace) Let $\mathcal{U} \subseteq \mathbb{R}^d$ be a subspace. The *orthogonal projector* onto $\mathcal{U}$ is the map $P_{\mathcal{U}} : \mathbb{R}^d \to \mathbb{R}^d$ defined by

$$P_{\mathcal{U}}(x) \in \arg\min_{u \in \mathcal{U}} \|x - u\|_2.$$

The minimizer exists and is unique, and $P_{\mathcal{U}}$ is linear.

Proposition B.3.1 (Characterization of Orthogonal Projectors) *A matrix $P \in \mathbb{R}^{d\times d}$ is an orthogonal projector iff*

$$P^2 = P \quad \textit{and} \quad P^\top = P.$$

In that case, range$(P) = \mathcal{U}$ *and* null$(P) = \mathcal{U}^\perp$ *for some subspace* $\mathcal{U}$.

$P = A(A^\top A)^{-1}A^\top$ (Full Column Rank)

Let $A \in \mathbb{R}^{d\times k}$ have full column rank, and let $\mathcal{U} = \text{range}(A)$. Then the orthogonal projector onto $\mathcal{U}$ is

$$P_{\mathcal{U}} = A(A^\top A)^{-1}A^\top.$$

Proposition B.3.2 (Projector from a Full-Rank Spanning Matrix) *If $A \in \mathbb{R}^{d\times k}$ has full column rank, then*

$$P := A(A^\top A)^{-1}A^\top$$

is the orthogonal projector onto $\mathcal{U} = \text{range}(A)$.

Proof $P^\top = P$ since $(A^\top A)^{-1}$ is symmetric, and

$$P^2 = A(A^\top A)^{-1}A^\top A(A^\top A)^{-1}A^\top = A(A^\top A)^{-1}(A^\top A)(A^\top A)^{-1}A^\top = P.$$

Also $Px \in \text{range}(A)$ for all x, and $P(Az) = Az$ for all z, so P fixes range(A) and hence projects onto it. □

$P = QQ^\top$ for Orthonormal Basis Q

If we have an orthonormal basis, the formula simplifies.

Proposition B.3.3 (Projector from an Orthonormal Basis) *Let $Q \in \mathbb{R}^{d\times k}$ have orthonormal columns ($Q^\top Q = I_k$) and let* $\mathcal{U} = \text{range}(Q)$. *Then the orthogonal projector onto $\mathcal{U}$ is*

$$P_{\mathcal{U}} = QQ^\top.$$

Proof $P^\top = (QQ^\top)^\top = QQ^\top$ and $P^2 = QQ^\top QQ^\top = Q(Q^\top Q)Q^\top = QQ^\top$. □

Remark B.3.1 (Numerical Advantage) Computing $QQ^\top x$ avoids inverting $A^\top A$ and is typically more stable. In practice, one obtains Q via QR factorization or Gram–Schmidt/Householder orthogonalization.

Residual Characterization: $r = x - Px \in \mathcal{U}^\perp$

The defining optimality condition of orthogonal projection is orthogonality of the residual.

Proposition B.3.4 (Residual Orthogonality) *Let $\mathcal{U} \subseteq \mathbb{R}^d$ be a subspace and let $P_\mathcal{U}$ be its orthogonal projector. For any $x \in \mathbb{R}^d$, define*

$$u := P_\mathcal{U} x \in \mathcal{U}, \qquad r := x - u = (I - P_\mathcal{U})x.$$

Then

$$r \in \mathcal{U}^\perp, \qquad \text{i.e.} \qquad \langle r, u' \rangle = 0 \ \forall u' \in \mathcal{U}.$$

Consequently, Pythagoras holds:

$$\|x\|_2^2 = \|u\|_2^2 + \|r\|_2^2.$$

Remark B.3.2 (Normal Equations as Orthogonality) In least squares $\min_w \|Xw - y\|_2^2$, the fitted vector is $\widehat{y} = P_{\mathcal{C}(X)} y$ and the residual $r = y - \widehat{y}$ satisfies $X^\top r = 0$, which is exactly the condition $r \in \mathcal{C}(X)^\perp = \mathcal{N}(X^\top)$.

B.3.1.2 Principal Angles and Subspace Distances (Reference)

Subspace distances quantify how much two projection operators disagree. They are fundamental in perturbation theory: when a matrix is perturbed (e.g., covariance estimated from samples), we need to know how much the leading eigenspace rotates. Principal angles provide a basis-invariant parametrization of subspace misalignment, and $\sin\Theta$ theorems bound these angles by $\|E\|_2$ divided by a spectral gap.

Eigenspace Perturbation and $\sin\Theta$ Bounds (Reference)

Let A be symmetric with an invariant subspace $\mathcal{U}$ (e.g., top-k eigenvectors), and let $\widehat{A} = A + E$ with corresponding subspace $\widehat{\mathcal{U}}$. Then Davis–Kahan-type results yield

$$\|\sin\Theta(\widehat{\mathcal{U}}, \mathcal{U})\|_2 \lesssim \frac{\|E\|_2}{\text{gap}},$$

where gap is the separation between the eigenvalues defining $\mathcal{U}$ and the rest. Via projector metrics, this implies

$$\|P_{\widehat{\mathcal{U}}} - P_{\mathcal{U}}\|_2 \lesssim \frac{\|E\|_2}{\text{gap}}.$$

(See Chap. 4 for precise statements and constants.)

Stability under Spectral Gaps (Reference)

The qualitative rule is:

- **Large gap** $\Rightarrow$ eigenspace is stable (small rotation under perturbations).
- **Small/zero gap** $\Rightarrow$ eigenspace is fragile (can rotate substantially even under tiny perturbations).

In PCA, this means leading principal directions are reliable only when leading eigenvalues are well separated. In numerical algorithms, this explains why power/subspace iteration converges quickly when there is an eigengap.

B.3.2 Projections onto Convex Sets

Orthogonal projection onto a subspace is linear and admits closed-form formulas. Projection onto a general *closed convex* set is typically nonlinear, yet it retains two powerful properties: (1) it exists and is unique, and (2) it is non-expansive (does not increase distances). These facts are the geometric foundation of projected-gradient methods, feasibility algorithms, and proximal optimization. Proximal operators extend projection by allowing a penalty term; they formalize shrinkage/thresholding as generalized projections.

B.3.2.1 Existence/Uniqueness and Non-expansiveness

Projection Theorem for Closed Convex C

Let $C \subseteq \mathbb{R}^d$ be nonempty.

Definition B.3.2 (Euclidean Projection) The (Euclidean) projection of $x \in \mathbb{R}^d$ onto C is

$$\Pi_C(x) \in \arg\min_{z \in C} \|z - x\|_2.$$

Theorem B.3.1 (Projection Theorem) *If $C \subseteq \mathbb{R}^d$ is nonempty, closed, and convex, then for every $x \in \mathbb{R}^d$ there exists a* unique *point $\Pi_C(x) \in C$ that minimizes $\|z - x\|_2$ over $z \in C$.*

Remark B.3.3 (Why Closedness and Convexity Matter) Closedness ensures the minimum is attained (no "missing boundary" issues), while convexity ensures uniqueness (the objective $\|z - x\|_2^2$ is strictly convex in z and C contains all line segments between points).

Variational Inequality: $\langle x - \Pi_C(x), z - \Pi_C(x)\rangle \leq 0$

The projection point is characterized by a first-order geometric condition.

Theorem B.3.2 (Variational Inequality and Pythagorean Inequality) *Let C be nonempty, closed, and convex and let $p = \Pi_C(x)$. Then:*

*1. (*Variational inequality*) For all $z \in C$,*

$$\langle x - p,\ z - p\rangle \leq 0.$$

*2. (*Pythagorean-type inequality*) For all $z \in C$,*

$$\|x - z\|_2^2 \geq \|x - p\|_2^2 + \|p - z\|_2^2.$$

Proof Consider $\varphi(t) = \|x - (p + t(z - p))\|_2^2$ for $t \in [0, 1]$. Convexity gives $p + t(z - p) \in C$, so minimality of p implies $\varphi(t) \geq \varphi(0)$ and hence $\varphi'(0) \geq 0$. Compute

$$\varphi(t) = \|x - p - t(z - p)\|_2^2 = \|x - p\|_2^2 - 2t\langle x - p,\ z - p\rangle + t^2\|z - p\|_2^2,$$

so $\varphi'(0) = -2\langle x - p,\ z - p\rangle \geq 0$, proving (1). For (2), expand:

$$\|x - z\|_2^2 = \|(x - p) + (p - z)\|_2^2 = \|x - p\|_2^2 + \|p - z\|_2^2 + 2\langle x - p,\ p - z\rangle,$$

and apply (1) with z to get $\langle x - p,\ p - z\rangle \geq 0$. □

Theorem B.3.3 (Non-expansiveness of Projection) *If C is nonempty, closed, and convex, then for all $x, y \in \mathbb{R}^d$,*

$$\|\Pi_C(x) - \Pi_C(y)\|_2 \leq \|x - y\|_2.$$

Proof Let $p = \Pi_C(x)$ and $q = \Pi_C(y)$. Apply the variational inequality (Theorem B.3.2(1)) twice:

$$\langle x - p,\ q - p\rangle \leq 0 \quad \text{and} \quad \langle y - q,\ p - q\rangle \leq 0.$$

Add them to obtain $\langle x - y,\ p - q\rangle \geq \|p - q\|_2^2$. Then by Cauchy–Schwarz,

$$\|p - q\|_2^2 \leq \langle x - y,\ p - q\rangle \leq \|x - y\|_2 \, \|p - q\|_2,$$

and cancel $\|p - q\|_2$ (or handle the zero case) to conclude $\|p - q\|_2 \leq \|x - y\|_2$. □

Remark B.3.4 (Stability Message) Projection is a "safe" operation: it cannot amplify perturbations. This is why projected-gradient and feasibility methods are numerically robust.

B.3.2.2 Prox Operators as Generalized Projections

Projections enforce hard constraints ($z \in C$). Proximal operators enforce *soft constraints* via penalties. They are the workhorse of nonsmooth optimization and appear throughout ML in sparse learning, total-variation denoising, and regularized regression. Many famous update rules (soft-thresholding, shrinkage) are exactly proximal maps.

$\mathrm{prox}_{\lambda\Omega}(x) = \arg\min_z \frac{1}{2}\|z - x\|_2^2 + \lambda\Omega(z)$

Definition B.3.3 (Proximal Operator) Let $\Omega : \mathbb{R}^d \to (-\infty, \infty]$ be a proper function (typically closed and convex) and let $\lambda > 0$. The proximal operator is

$$\mathrm{prox}_{\lambda\Omega}(x) := \arg\min_{z\in\mathbb{R}^d} \left\{ \frac{1}{2}\|z - x\|_2^2 + \lambda\, \Omega(z) \right\},$$

when the minimizer exists and is unique.

Remark B.3.5 (Projection as a Special Prox) If $\Omega = \iota_C$ is the indicator function of a set C,

$$\iota_C(z) = \begin{cases} 0, & z \in C, \\ +\infty, & z \notin C, \end{cases}$$

then $\mathrm{prox}_{\lambda\iota_C}(x) = \Pi_C(x)$ for every $\lambda > 0$. Thus, projection is a proximal operator.

Remark B.3.6 (First-Order Optimality for Prox (Convex Case)) If Ω is convex, then $z = \mathrm{prox}_{\lambda\Omega}(x)$ satisfies

$$0 \in z - x + \lambda\, \partial\Omega(z),$$

equivalently $(x - z)/\lambda \in \partial\Omega(z)$. This is the algebraic characterization behind shrinkage rules.

Soft-Thresholding and Shrinkage Templates

Two canonical proximal maps are used throughout ML.

Proposition B.3.5 (Soft-Thresholding: Prox of ℓ_1) *Let $\Omega(z) = \|z\|_1$ and $\lambda > 0$. Then for $x \in \mathbb{R}^d$,*

$$\operatorname{prox}_{\lambda\|\cdot\|_1}(x) = \big(\operatorname{sign}(x_i)\max\{|x_i| - \lambda, 0\}\big)_{i=1}^d.$$

Proposition B.3.6 (Ridge Shrinkage: Prox of $\frac{1}{2}\|\cdot\|_2^2$) *Let $\Omega(z) = \frac{1}{2}\|z\|_2^2$ and $\lambda > 0$. Then*

$$\operatorname{prox}_{\lambda\Omega}(x) = \frac{1}{1+\lambda}\,x.$$

Remark B.3.7 (Geometric Intuition) Soft-thresholding produces exact zeros due to the "corners" of the ℓ_1 geometry, inducing sparsity. Ridge shrinkage rescales smoothly because the ℓ_2 geometry is round, inducing shrinkage without sparsity.

Remark B.3.8 (Non-expansiveness (Reference)) For closed convex Ω, proximal operators are firmly non-expansive, hence in particular non-expansive:

$$\|\operatorname{prox}_{\lambda\Omega}(x) - \operatorname{prox}_{\lambda\Omega}(y)\|_2 \le \|x - y\|_2.$$

This is a key fact used in convergence proofs of proximal-gradient methods.

Appendix C
Probability Reference

C.1 Common Distributions at a Glance

This appendix records frequently used probability families and their basic formulas. Throughout, $\mathbb{P}$ denotes probability, $\mathbb{E}$ expectation, and Var variance. All distributions below are stated with explicit parameter conventions.

C.1.1 Discrete Families

C.1.1.1 Bernoulli/Binomial/Multinomial/Poisson

PMFs, Means, Variances

See Table C.1.

Remark C.1.1 (Multinomial as Dependent Counts) If $X = (X_1, \ldots, X_K) \sim \text{Multinomial}(n, p)$, then $\sum_{i=1}^{K} X_i = n$ almost surely, so the coordinates are not independent. The covariance identity $\text{Cov}(X_i, X_j) = -np_i p_j$ for $i \neq j$ reflects this "fixed-sum" dependence.

Poisson Limit of Binomial (Reference)

A classical scaling limit turns rare Bernoulli trials into Poisson counts.

P. Singh, B. Raman, *Foundations of Machine Learning and AI*, Studies in Big Data 199, https://doi.org/10.1007/978-3-032-30336-3

Table C.1 Discrete distributions: PMF, mean, variance (parameter conventions used in this book)

Family	Parameters	Support	PMF $\mathbb{P}(X = x)$	Mean/Variance
Bernoulli	$p \in (0, 1)$	$\{0, 1\}$	$p^x(1-p)^{1-x}$	$\mathbb{E}[X] = p,\ \mathrm{Var}(X) = p(1-p)$
Binomial	$n \in \mathbb{N},\ p \in (0, 1)$	$\{0, 1, \dots, n\}$	$\binom{n}{x} p^x (1-p)^{n-x}$	$\mathbb{E}[X] = np,\ \mathrm{Var}(X) = np(1-p)$
Multinomial	$n \in \mathbb{N},\ p \in \Delta^{K-1}$	$\{x \in \mathbb{N}^K : \sum_i x_i = n\}$	$\frac{n!}{\prod_{i=1}^K x_i!} \prod_{i=1}^K p_i^{x_i}$	$\mathbb{E}[X_i] = np_i$, $\mathrm{Var}(X_i) = np_i(1-p_i),\ \mathrm{Cov}(X_i, X_j) = -np_i p_j$
Poisson	$\lambda > 0$	$\{0, 1, 2, \dots\}$	$e^{-\lambda} \frac{\lambda^x}{x!}$	$\mathbb{E}[X] = \lambda,\ \mathrm{Var}(X) = \lambda$

Theorem C.1.1 (Poisson Limit of Binomial) *Let $X_n \sim \mathrm{Binomial}(n, p_n)$ with $p_n \to 0$ and $np_n \to \lambda > 0$ as $n \to \infty$. Then for every fixed $k \in \{0, 1, 2, \dots\}$,*

$$\mathbb{P}(X_n = k) \longrightarrow e^{-\lambda} \frac{\lambda^k}{k!},$$

i.e. $X_n \Rightarrow \mathrm{Poisson}(\lambda)$ in distribution.

Proof For fixed k,

$$\mathbb{P}(X_n = k) = \binom{n}{k} p_n^k (1-p_n)^{n-k} = \frac{n(n-1)\cdots(n-k+1)}{k!} p_n^k (1-p_n)^n (1-p_n)^{-k}.$$

Since $np_n \to \lambda$ and $p_n \to 0$,

$$\frac{n(n-1)\cdots(n-k+1)}{n^k} \to 1, \qquad (1-p_n)^{-k} \to 1, \qquad (1-p_n)^n \to e^{-\lambda}.$$

Also $n^k p_n^k = (np_n)^k \to \lambda^k$. Combining gives the claimed limit. □

C.1.1.2 Geometric/Negative Binomial (Optional)

Waiting-Time Interpretations and Memorylessness

These distributions model waiting times in Bernoulli trials with success probability p.

Definition C.1.1 (Geometric Distribution: Trials-Until-First-Success) Let $(B_t)_{t\geq 1}$ be i.i.d. Bernoulli(p) trials. Define T as the index of the first success:

$$T := \min\{t \geq 1 : B_t = 1\}.$$

Then T has the *Geometric*(p) distribution on $\{1, 2, 3, \dots\}$ with PMF

$$\mathbb{P}(T = t) = (1 - p)^{t-1} p, \qquad t = 1, 2, \dots$$

Proposition C.1.1 (Memorylessness of Geometric(p)) *If* $T \sim$ Geometric(p) *on* $\{1, 2, \dots\}$, *then for all integers* $m, n \geq 1$,

$$\mathbb{P}(T > m + n \mid T > m) = \mathbb{P}(T > n) = (1 - p)^n.$$

Proof $\mathbb{P}(T > t) = (1 - p)^t$ for $t \geq 0$. Hence

$$\mathbb{P}(T > m + n \mid T > m) = \frac{(1 - p)^{m+n}}{(1 - p)^m} = (1 - p)^n.$$

□

Remark C.1.2 (Alternative Geometric Convention) Some sources define G as the number of failures before the first success, supported on $\{0, 1, 2, \dots\}$ with $\mathbb{P}(G = g) = (1 - p)^g p$. In that convention, $T = G + 1$.

Mean/Variance and Parameter Conventions

Proposition C.1.2 (Mean/Variance: Geometric(p)) *If* $T \sim$ Geometric(p) *on* $\{1, 2, \dots\}$, *then*

$$\mathbb{E}[T] = \frac{1}{p}, \qquad \mathrm{Var}(T) = \frac{1 - p}{p^2}.$$

If $G = T - 1$ *is the failures-before-success version on* $\{0, 1, \dots\}$, *then* $\mathbb{E}[G] = \frac{1-p}{p}$ *and* $\mathrm{Var}(G) = \frac{1-p}{p^2}$.

Definition C.1.2 (Negative Binomial: Failures Before r Successes) Fix $r \in \mathbb{N}$ and $p \in (0, 1)$. Let $(B_t)_{t\geq 1}$ be i.i.d. Bernoulli(p) trials. Let F be the number of failures observed before the r-th success. Then F has the *Negative Binomial*(r, p) distribution on $\{0, 1, 2, \dots\}$ with PMF

$$\mathbb{P}(F = f) = \binom{f + r - 1}{r - 1}(1 - p)^f p^r, \qquad f = 0, 1, 2, \dots$$

Remark C.1.3 (Alternative Negative Binomial Convention) Some sources define the random variable as the *total number of trials* T_r needed to obtain r successes, supported on $\{r, r+1, \dots\}$. Under our convention, $T_r = F + r$ and thus T_r has PMF

$$\mathbb{P}(T_r = t) = \binom{t-1}{r-1} p^r (1-p)^{t-r}, \qquad t = r, r+1, \dots$$

Proposition C.1.3 (Mean/Variance: Negative Binomial(r, p)) *If $F \sim$ NegBin (r, p) is the number of failures before the r-th success, then*

$$\mathbb{E}[F] = \frac{r(1-p)}{p}, \qquad \mathrm{Var}(F) = \frac{r(1-p)}{p^2}.$$

Equivalently, for $T_r = F + r$ (total trials until r successes),

$$\mathbb{E}[T_r] = \frac{r}{p}, \qquad \mathrm{Var}(T_r) = \frac{r(1-p)}{p^2}.$$

Remark C.1.4 (Geometric as a Special Case) Geometric(p) is the special case $r = 1$ of Negative Binomial: the number of failures before the first success.

C.1.2 Continuous Families

This section lists continuous distributions most frequently used in ML/AI: Gaussian for noise models and quadratic objectives, Exponential/Gamma for waiting times and rate modeling, and Beta/Dirichlet for probabilities on [0, 1] and on the simplex. We include MGFs (where convenient), closure properties, and conjugacy statements that reappear in Bayesian estimation.

C.1.2.1 Gaussian/Exponential/Gamma

Gaussian (Normal) Distribution

Definition C.1.3 (Univariate Gaussian) A random variable $X \sim \mathcal{N}(\mu, \sigma^2)$ has density

$$f(x) = \frac{1}{\sqrt{2\pi\sigma^2}} \exp\left(-\frac{(x-\mu)^2}{2\sigma^2}\right), \qquad x \in \mathbb{R}.$$

Definition C.1.4 (Multivariate Gaussian) A random vector $X \in \mathbb{R}^d$ is $X \sim \mathcal{N}(\mu, \Sigma)$ with $\Sigma \succ 0$ if its density is

$$f(x) = \frac{1}{(2\pi)^{d/2} \det(\Sigma)^{1/2}} \exp\left(-\frac{1}{2}(x-\mu)^\top \Sigma^{-1}(x-\mu)\right), \qquad x \in \mathbb{R}^d.$$

Proposition C.1.4 (MGF and Closure: Gaussian) *If $X \sim \mathcal{N}(\mu, \sigma^2)$, its MGF is*

$$M_X(t) = \mathbb{E}[e^{tX}] = \exp\left(\mu t + \frac{1}{2}\sigma^2 t^2\right), \qquad t \in \mathbb{R}.$$

If $X_i \sim \mathcal{N}(\mu_i, \sigma_i^2)$ are independent, then $\sum_i X_i \sim \mathcal{N}(\sum_i \mu_i, \sum_i \sigma_i^2)$. More generally, if $X \sim \mathcal{N}(\mu, \Sigma)$ and A is a matrix and b a vector, then

$$AX + b \sim \mathcal{N}(A\mu + b,\ A\Sigma A^\top).$$

Remark C.1.5 (Quadratic Forms and Mahalanobis Geometry) Gaussian log-likelihoods involve $\log\det(\Sigma)$ and the Mahalanobis term $(x-\mu)^\top \Sigma^{-1}(x-\mu)$, which is why log-det and trace calculus identities appear repeatedly in ML.

Exponential Distribution

Definition C.1.5 (Exponential) $X \sim \mathrm{Exp}(\lambda)$ with rate $\lambda > 0$ has density

$$f(x) = \lambda e^{-\lambda x}, \qquad x \geq 0.$$

It satisfies $\mathbb{E}[X] = 1/\lambda$ and $\mathrm{Var}(X) = 1/\lambda^2$.

Proposition C.1.5 (MGF and Memorylessness: Exponential) *If $X \sim \mathrm{Exp}(\lambda)$, then for $t < \lambda$,*

$$M_X(t) = \mathbb{E}[e^{tX}] = \frac{\lambda}{\lambda - t}.$$

Moreover, X is memoryless: for all $s, t \geq 0$,

$$\mathbb{P}(X > t + s \mid X > t) = \mathbb{P}(X > s) = e^{-\lambda s}.$$

Gamma Distribution

Definition C.1.6 (Gamma (Shape–Rate Parameterization)) $X \sim \mathrm{Gamma}(\alpha, \lambda)$ with shape $\alpha > 0$ and rate $\lambda > 0$ has density

$$f(x) = \frac{\lambda^\alpha}{\Gamma(\alpha)} x^{\alpha-1} e^{-\lambda x}, \qquad x \geq 0.$$

Its mean and variance are

$$\mathbb{E}[X] = \frac{\alpha}{\lambda}, \qquad \mathrm{Var}(X) = \frac{\alpha}{\lambda^2}.$$

Proposition C.1.6 (MGF and Closure: Gamma) *If* $X \sim \mathrm{Gamma}(\alpha, \lambda)$*, then for* $t < \lambda$*,*

$$M_X(t) = \left(\frac{\lambda}{\lambda - t}\right)^\alpha .$$

If $X_1 \sim \mathrm{Gamma}(\alpha_1, \lambda)$ *and* $X_2 \sim \mathrm{Gamma}(\alpha_2, \lambda)$ *are independent with the same rate, then* $X_1 + X_2 \sim \mathrm{Gamma}(\alpha_1 + \alpha_2, \lambda)$*. If* $c > 0$*, then* $cX \sim \mathrm{Gamma}(\alpha, \lambda/c)$ *(rate rescales).*

Exponential as Gamma Special Case

Remark C.1.6 (Embedding) $\mathrm{Exp}(\lambda) = \mathrm{Gamma}(\alpha = 1, \lambda)$. More generally, if $\alpha \in \mathbb{N}$, $\mathrm{Gamma}(\alpha, \lambda)$ is the distribution of the sum of α independent $\mathrm{Exp}(\lambda)$ variables (Erlang distribution).

C.1.2.2 Beta/Dirichlet and Simplex Geometry

Beta and Dirichlet distributions are distributions on probability parameters: Beta on $[0, 1]$ and Dirichlet on the simplex Δ^{K-1}. They are conjugate priors for Bernoulli/Binomial and Multinomial models, respectively, making posterior updates algebraic. Geometrically, the simplex constraint $\sum_i p_i = 1$ is built into the density via the normalization constant (a Beta function/multivariate Beta function).

Beta Distribution

Definition C.1.7 (Beta) $P \sim \mathrm{Beta}(\alpha, \beta)$ with $\alpha, \beta > 0$ has density on $p \in (0, 1)$

$$f(p) = \frac{1}{\mathrm{B}(\alpha, \beta)}\, p^{\alpha-1}(1-p)^{\beta-1}, \qquad \mathrm{B}(\alpha, \beta) = \frac{\Gamma(\alpha)\Gamma(\beta)}{\Gamma(\alpha+\beta)}.$$

Its mean and variance are

$$\mathbb{E}[P] = \frac{\alpha}{\alpha+\beta}, \qquad \mathrm{Var}(P) = \frac{\alpha\beta}{(\alpha+\beta)^2(\alpha+\beta+1)}.$$

Dirichlet Distribution

Definition C.1.8 (Dirichlet) Let $\alpha = (\alpha_1, \dots, \alpha_K)$ with $\alpha_i > 0$. A random vector $P = (P_1, \dots, P_K)$ has the Dirichlet(α) distribution on the simplex

$$\Delta^{K-1} := \left\{ p \in \mathbb{R}^K : \ p_i \geq 0, \ \sum_{i=1}^{K} p_i = 1 \right\}$$

if its density is

$$f(p) = \frac{1}{\mathrm{B}(\alpha)} \prod_{i=1}^{K} p_i^{\alpha_i - 1}, \qquad \mathrm{B}(\alpha) = \frac{\prod_{i=1}^{K} \Gamma(\alpha_i)}{\Gamma\left(\sum_{i=1}^{K} \alpha_i\right)}.$$

Its mean is

$$\mathbb{E}[P_i] = \frac{\alpha_i}{\alpha_0}, \qquad \alpha_0 := \sum_{i=1}^{K} \alpha_i.$$

Remark C.1.7 (Beta as Dirichlet Special Case) For $K = 2$, Dirichlet(α_1, α_2) induces $P_1 \sim \text{Beta}(\alpha_1, \alpha_2)$ with $P_2 = 1 - P_1$.

Conjugacy with Bernoulli/Binomial/Multinomial

Proposition C.1.7 (Beta–Binomial Conjugacy) *Let* $P \sim \text{Beta}(\alpha, \beta)$ *and conditional on* $P = p$ *let* $X \sim \text{Binomial}(n, p)$*. Then the posterior is*

$$P \mid X = x \sim \text{Beta}(\alpha + x, \ \beta + n - x).$$

In particular, for Bernoulli(p) (the case $n = 1$*), observing* $X \in \{0, 1\}$ *updates* (α, β) *by adding* $(X, 1 - X)$*.*

Proposition C.1.8 (Dirichlet–Multinomial Conjugacy) *Let* $P \sim \text{Dirichlet}(\alpha)$ *and conditional on* $P = p$ *let* $X = (X_1, \dots, X_K) \sim \text{Multinomial}(n, p)$*. Then the posterior is*

$$P \mid X = x \sim \text{Dirichlet}(\alpha_1 + x_1, \dots, \alpha_K + x_K).$$

Remark C.1.8 (Pseudo-Count Interpretation) The parameters α_i act as prior counts: observing x_i additional occurrences adds directly to α_i. The total concentration α_0 controls how strongly the prior resists updates.

Simplex Constraints and Normalization Constants (Reference)

Remark C.1.9 (Normalization via (Multi)variate Beta Function) The normalization constants $\mathrm{B}(\alpha, \beta)$ and $\mathrm{B}(\alpha)$ ensure densities integrate to 1 over $(0, 1)$ and Δ^{K-1}, respectively. They are expressed in terms of Gamma functions:

$$\mathrm{B}(\alpha, \beta) = \frac{\Gamma(\alpha)\Gamma(\beta)}{\Gamma(\alpha + \beta)}, \qquad \mathrm{B}(\alpha) = \frac{\prod_{i=1}^{K} \Gamma(\alpha_i)}{\Gamma(\sum_i \alpha_i)}.$$

This makes Beta/Dirichlet analytically tractable and is the reason conjugacy updates remain algebraic.

C.2 Expectation and Variance Formulas

Expectation and variance identities are the algebraic backbone of statistical learning theory and probabilistic modeling. The tower property (law of total expectation) formalizes iterated averaging and underlies many conditioning tricks. The law of total variance decomposes uncertainty into "noise within groups" plus "uncertainty of group means," a conceptual analogue of bias–variance and explained vs. unexplained variability. Matrix covariance identities connect probability to linear algebra: covariance transforms quadratically under linear maps, and trace identities translate second moments into scalar summaries (total energy/variance).

C.2.1 Useful Identities

C.2.1.1 Law of Total Expectation and Total Variance

$\mathbb{E}[X] = \mathbb{E}[\mathbb{E}[X \mid Y]]$

Theorem C.2.1 (Law of Total Expectation (Tower Property)) *Let X be an integrable random variable and Y any random variable (or, more generally, any σ-field). Then*

$$\mathbb{E}[X] = \mathbb{E}\big[\mathbb{E}[X \mid Y]\big].$$

Remark C.2.1 (Practical Meaning) Compute expectation by conditioning on Y first, then averaging over Y. This is frequently used to simplify expectations when X has a simple conditional distribution given Y.

$\mathrm{Var}(X) = \mathbb{E}[\mathrm{Var}(X \mid Y)] + \mathrm{Var}(\mathbb{E}[X \mid Y])$

Theorem C.2.2 (Law of Total Variance) *Let X be square-integrable and Y any random variable. Then*

$$\mathrm{Var}(X) = \mathbb{E}\big[\mathrm{Var}(X \mid Y)\big] + \mathrm{Var}\big(\mathbb{E}[X \mid Y]\big).$$

Proof Write

$$\mathrm{Var}(X) = \mathbb{E}[X^2] - \big(\mathbb{E}[X]\big)^2.$$

By the tower property,

$$\mathbb{E}[X^2] = \mathbb{E}\big[\mathbb{E}[X^2 \mid Y]\big], \qquad \mathbb{E}[X] = \mathbb{E}\big[\mathbb{E}[X \mid Y]\big].$$

Also, by definition,

$$\mathrm{Var}(X \mid Y) = \mathbb{E}[X^2 \mid Y] - \big(\mathbb{E}[X \mid Y]\big)^2.$$

Taking expectations yields

$$\mathbb{E}[\mathrm{Var}(X \mid Y)] = \mathbb{E}[X^2] - \mathbb{E}\big[\big(\mathbb{E}[X \mid Y]\big)^2\big].$$

Finally,

$$\mathrm{Var}(\mathbb{E}[X \mid Y]) = \mathbb{E}\big[\big(\mathbb{E}[X \mid Y]\big)^2\big] - \big(\mathbb{E}[X]\big)^2,$$

and summing the last two displays gives $\mathrm{Var}(X)$. □

Remark C.2.2 (Interpretation: Within + Between) $\mathbb{E}[\mathrm{Var}(X \mid Y)]$ is the average noise within each Y-slice, while $\mathrm{Var}(\mathbb{E}[X \mid Y])$ is the variance of the conditional mean across Y. This decomposition is a probabilistic analogue of explained vs. residual variance.

C.2.1.2 Covariance Identities and Matrix Forms

$\mathrm{Cov}(AX) = A\,\mathrm{Cov}(X)\,A^\top$

Definition C.2.1 (Covariance Matrix) For a random vector $X \in \mathbb{R}^d$ with mean $\mu = \mathbb{E}[X]$, its covariance is

$$\Sigma := \mathrm{Cov}(X) := \mathbb{E}\big[(X - \mu)(X - \mu)^\top\big] \in \mathbb{R}^{d \times d}.$$

Proposition C.2.1 (Covariance under Affine Maps) *Let $X \in \mathbb{R}^d$ have covariance Σ and let $Y = AX + b$ with $A \in \mathbb{R}^{k \times d}$ and $b \in \mathbb{R}^k$. Then*

$$\mathrm{Cov}(Y) = A\,\Sigma\,A^\top.$$

In particular, translation by b does not affect covariance.

Proof Let $\mu = \mathbb{E}[X]$ so that $\mathbb{E}[Y] = A\mu + b$. Then

$$Y - \mathbb{E}[Y] = A(X - \mu),$$

hence

$$\mathrm{Cov}(Y) = \mathbb{E}\big[A(X-\mu)(X-\mu)^\top A^\top\big] = A\,\mathbb{E}\big[(X-\mu)(X-\mu)^\top\big]A^\top = A\Sigma A^\top.$$

□

Remark C.2.3 (Scalar Special Case) If $a \in \mathbb{R}^d$ and $Z = a^\top X$, then $\mathrm{Var}(Z) = a^\top \Sigma a$. Thus quadratic forms are variances of linear functionals.

Trace Forms: $\mathbb{E}\|X\|_2^2 = \mathrm{tr}(\mathrm{Cov}(X)) + \|\mathbb{E}X\|_2^2$

Proposition C.2.2 (Second-Moment Trace Identity) *Let $X \in \mathbb{R}^d$ have mean $\mu = \mathbb{E}[X]$ and covariance $\Sigma = Cov(X)$. Then*

$$\mathbb{E}\|X\|_2^2 = \mathrm{tr}(\Sigma) + \|\mu\|_2^2.$$

Equivalently,

$$\mathrm{tr}(\Sigma) = \mathbb{E}\|X - \mu\|_2^2.$$

Proof Expand:

$$\mathbb{E}\|X\|_2^2 = \mathbb{E}[X^\top X] = \mathbb{E}\big[\mathrm{tr}(XX^\top)\big] = \mathrm{tr}\big(\mathbb{E}[XX^\top]\big).$$

Also,

$$\mathbb{E}[XX^\top] = \mathbb{E}[(X-\mu+\mu)(X-\mu+\mu)^\top] = \mathbb{E}[(X-\mu)(X-\mu)^\top] + \mu\mu^\top = \Sigma + \mu\mu^\top.$$

Taking trace gives

$$\mathbb{E}\|X\|_2^2 = \mathrm{tr}(\Sigma) + \mathrm{tr}(\mu\mu^\top) = \mathrm{tr}(\Sigma) + \|\mu\|_2^2,$$

since $\mathrm{tr}(\mu\mu^\top) = \mu^\top\mu$. □

Remark C.2.4 (Total Variance as Trace) $\mathrm{tr}(\Sigma) = \sum_{i=1}^{d} \mathrm{Var}(X_i)$ is the total marginal variance. In PCA, $\mathrm{tr}(\widehat{\Sigma})$ equals the total explained variance available in the data, and the top-k eigenvalues capture a fraction of this trace.

C.2.2 Conditional Expectation Tricks

Conditional expectation is both a probabilistic object and a geometric one: it is an L^2-projection onto the space of functions measurable with respect to available information. This viewpoint underlies optimal prediction, regression, and many proof shortcuts in learning theory. We record (i) the tower property in sigma-field form and the minimal measurability facts, and (ii) the squared-loss optimality (best predictor) and the orthogonality principle that parallels least squares geometry.

C.2.2.1 Tower Property and Conditioning on Sigma-Fields (Optional)

Sigma-Field Notation and Measurability Checks (Reference)

Definition C.2.2 (σ-Field (Sigma-Algebra)) Let Ω be a sample space. A collection $\mathcal{G} \subseteq 2^{\Omega}$ is a *σ-field* if: (i) $\Omega \in \mathcal{G}$, (ii) $E \in \mathcal{G} \Rightarrow E^c \in \mathcal{G}$, (iii) if $E_1, E_2, \dots \in \mathcal{G}$ then $\bigcup_{i=1}^{\infty} E_i \in \mathcal{G}$.

Definition C.2.3 (Measurability) A random variable $Z : \Omega \to \mathbb{R}$ is $\mathcal{G}$-*measurable* if $\{\omega : Z(\omega) \leq t\} \in \mathcal{G}$ for all $t \in \mathbb{R}$. Equivalently, Z is measurable with respect to the information encoded by $\mathcal{G}$.

Definition C.2.4 (Conditional Expectation Given a σ-Field) Let X be integrable and let $\mathcal{G}$ be a σ-field. A random variable Z is a version of $\mathbb{E}[X \mid \mathcal{G}]$ if:

1. Z is $\mathcal{G}$-measurable, and
2. for all $G \in \mathcal{G}$,

$$\mathbb{E}[X \, \mathbf{1}_G] = \mathbb{E}[Z \, \mathbf{1}_G].$$

Remark C.2.5 (Conditioning on a Random Variable) For a random variable Y, $\mathbb{E}[X \mid Y]$ is shorthand for $\mathbb{E}[X \mid \sigma(Y)]$, where $\sigma(Y)$ is the σ-field generated by Y.

Theorem C.2.3 (Tower Property (Sigma-Field Form)) *Let X be integrable and let $\mathcal{H} \subseteq \mathcal{G}$ be σ-fields. Then*

$$\mathbb{E}[\mathbb{E}[X \mid \mathcal{G}] \mid \mathcal{H}] = \mathbb{E}[X \mid \mathcal{H}].$$

In particular, taking expectations yields $\mathbb{E}[\mathbb{E}[X \mid \mathcal{G}]] = \mathbb{E}[X]$.

Remark C.2.6 (Quick Measurability Check) If Z is $\mathcal{G}$-measurable, then $\mathbb{E}[Z \mid \mathcal{G}] = Z$ (a.s.). This is often the fastest way to simplify nested conditionals.

C.2.2.2 Best Prediction under Squared Loss

Projection Viewpoint: $\mathbb{E}[Y \mid X]$ Minimizes $\mathbb{E}(Y - g(X))^2$

In $L^2(\Omega)$, squared loss corresponds to Euclidean distance. The set of predictors $g(X)$ (measurable functions of X) forms a closed subspace (under mild integrability conditions), and conditional expectation is the orthogonal projection of Y onto this subspace. This is the probabilistic analogue of least squares projection onto a column space.

Theorem C.2.4 (Conditional Expectation as L^2-Projection) *Let Y be square-integrable and let $\mathcal{G}$ be a σ-field. Then $\mathbb{E}[Y \mid \mathcal{G}]$ is the unique minimizer (up to a.s. equality) of*

$$\min_{Z\ \mathcal{G}\text{-measurable}} \mathbb{E}\big[(Y - Z)^2\big].$$

Equivalently, for a random variable X, the function $g^\star(X) = \mathbb{E}[Y \mid X]$ minimizes

$$\min_{g} \mathbb{E}\big[(Y - g(X))^2\big]$$

over all measurable functions g with $\mathbb{E}[g(X)^2] < \infty$.

Remark C.2.7 (Uniqueness is a.s.) Two predictors that differ only on a probability-zero event are equally valid; hence conditional expectations are defined up to almost sure equality.

Orthogonality Principle for Regression Residuals (Reference)

The projection theorem yields the fundamental orthogonality condition.

Theorem C.2.5 (Orthogonality Principle) *Let $Z^\star = \mathbb{E}[Y \mid \mathcal{G}]$ and define the residual $R := Y - Z^\star$. Then for every square-integrable $\mathcal{G}$-measurable random variable W,*

$$\mathbb{E}[R\, W] = 0.$$

Equivalently,

$$\mathbb{E}\big[(Y - \mathbb{E}[Y \mid \mathcal{G}])\, W\big] = 0 \quad \forall\, W \in L^2(\mathcal{G}).$$

Proof sketch By the defining property of conditional expectation, for any $G \in \mathcal{G}$, $\mathbb{E}[R\,\mathbf{1}_G] = \mathbb{E}[Y\mathbf{1}_G] - \mathbb{E}[Z^\star \mathbf{1}_G] = 0$. Extend from indicators to simple functions and then to all $W \in L^2(\mathcal{G})$ by density. □

Remark C.2.8 (ML Interpretation: Residual Uncorrelated with Features) Take $\mathcal{G} = \sigma(X)$ and $W = g(X)$. Then

$$\mathbb{E}\big[(Y - \mathbb{E}[Y \mid X])\, g(X)\big] = 0,$$

meaning the optimal residual is uncorrelated with every function of the features. In linear regression, this reduces to the sample normal equations $X^\top r = 0$ when $g(X)$ ranges over linear functions.

Appendix D
Concentration and Inequality Toolbox

D.1 Inequality Templates

Many ML generalization and statistical guarantees reduce to bounding tail probabilities of random quantities (empirical means, losses, gradients). A small set of inequalities is used repeatedly: Markov (for nonnegative variables), Chebyshev (for second-moment control), and Jensen (for convexity-based bounds). These are "templates": each has a standard derivation and a standard usage pattern. Equally important are the pitfalls: heavy tails can invalidate variance-based bounds and dependence can destroy i.i.d. concentration logic.

D.1.1 Markov, Chebyshev, Jensen

D.1.1.1 Standard Derivations and When to Use

Markov on Nonnegative RVs; Chebyshev via Second Moment

Theorem D.1.1 (Markov Inequality) *If* $X \geq 0$ *almost surely and* $\mathbb{E}[X] < \infty$, *then for any* $t > 0$,

$$\mathbb{P}(X \geq t) \leq \frac{\mathbb{E}[X]}{t}.$$

Proof Let $\mathbf{1}\{X \geq t\}$ be the indicator of the event $\{X \geq t\}$. Since $X \geq t\,\mathbf{1}\{X \geq t\}$, taking expectations yields

$$\mathbb{E}[X] \geq t\,\mathbb{E}[\mathbf{1}\{X \geq t\}] = t\,\mathbb{P}(X \geq t).$$

Divide by t. □

P. Singh, B. Raman, *Foundations of Machine Learning and AI*, Studies in Big Data 199, https://doi.org/10.1007/978-3-032-30336-3

Remark D.1.1 (When to Use Markov) Markov is used when you can bound a nonnegative expectation but not higher moments, or as the first step of Chernoff bounds by applying Markov to $e^{\lambda X}$.

Theorem D.1.2 (Chebyshev Inequality) *Let X have finite mean* $\mu = \mathbb{E}[X]$ *and finite variance* $\sigma^2 = \mathrm{Var}(X)$. *Then for any* $t > 0$,

$$\mathbb{P}(|X - \mu| \geq t) \leq \frac{\sigma^2}{t^2}.$$

Proof Apply Markov to the nonnegative variable $(X - \mu)^2$:

$$\mathbb{P}(|X - \mu| \geq t) = \mathbb{P}\big((X - \mu)^2 \geq t^2\big) \leq \frac{\mathbb{E}[(X - \mu)^2]}{t^2} = \frac{\sigma^2}{t^2}.$$

□

Remark D.1.2 (When to Use Chebyshev) Chebyshev is the default tool when only a variance bound is available. It gives polynomial $(1/t^2)$ tails, so it is typically loose compared to sub-Gaussian bounds, but it works without boundedness assumptions.

Jensen for Convex Functions and Risk Bounds

Theorem D.1.3 (Jensen Inequality) *Let* $\varphi : \mathbb{R} \to \mathbb{R}$ *be convex and let X be integrable. Then*

$$\varphi(\mathbb{E}[X]) \leq \mathbb{E}[\varphi(X)].$$

More generally, for $X \in \mathbb{R}^d$ *and convex* $\varphi : \mathbb{R}^d \to \mathbb{R}$, *the same inequality holds.*

Remark D.1.3 (When to Use Jensen) Jensen is used to swap expectation and convex transformations. Typical ML uses include:

- upper-bounding risk or log-likelihood terms,
- proving $\log \mathbb{E}[e^X] \geq \mathbb{E}[X]$ (via convexity of e^x or concavity of log),
- bounding generalization quantities by convex surrogates.

Remark D.1.4 (Concave Version) If ψ is concave, then $\psi(\mathbb{E}[X]) \geq \mathbb{E}[\psi(X)]$. A common example is $\psi = \log$, giving $\log \mathbb{E}[X] \geq \mathbb{E}[\log X]$ for $X > 0$.

D.1.1.2 Common Pitfalls and Counterexamples

Heavy Tails: Variance May Be Infinite

Pitfall Chebyshev requires finite variance. Heavy-tailed distributions can have infinite variance, making Chebyshev inapplicable.

Example D.1.1 (Infinite Variance: Pareto Tail) Let X have a Pareto(x_m, α) distribution with $x_m = 1$ and shape $\alpha \in (1, 2]$:

$$\mathbb{P}(X \geq t) = t^{-\alpha}, \qquad t \geq 1.$$

Then $\mathbb{E}[X] < \infty$ for $\alpha > 1$, but $\mathrm{Var}(X) = \infty$ for $\alpha \leq 2$. Hence Chebyshev cannot be applied (variance-based concentration fails), and sample averages may concentrate very slowly.

Remark D.1.5 (ML Implication) If losses/gradients have heavy tails, variance-based generalization and SGD noise models can be misleading. Robust methods or truncation/median-of-means style estimators may be required.

Dependence Breaks Naive Concentration

Pitfall Many concentration bounds assume i.i.d. samples. Strong dependence can invalidate such bounds.

Example D.1.2 (Perfect Dependence: No $1/n$ Variance Reduction) Let $X_1 = \cdots = X_n = Z$ almost surely, with $\mathrm{Var}(Z) = \sigma^2 > 0$. Then the sample mean $\bar{X} = \frac{1}{n}\sum_{i=1}^{n} X_i = Z$ has variance $\mathrm{Var}(\bar{X}) = \sigma^2$, not σ^2/n. Thus, increasing n provides no concentration benefit under perfect dependence.

Remark D.1.6 (ML Implication) Time series, correlated minibatches, data augmentation with shared randomness, and graph data violate independence. One must use dependence-aware tools (mixing conditions, martingale inequalities, block arguments) or adjust evaluation protocols.

D.1.2 Chernoff Bounding Method

Chernoff bounds are the workhorse of modern concentration: they convert tail probability questions into optimization problems over a scalar parameter $\lambda > 0$ using the moment generating function (MGF). The key move is Markov's inequality applied to $e^{\lambda X}$. Optimizing over λ yields exponentially decaying bounds whenever the log-MGF (cumulant generating function, CGF) is finite in a neighborhood of 0. Hoeffding- and Bernstein-type inequalities are standard outcomes of this method for bounded and sub-exponential variables.

D.1.2.1 MGF/CGF and Optimization of Parameters

Chernoff: $\mathbb{P}(X \geq t) \leq \inf_{\lambda>0} e^{-\lambda t}\mathbb{E}e^{\lambda X}$

Definition D.1.1 (MGF and CGF) For a random variable X, the *moment generating function* (MGF) is

$$M_X(\lambda) := \mathbb{E}[e^{\lambda X}],$$

whenever it is finite. The *cumulant generating function* (CGF) is

$$\psi_X(\lambda) := \log M_X(\lambda) = \log \mathbb{E}[e^{\lambda X}].$$

Theorem D.1.4 (Chernoff Bound (One-Sided)) *Let X be a random variable such that $M_X(\lambda) < \infty$ for some $\lambda > 0$. Then for any $t \in \mathbb{R}$,*

$$\mathbb{P}(X \geq t) \leq \inf_{\lambda>0} \exp\big(-\lambda t\big)\,\mathbb{E}[e^{\lambda X}] = \inf_{\lambda>0} \exp\big(\psi_X(\lambda) - \lambda t\big).$$

Similarly,

$$\mathbb{P}(X \leq t) \leq \inf_{\lambda<0} \exp\big(\psi_X(\lambda) - \lambda t\big).$$

Proof Fix $\lambda > 0$. Then

$$\mathbb{P}(X \geq t) = \mathbb{P}(e^{\lambda X} \geq e^{\lambda t}) \leq e^{-\lambda t}\,\mathbb{E}[e^{\lambda X}]$$

by Markov's inequality applied to the nonnegative variable $e^{\lambda X}$. Taking the infimum over $\lambda > 0$ yields the result. □

Remark D.1.7 (Centered Form (Common in ML)) Apply Theorem D.1.4 to $X - \mathbb{E}[X]$ to bound deviations:

$$\mathbb{P}(X - \mathbb{E}[X] \geq t) \leq \inf_{\lambda>0} \exp\big(\psi_{X-\mathbb{E}X}(\lambda) - \lambda t\big).$$

This is the standard starting point for concentration of empirical means.

CGF Convexity and Saddle-Point Intuition (Reference)

Proposition D.1.1 (Convexity of the CGF) *Whenever $M_X(\lambda)$ is finite on an open interval, $\psi_X(\lambda) = \log \mathbb{E}[e^{\lambda X}]$ is convex in λ on that interval.*

Proof By Hölder (or by direct differentiation when justified), $\lambda \mapsto \log \mathbb{E}[e^{\lambda X}]$ is a log-moment of an exponential family and is convex. A standard short argument uses

Hölder: for $\theta \in (0, 1)$ and λ_1, λ_2,

$$\mathbb{E}\Big[e^{(\theta\lambda_1+(1-\theta)\lambda_2)X}\Big] = \mathbb{E}\Big[(e^{\lambda_1 X})^{\theta}(e^{\lambda_2 X})^{1-\theta}\Big] \le \Big(\mathbb{E}[e^{\lambda_1 X}]\Big)^{\theta}\Big(\mathbb{E}[e^{\lambda_2 X}]\Big)^{1-\theta}.$$

Taking logs yields convexity. □

Remark D.1.8 (Saddle-Point/Legendre-Transform Intuition) The optimized exponent

$$\inf_{\lambda>0}\big(\psi_X(\lambda) - \lambda t\big)$$

is (up to sign) a Legendre–Fenchel transform of ψ_X. The minimizing $\lambda^\star$ (when it exists) satisfies $\psi_X'(\lambda^\star) = t$ (a "saddle-point" condition), linking tail events to tilted (exponentially reweighted) distributions. This viewpoint is the conceptual foundation of large deviations.

D.1.2.2 Hoeffding/Bernstein-Style Outcomes (Reference)

Hoeffding and Bernstein inequalities are standard Chernoff outcomes for sums of independent random variables. Hoeffding uses only boundedness; Bernstein uses boundedness plus variance and yields variance-sensitive behavior. For sub-exponential variables, Bernstein-style bounds remain valid with appropriate parameters. These results are used constantly for high-probability confidence intervals and generalization bounds.

Bounded vs. Sub-exponential Cases; Variance-Sensitive Forms

Theorem D.1.5 (Hoeffding Inequality (Bounded Independent Sum)) *Let $X_1, \dots, X_n$ be independent with $a_i \le X_i \le b_i$ almost surely, and let $S = \sum_{i=1}^n (X_i - \mathbb{E}[X_i])$. Then for any $t > 0$,*

$$\mathbb{P}(S \ge t) \le \exp\left(-\frac{2t^2}{\sum_{i=1}^n (b_i - a_i)^2}\right), \qquad \mathbb{P}(|S| \ge t) \le 2\exp\left(-\frac{2t^2}{\sum_{i=1}^n (b_i - a_i)^2}\right).$$

Theorem D.1.6 (Bernstein Inequality (Bounded Independent Sum)) *Let $X_1, \dots, X_n$ be independent, mean-zero, and satisfy $|X_i| \le M$ a.s. Let $v = \sum_{i=1}^n \mathrm{Var}(X_i)$ and $S = \sum_{i=1}^n X_i$. Then for any $t > 0$,*

$$\mathbb{P}(S \ge t) \le \exp\left(-\frac{t^2}{2(v + Mt/3)}\right), \qquad \mathbb{P}(|S| \ge t) \le 2\exp\left(-\frac{t^2}{2(v + Mt/3)}\right).$$

Remark D.1.9 (Two Regimes in Bernstein) If $t \ll v/M$, the exponent behaves like $-t^2/(2v)$ (sub-Gaussian regime). If $t \gg v/M$, it behaves like $-3t/(2M)$ (sub-exponential tail regime). This "variance-sensitive" transition is why Bernstein is often sharper than Hoeffding.

High-Probability Confidence Bounds in ML Notation

A common ML use-case is bounding the deviation of an empirical mean $\bar{X} = \frac{1}{n}\sum_{i=1}^{n} X_i$.

Corollary D.1.1 (Hoeffding-Style Mean Bound) *If $X_1, \ldots, X_n$ are i.i.d. with $a \le X_i \le b$ a.s., then for any $\delta \in (0, 1)$, with probability at least $1 - \delta$,*

$$\left|\bar{X} - \mathbb{E}[X_1]\right| \le (b-a)\sqrt{\frac{\log(2/\delta)}{2n}}.$$

Corollary D.1.2 (Bernstein-Style Mean Bound) *If $X_1, \ldots, X_n$ are i.i.d. mean-zero, $|X_i| \le M$ a.s., and $\mathrm{Var}(X_i) = \sigma^2$, then for any $\delta \in (0, 1)$, with probability at least $1 - \delta$,*

$$\left|\frac{1}{n}\sum_{i=1}^{n} X_i\right| \le \sqrt{\frac{2\sigma^2 \log(2/\delta)}{n}} + \frac{2M\log(2/\delta)}{3n}.$$

Remark D.1.10 (How to Read these in ML) Bounds are typically stated as

$$\text{empirical error} \le \text{population error} + O\left(\sqrt{\frac{\log(1/\delta)}{n}}\right),$$

possibly with variance and boundedness constants. The parameter δ is the failure probability (confidence level).

D.2 Sub-Gaussian/Sub-exponential Reference

Sub-Gaussian and sub-exponential classes provide a calibrated language for tail behavior that is stronger than finite-variance assumptions and flexible enough for ML applications. Many modern concentration results can be stated cleanly using these classes, and their closure properties explain why sums/averages of light-tailed variables concentrate well. Orlicz norms ψ_2 and ψ_1 provide convenient, quantitative summaries of tail heaviness and are frequently used to turn probabilistic statements into norm inequalities.

D.2.1 *Definitions and Closure Properties*

D.2.1.1 Sums, Scaling, and Maxima (Reference)

Sub-Gaussian Sums Remain Sub-Gaussian; Parameter Scaling Rules

We state standard (equivalent) definitions and then record closure rules.

Definition D.2.1 (Sub-Gaussian Random Variable (MGF Form)) A mean-zero random variable X is *sub-Gaussian* with parameter σ^2 if

$$\mathbb{E}[e^{\lambda X}] \leq \exp\left(\frac{\sigma^2\lambda^2}{2}\right) \qquad \forall \lambda \in \mathbb{R}.$$

Definition D.2.2 (Sub-exponential Random Variable (MGF Form)) A mean-zero random variable X is *sub-exponential* with parameters (ν^2, b) if

$$\mathbb{E}[e^{\lambda X}] \leq \exp\left(\frac{\nu^2\lambda^2}{2}\right) \qquad \forall |\lambda| < \frac{1}{b}.$$

Theorem D.2.1 (Tail Bounds)

1. *If X is sub-Gaussian with parameter σ^2, then for all $t > 0$,*

$$\mathbb{P}(|X| \geq t) \leq 2\exp\left(-\frac{t^2}{2\sigma^2}\right).$$

2. *If X is sub-exponential with parameters (ν^2, b), then for all $t > 0$,*

$$\mathbb{P}(|X| \geq t) \leq 2\exp\left(-\frac{1}{2}\min\left\{\frac{t^2}{\nu^2}, \frac{t}{b}\right\}\right).$$

Remark D.2.1 (Two-Regime Behavior) Sub-exponential tails behave like Gaussian for moderate deviations ($t \lesssim \nu^2/b$) and like exponential for large deviations ($t \gtrsim \nu^2/b$).

Proposition D.2.1 (Scaling) *If X is sub-Gaussian with parameter σ^2, then aX is sub-Gaussian with parameter $a^2\sigma^2$. If X is sub-exponential with parameters (ν^2, b), then aX is sub-exponential with parameters $(a^2\nu^2, |a|b)$.*

Proposition D.2.2 (Sums of Independent Sub-Gaussians) *Let $X_1, \ldots, X_n$ be independent mean-zero sub-Gaussian variables with parameters σ_i^2. Then $S =$*

$\sum_{i=1}^{n} X_i$ is sub-Gaussian with parameter $\sum_{i=1}^{n} \sigma_i^2$, i.e.

$$\mathbb{E}[e^{\lambda S}] \leq \exp\left(\frac{\lambda^2}{2} \sum_{i=1}^{n} \sigma_i^2\right).$$

Consequently,

$$\mathbb{P}(|S| \geq t) \leq 2 \exp\left(-\frac{t^2}{2 \sum_{i=1}^{n} \sigma_i^2}\right).$$

Proof Independence gives

$$\mathbb{E}[e^{\lambda S}] = \prod_{i=1}^{n} \mathbb{E}[e^{\lambda X_i}] \leq \prod_{i=1}^{n} \exp\left(\frac{\sigma_i^2 \lambda^2}{2}\right) = \exp\left(\frac{\lambda^2}{2} \sum_{i=1}^{n} \sigma_i^2\right).$$

Then apply Theorem D.2.1(1). □

Proposition D.2.3 (Sums of Independent Sub-exponentials) *Let $X_1, \ldots, X_n$ be independent mean-zero sub-exponential variables with parameters (ν_i^2, b_i). Then $S = \sum_{i=1}^{n} X_i$ is sub-exponential with parameters*

$$\left(\nu^2, b\right) = \left(\sum_{i=1}^{n} \nu_i^2,\ \max_{i \in [n]} b_i\right),$$

up to universal constants (i.e. there exists $c > 0$ such that S is sub-exponential with $(c\nu^2, cb)$).

Maxima Bounds and Union Bounds (Reference)

A standard pattern in ML is bounding the maximum of many random quantities by combining tail bounds with a union bound.

Proposition D.2.4 (Union Bound) *For events $E_1, \ldots, E_m$,*

$$\mathbb{P}\left(\bigcup_{i=1}^{m} E_i\right) \leq \sum_{i=1}^{m} \mathbb{P}(E_i).$$

Proposition D.2.5 (Max of Sub-Gaussians (High-Probability Bound)) *Let $X_1, \ldots, X_m$ be sub-Gaussian with a common parameter σ^2 (not necessarily* *independent). Then for any $\delta \in (0, 1)$, with probability at least $1 - \delta$,*

$$\max_{i \in [m]} |X_i| \le \sigma \sqrt{2 \log \left(\frac{2m}{\delta} \right)}.$$

Proof By Theorem D.2.1(1), $\mathbb{P}(|X_i| \ge t) \le 2 \exp(-t^2/(2\sigma^2))$. By the union bound,

$$\mathbb{P} \left(\max_i |X_i| \ge t \right) \le \sum_{i=1}^{m} \mathbb{P}(|X_i| \ge t) \le 2m \exp\left(-\frac{t^2}{2\sigma^2} \right).$$

Set the right-hand side equal to δ and solve for t. □

Remark D.2.2 (How This Is Used) This template is used to turn a pointwise concentration bound into a uniform bound over m coordinates, hypotheses, or grid points. The logarithmic factor $\log m$ is the price of uniformity via a union bound.

D.2.1.2 ψ_2/ψ_1 Orlicz Norms (Optional)

Definitions and Equivalences (Reference)

Definition D.2.3 (Orlicz Norms ψ_2 and ψ_1) For a random variable X define

$$\|X\|_{\psi_2} := \inf \left\{ c > 0 : \ \mathbb{E} \exp\left(\frac{X^2}{c^2} \right) \le 2 \right\},$$

$$\|X\|_{\psi_1} := \inf \left\{ c > 0 : \ \mathbb{E} \exp\left(\frac{|X|}{c} \right) \le 2 \right\}.$$

Remark D.2.3 (Interpretation) $\|X\|_{\psi_2}$ is finite iff X is sub-Gaussian (up to constants), and $\|X\|_{\psi_1}$ is finite iff X is sub-exponential (up to constants). These norms quantify tail heaviness in a way that composes nicely under sums and maxima.

Proposition D.2.6 (Equivalence to Tail Behavior (Up to Constants)) *There exist universal constants $c_1, c_2 > 0$ such that:*

1. *If $\|X\|_{\psi_2} \le K$, then $\mathbb{P}(|X| \ge t) \le 2 \exp(-c_1 t^2/K^2)$ for all $t \ge 0$.*
2. *If $\|X\|_{\psi_1} \le K$, then $\mathbb{P}(|X| \ge t) \le 2 \exp(-c_2 t/K)$ for all $t \ge 0$.*

Conversely, such tail bounds imply finiteness of the corresponding Orlicz norms (with different constants).

Practical Uses in Tail Bounds (Reference)

Remark D.2.4 (Closure Rules (Typical))

- (Scaling) $\|aX\|_{\psi_2} = |a|\|X\|_{\psi_2}$ and $\|aX\|_{\psi_1} = |a|\|X\|_{\psi_1}$.
- (Sum, independent) If X_i are independent mean-zero sub-Gaussian, then

$$\left\|\sum_{i=1}^{n} X_i\right\|_{\psi_2} \lesssim \left(\sum_{i=1}^{n} \|X_i\|_{\psi_2}^2\right)^{1/2}.$$

  If X_i are independent sub-exponential, then

$$\left\|\sum_{i=1}^{n} X_i\right\|_{\psi_1} \lesssim \sum_{i=1}^{n} \|X_i\|_{\psi_1}.$$

- (Max) For sub-Gaussian X_i,

$$\left\|\max_{i\in[m]} |X_i|\right\|_{\psi_2} \lesssim \max_{i\in[m]} \|X_i\|_{\psi_2} \sqrt{\log m}.$$

Here $\lesssim$ hides universal constants.

Remark D.2.5 (Why Orlicz Norms Are Convenient) They allow many concentration statements to be written as norm inequalities rather than repeated MGF computations, and they integrate smoothly with union bounds and chaining arguments.

D.3 Uniform Convergence Preview References

Uniform convergence is the classical route to generalization: show that empirical risk uniformly approximates population risk over a hypothesis class. Two main toolkits quantify the size/complexity of the class: covering numbers (metric entropy) and Rademacher complexity (data-dependent complexity). This section provides reference-level definitions and the core lemmas used later in the book. We emphasize how these quantities translate to generalization error with explicit confidence parameters and the origin of logarithmic factors.

D.3.1 Covering Numbers and Rademacher Complexity (Reference)

D.3.1.1 Definitions and Key Lemmas (Reference)

Covering Numbers and Metric Entropy

Definition D.3.1 (ε-Cover and Covering Number) Let $(\mathcal{F}, d)$ be a metric space and let $S \subseteq \mathcal{F}$. A set $\mathcal{N} \subseteq \mathcal{F}$ is an *ε-cover* of S if for every $f \in S$ there exists $g \in \mathcal{N}$ such that $d(f, g) \leq \varepsilon$. The *covering number* is

$$\mathcal{N}(S, d, \varepsilon) := \min\{|\mathcal{N}| : \mathcal{N} \text{ is an } \varepsilon\text{-cover of } S\},$$

and the *metric entropy* is $\log \mathcal{N}(S, d, \varepsilon)$.

Remark D.3.1 (Typical Metric in Learning) For supervised learning with loss ℓ, a common choice is the empirical L_2 (or L_∞) metric on a sample $S = (x_1, \ldots, x_n)$:

$$d_S(f, g) := \left(\frac{1}{n}\sum_{i=1}^{n}(f(x_i) - g(x_i))^2\right)^{1/2}.$$

Covering numbers with respect to d_S quantify how many "distinct" predictors the class contains on the observed sample.

Rademacher Complexity

Definition D.3.2 (Empirical and Expected Rademacher Complexity) Let $\mathcal{F}$ be a class of real-valued functions on $\mathcal{X}$ and let $S = (x_1, \ldots, x_n)$. Let $\varepsilon_1, \ldots, \varepsilon_n$ be i.i.d. Rademacher random variables ($\mathbb{P}(\varepsilon_i = \pm 1) = 1/2$). The *empirical Rademacher complexity* is

$$\widehat{\mathfrak{R}}_S(\mathcal{F}) := \mathbb{E}_\varepsilon\left[\sup_{f \in \mathcal{F}} \frac{1}{n}\sum_{i=1}^{n} \varepsilon_i f(x_i)\right].$$

The *expected Rademacher complexity* is

$$\mathfrak{R}_n(\mathcal{F}) := \mathbb{E}_S\big[\widehat{\mathfrak{R}}_S(\mathcal{F})\big],$$

where the outer expectation is over the sample S drawn from the data distribution.

Remark D.3.2 (Geometric Interpretation) The quantity $\sup_{f \in \mathcal{F}} \sum_i \varepsilon_i f(x_i)$ measures how well the function class can correlate with random ± 1 noise on the sample. Smaller complexity means the class cannot fit random noise easily, implying better generalization.

Massart Lemma, Contraction Lemma (Reference)

We record two standard lemmas frequently used to bound Rademacher complexity.

Lemma D.3.1 (Massart's Finite Class Lemma) *Let $\mathcal{F}$ be a finite class of functions $f : \mathcal{X} \to \mathbb{R}$ and let $S = (x_1, \ldots, x_n)$. Define $v_f := (f(x_1), \ldots, f(x_n)) \in \mathbb{R}^n$ and assume $\|v_f\|_2 \le R$ for all $f \in \mathcal{F}$. Then*

$$\widehat{\mathfrak{R}}_S(\mathcal{F}) \le \frac{R}{n}\sqrt{2\log|\mathcal{F}|}.$$

Lemma D.3.2 (Contraction Lemma (Lipschitz Composition)) *Let $\phi : \mathbb{R} \to \mathbb{R}$ be L-Lipschitz, i.e. $|\phi(a) - \phi(b)| \le L|a - b|$. Let $\mathcal{F}$ be a class of real-valued functions on $\mathcal{X}$ and define $\phi \circ \mathcal{F} = \{\phi \circ f : f \in \mathcal{F}\}$. Then for any sample S,*

$$\widehat{\mathfrak{R}}_S(\phi \circ \mathcal{F}) \le L\,\widehat{\mathfrak{R}}_S(\mathcal{F}).$$

More generally, for coordinatewise Lipschitz maps acting on vectors, an analogous contraction bound holds.

Remark D.3.3 (Why These Lemmas Matter) Massart gives a $\sqrt{\log|\mathcal{F}|/n}$ rate for finite classes. Contraction allows bounding complexity after passing through Lipschitz losses or activations. Together they form the standard pipeline for generalization bounds via Rademacher complexity.

Dudley Integral (Optional Reference)

A deeper result links Rademacher complexity to covering numbers.

Theorem D.3.1 (Dudley Entropy Integral (Informal Statement)) *Let $\mathcal{F}$ be uniformly bounded on the sample S and consider the empirical L_2 metric d_S. Then, up to universal constants,*

$$\widehat{\mathfrak{R}}_S(\mathcal{F}) \lesssim \inf_{\alpha>0}\left(\alpha + \frac{1}{\sqrt{n}}\int_{\alpha}^{\mathrm{diam}(\mathcal{F})} \sqrt{\log\mathcal{N}(\mathcal{F}, d_S, \varepsilon)}\,d\varepsilon\right).$$

Remark D.3.4 (Interpretation) The integral accumulates complexity across scales: large-scale structure and fine-scale structure both contribute. This is a precursor to chaining arguments and modern complexity measures.

D.3.1.2 How Bounds Translate to Generalization (Reference)

ERM and Uniform Convergence Pipeline (Reference)

Pipeline Let $\mathcal{F}$ be a hypothesis class and let ℓ be a loss. Define population risk $R(f) = \mathbb{E}[\ell(f(X), Y)]$ and empirical risk $\widehat{R}_S(f) = \frac{1}{n}\sum_{i=1}^{n} \ell(f(x_i), y_i)$. Uniform convergence aims to prove

$$\sup_{f\in\mathcal{F}} |R(f) - \widehat{R}_S(f)| \le \varepsilon(n, \delta) \quad \text{with probability at least } 1-\delta.$$

If $\widehat{f} \in \arg\min_{f\in\mathcal{F}} \widehat{R}_S(f)$ is an ERM solution and $f^\star \in \arg\min_{f\in\mathcal{F}} R(f)$, then

$$R(\widehat{f}) - R(f^\star) \le 2 \sup_{f\in\mathcal{F}} |R(f) - \widehat{R}_S(f)|.$$

Thus bounding the uniform deviation yields an excess-risk bound.

Remark D.3.5 (Why the Factor 2 Appears) The factor arises from adding/ subtracting $\widehat{R}_S(\widehat{f})$ and $\widehat{R}_S(f^\star)$ and using optimality of $\widehat{f}$ for $\widehat{R}_S$. This is the standard ERM-to-risk comparison step.

Confidence Parameters and Log Factors (Reference)

High-probability bounds usually take the form

$$\varepsilon(n, \delta) \;=\; O(\text{complexity term}) + O\left(\sqrt{\frac{\log(1/\delta)}{n}}\right),$$

where the $\log(1/\delta)$ factor comes from tail inequalities (Chernoff/Hoeffding/ Bernstein) and union bounds, and the complexity term is captured by either:

- **Covering numbers:** $\log \mathcal{N}(\mathcal{F}, d_S, \varepsilon)$ (metric entropy),
- **Rademacher complexity:** $\widehat{\mathfrak{R}}_S(\mathcal{F})$ or $\mathfrak{R}_n(\mathcal{F})$,
- **VC-type dimensions:** in classification settings (covered later).

Remark D.3.6 (The Origin of $\log m$ and $\log(1/\delta)$)

- $\log(1/\delta)$ appears when converting an exponential tail bound into a failure probability δ.
- $\log m$ appears when applying a union bound over m hypotheses or grid points (or when bounding finite class complexity via Massart).

Understanding where these logs come from helps diagnose whether a bound is genuinely tight or merely a proof artifact.

Appendix E
Optimization Reference

E.1 Convexity and Smoothness Lemmas

Most first-order optimization theory in ML rests on two structural conditions on $f : \mathbb{R}^d \to \mathbb{R}$: *smoothness* (Lipschitz gradients) and *strong convexity* (quadratic lower curvature). Smoothness yields descent lemmas and step-size rules; strong convexity yields uniqueness and linear convergence. Co-coercivity (Baillon–Haddad) is a powerful equivalence in the convex smooth setting that underpins many modern analyses (proximal methods, gradient mapping bounds). We record the equivalent characterizations and the key consequences used throughout the book.

E.1.1 Descent Lemmas and Co-coercivity (Reference)

E.1.1.1 Equivalent Characterizations of L-Smoothness

Quadratic Upper Bound and Lipschitz Gradient Equivalence

Definition E.1.1 (**L-Smoothness**) A differentiable function $f : \mathbb{R}^d \to \mathbb{R}$ is L-*smooth* if its gradient is L-Lipschitz:

$$\|\nabla f(x) - \nabla f(y)\|_2 \leq L\|x - y\|_2 \qquad \forall x, y \in \mathbb{R}^d.$$

Theorem E.1.1 (Descent Lemma/Quadratic Upper Bound) *If f is L-smooth, then for all $x, y \in \mathbb{R}^d$,*

$$f(y) \leq f(x) + \langle \nabla f(x),\ y - x\rangle + \frac{L}{2}\|y - x\|_2^2.$$

P. Singh, B. Raman, *Foundations of Machine Learning and AI*, Studies in Big Data 199, https://doi.org/10.1007/978-3-032-30336-3

Remark E.1.1 (How It Is Used) Setting $y = x - \eta\nabla f(x)$ yields the standard descent guarantee for gradient descent:

$$f(x - \eta\nabla f(x)) \le f(x) - \eta\left(1 - \frac{L\eta}{2}\right)\|\nabla f(x)\|_2^2,$$

so any $\eta \in (0, 2/L)$ gives a decrease (and $\eta = 1/L$ is the canonical safe step).

Proposition E.1.1 (Equivalent Forms of L-Smoothness (Reference)) *Assume f is differentiable. The following implications hold (and are equivalent under mild regularity):*

1. $\|\nabla f(x) - \nabla f(y)\|_2 \le L\|x - y\|_2$ *for all* x, y.
2. $f(y) \le f(x) + \langle\nabla f(x), y - x\rangle + \frac{L}{2}\|y - x\|_2^2$ *for all* x, y.
3. *(If f is twice differentiable)* $\nabla^2 f(x) \preceq LI$ *for all* x.

Remark E.1.2 (Hessian Interpretation) For C^2 functions, L-smoothness means curvature is globally bounded above by L in every direction: $u^\top\nabla^2 f(x)u \le L\|u\|_2^2$.

Baillon–Haddad (Co-coercivity) Statement (Reference)

Co-coercivity is a strengthening of monotonicity for gradients of convex smooth functions.

Theorem E.1.2 (Baillon–Haddad (Co-coercivity)) *If f is convex and L-smooth, then for all $x, y \in \mathbb{R}^d$,*

$$\langle\nabla f(x) - \nabla f(y),\ x - y\rangle \ \ge\ \frac{1}{L}\|\nabla f(x) - \nabla f(y)\|_2^2.$$

Equivalently, ∇f is $(1/L)$-co-coercive.

Remark E.1.3 (Why It Matters) Co-coercivity allows replacing distances in parameter space by gradient differences and is central in convergence proofs for gradient descent, proximal-gradient methods, and operator-splitting algorithms.

E.1.1.2 Strong Convexity Consequences

PL Inequalities vs. Strong Convexity (Reference)

Definition E.1.2 (μ-Strong Convexity) A differentiable function f is *μ-strongly convex* if for all $x, y \in \mathbb{R}^d$,

$$f(y) \ge f(x) + \langle\nabla f(x), y - x\rangle + \frac{\mu}{2}\|y - x\|_2^2.$$

If f is twice differentiable, this is equivalent to $\nabla^2 f(x) \succeq \mu I$ for all x.

Definition E.1.3 (Polyak–Łojasiewicz (PL) Condition) A differentiable function f satisfies the *PL condition* with parameter $\mu > 0$ if for all x,

$$\frac{1}{2}\|\nabla f(x)\|_2^2 \geq \mu\big(f(x) - f^\star\big), \qquad f^\star := \inf_x f(x).$$

Remark E.1.4 (Relationship) Strong convexity $\Rightarrow$ PL (for L-smooth functions), but PL does not imply convexity. Thus PL captures a class of problems where gradient descent enjoys linear convergence even in some nonconvex settings.

Error Bounds: Function Gap $\leftrightarrow$ Distance to Optimum

Let $x^\star$ be the unique minimizer of a μ-strongly convex function.

Proposition E.1.2 (Quadratic Growth/Distance–Gap Bounds) *If f is μ-strongly convex and has minimizer $x^\star$, then for all x,*

$$f(x) - f(x^\star) \geq \frac{\mu}{2}\|x - x^\star\|_2^2.$$

If additionally f is L-smooth, then

$$f(x) - f(x^\star) \leq \frac{L}{2}\|x - x^\star\|_2^2.$$

Remark E.1.5 (Condition Number and Rates) For functions that are both μ-strongly convex and L-smooth, the condition number $\kappa := L/\mu$ governs the linear rate of gradient descent:

$$f(x_t) - f^\star \leq \left(1 - \frac{\mu}{L}\right)^t (f(x_0) - f^\star)$$

for step size $\eta = 1/L$ (standard result). This is the optimization analogue of conditioning in numerical linear algebra.

E.2 Convergence Proof Templates

Convergence proofs in first-order optimization are highly modular. Most proofs follow a small set of skeletons built on: (i) the descent lemma for smoothness, (ii) convexity/strong convexity inequalities, and (iii) telescoping sums over iterations. For SGD, the extra ingredient is handling noise via conditional expectation and bounding a variance term, which leads to a steady-state "noise floor." This section records reference-level proof templates that can be reused throughout the book.

E.2.1 GD/SGD Convergence Skeletons

E.2.1.1 Convex Case: Rates and Step-Size Choices

$O(1/\sqrt{T})$ and $O(1/T)$ Template Arguments

Assume $f : \mathbb{R}^d \to \mathbb{R}$ is convex and L-smooth, and let $x^\star \in \arg\min f$. Gradient descent (GD) is

$$x_{t+1} = x_t - \eta \nabla f(x_t).$$

Lemma E.2.1 (One-Step Descent Inequality) *If f is L-smooth and $\eta \in (0, 2/L)$, then*

$$f(x_{t+1}) \le f(x_t) - \eta \left(1 - \frac{L\eta}{2}\right) \|\nabla f(x_t)\|_2^2.$$

In particular, for $\eta = 1/L$,

$$f(x_{t+1}) \le f(x_t) - \frac{1}{2L} \|\nabla f(x_t)\|_2^2.$$

Lemma E.2.2 (Distance Recursion (Convex Smooth)) *Let f be convex. For any $\eta > 0$,*

$$\|x_{t+1} - x^\star\|_2^2 = \|x_t - x^\star\|_2^2 - 2\eta \langle \nabla f(x_t), x_t - x^\star \rangle + \eta^2 \|\nabla f(x_t)\|_2^2.$$

By convexity, $\langle \nabla f(x_t), x_t - x^\star \rangle \ge f(x_t) - f(x^\star)$. Thus,

$$2\eta\big(f(x_t) - f^\star\big) \le \|x_t - x^\star\|_2^2 - \|x_{t+1} - x^\star\|_2^2 + \eta^2 \|\nabla f(x_t)\|_2^2.$$

Remark E.2.1 (How $O(1/T)$ Arises) If f is convex and L-smooth and we choose $\eta = 1/L$, then one can combine Lemma E.2.1 and Lemma E.2.2 to obtain a telescoping bound on $\sum_{t=0}^{T-1} \big(f(x_t) - f^\star\big)$ and conclude

$$f(\bar{x}_T) - f^\star \le \frac{L\|x_0 - x^\star\|_2^2}{2T}, \qquad \bar{x}_T := \frac{1}{T} \sum_{t=0}^{T-1} x_t,$$

i.e. an $O(1/T)$ rate for the *averaged iterate*.

Remark E.2.2 (How $O(1/\sqrt{T})$ Arises for SGD) For SGD with unbiased gradients and bounded second moments, a similar telescoping argument yields

$$\mathbb{E}[f(\bar{x}_T)] - f^\star \le O\left(\frac{1}{\eta T}\right) + O(\eta),$$

and choosing $\eta \asymp 1/\sqrt{T}$ balances the terms to give $O(1/\sqrt{T})$. This is the canonical convex SGD tradeoff: optimization error decreases with smaller η, noise error increases with smaller η.

Averaging vs. Last-Iterate Discussion (Reference)

Remark E.2.3 (Averaging Is a Proof Device with Practical Variants) For general convex (not strongly convex) objectives, the cleanest rates are usually obtained for an averaged iterate $\bar{x}_T = \frac{1}{T}\sum_{t=0}^{T-1} x_t$ (or weighted averages). Last-iterate guarantees may require additional assumptions or more delicate arguments. In practice, Polyak/Ruppert averaging for SGD is a robust technique; for GD, last-iterate can also converge at $O(1/T)$ for smooth convex, but averaging remains the simplest reference bound.

E.2.1.2 Strongly Convex Case: Linear Rates

$(1 - \mu/L)^t$ Contraction Template

Assume f is μ-strongly convex and L-smooth, so $\kappa = L/\mu$ is the condition number. Then GD with $\eta = 1/L$ satisfies a linear (geometric) contraction.

Theorem E.2.1 (GD Linear Rate (Reference Form)) *Let f be μ-strongly convex and L-smooth with minimizer $x^\star$. Then gradient descent with step size $\eta = 1/L$ satisfies*

$$f(x_t) - f^\star \le \left(1 - \frac{\mu}{L}\right)^t \left(f(x_0) - f^\star\right),$$

and also

$$\|x_t - x^\star\|_2^2 \le \left(1 - \frac{\mu}{L}\right)^t \|x_0 - x^\star\|_2^2.$$

Remark E.2.4 (Proof Skeleton) Combine the descent lemma (upper quadratic bound) with strong convexity (lower quadratic bound) to relate $\|\nabla f(x_t)\|_2^2$ to $f(x_t) - f^\star$ (PL-type inequality), then plug into the one-step descent inequality to get contraction. Telescoping is replaced by geometric decay.

Noise Floor and Steady-State Error for SGD (Reference)

In strongly convex problems, SGD can converge linearly at first, but noise prevents convergence to the exact minimizer with constant step size. Instead, iterates enter a neighborhood whose radius scales with the step size and the noise variance. This is the "noise floor." Decaying step sizes can drive the error to zero at slower rates.

Remark E.2.5 (Canonical Steady-State Form) For μ-strongly convex, L-smooth f, and unbiased stochastic gradients with bounded variance $\mathbb{E}\|\widehat{g}_t - \nabla f(x_t)\|_2^2 \leq \sigma^2$, constant-step SGD with $\eta \leq c/L$ typically yields

$$\mathbb{E}[f(x_t) - f^\star] \leq (1 - \mu\eta)^t \, (f(x_0) - f^\star) + O\left(\frac{\eta\sigma^2}{\mu}\right).$$

The second term is the noise floor; choosing smaller η reduces it but slows the transient rate.

Remark E.2.6 (Decaying Step Sizes) With $\eta_t \asymp 1/t$, one can achieve $\mathbb{E}[f(x_t) - f^\star] = O(1/t)$ under standard assumptions, mirroring the deterministic strongly convex rate up to constants, but without a fixed noise floor.

E.2.2 Proximal and ADMM Reference

Many ML objectives are *composite*: a smooth data-fit term plus a nonsmooth regularizer (e.g., least squares + ℓ_1 for sparsity, logistic loss + group norms, TV penalties). Proximal-gradient methods (ISTA/FISTA) solve these efficiently by alternating a gradient step on the smooth part with a proximal step on the nonsmooth part. ADMM is a splitting method that handles constraints and separable structure by alternating minimizations of an augmented Lagrangian and updating dual variables; it is especially useful for large-scale, distributed, or structured problems.

E.2.2.1 Proximal Gradient: Composite Objectives

Composite Form $f + g$, Proximal Descent Inequality (Reference)

Consider the composite problem

$$\min_{x \in \mathbb{R}^d} \; F(x) := f(x) + g(x),$$

where f is convex and L-smooth and g is closed, convex, possibly nonsmooth.

Definition E.2.1 (Proximal Gradient Step) Given step size $\eta > 0$, the proximal-gradient update is

$$x_{t+1} := \operatorname{prox}_{\eta g}\big(x_t - \eta \nabla f(x_t)\big),$$

where

$$\operatorname{prox}_{\eta g}(v) := \arg\min_{z\in\mathbb{R}^d} \left\{ \frac{1}{2}\|z - v\|_2^2 + \eta\, g(z) \right\}.$$

Definition E.2.2 (Proximal-Gradient Mapping) The proximal-gradient mapping is

$$G_\eta(x) := \frac{1}{\eta}\left(x - \operatorname{prox}_{\eta g}\big(x - \eta\nabla f(x)\big)\right).$$

It generalizes $\nabla f(x)$: if $g \equiv 0$, then $G_\eta(x) = \nabla f(x)$.

Proposition E.2.1 (Basic Descent Inequality (Reference)) *Assume f is convex and L-smooth and take $\eta \in (0, 1/L]$. Let $x^+ = \operatorname{prox}_{\eta g}(x - \eta\nabla f(x))$. Then*

$$F(x^+) \le F(x) - \frac{\eta}{2}\|G_\eta(x)\|_2^2.$$

In particular, $F(x_t)$ is nonincreasing along proximal-gradient iterates.

Remark E.2.7 (Stationarity Criterion) A point $x^\star$ is optimal iff $0 \in \nabla f(x^\star) + \partial g(x^\star)$. Equivalently, $G_\eta(x^\star) = 0$ for any $\eta > 0$ (fixed point of the prox-gradient map). Thus $\|G_\eta(x)\|$ is a natural stopping metric.

ISTA/FISTA Templates (Reference)

ISTA is proximal gradient with constant step size $\eta \le 1/L$. It achieves the basic convex rate $F(x_t) - F^\star = O(1/t)$ under standard assumptions.

Algorithm 4 ISTA (proximal gradient)

Require: $x_0 \in \mathbb{R}^d$, step size $\eta \le 1/L$
1: **for** $t = 0, 1, 2, \ldots$ **do**
2: $\quad x_{t+1} \leftarrow \operatorname{prox}_{\eta g}(x_t - \eta\nabla f(x_t))$
3: **end for**

Remark E.2.8 (FISTA Acceleration (Reference)) FISTA adds Nesterov-style momentum: it maintains an auxiliary sequence y_t and extrapolates using a scalar t_k sequence. Under convexity and L-smoothness, it achieves $F(x_t) - F^\star = O(1/t^2)$.

Full details are deferred to the optimization chapter; here we record the standard template:

$$x_{t+1} = \mathrm{prox}_{\eta g}(y_t - \eta \nabla f(y_t)),$$

$$\beta_{t+1} = \frac{t_t - 1}{t_{t+1}}, \quad t_{t+1} = \frac{1 + \sqrt{1 + 4t_t^2}}{2},$$

$$y_{t+1} = x_{t+1} + \beta_{t+1}(x_{t+1} - x_t).$$

E.2.2.2 ADMM: Augmented Lagrangian and Residual Criteria

Primal/Dual Residual Definitions and Stopping Rules

ADMM targets problems of the form

$$\min_{x,z} \ f(x) + g(z) \quad \text{s.t.} \quad Ax + Bz = c,$$

with convex f, g and matrices A, B.

Definition E.2.3 (Augmented Lagrangian) For penalty $\rho > 0$ and dual variable y, the augmented Lagrangian is

$$\mathcal{L}_\rho(x, z, y) = f(x) + g(z) + y^\top (Ax + Bz - c) + \frac{\rho}{2} \|Ax + Bz - c\|_2^2.$$

In scaled form with $u = y/\rho$,

$$\mathcal{L}_\rho(x, z, u) = f(x) + g(z) + \frac{\rho}{2} \|Ax + Bz - c + u\|_2^2 - \frac{\rho}{2} \|u\|_2^2.$$

Algorithm 5 ADMM (scaled form)

Require: x^0, z^0, u^0, penalty $\rho > 0$
1: **for** $t = 0, 1, 2, \ldots$ **do**
2: $\quad x^{t+1} \leftarrow \arg\min_x \left(f(x) + \frac{\rho}{2} \|Ax + Bz^t - c + u^t\|_2^2 \right)$
3: $\quad z^{t+1} \leftarrow \arg\min_z \left(g(z) + \frac{\rho}{2} \|Ax^{t+1} + Bz - c + u^t\|_2^2 \right)$
4: $\quad u^{t+1} \leftarrow u^t + (Ax^{t+1} + Bz^{t+1} - c)$
5: **end for**

Definition E.2.4 (Primal and Dual Residuals) The *primal residual* and *dual residual* at iteration $t+1$ are

$$r^{t+1} := Ax^{t+1} + Bz^{t+1} - c, \qquad s^{t+1} := \rho A^\top B\,(z^{t+1} - z^t),$$

in the standard two-block ADMM form (the precise dual residual expression depends on scaling and splitting; this is the common consensus-like form).

Remark E.2.9 (Stopping Rules (Reference)) A standard criterion is to stop when

$$\|r^t\|_2 \le \varepsilon_{\text{pri}} \quad \text{and} \quad \|s^t\|_2 \le \varepsilon_{\text{dual}},$$

where tolerances combine absolute and relative parts, e.g.,

$$\varepsilon_{\text{pri}} = \sqrt{m}\,\varepsilon_{\text{abs}} + \varepsilon_{\text{rel}} \max\{\|Ax^t\|_2, \|Bz^t\|_2, \|c\|_2\},$$

and similarly for $\varepsilon_{\text{dual}}$.

Consensus and Splitting Forms (Reference)

Remark E.2.10 (Consensus Form) A widely used special case is consensus:

$$\min_x \sum_{i=1}^N f_i(x) \quad \Longleftrightarrow \quad \min_{x_1,\ldots,x_N,z} \sum_{i=1}^N f_i(x_i) \text{ s.t. } x_i = z\ \forall i,$$

which is ADMM-friendly and supports distributed optimization (each x_i updated locally, then averaged into z).

Remark E.2.11 (Splitting for Structured Regularizers) If g is complicated but proximal-friendly after introducing an auxiliary variable (e.g., $z = Dx$ for TV regularization), ADMM splits $f(x) + g(Dx)$ into separate x and z updates:

$$\min_x f(x) + g(Dx) \implies \min_{x,z}\ f(x) + g(z) \text{ s.t. } z = Dx.$$

This is the standard route for constrained and nonsmooth ML objectives.

E.3 Nonconvex Reference

Nonconvex objectives are the norm in deep learning. Global optimality is usually intractable to certify, so the central questions become: (i) what *stationarity* notion should we target, (ii) how do we measure progress, and (iii) how do algorithms behave near saddles? This reference records standard first-order stationarity measures (gradient norms and gradient mappings) and the high-level intuition behind saddle escaping (negative curvature + perturbations/noise).

E.3.1 Stationarity Measures and Gradient Norms

E.3.1.1 First-Order Stationarity and Approximate Solutions

$\|\nabla f(x)\| \leq \varepsilon$ as Stopping Criterion

For unconstrained smooth optimization, the basic first-order necessary condition for a local minimizer $x^\star$ is $\nabla f(x^\star) = 0$. In algorithms, we therefore measure approximate stationarity by the gradient norm.

Definition E.3.1 (ε-First-Order Stationary Point) Let $f : \mathbb{R}^d \to \mathbb{R}$ be differentiable. A point x is an *ε-first-order stationary point* (FOSP) if

$$\|\nabla f(x)\|_2 \leq \varepsilon.$$

Remark E.3.1 (Why This Is Meaningful) If $\|\nabla f(x)\|$ is small, then the linearization $f(x + \Delta) \approx f(x) + \langle \nabla f(x), \Delta \rangle$ suggests no direction yields large immediate descent for small Δ. Thus gradient norm is the standard termination metric for first-order methods in nonconvex ML.

Gradient Mapping for Constrained/Prox Problems (Reference)

For constrained or composite problems, $\nabla f(x)$ alone is not the right stationarity measure. The proximal-gradient mapping generalizes it.

Definition E.3.2 (Gradient Mapping (Composite/Proximal Stationarity)) Consider the composite objective

$$F(x) = f(x) + g(x),$$

where f is differentiable and g is proper (often convex, possibly nonsmooth). For step size $\eta > 0$, define the proximal-gradient mapping

$$G_\eta(x) := \frac{1}{\eta}\left(x - \mathrm{prox}_{\eta g}(x - \eta \nabla f(x))\right).$$

Remark E.3.2 (Stationarity Equivalence (Reference)) Under standard regularity (e.g., g closed convex), a point $x^\star$ satisfies the first-order optimality condition

$$0 \in \nabla f(x^\star) + \partial g(x^\star)$$

iff $G_\eta(x^\star) = 0$ for any $\eta > 0$. Hence $\|G_\eta(x)\|$ is the natural stopping criterion for proximal methods. If $g = \iota_C$ is an indicator of a convex set C, then $\text{prox}_{\eta g} = \Pi_C$ and G_η becomes the projected-gradient mapping.

E.3.2 Saddle-Escaping Intuition (Reference)

In nonconvex landscapes, critical points can be local minima, local maxima, or saddle points. Many saddles in high-dimensional problems are *strict saddles*: the Hessian has a negative eigenvalue. In such regions, there exists a direction of negative curvature along which the function decreases quadratically. Gradient descent can be slow to leave flat saddles, but small perturbations (explicit noise, stochastic gradients) can help discover negative-curvature directions and escape.

Negative Curvature Directions and Perturbation Arguments (Reference)

Definition E.3.3 (Second-Order Stationarity and Strict Saddles) Let f be twice differentiable. A point x is an (ε, γ)-*second-order stationary point* if

$$\|\nabla f(x)\|_2 \leq \varepsilon \quad \text{and} \quad \lambda_{\min}(\nabla^2 f(x)) \geq -\gamma.$$

A *strict saddle* is a critical point x with $\nabla f(x) = 0$ and $\lambda_{\min}(\nabla^2 f(x)) < 0$.

Remark E.3.3 (Local Decrease under Negative Curvature) If v is a unit eigenvector of $\nabla^2 f(x)$ with eigenvalue $\lambda_{\min} < 0$, then for small α, a second-order Taylor approximation gives

$$f(x + \alpha v) \approx f(x) + \alpha \langle \nabla f(x), v \rangle + \frac{\alpha^2}{2} \lambda_{\min},$$

so when $\nabla f(x) \approx 0$ and $\lambda_{\min} < 0$, moving along v decreases f by order $\alpha^2 |\lambda_{\min}|$. This is the geometric reason strict saddles are escapable.

Why Noise Can Help in Practice (Reference)

Remark E.3.4 (SGD as Implicit Perturbation) Stochastic gradients can be written as

$$\widehat{g}(x) = \nabla f(x) + \xi,$$

where ξ is a noise term (mean zero under unbiased sampling). Near saddles with small gradients, the noise can dominate and push iterates into directions that reveal negative curvature, helping the algorithm leave the saddle region. This is one mechanism behind the empirical observation that SGD "escapes saddles" better than deterministic GD.

Remark E.3.5 (Perturbed Gradient Methods (Reference)) A standard theoretical tool is to add occasional random perturbations to the iterate when gradient norms are small. Under smoothness and strict-saddle assumptions, one can show convergence to approximate second-order stationary points in polynomial time. We omit full proofs here; the key geometric idea is that randomness finds descent directions when gradients provide insufficient guidance.

Appendix F
Notation Index and Quick Reference

F.1 Core Symbols Used Across Chapters

This appendix lists the core symbols used throughout the book. Notation is presented in the form: *symbol → meaning (and common variants)*. Whenever a symbol has multiple common conventions in the literature, the convention used in this book is stated explicitly.

F.1.1 Sets, Spaces, and Operators

F.1.1.1 $\mathcal{X}, \mathcal{Y}, \mathcal{H}$; arg min; Indicators

Basic Set-Builder Notation; Events as Sets

- $\mathcal{X}$: input/feature space (e.g., $\mathbb{R}^d$, sequences, graphs).
- $\mathcal{Y}$: output/label space (e.g., $\{1, \ldots, K\}$, $\mathbb{R}$, structured outputs).
- Θ: parameter space (often $\mathbb{R}^p$ or a constrained set).
- $\mathcal{H}$: hypothesis class (set of candidate predictors), typically $\mathcal{H} \subseteq \{h : \mathcal{X} \to \mathcal{Y}\}$.
- $A \subseteq B$: subset; $A \subsetneq B$: strict subset; A^c: complement (in a stated universe Ω).
- $\{x \in A : P(x)\}$: set-builder notation (elements of A satisfying predicate P).
- Ω: sample space (probability); $E \subseteq \Omega$: event.
- $\sigma(\cdot)$: sigma-field generated by a random variable or collection (measure-theoretic conditioning).

P. Singh, B. Raman, *Foundations of Machine Learning and AI*, Studies in Big Data 199, https://doi.org/10.1007/978-3-032-30336-3

Indicator $\mathbf{1}\{\cdot\}$ and Support $\mathrm{supp}(\cdot)$

- $\mathbf{1}\{E\}$ (or $\mathbb{I}\{E\}$): indicator of event/condition E; equals 1 if E holds, else 0.
- $\mathrm{supp}(x)$ for $x \in \mathbb{R}^d$: index set of nonzeros, $\mathrm{supp}(x) = \{i \in [d] : x_i \neq 0\}$.
- $\mathrm{supp}(X)$ for a discrete random variable X: $\{x : \mathbb{P}(X = x) > 0\}$.
- $\mathrm{supp}(p)$ for a density p: typically $\overline{\{x : p(x) > 0\}}$ (closure), when needed.
- $\min_{x\in S} f(x)$, $\max_{x\in S} f(x)$: minimum/maximum value (when attained).
- $\inf_{x\in S} f(x)$, $\sup_{x\in S} f(x)$: infimum/supremum (always defined for extended reals).
- $\arg\min_{x\in S} f(x)$: set of minimizers, $\{x^\star \in S : f(x^\star) = \inf_{x\in S} f(x)\}$.
- $\arg\max_{x\in S} f(x)$: set of maximizers, $\{x^\star \in S : f(x^\star) = \sup_{x\in S} f(x)\}$.

F.1.1.2 Inner Products, Norms, Projections, Trace

$\langle\cdot,\cdot\rangle$, $\|\cdot\|_p$, Dual Norm $\|\cdot\|_*$

- $\langle x, y\rangle$: Euclidean inner product $x^\top y$ (unless explicitly stated otherwise).
- $\|x\|_2$: Euclidean norm $\sqrt{\langle x, x\rangle}$.
- $\|x\|_p$: ℓ_p norm, $\|x\|_p = (\sum_i |x_i|^p)^{1/p}$ for $1 \le p < \infty$; $\|x\|_\infty = \max_i |x_i|$.
- $\|A\|_2$: spectral/operator norm (largest singular value).
- $\|A\|_F$: Frobenius norm, $\|A\|_F = \sqrt{\sum_{i,j} A_{ij}^2}$.
- Dual norm: for a norm $\|\cdot\|$, $\|y\|_* = \sup_{\|x\|\le 1}\langle x, y\rangle$.

Projectors P, Projection Π_C, Prox $\mathrm{prox}_{\lambda\Omega}$

- P: projector matrix. Orthogonal projector satisfies $P^2 = P$ and $P^\top = P$.
- $P_{\mathcal{U}}$: orthogonal projector onto subspace $\mathcal{U}$.
- $\Pi_C(x)$: Euclidean projection of x onto a set C, $\Pi_C(x) \in \arg\min_{z\in C}\|z - x\|_2$ (unique if C closed convex).
- $\mathrm{prox}_{\lambda\Omega}(x)$: proximal operator, $\arg\min_z \frac{1}{2}\|z - x\|_2^2 + \lambda\Omega(z)$ (unique when Ω is closed convex).
- $G_\eta(x)$: proximal-gradient mapping, $G_\eta(x) = \frac{1}{\eta}\big(x - \mathrm{prox}_{\eta g}(x - \eta\nabla f(x))\big)$.

Trace tr, Determinant det, Spectral Radius ρ

- $\mathrm{tr}(A)$: trace of a square matrix A, $\sum_i A_{ii}$; invariant under similarity.
- $\det(A)$: determinant of a square matrix A; volume scaling factor of the linear map.
- $\log\det(A)$: log-determinant (computed stably via Cholesky for SPD matrices).
- $\rho(A)$: spectral radius, $\rho(A) = \max\{|\lambda| : \lambda \in \mathrm{spec}(A)\}$.
- $\mathrm{spec}(A)$: spectrum (set/multiset of eigenvalues).

F.1.2 Probability and Statistics

F.1.2.1 Random Variables, Expectations, Variances, Covariances

$X, Y, \mathbb{E}$, Var, Cov, Σ

- X, Y: random variables (scalar) or random vectors (context clarifies). In supervised learning, typically $(X, Y) \sim \mathcal{D}$ denotes a random input–output pair drawn from a data-generating distribution.
- x, y: realizations (observed values) of X, Y.
- $\mathbb{P}(E)$: probability of event E.
- $\mathbb{E}[Z]$: expectation of a random variable Z (assumed finite when used).
- $\mathrm{Var}(Z)$: variance, $\mathrm{Var}(Z) = \mathbb{E}[(Z - \mathbb{E}Z)^2]$ (when finite).
- $\mathrm{Cov}(Z, W)$: covariance, $\mathbb{E}[(Z - \mathbb{E}Z)(W - \mathbb{E}W)]$.
- For a random vector $X \in \mathbb{R}^d$:

$$\mu := \mathbb{E}[X] \in \mathbb{R}^d, \qquad \Sigma := \mathrm{Cov}(X) := \mathbb{E}\big[(X - \mu)(X - \mu)^\top\big] \in \mathbb{R}^{d \times d}.$$

- $\widehat{\mu}, \widehat{\Sigma}$: empirical (sample) mean and covariance computed from data. Common convention used in this book for centered data matrix $X_c \in \mathbb{R}^{n \times d}$:

$$\widehat{\Sigma} := \frac{1}{n} X_c^\top X_c \quad \text{(uncorrected, for linear-algebra convenience).}$$

IID, Conditional Expectation, Likelihood $p(y \mid x)$

- i.i.d.: independent and identically distributed. For samples (X_i, Y_i),

$$(X_i, Y_i) \text{ i.i.d.} \sim \mathcal{D} \quad \Longleftrightarrow \quad (X_i, Y_i) \text{ independent and each has distribution } \mathcal{D}.$$

- Conditional expectation:

$$\mathbb{E}[Z \mid Y] \quad \text{or} \quad \mathbb{E}[Z \mid \mathcal{G}]$$

is the expectation of Z given information Y (or sigma-field $\mathcal{G}$). Key tower property: $\mathbb{E}[Z] = \mathbb{E}[\mathbb{E}[Z \mid Y]]$.
- Likelihood: for a parametric model $p_\theta(y \mid x)$, the likelihood of θ given data $(x_i, y_i)_{i=1}^n$ is

$$L(\theta) := \prod_{i=1}^n p_\theta(y_i \mid x_i), \qquad \ell(\theta) := \log L(\theta) = \sum_{i=1}^n \log p_\theta(y_i \mid x_i).$$

- $p(x)$, $p(y \mid x)$, $p(x, y)$: density/PMF notation (context determines discrete vs. continuous).

F.1.2.2 Likelihood, Posterior, Risk, and Loss

Risk $R(h)$, Empirical Risk $\widehat{R}_n(h)$, Loss $\ell(\cdot, \cdot)$

- $\ell(\hat{y}, y)$ or $\ell(h(x), y)$: loss function measuring prediction error. Common examples: squared loss $(\hat{y} - y)^2$, logistic loss, hinge loss, cross-entropy.
- (Population) risk:

$$R(h) := \mathbb{E}_{(X,Y)\sim\mathcal{D}}\big[\ell(h(X), Y)\big].$$

- Empirical risk on a sample $S = \{(x_i, y_i)\}_{i=1}^n$:

$$\widehat{R}_n(h) := \frac{1}{n}\sum_{i=1}^{n} \ell(h(x_i), y_i).$$

- Empirical risk minimization (ERM):

$$\widehat{h} \in \arg\min_{h\in\mathcal{H}} \widehat{R}_n(h), \qquad h^\star \in \arg\min_{h\in\mathcal{H}} R(h).$$

- Generalization gap: $R(h) - \widehat{R}_n(h)$. Excess risk: $R(\widehat{h}) - R(h^\star)$.

MLE/MAP, Posterior, Evidence (Reference)

- Maximum likelihood estimator (MLE):

$$\widehat{\theta}_{\text{MLE}} \in \arg\max_{\theta} \prod_{i=1}^{n} p_\theta(y_i \mid x_i) \iff \widehat{\theta}_{\text{MLE}} \in \arg\max_{\theta} \sum_{i=1}^{n} \log p_\theta(y_i \mid x_i).$$

- Prior: $\pi(\theta)$, a density/PMF encoding prior belief on parameters.
- Posterior (Bayes rule):

$$p(\theta \mid \mathcal{D}) \propto p(\mathcal{D} \mid \theta)\, \pi(\theta).$$

- Maximum a posteriori (MAP):

$$\widehat{\theta}_{\text{MAP}} \in \arg\max_{\theta}\ p(\mathcal{D} \mid \theta)\pi(\theta) \iff \arg\max_{\theta}\ \big(\log p(\mathcal{D} \mid \theta) + \log \pi(\theta)\big).$$

- Evidence / marginal likelihood:

$$p(\mathcal{D}) = \int p(\mathcal{D} \mid \theta)\pi(\theta)\, d\theta,$$

used for model comparison and Bayesian hyperparameter selection.

F.1.3 Optimization and Computation

F.1.3.1 Gradients, Hessians, Step Sizes, Tolerances

∇f, $\nabla^2 f$, Lipschitz L, Strong Convexity μ

- $f : \mathbb{R}^d \to \mathbb{R}$: objective (loss/risk/energy) function.
- $\nabla f(x) \in \mathbb{R}^d$: gradient at x (column-vector convention).
- $\nabla^2 f(x) \in \mathbb{R}^{d\times d}$: Hessian at x (when it exists).
- L: smoothness constant (Lipschitz gradient constant). Typical meaning:

$$\|\nabla f(x) - \nabla f(y)\|_2 \le L\|x - y\|_2 \quad \forall x, y.$$

  Equivalently (for C^2 functions), $\nabla^2 f(x) \preceq LI$ for all x.
- μ: strong convexity constant. Typical meaning:

$$f(y) \ge f(x) + \langle \nabla f(x), y - x\rangle + \frac{\mu}{2}\|y - x\|_2^2 \quad \forall x, y,$$

  equivalently (for C^2) $\nabla^2 f(x) \succeq \mu I$ for all x.
- $\kappa := L/\mu$: condition number for μ-strongly convex, L-smooth objectives (governs linear convergence rates).

Step Size η, Tolerance ε, Failure Prob. δ

- η (or α): step size / learning rate in GD/SGD and related methods. Common safe regime for L-smooth GD: $\eta \in (0, 2/L)$ (often $\eta = 1/L$ in theory).
- ε: accuracy tolerance. Common uses:
  - optimization accuracy: $f(x) - f^\star \le \varepsilon$,
  - stationarity: $\|\nabla f(x)\|_2 \le \varepsilon$,
  - constraint/feasibility: $\|r\| \le \varepsilon$.
- $\delta \in (0, 1)$: failure probability / confidence parameter in high-probability bounds. Statements "with probability at least $1 - \delta$" are standard; $\log(1/\delta)$ frequently appears in bounds.

F.1.3.2 Complexities: $n, d, p, T, \varepsilon, \delta$

Sample Size n, Dimension d, Parameters p, Iterations T

- n: number of samples (data points) in a dataset.
- d: ambient feature dimension (input dimension) or embedding dimension (context clarifies).
- p: number of parameters of a model (weights/biases), or a complexity parameter (e.g., number of features) depending on context.
- T: number of iterations/steps (optimization) or time horizon (dynamical systems); context clarifies.
- $\mathrm{nnz}(A)$: number of nonzeros in a sparse matrix A (used in sparse cost models).

Asymptotics: $O(\cdot)$, $\Theta(\cdot)$, $o(\cdot)$; $\widetilde{O}(\cdot)$

- $f(n) = O(g(n))$: $\exists C, n_0$ such that $|f(n)| \le Cg(n)$ for all $n \ge n_0$.
- $f(n) = \Theta(g(n))$: both $f(n) = O(g(n))$ and $f(n) = \Omega(g(n))$; i.e. f and g grow at the same rate up to constants.
- $f(n) = o(g(n))$: $\lim_{n\to\infty} f(n)/g(n) = 0$ (strictly smaller order).
- $\widetilde{O}(g(n))$: soft-O notation, hides polylog factors:

$$\widetilde{O}(g(n)) = O\big(g(n) \cdot (\log n)^{O(1)}\big),$$

  and similarly for $\widetilde{\Theta}$, $\widetilde{\Omega}$.

Remark F.1.1 (Which Variable Tends to Infinity) Always interpret asymptotics with respect to the regime under discussion: $n \to \infty$ (more data), $d \to \infty$ (high-dimensional), $T \to \infty$ (more iterations), or $\varepsilon \downarrow 0$ (higher accuracy). The same expression can have very different meaning under different regimes.

Bibliography

Arlot, S., Celisse, A.: A survey of cross-validation procedures for model selection. Stat. Surv. **4**, 40–79 (2010)
Aronszajn, N.: Theory of reproducing kernels. Trans. Am. Math. Soc. **68**(3), 337–404 (1950)
Axler, S.: Linear Algebra Done Right, 3rd edn. Springer, Cham (2015)
Benjamini, Y., Hochberg, Y.: Controlling the false discovery rate: a practical and powerful approach to multiple testing. J. R. Stat. Soc.: Ser. B **57**(1), 289–300 (1995)
Berger, J.O.: Statistical Decision Theory and Bayesian Analysis. Springer, New York (1985)
Bhatia, R.: Positive Definite Matrices. Princeton University Press, Princeton (2007)
Bickel, P.J., Doksum, K.A.: Mathematical Statistics: Basic Ideas and Selected Topics, Volume I, 2nd edn. Chapman and Hall/CRC, Boca Raton (2015)
Bishop, C.M.: Pattern Recognition and Machine Learning. Springer, New York (2006)
Bottou, L., Curtis, F.E., Nocedal, J.: Optimization methods for large-scale machine learning. SIAM Rev. **60**(2), 223–311 (2018)
Boucheron, S., Lugosi, G., Massart, P.: Concentration Inequalities: A Non-Asymptotic Theory of Independence. Oxford University Press, Oxford (2013)
Boyd, S., Vandenberghe, L.: Convex Optimization. Cambridge University Press, Cambridge (2004)
Casella, G., Berger, R.L.: Statistical Inference. Cengage Learning, 2nd edn. (2002)
Cormen, T.H., Leiserson, C.E., Rivest, R.L., Stein, C.: Introduction to Algorithms, 4th edn. MIT Press, Cambridge (2022)
Duchi, J., Hazan, E., Singer, Y.: Adaptive subgradient methods for online learning and stochastic optimization. J. Mach. Learn. Res. **12**(null), 2121–2159 (2011). ISSN 1532-4435
Durrett, R.: Probability: Theory and Examples, 5th edn. Cambridge University Press, Cambridge (2019)
Efron, B., Tibshirani, R.J.: An Introduction to the Bootstrap. Chapman & Hall, Boca Raton (1993)
Gelman, A., Carlin, J.B., Stern, H.S., Dunson, D.B., Vehtari, A., Rubin, D.B.: Bayesian Data Analysis, 3rd edn. Chapman and Hall/CRC, Boca Raton (2014)
Gneiting, T., Raftery, A.E.: Strictly proper scoring rules, prediction, and estimation. J. Am. Stat. Assoc. **102**(477), 359–378 (2007)
Golub, G.H., Van Loan, C.F.: Matrix Computations, 4th edn. Johns Hopkins University Press, Baltimore (2013)
Grimmett, G., Stirzaker, D.: Probability and Random Processes, 4th edn. Oxford University Press, Oxford (2020)
Halko, N., Martinsson, P.G., Tropp, J.A.: Finding structure with randomness: probabilistic algorithms for constructing approximate matrix decompositions. SIAM Rev. **53**(2), 217–288 (2011)

P. Singh, B. Raman, *Foundations of Machine Learning and AI*, Studies in Big Data 199, https://doi.org/10.1007/978-3-032-30336-3

Halmos, P.R.: Naive Set Theory. Springer, New York (1960)

Hastie, T., Tibshirani, R., Friedman, J.: The Elements of Statistical Learning: Data Mining, Inference, and Prediction, 2nd edn. Springer, New York (2009)

Higham, N.J.: Accuracy and Stability of Numerical Algorithms, 2nd edn. SIAM, Philadelphia (2002)

Hoerl, A.E., Kennard, R.W.: Ridge regression: biased estimation for nonorthogonal problems. Technometrics **42**(1), 80–86 (2000). ISSN 0040-1706. https://doi.org/10.2307/1271436.

Horn, R.A., Johnson, C.R.: Matrix Analysis, 2 edn. Cambridge University Press, Cambridge (2012)

Jolliffe, I.T.: Principal Component Analysis, 2 edn. Springer, New York (2002)

Kolda, T.G., Bader, B.W.: Tensor decompositions and applications. SIAM Rev. **51**(3), 455–500 (2009). ISSN 0036-1445. https://doi.org/10.1137/07070111X

Lehmann, E.L., Romano, J.P.: Testing Statistical Hypotheses, 3rd edn. Springer, Berlin (2005)

Magnus, J.R., Neudecker, H.: Matrix Differential Calculus with Applications in Statistics and Econometrics, 3rd edn. Wiley, Hoboken (2019)

McCullagh, P., Nelder, J.A.: Generalized Linear Models, 2 edn. Chapman and Hall, London (1989)

Nocedal, J., Wright, S.J.: Numerical Optimization, 2 edn. Springer, New York (2006)

Parikh, N., Boyd, S.: Proximal algorithms. Found. Trends Optim. **1**(3), 123–231 (2014)

Rahimi, A., Recht, B.: Random features for large-scale kernel machines. In: Proceedings of the 21st International Conference on Neural Information Processing Systems, NIPS'07, pp. 1177–1184, Red Hook, NY. Curran Associates Inc., Red Hook (2007). ISBN 9781605603520

Ross, S.M.: A First Course in Probability, 10 edn. Pearson, London (2023)

Saad, Y.: Iterative Methods for Sparse Linear Systems, 2nd edn. SIAM, Philadelphia (2003)

Schölkopf, B., Smola, A.J.: Learning with Kernels: Support Vector Machines, Regularization, Optimization, and Beyond. MIT Press, Cambridge (2002)

Schölkopf, B., Smola, A., Müller, K.-R.: Nonlinear component analysis as a kernel eigenvalue problem. Neural Comput. **10**(5), 1299–1319 (1998)

Seber, G.A.F., Lee, A.J.: Linear Regression Analysis, 2nd edn. Wiley, Hoboken (2003)

Strang, G.: Linear Algebra and Learning from Data. Wellesley-Cambridge Press, Wellesley (2019)

Tibshirani, R.: Regression shrinkage and selection via the lasso. J. R. Stat. Soc.: Ser. B (Methodol.) **58**(1), 267–288 (1996)

Trefethen, L.N., Bau, D. III: Numerical Linear Algebra. SIAM, Philadelphia (1997)

van der Vaart, A., Wellner, J.: Weak Convergence and Empirical Processes: With Applications to Statistics. Springer, New York (1996)

Vershynin, R.: High-Dimensional Probability: An Introduction with Applications in Data Science. Cambridge University Press, Cambridge (2018)

von Luxburg, U.: A tutorial on spectral clustering. Stat. Comput. **17**(4), 395–416 (2007)

Wainwright, M.J.: High-Dimensional Statistics: A Non-Asymptotic Viewpoint. Cambridge University Press, Cambridge (2019)

Williams, D.: Probability with Martingales. Cambridge University Press, Cambridge (1991)

FSC
www.fsc.org
MIX
Papier | Fördert gute Waldnutzung
FSC® C083411

Kolibri 360 GmbH
Ferdinand-Jühlke-Straße 7
99095 Erfurt, Deutschland
produktsicherheit@kolibri360.de